Genetics and Molecular Biology of Streptococci, Lactococci, and Enterococci

Genetics and Molecular Biology of Streptococci, Lactococci, and Enterococci

Editors

Gary M. Dunny
University of Minnesota, St. Paul, Minnesota

P. Patrick Cleary
University of Minnesota Medical School, Minneapolis, Minnesota

Larry L. McKay
University of Minnesota, St. Paul, Minnesota

American Society for Microbiology
Washington, D.C.

Library of Congress Cataloging-in-Publication Data

Genetics and molecular biology of streptococci, lactococci, and enterococci/editors Gary
M. Dunny, P. Patrick Cleary, Larry L. McKay.
    p. cm.
  Includes bibliographical references and indexes.
  ISBN 1-55581-034-9
  1. Streptococcus—Genetics—Congresses. 2. Lactococcus—Genetics—Congresses. 3.
Enterococcus—Genetics—Congresses. 4. Molecular genetics—Congresses. I. Dunny,
Gary M. II. Cleary, P. Patrick. III. McKay, Larry L.
QR82.S78G46   1991
589.9'5—dc20   91-2846
                                                                                  CIP

Cover photograph: Immunogold labeling of pheromone-induced aggregation substance of
Enterococcus faecalis. *E. faecalis* cells carrying the conjugative, pheromone-inducible plasmid pAD1 were reacted with colloidal gold-labeled antibody specific for the Asa1 surface
protein which mediates formation of mating aggregates. The labeled cells were fixed and
visualized by scanning electron microscopy. (Photograph courtesy of Gerhard Wanner and
Reinhard Wirth, University of Munich.)

Contents

III. Lactococci: Molecular Biology and Biotechnology

IV. Structure and Evolution of the M-Protein Gene Family

V. Extracellular Products of Pathogenic Streptococci: Genetics and Regulation

VI. Molecular Biology of Oral Streptococci

Introduction

Gram-positive cocci have been more resistant than enteric bacteria to the tools of molecular biology. Scientific activity in the past decade, however, has given rise to new vectors, systems of transformation, and other molecular approaches tailored to investigating important questions unique to these bacteria. The Third International ASM Conference on Streptococcal Genetics, held at the Hyatt Regency Hotel in Minneapolis, Minn., from 6 to 9 June 1990, addressed many of these questions. Scientists attending the conference came from 14 countries around the world, reflecting the international nature of laboratories involved in investigations of streptococcal molecular biology. Oral presentations and posters were concerned with important theoretical questions of gene regulation and transfer and with practical questions ranging from mechanisms of virulence to strain optimization for dairy fermentations. Six sessions were convened: Gene Transfer, Molecular and Genetic Analysis of Pneumococci, Lactococci: Molecular Biology and Biotechnology, Structure and Evolution of the M-Protein Gene Family, Extracellular Products of Pathogenic Streptococci: Genetics and Regulation, and Molecular Biology of Oral Streptococci. Each oral session was accompanied by separate poster sessions. All attendees left the conference excited about future directions of their respective fields, armed with new ideas, and fully aware of recent advances in the molecular biology of these cocci. Chapters in this volume are based on oral and selected poster presentations, and they reflect the high caliber and informative nature of the science discussed during the conference.

The conference and chapters to follow exemplify the great biological diversity of this collection of gram-positive cocci. What other group of bacteria has such broad and important influence on human health and is also responsible for a nutritious food with delicate taste such as cheddar cheese? The products of these organisms can stimulate your taste buds, damage your heart, or dissolve the enamel of your teeth. It is true that taxonomists recently divided these streptococci into three different genera, *Streptococcus*, *Enterococcus*, and *Lactococcus*. The surprise, however, is that our creative activity and new understanding seem to bring those who study these creatures closer together. For example, who would have imagined or predicted that a peptidase from pathogenic streptococci (Cleary et al., this volume) which is specific for a complement protein would have significant genetic homology with a protease, important in cheese production (de Vos et al., this volume)? The comparison by Dr. Fischetti and colleagues of surface proteins from a variety of species has also revealed striking sequence similarities and demonstrates a common mode of anchorage of these proteins to the cell surface (Schneewind et al., this volume).

Lewis Thomas, a renowned medical essayist, noted that some microbes are uninformed to cause disease unless they themselves are infected with an extrachromosomal element.[†] Streptococci and lactococci are magnificent examples of this life-style. Mobile elements, plasmids, and bacteriophages encode a variety of virulence factors, fermentative enzymes, and other accessory macromolecules which permit them to adapt to their unique environmental niches. He also discussed and speculated on the impact of pheromones on human interpersonal relations. No one is certain whether such molecules affect human personal or sexual behavior, but it is clear from session I (Gene Transfer) that pheromones have a dramatic impact on the steamy sex life of enterococci. Perhaps Dr. Thomas's speculations should be given more serious consideration.

Of course, molecular genetics began with *Streptococcus pneumoniae*. The renewed burst of interest in this organism will surely lay aside many old ideas. It was made clear in session II (Molecular and Genetic Analysis of Pneumococci) that factors other than the polysaccharide capsule are important to the virulence of the organisms (Yother et al., this volume).

This volume clearly reveals the sophistication that streptococcal genetics has attained. The boundaries of knowledge have advanced dramatically in the past 4 years, and progress over the next 3 or 4 years will be the topic of a future conference. The development of new and improved vaccines that prevent streptococcal disease is just over the horizon, and the application of biotechnology to fermentation and food industries has matured and is about to blossom.

The American Society for Microbiology and the conference organizers gratefully acknowledge the generous financial support from the following organizations: The BOC Group; BSN Groupe;

[†]Thomas, L. 1974. *The Lives of a Cell: Notes of a Biology Watcher*. Bantam Books/The Viking Press, New York.

Connaught Laboratories; Ecolab; General Mills, Inc.; Land O'Lakes; Microbial Genetics/Pioneer Hi-Bred International; Microlife Technics; Marschall Products, Rhone Poulenc, Inc.; Nestle S.A.; The Nutrasweet Company; R&D Systems, Inc.; Lifecore Biomedical, Inc.; Bristol-Myers Squibb Company; National Institutes of Health; and National Science Foundation.

P. PATRICK CLEARY
University of Minnesota
Minneapolis, Minn.

I. Gene Transfer

Enterococcus faecalis Hemolysin/Bacteriocin Plasmid pAD1: Regulation of the Pheromone Response

DON B. CLEWELL,[1,2] LINDA T. PONTIUS,[2] KEITH E. WEAVER,[1†] FLORENCE Y. AN,[1] YASUYOSHI IKE,[1‡] AKINORI SUZUKI,[3] AND JIRO NAKAYAMA[3]

Department of Biologic and Materials Sciences, School of Dentistry,[1] and Department of Microbiology and Immunology, School of Medicine,[2] The University of Michigan, Ann Arbor, Michigan 48109, and Department of Agricultural Chemistry, University of Tokyo, Bunkyo-ku, Tokyo 113, Japan[3]

The conjugative plasmid pAD1 (60 kb), originally identified in *Enterococcus faecalis* DS16 (4, 26), encodes a hemolysin/bacteriocin activity shown to contribute to virulence in a mouse model (17). Closely related hemolysin plasmids have been identified in clinical isolates derived from human parenteral infections (15, 18). Analyses of pAD1 conjugation, particularly with respect to the related response to the peptide sex pheromone cAD1 (5), have been ongoing in our laboratory. Plasmid-free strains of *E. faecalis* excrete a number of sex pheromones specific for potential donors with different classes of plasmids (5, 7, 8). Donors are induced to synthesize a surface protein(s) referred to as aggregation substance (AS) that facilitates formation of mating aggregates. AS binds to a substance on the recipient surface called binding substance, which may correspond to lipoteichoic acid (10). Donor cells, which also have lipoteichoic acid, undergo a self-aggregation (clumping) when exposed to pheromone in the absence of recipients.

Several inducible pAD1-related surface proteins have been identified in Western immunoblot analyses utilizing polyclonal antiserum raised against gluteraldehyde-fixed cells previously induced with pheromone (10, 28). The inducible proteins range in size from 52 to about 190 kDa (10, 28, 30). Electron microscopic analyses by Wirth and colleagues (13, 27), using immunogold labeling and antiserum raised specifically against the predominant inducible 74-kDa surface protein, have recently revealed a dense layer of "hairs." Interestingly, the hairs seem to appear only on that portion of the surface corresponding to old wall, i.e., wall not being synthesized at the time of pheromone exposure (see Wirth et al., this volume).

When a given plasmid (e.g., pAD1) enters the recipient, it shuts down the synthesis of the related pheromone (e.g., cAD1), but different pheromones (e.g., cPD1 and cAM373) continue to be elaborated. *E. faecalis* isolates have been found to harbor as many as three different conjugative plasmids, each of which encodes a response to a different pheromone (6, 21). Bacteria containing pAD1 specifically excrete a different peptide, iAD1, which acts as a competitive inhibitor of cAD1 (16); thus, potential recipient cells must be reasonably close to donors in order that cAD1 outcompete iAD1 so that induction can occur. Both cAD1 and iAD1 have been purified and characterized (19, 20); both are hydrophobic octapeptides and exhibit 50% homology with each other.

Organization of pAD1 Conjugation-Related Genes

At least half of pAD1 corresponds to a contiguous segment devoted to various aspects of the mating event and includes determinants referred to as *traA* and *traB* as well as regions designated C through H (5, 9, 14) (Fig. 1). Genetic analyses involved the generation of insertion mutations by using the transposon Tn917 or Tn917lac (which generates *lacZ* transcriptional fusions) (31). Insertions in regions G and H affect plasmid transfer but not aggregation, whereas insertions in region F specifically affect the production of pheromone-inducible surface proteins and alter the ability to aggregate. Tn917lac fusions located counterclockwise to the H region as far as position 44 kb on the map have recently been found to undergo pheromone-induced production of β-galactosidase (Fig. 1); however, a specific phenotype(s) related to this region is not yet evident.

†Present address: Department of Microbiology, School of Medicine, University of South Dakota, Vermillion, SD 57069.
‡Present address: Department of Microbiology, Gunma University School of Medicine, Maebashi City, Gunma 371, Japan.

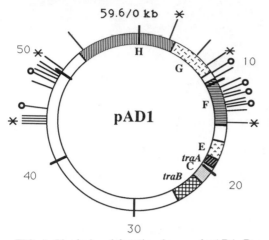

FIG. 1. Physical and functional map of pAD1. Regions important for transfer are indicated as shaded or open boxes and are labeled *traA*, *traB*, and C through H. The lines located at various points outside the circle represent Tn*917lac* insertions. Those with small open circles on the ends represent insertions that are oriented such that *lacZ* would have to be transcribed clockwise; those with no circles or with an asterisk are oriented counterclockwise. All counterclockwise-oriented insertions resulted in β-galactosidase expression as a result of exposure to pheromone, whereas the clockwise inserts did not respond. The asterisks indicate insertions that were tested for switching on related to the phase variation phenomenon; these were all found to switch on β-galactosidase production at a frequency of 10^{-3} to 10^{-4}.

Examination of numerous transposon fusions suggests that inducible transcription occurs counterclockwise over the entire region from about 18 to 44 kb (Fig. 1). It is likely that more than one promoter is involved over this span, since some insertions do not result in polar mutations. For example, Tn*917* insertions in the F region, which affect aggregation, do not reduce plasmid transfer (related to the G and H regions) if the matings are performed on filter membranes.

The remaining determinants and regions (Fig. 2) are involved in regulation and are located within the *Eco*RI B fragment, which also contains the necessary functions for plasmid replication (29). The E-region product(s) regulates the response in a positive manner, as mutations here knock out the ability to aggregate or transfer DNA (9). Entry exclusion is also eliminated in such mutations (L. T. Pontius, unpublished data). The *traA* product represses the E-region determinant(s) and is also involved in transduction of the exogenous signal (pheromone) (30). Mutants in *traA* constitutively clump and transfer plasmid DNA at high frequency in short (10-min) matings. (It generally takes more than 30 min of prior pheromone exposure to induce wild-type donors to transfer at high frequency in 10-min matings.) The β-galactosidase activity of Tn*917lac* fusions in the E region is negatively regulated by the *traA* product; derivatives with E-region fusions as well as a mutation in *traA* constitutively express the enzyme. The *traB* product is involved with shutdown of endogenous cAD1; however, it appears to require help from the E region, the C region, or both (30). The C-region product(s) includes an activity that enhances the binding of exogenous cAD1 or excreted iAD1 to the receptor. Interestingly, the activity of the related C protein is host specific; it increases binding in *E. faecalis* OG1X but not in the nonisogenic strain FA2-2 (30).

Transcriptional fusions (Tn*917lac*) in *traA*, *traB*, and region C indicate that expression occurs constitutively and at a very low level (28). Although *traB* and region C are transcribed to about the same degree and are in the same orientation, they are probably expressed from different promoters, since transposon mutations in *traB* are not polar.

The iAD1 Determinant, *iad*

Figure 3 shows a DNA sequence located near the 5' end of the E region (also see Fig. 2) and containing a short open reading frame corresponding to the iAD1 structural gene *iad*. The

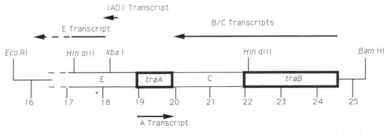

FIG. 2. Map of the pheromone response regulatory region of pAD1. This map is based on maps previously reported (5, 30); the arrows indicate transcription units and their orientation based on fusion studies using Tn*917lac*. The location of the iAD1 determinant was based on sequence data shown in Fig. 3. The numbers indicate the pAD1 map position in kilobases.

iAD1 peptide (Leu-Phe-Val-Val-Thr-Leu-Val-Gly) can be seen at the carboxyl terminus of what appears to be a 22-amino-acid precursor. The first of two methionines was chosen as the likely translational start site because of the optimally located Shine-Dalgarno (SD) ribosome binding sequence. Two possible promoters are also observed (overlapping) upstream, with spacings between their −10 and −35 hexamers of 18 and 20 nucleotides. The closest apparent transcription terminator is well over 300 nucleotides downstream from the *iad* sequence and corresponds to a 17-bp inverted repeat sequence (designated C) followed by five T's. (The inverted repeats would exhibit a ΔG of −34.4 kcal [ca. −144 kJ].) Other inverted repeats, designated A and B, are also

```
                                              50
TTGGTTTTTTTATTATGTAATAAATTTTTTTGATGAAAAAGCGCAAATTT

                                             100
                                               *
TTGCATTTTTGTTTATTTTTATTTTAATCTATGCTATTATTAATTTGTAA
 -35     -35                  -10       -10
                                             150
GTTAAGTTTAAATAAGAGGAGAGCTATTAGAATGAGCAAACGAGCTATGA
              S.D.                  M  S  K  R  A  M

                                             200
                                               *
AAAAAATTATTCCATTGATAACTTTATTTGTTGTCACACTTGTAGGATAA
K  K  I  I  P  L  I  T  L  F  V  V  T  L  V  G  *
                              iad         AA12
                                              ▽   250
TTAGTTTAGAATCTCGTAGTTACCTTGATATAGCAATTATCTGAGAGATA
               - - - - - - - ->
                        A
                                             300
                                               *
AATCTTTTTACTTTTTTAAACTTAGTGCGATATAAAGGCAACTTACAAGA
                        <- - - - -
                              A
                                             350
CAATTAACAAAATAAGAACCGACTGCCATAGGACGGGAATCCTAGAGGAC

                                             400
                                               *
AGTTAAACAATTCATGCTATACCCATGAACTATACTCGGTTCTCGTTTGT
                        - - - - ->
                          B
                                             450
TGCAACATTAGTTACAACGTATAGTATAACAATTTTTTATGTAAAATTCT

                                             500
                                               *
AGACTTTTTTAAACTCCTTTATTTGTCTAGGAAAAGTTTTTACAGTGAAT
Xba I
                                             550
TGTTTTAATTAGTTGTATAAATGTTGGAGCAGCGGGGAATGTATACAGTT
                                    M  Y  T  V
              S.D.
                                        <-  -
                                      B
                           - - - - - - ->   600
                       C  AA43               *
                              ▽
CATGTATATATTCCCCGCTTTTTTGTTGTCTGTTAACTTGTTGAAAAAAT
H  V  Y  I  P  R  F  F  V  V  C  *
- - - - - - - -
     <- - - - -   - - -
            C
```

present between the translational stop site and the putative transcriptional terminator; their significance is not known.

Within the presumed transcription terminator, there is a possible SD sequence preceding a short open reading frame (15 amino acid residues). If this is in fact an SD sequence, the level of translation through it might exert an effect on transcription termination and affect readthrough into the E region. Since the apparent iAD1 activity in culture filtrates is similar for uninduced wild-type and *traA* (fully induced) mutants, it would appear that iAD1 expression is constitutive. The behavior of the Tn917lac fusion indicated in Fig. 3 as AA12 is consistent with this, as β-galactosidase expression is high in the absence of pheromone. Interestingly, the fusion designated AA43 immediately downstream of the transcription terminator exhibits expression (high level) only in the presence of pheromone, as do fusions farther downstream in region E (not shown). Since these are negatively regulated by the *traA* product (TraA), it is possible that expression of the E-region product(s) depends on readthrough of the transcription terminator of *iad*. Thus, TraA may exhibit its negative control by binding to DNA at or near the transcription terminator and facilitating termination.

A chimeric plasmid consisting of pBluescript (Stratagene, San Diego, Calif.) containing within it the region defined by the *Hin*dIII-*Xba*I fragment seen in Fig. 2 gave rise to iAD1 expression in *Escherichia coli* DH5α. Culture filtrates of late-exponential-phase cells grown in L broth contained activity at a titer of 32. Expression occurred equally well when the fragment was in either orientation, implying that *iad* utilized its

FIG. 3. Nucleotide sequence of the region possessing the *iad* determinant (see Fig. 2 for indication of region). Two open reading frames based on the positions of SD sequences are indicated with the inferred amino acid residues. The first corresponds to *iad* and is preceded by two possible promoters (−10 and −35 sequences); the second reflects a short reading frame beginning within the apparent *iad* transcription termination site. The latter is indicated by dashed arrows showing inverted repeats designated C (17 bp with one mismatch; $\Delta G = -34.4$ kcal [ca. −144 kJ]). Two other pairs of inverted repeats are also noted and indicated as A (24 bp each with four mismatchs and one unpaired; $\Delta G = -19.2$ kcal [ca. −80.3 kJ]) and B (17 bp with one mismatch; $\Delta G = -9.1$ kcal [ca. 30 kJ]), with one of the B repeats occurring within the putative downstream transcription terminator. The two triangles indicated as AA12 and AA43 represent Tn917lac fusions. AA12 expressed β-galactosidase constitutively, whereas AA43 expressed it only in the presence of cAD1.

own promoter. When iAD1 activity was examined by reverse-phase high-pressure liquid chromatography (2), it was found to exhibit a retention time essentially identical to that of authentic iAD1 (chemically synthesized); thus, processing of the inferred 22-amino-acid precursor must occur in this host.

The entire putative iAD1 precursor (Fig. 3) resembles a signal sequence (22, 24); four of the first eight residues in the amino-terminus end are positively charged, while the remaining residues are very hydrophobic. The precursor sequence is probably important for insertion into or transport through the membrane. It is interesting that despite the presence of the four charged residues, along with the amino-terminal charge, an α-helical wheel projection (24) depicts a hydrophilic and hydrophobic side (not shown). Although N-terminal amphipathic signal peptides are known to occur in prokaryotes, their significance is not yet clear; in eukaryotes, such structures appear important for targeting to mitochondria (24).

Phase Variation

When E. faecalis strains harboring pAD1 are plated on solid media containing synthetic pheromone, the resulting colonies exhibit a characteristic dry morphology presumed to represent an induced aggregative state (5). Without cAD1 in the plate, the colonies are smooth; however, dry colonies appear spontaneously at a frequency of 10^{-3} to 10^{-4}. Subcultures continue to give rise to dry colonies, but normal colonies appear, again at a frequency of 10^{-3} to 10^{-4}. Cultures of cells from dry colonies appear to be "turned on" with respect to the mating response; they exhibit clumping in liquid culture and donate plasmid at high frequency in short (10-min) broth matings, and they generally behave much like traA mutants (Table 1).

The reversible switching phenomenon affects expression of regions within E, F, G, and H and even counterclockwise to H, since colonies representing appropriate transcriptional fusions here appear blue on 5-bromo-4-choro-3-indolyl-β-D-glactoside (X-Gal) plates devoid of pheromone at a similar frequency (see Fig. 1). Curing and reconstruction experiments have shown that the switch occurs on the plasmid and not on the bacterial chromosome. A pAD1 derivative deleted

of all DNA except the EcoRI B fragment and also containing a pheromone-inducible Tn917lac in the E region exhibits a reversible phase variation relating to β-galactosidase expression. The specific nature of the switch is not yet known but probably involves a reversible change affecting the ability of TraA to regulate the E region. Whether the change involves a DNA rearrangement (e.g., an inversion related to one of the sets of inverted repeats downstream from iad [Fig. 3]), a frameshift, or some other event, it should be revealed in the very near future.

Effect of Aeration on Pheromone Expression

It has recently been observed that the secretion of cAD1 is greatly affected by the degree of aeration during bacterial growth. This is illustrated in Table 2, which shows that culture filtrates from plasmid-free E. faecalis strains (both OG1X and FA2-2) grown without shaking (stagnant) contain 8 to 16 times as much cAD1 activity as do cells grown with vigorous shaking. A culture placed inside an anaerobic GasPak (American Scientific Products, McGaw Park, Ill.) and shaken exhibited the same elevated level of activity as did a similarly grown but not shaken culture. In contrast, the two unrelated sex pheromones cPD1 and cAM373 were not affected by aeration. It therefore appears that the presence of oxygen has a significant negative effect on the expression of cAD1.

The phenomenon probably also relates to an observation made for certain behavioral aspects of traB mutants. As noted earlier, the traB product is believed to contribute to the shutdown of endogenous cAD1 (with the help of factors related to the E and/or C region); cAD1 was barely, if at all, detected in culture filtrates of traB insertion derivatives. These mutants were originally characterized by their derepressed state and the ability to transfer plasmid at an elevated frequency in broth matings (15). In this regard they resembled traA mutants; however, their colony morphology on solid media differed somewhat from that of traA variants. Rather than the dry colony appearance typical of traA mutants, traB mutants gave rise to "rings" within the colony. If synthetic cAD1 was included in the plates, then the colonies appeared dry. Interestingly, when plate cultures containing traB variants were

TABLE 1. Properties of "switched-on" variants[a]

Type	Colony morphology	Appearance in broth suspension	Production of novel surface proteins	Transfer frequency in 10-min matings (approx)
Normal	Normal	Dispersed	Inducible	10^{-7}
Switched on	Dry	Clumping	Constitutive	10^{-3}
Revertant	Normal	Dispersed	Inducible	10^{-7}

[a]In all variants, cAD1 excretion is not detected and iAD1 excretion is normal.

TABLE 2. Effect of aeration on *E. faecalis* pheromone titers[a]

Strain	cAD1		cPD1		cAM373	
	Stagnant	Aerated	Stagnant	Aerated	Stagnant	Aerated
OG1X	256	16	32	32	32	32
FA2-2	32	4	32	32	16	16

[a] Pheromone titer is defined as the highest dilution of culture filtrate still able to induce an aggregation response in a strain bearing the plasmid for which that pheromone is specific. Pheromones cAD1, cPD1, and cAM373 are specific for plasmids pAD1, pPD1, and pAM373, respectively. Plasmid-containing strains DS16, 39-5, and FA373 were used as responders in microtiter assays for cAD1, cPD1, and cAM373, respectively. Titers shown are representative of at least two independent experiments.

grown in the GasPak, then the colonies appeared dry even in the absence of exogenous cAD1. The interpretation here is that a small amount of cAD1 is in fact produced in *traB* mutants and is attempting to self-induce the mating response; the ring appearance of the colony may represent a cyclic effect related to environmental changes at different locations within the growing colony. When grown anaerobically (GasPak), the cells probably produce more cAD1, causing a more efficient self-induction and resulting in a dry colony morphology. Tn917 insertions in *traB* also transferred plasmid DNA in short (10-min) matings at frequencies at least 2 orders of magnitude better if they were grown without shaking in comparison with shaken cultures.

Concluding Remarks

Although significant progress is being made in understanding the processes involved in the pAD1 mating response, much remains to be learned. At least two mechanisms are evident for activating conjugation: a physiological one involving an external ligand (pheromone), and a genetic event relating to an alteration at a site in or near the regulatory region on the plasmid. The latter process may be useful in bacterial hosts that are incapable of responding to exogenous cAD1, perhaps as a result of an absence or dysfunction of the receptor. It is conceivable that the receptor is actually determined by the chromosome and acts to feedback regulate the production of cAD1 in plasmid-free cells. The plasmid might "tap in" to such a receptor (e.g., via *traA*) to regulate the mating response. Thus, a particular host not having such a receptor would be unable to support a pheromone response; however, plasmid transfer might still be possible as a result of the switch.

Another interesting point relates to the recent finding by Galli et al. (12a) that the AS protein of pAD1 contains the amino acid motifs Arg-Gly-Asp-Ser and Arg-Gly-Asp-Val (each appears once), which have been proposed to play a crucial role (in other systems) in adherence to eukaryotic cells (23). Since there is evidence that pAD1 and related plasmids may contribute to virulence, it is conceivable that AS plays a role in colonization.

The absence of pheromone in such situations would mean that AS expression might depend on a switching mechanism, much like the case in turning on production of type 1 fimbriae necessary for colonization of certain pathogenic strains of *Escherichia coli* (11).

It has been speculated (1) that the peptide sex pheromones produced by plasmid-free enterococci might have functions independent of the conjugation phenomenon and that plasmids evolved to take advantage of their potential use as intercellular signals. Recent reports that certain pheromones exhibit potent neutrophil chemotaxis activity (12, 25) raise the possibility that these substances could influence the course of a related bacterial infection by enhancing or modifying the host response. Perhaps some peptides could act to competitively inhibit the activity of certain chemotactic factors; such compounds might then act as virulence factors. It is interesting that essentially all strains of *Staphylococcus aureus* produce an activity similar to that of the *E. faecalis* pheromone cAM373, whereas strains of coagulase-negative staphylococci (less pathogenic) do not (3). The *S. aureus* peptide differs in structure from the enterococcal peptide only at the carboxy-terminal residue (J. Nakayama and A. Suzuki, unpublished data); thus, it will be interesting to determine how neutrophils react in comparison with the chemotactic activity exhibited by the enterococcal structure. In the case of the pAD1-related peptides, it was found that iAD1 exhibited chemotactic activity with human neutrophils, whereas cAD1 did not (12, 25).

This study was supported by Public Health Service grants GM33956 and DE02731 from the National Institutes of Health and a grant from the Ministry of Education, Science and Culture of Japan.

LITERATURE CITED

1. **Clewell, D. B.** 1981. Plasmids, drug resistance, and gene transfer in the genus *Streptococcus*. *Microbiol. Rev.* **45:**409–436.
2. **Clewell, D. B., F. Y. An, M. Mori, Y. Ike, and A. Suzuki.** 1987. *Streptococcus faecalis* sex pheromone (cAD1) response: evidence that the peptide inhibitor excreted by pAD1-containing cells may be plasmid determined. *Plasmid* **17:**65–68.
3. **Clewell, D. B., F. Y. An, B. A. White, and C. Gawron-**

Burke. 1985. *Streptococcus faecalis* sex pheromone (cAM373) also produced by *Staphylococcus aureus* and identification of a conjugative transposon (Tn*918*). *J. Bacteriol.* **143**:1063–1065.

4. **Clewell, D. B., P. K. Tomich, M. C. Gawron-Burke, A. E. Franke, Y. Yagi, and F. Y. An.** 1982. Mapping of *Streptococcus faecalis* plasmids pAD1 and pAD2 and studies relating to the transposition of Tn*917*. *J. Bacteriol.* **152**:1220–1230.

5. **Clewell, D. B., and K. E. Weaver.** 1989. Sex pheromones and plasmid transfer in *Enterococcus faecalis* (a review). *Plasmid* **21**:175–184.

6. **Clewell, D. B., Y. Yagi, Y. Ike, R. A. Craig, B. L. Brown, and F. An.** 1982. Sex pheromones in *Streptococcus faecalis*: multiple pheromone systems in strain DS5, similarities of pAD1 and pAMγ1, and mutants of pAD1 altered in conjugative properties, p. 97–100. In D. Schlessinger (ed.), *Microbiology—1982*. American Society for Microbiology, Washington, D.C.

7. **Dunny, G. M., B. L. Brown, and D. B. Clewell.** 1978. Induced cell aggregation and mating in *Streptococcus faecalis*: evidence for a bacterial sex pheromone. *Proc. Natl. Acad. Sci. USA* **75**:3479–3483.

8. **Dunny, G. M., R. A. Craig, R. Carron, and D. B. Clewell.** 1979. Plasmid transfer in *Streptococcus faecalis*. Production of multiple sex pheromones by recipients. *Plasmid* **2**:454–465.

9. **Ehrenfeld, E. E., and D. B. Clewell.** 1987. Transfer functions of the *Streptococcus faecalis* plasmid pAD1: organization of plasmid DNA encoding response to sex pheromone. *J. Bacteriol.* **169**:3473–3481.

10. **Ehrenfeld, E. E., R. E. Kessler, and D. B. Clewell.** 1986. Identification of pheromone-induced surface proteins in *Streptococcus faecalis* and evidence of a role for lipoteichoic acid in formation of mating aggregates. *J. Bacteriol.* **168**:6–12.

11. **Eisenstein, B. I.** 1981. Phase variation of type I fimbriae in *Escherichia coli* is under transcriptional control. *Science* **214**:337–339.

12. **Ember, J., and T. E. Hugli.** 1989. Characterization of the human neutrophil response to sex pheromones from *Streptococcus faecalis*. *Am. J. Pathol.* **134**:797–805.

12a. **Galli, D., F. Lottspeich, and R. Wirth.** 1990. Sequence analysis of *Enterococcus faecalis* aggregation substance encoded by the sex pheromone plasmid pAD1. *Mol. Microbiol.* **4**:895–904.

13. **Galli, D., R. Wirth, and G. Wanner.** 1989. Identification of aggregation substances of *Enterococcus faecalis* cells after induction by sex pheromones. *Arch. Microbiol.* **151**:486–490.

14. **Ike, Y., and D. B. Clewell.** 1984. Genetic analysis of the pAD1 pheromone response in *Streptococcus faecalis*, using transposon Tn*917* as an insertional mutagen. *J. Bacteriol.* **158**:777–783.

15. **Ike, Y., and D. B. Clewell.** 1987. High incidence of hemolysin production by *Streptococcus faecalis* strains associated with human parenteral infections: structure of hemolysin plasmids, p. 159–164. In J. Ferretti and R. Curtiss (ed.), *Streptococcal Genetics*. American Society for Microbiology, Washington, D.C.

16. **Ike, Y., R. C. Craig, B. A. White, Y. Yagi, and D. B. Clewell.** 1983. Modification of *Streptococcus faecalis* sex pheromone after acquisition of plasmid DNA. *Proc. Natl. Acad. Sci. USA* **80**:5369–5373.

17. **Ike, Y., H. Hashimoto, and D. B. Clewell.** 1984. Hemo-

lysin of *Streptococcus faecalis* subspecies *zymogenes* contributes to virulence in mice. *Infect. Immun.* **45**:528–530.

18. **Ike, Y., H. Hashimoto, and D. B. Clewell.** 1987. High incidence of hemolysin production by *Enterococcus (Streptococcus) faecalis* strains associated with human parenteral infection. *J. Clin. Microbiol.* **25**:1524–1528.

19. **Mori, M., A. Isogai, Y. Sakagami, M. Fujino, C. Kitada, D. B. Clewell, and A. Suzuki.** 1986. Isolation and structure of *Streptococcus faecalis* sex pheromone inhibitor, iAD1, that is excreted by donor strains harboring plasmid pAD1. *Agric. Biol. Chem.* **50**:539–541.

20. **Mori, M., H. Tanaka, Y. Sakagami, A. Isogai, M. Fujino, C. Kitada, D. B. Clewell, and A. Suzuki.** 1984. Isolation and structure of the bacterial sex pheromone, cAD1, that induces plasmid transfer in *Streptococcus faecalis*. *FEBS Lett.* **178**:97–100.

21. **Murray, B. E., F. An, and D. B. Clewell.** 1988. Plasmids and pheromone response of the β-lactamase producer *Streptococcus (Enterococcus) faecalis* HH22. *Antimicrob. Agents Chemother.* **32**:547–551.

22. **Oliver, D. B.** 1987. Periplasm and protein secretion, p. 56–69. In F. C. Neidhardt, J. L. Ingraham, K. B. Low, B. Magasanik, M. Schaechter, and H. E. Umbarger (ed.), *Escherichia coli and Salmonella typhimurium: Cellular and Molecular Biology*. American Society for Microbiology, Washington, D.C.

23. **Ruoslahti, E.** 1988. Fibronectin and its receptors. *Annu. Rev. Biochem.* **57**:375–413.

24. **Saier, M. H., Jr., P. K. Werner, and M. Muller.** 1989. Insertion of proteins into bacterial membranes: mechanism, characteristics, and comparisons with eucaryotic process. *Microbiol. Rev.* **53**:333–366.

25. **Sannomiya, P., R. A. Craig, D. B. Clewell, A. Suzuki, M. Fujino, G. O. Till, and W. Morasco.** 1990. Characterization of a class of nonformylated *Enterococcus faecalis*-derived neutrophil chemotactic peptides: the sex pheromones. *Proc. Natl. Acad. Sci. USA* **87**:66–70.

26. **Tomich, P. K., F. Y. An, S. P. Damle, and D. B. Clewell.** 1979. Plasmid-related transmissibility and multiple drug resistance in *Streptococcus faecalis* subsp. *zymogenes* strain DS16. *Antimicrob. Agents Chemother.* **15**:828–830.

27. **Wanner, G., H. Formanek, D. Galli, and R. Wirth.** 1989. Localization of aggregation substances of *Enterococcus faecalis* after induction by sex pheromone. An ultrastructural comparison using immuno labelling, transmission and high resolution scanning electron microscopic techniques. *Arch. Microbiol.* **151**:491–497.

28. **Weaver, K. E., and D. B. Clewell.** 1988. Regulation of the pAD1 sex pheromone response in *Enterococcus faecalis*: construction and characterization of *lacZ* transcriptional fusions in a key control region of the plasmid. *J. Bacteriol.* **170**:4343–4352.

29. **Weaver, K. E., and D. B. Clewell.** 1989. Construction of *Enterococcus faecalis* pAD1 miniplasmids: identification of a minimal pheromone response regulatory region and evaluation of a novel pheromone-dependent growth inhibition. *Plasmid* **22**:106–119.

30. **Weaver, K. E., and D. B. Clewell.** 1990. Regulation of the pAD1 sex pheromone response in *Enterococcus faecalis*: effects of host strain and *traA*, *traB*, and C region mutants on expression of an E region pheromone-inducible *lacZ* fusion. *J. Bacteriol.* **172**:2633–2641.

31. **Youngman, P. J.** 1987. Plasmid vectors for recovering and exploiting Tn*917* transposition in *Bacillus* and other gram-positives, p. 79–103. In K. Hardy (ed.), *Plasmids: a Practical Approach*. IRL Press, Oxford.

Cell-Cell Interactions and Conjugal Gene Transfer Events Mediated by the Pheromone-Inducible Plasmid Transfer System and the Conjugative Transposon Encoded by *Enterococcus faecalis* Plasmid pCF10

G. M. DUNNY,[1†] J. W. CHUNG,[1,2] J. C. GALLO,[1] S.-M. KAO,[1] K. M. TROTTER,[1]
R. Z. KORMAN,[3‡] S. B. OLMSTED,[1,2] R. RUHFEL,[2] O. R. TORRES,[1] AND S. A. ZAHLER[3]

*Department of Microbiology, Immunology, and Parasitology, New York State College of Veterinary Medicine,[1]
and Section of Genetics and Development, Cornell University,[3] Ithaca, New York 14853, and Institute for
Advanced Studies in Biological Process Technology and Department of Microbiology, University of Minnesota,
St. Paul, Minnesota 55108[2]*

Conjugation in Streptococci and Related Organisms

All conjugal transfer events in bacteria require four critical steps: (i) direct contact between the donor and recipient cells; (ii) formation of a channel allowing for the transfer of DNA; (iii) transfer of plasmid or other DNA through the channel; and (iv) stabilization of the DNA in the new host, either via autonomous replication or by integration into the host genome. Although the streptococci and related organisms such as the lactococci and enterococci constitute a fairly closely related phylogenetic group of bacteria, they feature a remarkable variety of systems for carrying out the conjugal DNA transfer process. Genetic elements encoding conjugal transfer functions in the streptococcal family include large pheromone-inducible plasmids in *Enterococcus faecalis* (9,10), smaller broad-host-range plasmids (19), and conjugative transposons (7). Because these systems have been reviewed elsewhere (7, 9, 10, 19), they will not be described comprehensively here. Instead, this chapter will focus on two different conjugal transfer systems, both encoded by the 58-kb tetracycline resistance plasmid, pCF10, from *E. faecalis* (10). The striking diversity of cell interactions associated with conjugal gene transfer systems in gram-positive cocci will be illustrated by the discussion of our present understanding of the mechanism of transfer for each of the two pCF10-associated conjugation systems.

Pheromone-Inducible Conjugal Transfer of pCF10

When *E. faecalis* donor cells carrying pCF10 or certain other conjugative plasmids are mixed in broth with plasmid-free *E. faecalis* recipient cells, the efficiency of plasmid transfer in the mixed culture rises from about 10^{-6} during the first 30 to 45 min to about 10^{-1} after 2 to 3 h. Visible aggregation of the cells in the mating mixture is often apparent after several hours of mating. Exposure of the donor cells to culture filtrates of the recipient prior to mating induces rapid, efficient transfer and formation of aggregates upon mixing with recipients, and it can also induce aggregation of pure cultures of donors. It is now known that these phenomena result from the plasmid-determined response of the donor cells to a small peptide sex pheromone excreted by the recipient cells (9, 10). Our current understanding of the mechanisms by which the four stages of DNA transfer occur in the pCF10 pheromone-inducible system is summarized in Fig. 1 and in the following paragraphs.

Chemical signaling between recipient and donor cells. *E. faecalis* cells typically excrete a number of chromosomally encoded small, hydrophobic peptides, and various pheromone-inducible plasmids confer on their host cell the ability to recognize selected members of this set of peptides as pheromones (9, 11). Although the peptides show considerable similarity in amino acid sequence, there is very little cross-reactivity between them in terms of biological activity (9, 18). Thus, each different pheromone-inducible plasmid would appear to determine the production of a highly specific pheromone-sensing mechanism. (We have not investigated in detail the genetics of pheromone production nor the plasmid-encoded shutdown mechanisms encoded by pCF10 that prevent cells carrying the plasmid from producing cCF10. The results of some investigations of these phenomena in relation to the pAD1 system by Clewell and co-workers are presented in this volume and elsewhere [5, 8, 15].)

The very first form of cell interaction in the pCF10 system involves the transmission of a chemical signal in the form of a heptapeptide called cCF10 from the recipient to the donor. As

†Present address: Institute for Advanced Studies in Biological Process Technology and Department of Microbiology, University of Minnesota, Minneapolis, MN 55108.
‡Deceased.

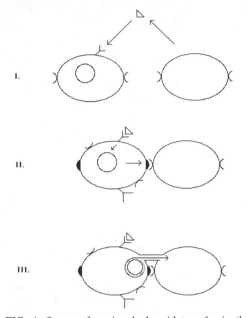

I.

II.

III.

FIG. 1. Stages of conjugal plasmid transfer in the pheromone-inducible conjugation system encoded by pCF10. (I) The first stage involves the synthesis of a diffusible heptapeptide pheromone, cCF10, by recipient cells. This pheromone is recognized by a plasmid-determined receptor on the donor cell. (II) The binding of one to five molecules of cCF10 by the donor cell initiates a signal transduction process resulting in expression of a number of plasmid-encoded transfer functions, including the synthesis of a surface protein, aggregation substance (AS), which can bind to binding substance (BS) on the recipient cell surface. (III) After the initial attachment of donor and recipient cells, via AS-BS binding, a channel permits the transfer of plasmid DNA from the donor to the recipient cell. The transfer is depicted as a unidirectional event occurring via a rolling-circle type of mechanism, but there is little direct evidence for such a mechanism at present. △, Pheromone; ⊥, pheromone receptor; ⟨, binding substance; ❫, aggregation substance.

shown by Mori et al. (18), one to five molecules of this compound are capable of inducing a clumping and mating response in a donor cell. We have determined that pCF10-containing cells can bind and remove cCF10 activity from culture medium, and we have recently identified a segment of the plasmid necessary for pheromone binding within the region of pCF10 encoding various negative control genes (see next paragraph). Current efforts are directed toward cloning this region to determine whether it encodes a pheromone receptor. Although the details of the signal transduction process are completely unknown at present, we do know that the outcome of the binding of a very small number of molecules to

the responder cell initiates a dramatic response at the levels of transcription and translation in this cell, as described below.

Genes and regulatory mechanisms involved in the pheromone response. A combination of transposon mutagenesis, molecular cloning, DNA sequencing, and transcriptional analysis has been used to identify the pCF10 genes involved in pheromone-inducible conjugation and to study the regulation of their expression. Our current understanding of the molecular organization of these genes is depicted in Fig. 2. Initial analysis of pCF10 by transposon mutagenesis (2) indicated that about half of the plasmid contained genes encoding pheromone-inducible conjugation functions. Additional information about this region was obtained from cloning and expression studies in both *E. faecalis* and *Escherichia coli* hosts and by further transposon mutagenesis of cloned fragments (3). These studies showed that a portion of the plasmid encompassing the contiguous *Eco*RI c and e fragments (the thick portion of line II in Fig. 2) encodes two surface proteins, Sec10 (formerly Tra130), which appears to be involved in a surface exclusion function (12), and Asc10 (formerly Tra150), which mediates mating aggregate formation (3; Wirth et al., this volume). A region to the left of the structural genes encoding the surface proteins encodes several positive regulatory factors, and negative regulatory genes are probably to the left of the *Eco*RI c fragment. The complete DNA sequence of the *Eco*RI c and e fragments has been determined recently (S.-M. Kao, Ph.D. dissertation, Cornell University, Ithaca, N.Y., 1990; Wirth et al., this volume; unpublished data). The sequence data, along with various genetic analyses, have resulted in the designation of genes in this system as noted in line IV of Fig. 2. Five of the 11 open reading frames revealed by sequence analysis correspond to regions identified by genetic studies as encoding either surface proteins or regulatory functions, and they have been given *prg* (pheromone-responsive gene) designations. Thus, the *prgA* and *prgB* genes encode the Sec10 and Asc10 proteins, while the *prgX*, *-R*, and *-S* genes have regulatory functions.

Transposon insertions into the region spanning *prgR* and *prgS* abolished expression of *prgB* (but not *prgA*) in *E. faecalis* but not in *E. coli* (3). This region was initially believed to contain a single regulatory gene that was designated R150, since it appeared to activate expression of the 150-kDa Asc10 protein (3). Subsequently, the sequence data clearly showed two open reading frames, which would be predicted to compose an operon expressed from a promoter at the 5' end of *prgR* (Kao, Ph.D. dissertation, 1990). The initial analyses of *prgX* by Tn5 mutagenesis indicated that this gene might be an activator of *prgA*

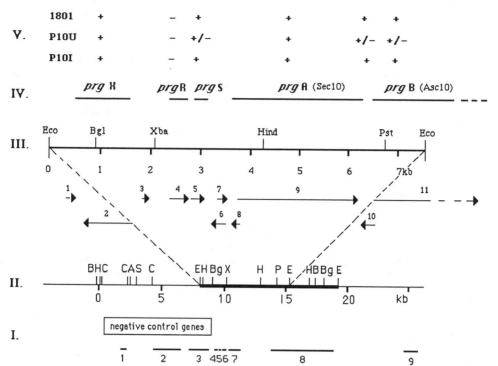

FIG. 2. Molecular organization and expression of pCF10 genes involved in pheromone-inducible conjugation. (I) Transfer region as determined by Tn917 insertional mutagenesis (2). Nine different tra segments, some probably containing multiple genes, were identified and grouped by location and phenotype by Christie and Dunny (2). (II) Restriction enzyme cleavage map of the transfer region, in relation to the various tra segments. The adjacent EcoRI c and e fragments encoding several important structural and regulatory genes are indicated by the heavy line. (III) Expanded view of the EcoRI c fragment, with the 11 open reading frames identified by DNA sequencing indicated by the arrows. (IV) Genes encoded by the EcoRI c fragment, as determined by transposon mutagenesis, DNA cloning, and expression studies (2, 3; see text). (V) Summary of Northern hybridization analysis of the transcription of this region of pCF10. Radiolabeled oligonucleotide probes corresponding to the positions indicated were hybridized to mRNA from pheromone-induced and uninduced cells carrying pCF10 to examine transcription. Symbols: +, transcript detected; −, no transcript; +/−, very low level of transcript detected. See text for further details.

expression (3), but further investigations have shown that certain transposon insertions into this region apparently are lethal in *E. faecalis* and result in extensive deletion of pCF10 DNA. Recent subcloning studies have shown that like *prgR* and -*S*, *prgX* actually is required for expression of *prgB*. Interestingly, expression of the *prgA* gene, which is located between *prgB* and *prgR*-*S*, is not dependent on these positive regulatory genes.

The distance between *prgB* and its regulatory genes (~3 to 6 kb; Fig. 2, lines III and IV) led us to predict that the cloned positive regulatory genes would function in *trans* to activate transcription of *prgB*. However, numerous attempts to activate *prgB* expression with all or part of the *prgX-R-S* region were unsuccessful when the regulatory and structural regions were cloned separately on compatible plasmids and were inserted into an *E. faecalis* host. Recently, however, ex-

periments involving the cloning of the two regions several kilobases apart on the same plasmid have shown some signs of activation if the cloned fragments are placed in the same relative orientation in which they exist in pCF10. These data are leading us to consider some sort of DNA looping model for the activation mechanism, but much more experimental work will be necessary to establish this. Computer analysis of the predicted amino acid sequences encoded by the three regulatory genes indicates that all three of the predicted proteins are basic and could thus bind DNA, but none of them show any significant similarity to any DNA-binding proteins with well-characterized structural motifs (17). The predicted *prgX* gene product does show significant similarity to the protein products of two positive regulatory genes in other organisms, the *Shigella* gene *virB* (1) and the *modulo* gene of *Drosophila* (16). Neither the structural motifs nor the target sites for the products of these two genes have been determined as yet (1, 16).

We have carried out transcriptional analysis of the genes located on the *Eco*RI c fragment by RNA dot blotting and Northern (RNA) blotting, using synthetic oligonucleotide probes (Fig. 2, line V). In cells carrying wild-type pCF10, the *prgX* and *prgA* genes were transcribed constitutively as monocistronic messages, whereas the *prgB* message (also apparently monocistronic) was produced at very low levels in the absence of pheromone and increased greatly as a result of pheromone induction. The *prgS* gene also showed inducible transcription. However, we have not been able to detect transcription of the *prgR* gene in *E. faecalis* cells, carrying either wild-type pCF10 or cloned fragments thereof, under any physiological conditions, using either of two different probes from different regions of the gene. This is somewhat surprising in view of the sequence data described above suggesting a putative single promoter for both *prgR* and *prgS*, 5′ to *prgR*. We are currently mapping both ends of this transcript to clarify the situation.

Our analysis of a negative regulatory region to the left of the *Eco*RI c fragment has not progressed as far as that of the positive control genes. However, several lines of evidence suggest that multiple negative control genes exist in this region and that these negative control functions can function in *trans* to block expression of aggregation and plasmid transfer in the absence of exogenous pheromone. The constitutive clumping phenotype conferred by the cloned pCF10 DNA represented by the heavy portion of line I in Fig. 2 was repressed when a second chimeric plasmid containing the negative control region was inserted into the same strain. When this strain was grown in the absence of selection for the second plasmid, it was cured spontaneously, restoring the clumping phenotype. At least one negative control gene, located in the tra 2 region (Fig. 2, line I), is involved in pheromone-binding activity and could encode the receptor. Our current model is that in the absence of pheromone, the negative control genes block the transcription of *prgB* and of additional transfer genes, possibly by interfering with the transcriptional activation functions of one or more genes in the *prgX-R-S* region. This interference could occur either via blocking of activator, e.g., *prgS* expression, by degradation of an activator gene product, or via direct competitive interference with activation near the 5′ end of the *prgB* gene.

Aggregation and transfer. Experimental evidence from several laboratories has indicated that synthesis of a pheromone-inducible adhesin, termed aggregation substance (AS), is necessary for efficient plasmid transfer in broth matings (3, 14, 22). Our results have indicated that this aggregation is mediated by a 150-kDa surface protein originally designated Tra150 (3) and now called Asc10, using the nomenclature proposed in this volume by Wirth et al. A protein of similar size, Asa1, has been implicated in aggregation in the pAD1 system by Galli and co-workers (14), and these two proteins share extensive homology at the DNA and protein levels (Wirth et al., this volume). The receptor on the recipient cell surface, binding substance (BS), is recognized and bound tightly by AS, enabling the stabilization of the mating pair or aggregate in liquid culture. BS is chromosomally determined, and since it is also expressed by donor cells, it is involved in self-aggregation of donor cultures exposed to pheromone in the absence of recipient cells (13, 20). There is published evidence (13) implicating lipoteichoic acid as BS in the pAD1 system. We recently generated BS⁻ mutants by transposon mutagenesis and carried out genetic and biochemical analysis of these strains (20). These results indicated that either lipoteichoic acid, a 105-kDa cell envelope protein, or both could comprise BS. The results also showed that aggregation via AS-AS binding does not occur and that the synthesis of AS on a donor cell following pheromone induction is not dependent on the presence of BS on the same cell.

We conducted genetic experiments to investigate the mechanism of formation of a mating channel between AS⁺ BS⁺ donors and BS⁻ recipient strains carrying chimeric plasmids conferring constitutive expression of various combinations of Asc10 and Sec10. We tested whether aggregation and plasmid transfer could occur between two *E. faecalis* cells when the normal orientation of the aggregation receptors was reversed, i.e., if the recipient cell was expressing AS but not BS and the donor cell was expressing BS (Table 1). Interestingly, AS-BS binding could

TABLE 1. Mating-pair formation mediated by reversed receptors[a]

Mating	Recipient cell receptors	Transconjugants/ donor
I	None	1×10^{-6}
II	AS, SE	7×10^{-5}
III	AS	2×10^{-4}
IV	SE	3×10^{-6}

[a]Fifteen-minute broth matings were carried out between pheromone-induced *E. faecalis* donor cells containing a pCF10:TN917 derivative showing a wild-type pheromone response and various derivatives of strain INY3000, a BS⁻ derivative of *E. faecalis* OG1SSp (20). (Under the conditions used in these experiments, transfer to a wild-type [BS⁺] recipient strain is normally about 10^{-2}.) The BS⁻ recipient strains carried plasmid pWM401 and chimeric derivatives containing cloned fragments of pCF10 encoding either the Asc10 aggregation substance protein (AS), the Sec10 surface exclusion protein (SE), or both. The donor cell expressed both AS and BS, but the AS on the donor cell could not promote binding to recipients since they did not express BS.

promote plasmid transfer in either orientation. This result suggests that AS acts as a bacterial grappling hook, whose function is to bring donor and recipient cells close enough together to allow for the formation of a mating channel between the two cells. We speculate that the formation of this channel involves additional, as yet unidentified plasmid-encoded gene products. If AS were directly involved in channel formation, it would be expected that this protein would have to be expressed in the donor cell in order to interact with both the recipient cell and the transferred DNA in the proper orientation. The expression of Sec10 in the recipient cell resulted in a slightly decreased frequency of transfer, consistent with the previously postulated role of this protein in surface exclusion (12). It should be noted that surface exclusion acts subsequent to aggregation (6, 12), probably by blocking mating channel formation, or the movement of DNA through the channel, also supporting the idea that aggregation and transfer are separable events.

In the case of the pheromone-inducible plasmids such as pCF10, it is generally assumed that the plasmid DNA transfer and the establishment of an autonomous replicon in the recipient cell constitute a unidirectional process similar to that involved in the conjugal transfer of the *E. coli* sex factor F (21). However very little investigation of these latter stages of transfer has been carried out, and there is almost no relevant experimental evidence available in relation to DNA transfer and subsequent events in this system.

Conjugal Transfer Events Mediated by Tn*925*

Plasmid pCF10 is somewhat unusual in that it carries two independent conjugal transfer systems. The tetracycline resistance determinant of this plasmid is located within a 15- to 17-kb region homologous to Tn*916* (2). For several years, we had no genetic evidence that this region actually functioned as a transposon. However, when *E. faecalis* cells were mated with *Bacillus subtilis* with selection for the transfer of tetracycline resistance (4), we obtained *Bacillus* transconjugants that contained insertions of the 17-kb Tn*916*-like DNA at random locations in the chromosome (but no other pCF10 DNA). These strains could serve as donors in subsequent matings with other *B. subtilis* or *E. faecalis* recipient strains. Further genetic and molecular analysis showed that a conjugative transposon similar to Tn*916* in both molecular structure and genetic properties, designated Tn*925*, was carried on pCF10 (4). We believe that when pCF10 is transferred between *E. faecalis* strains, the pheromone-inducible transfer system is so much more efficient than that of Tn*925* that the independent transfer of the latter element is never observed. However, pCF10

appears to be unable to replicate in *B. subtilis*, and this circumstance allowed us to detect the independent conjugative transposition of Tn*925*.

As illustrated by several chapters in this volume (Horaud et al.; Trieu-Cuot et al.; Scott; Clewell et al.), there is a great deal of interest in the transposition mechanism of conjugative transposons. However, relatively little is known about the cell-cell interactions involved in the conjugal transfer events mediated by these elements. We have carried out a number of genetic experiments investigating certain aspects of the conjugal transfer of Tn*925* that have not been examined in detail by other groups. Our investigations have focused on the ability of Tn*925* to promote the transfer of unlinked chromosomal genes and plasmids from the same donor cell. The results of a recent investigation (O. R. Torres, R. Z. Korman, S. A. Zahler, and G. M. Dunny, *Mol. Gen. Genet.*, in press) along with those of additional experiments will be summarized.

We carried out mating experiments between multiply marked strains of either *B. subtilis* or *E. faecalis* in which one of the members of the mating mixture contained Tn*925*. In the case of *B. subtilis*, two noncompetent strains derived from strain 168 and differentially marked at six well-characterized chromosomal loci (*urc*, *metB*, *trpC*, *ilvB*, *thrA*, and *hisA*) distributed over a wide portion of the chromosome were used to measure Tn*925*-mediated chromosomal mobilization. We observed that Tn*925* promoted extensive random recombination between all of these markers. This suggested that the mating interaction resulted in formation of a zygote that allowed for extensive interaction between the two chromosomes throughout their lengths. Although the chromosome of *E. faecalis* is much less well characterized and the transfer efficiency of Tn*925* is generally much lower in this species, we were able to demonstrate transfer of several chromosomal markers in this organism as well. We also have preliminary evidence for the mobilization of nonconjugative plasmids by Tn*925* in bacilli and in enterococci. Our chromosomal transfer studies indicated that the presence of the transposon was required for mobilization but that transfer of Tn*925* did not always occur in the case of chromosomal or plasmid mobilization. Moreover, it is not always possible to show that the transfer of genetic information has not been from the recipient (the strain originally lacking the transposon) to the transposon-containing donor strain. Taken together, these results suggest that at least in some cases, the conjugal transfer event mediated by Tn*925* genetically resembles a fusion of donor and recipient cells, creating at least transiently a diploid cell in which extensive chromosomal interactions can occur (Fig. 3). Our future investigations of this system will focus on further

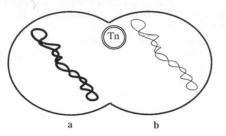

a   b

FIG. 3. Model for cell interactions in the transfer of the conjugative transposon Tn*925*. Cells have undergone a type of fusion event in which a diploid zygote is formed, at least temporarily, enabling substantial genetic recombination to occur between the two cells. The putative transposition intermediate is shown as an excised circular molecule in accordance with current models of transposition of these elements (see the chapters in this volume by Scott, Trieu-Cuot et al., and Clewell et al.). See text for further discussion.

genetic and physical characterization of the bacterial products of this novel mating interaction and on the potential of conjugative transposons to promote extensive genetic exchange among a wide variety of bacterial hosts.

Conclusions

From the preceding discussion, it is obvious that a remarkable variety of genetic interactions can be promoted by genes carried on a single *E. faecalis* plasmid. It will be of great interest to examine other, less well characterized gene transfer systems of gram-positive bacteria to determine what other novel genetic exchange phenomena have evolved in these organisms. In addition, there are many important questions that remain about the biology of the transfer systems discussed here and the roles that they play in the evolution of their bacterial hosts.

We thank Ann Viksnins and Vickie Johncox for excellent assistance with DNA sequence analysis.

This work was supported by Public Health Service grant AI19310 from the National Institutes of Health to G.M.D. and by a grant from the Cornell Biotechnology Program to S.A.Z. and G.D. S.-M.K. was supported by a graduate fellowship from the Cornell Biotechnology Program, and K.M.T. was the recipient of a graduate research award from Sigma Xi.

LITERATURE CITED

1. **Adler, B., C. Sasakawa, T. Tobe, S. Makino, K. Komatsu, and M. Yoshikawa.** 1989. A dual transcriptional activation system for the 230 kb plasmid genes coding for virulence-associated antigens of *Shigella flexneri. Mol. Microbiol.* **3:**627–635.
2. **Christie, P. J., and G. M. Dunny.** 1986. Identification of regions of the *Streptococcus faecalis* plasmid pCF-10 that encode antibiotic resistance and pheromone response functions. *Plasmid* **15:**230–241.
3. **Christie, P. J., S.-M. Kao, J. C. Adsit, and G. M. Dunny.** 1988. Cloning and expression of genes encoding pheromone-inducible antigens of *Enterococcus (Streptococcus) faecalis. J. Bacteriol.* **170:**5161–5168.
4. **Christie, P. J., R. Z. Korman, S. A. Zahler, J. C. Adsit, and G. M. Dunny.** 1987. Two conjugation systems associated with *Streptococcus faecalis* plasmid pCF10: identification of a conjugative transposon that transfers between *S. faecalis* and *Bacillus subtilis. J. Bacteriol.* **169:**2529–2536.
5. **Clewell, D. B., F. Y. An, M. Mori, Y. Ike, and A. Suzuki.** 1987. *Streptococcus faecalis* pheromone cAD1 response: evidence that the peptide inhibitor excreted by pAD1-containing cells may be plasmid determined. *Plasmid* **17:**65–68.
6. **Clewell, D. B., and B. L. Brown.** 1980. Sex pheromone cAD1 in *Streptococcus faecalis*: induction of a function related to plasmid transfer. *J. Bacteriol.* **143:**1063–1065.
7. **Clewell, D. B., and M. C. Gawron-Burke.** 1986. Conjugative transposons and the dissemination of antibiotic resistance in streptococci. *Annu. Rev. Microbiol.* **40:**653–659.
8. **Clewell, D. B., L. T. Pontius, F. Y. An, Y. Ike, A. Suzuki, and J. Nakayama.** 1990. Nucleotide sequence of the sex pheromone inhibitor (iAD1) determinant of *Enterococcus faecalis* conjugative plasmid pAD1. *Plasmid* **24:**156–161.
9. **Clewell, D. B., and K. E. Weaver.** 1989. Sex pheromones and plasmid transfer in *Enterococcus faecalis*: a review. *Plasmid* **21:**175–184.
10. **Dunny, G. M.** 1990. Genetic functions and cell-cell interactions in the pheromone-inducible plasmid transfer system of *Enterococcus faecalis. Mol. Microbiol.* **4:**689–696.
11. **Dunny, G. M., R. A. Craig, R. L. Carron, and D. B. Clewell.** 1979. Plasmid transfer in *Streptococcus faecalis*: production of multiple pheromones by recipients. *Plasmid* **2:**454–465.
12. **Dunny, G. M., D. L. Zimmerman, and M. L. Tortorello.** 1985. Induction of surface exclusion (entry exclusion) by *Streptococcus faecalis* sex pheromones: use of monoclonal antibodies to identify an inducible surface antigen involved in the exclusion process. *Proc. Natl. Acad. Sci. USA* **82:**8582–8586.
13. **Ehrenfeld, E. E., R. E. Kessler, and D. B. Clewell.** 1986. Identification of pheromone-induced surface proteins in *Streptococcus faecalis* and evidence of a role for lipoteichoic acid in the formation of mating aggregates. *J. Bacteriol.* **168:**6–12.
14. **Galli, D., F. Lottspeich, and R. Wirth.** 1990. Sequence analysis of *Enterococcus faecalis* aggregation substance encoded by the sex pheromone plasmid pAD1. *Mol. Microbiol.* **4:**895–904.
15. **Ike, Y., R. Craig, B. White, Y. Yagi, and D. B. Clewell.** 1983. Modification of a *Streptococcus faecalis* sex pheromone after acquisition of plasmid DNA. *Proc. Natl. Acad. Sci. USA* **80:**5369–5373.
16. **Krejci, E., V. Garzino, C. Mary, N. Bennani, and J. Pradel.** 1989. *Modulo*, a new maternally expressed *Drosophila* gene, encodes a DNA-binding protein with distinct acidic and basic regions. *Nucleic Acids Res.* **17:**8101–8116.
17. **Mitchell, P. M., and R. Tjian.** 1989. Transcriptional regulation in mammalian cells by sequence-specific DNA binding proteins. *Science* **245:**371–378.
18. **Mori, M., Y. Sakagami, Y. Ishii, A. Isogai, C. Kitada, M. Fujino, J. C. Adsit, G. M. Dunny, and A. Suzuki.** 1988. Structure of cCF10, a peptide sex pheromone which induces conjugative transfer of the *Streptococcus faecalis* tetracycline resistance plasmid, pCF10. *J. Biol. Chem.* **263:**14574–14578.
19. **Schaberg, D. R., and M. J. Zervos.** 1986. Intergeneric and interspecies gene exchange in gram-positive cocci. *Antimicrob. Agents Chemother.* **30:**817–822.
20. **Trotter, K. M., and G. M. Dunny.** 1990. Mutants of *Enterococcus faecalis* deficient as recipients in mating with

donors carrying pheromone-inducible plasmids. *Plasmid* **24**:57–67.

21. **Willets, N., and B. Wilkins.** 1984. Processing of plasmid DNA during bacterial conjugation. *Microbiol. Rev.* **48**:24–41.

22. **Yagi, Y., R. E. Kessler, J. H. Shaw, D. E. Lopatin, F. Y. An, and D. B. Clewell.** 1983. Plasmid content of *Streptococcus faecalis* strain 39-5 and identification of a pheromone (cPD1)-induced surface antigen. *J. Gen. Microbiol.* **129**: 1207–1215.

Variability of Chromosomal Genetic Elements in Streptococci

THEA HORAUD, GILDA DE CESPÉDÈS, DOMINIQUE CLERMONT, FELICIA DAVID, AND FRANÇOISE DELBOS

Laboratoire des Staphylocoques et des Streptocoques, Institut Pasteur, 75724 Paris Cedex 15, France

For the last 15 years, we have been concerned with the study of the genetic basis of antibiotic resistance in pathogenic streptococci. The results obtained in our laboratory with 152 antibiotic-resistant strains, belonging to streptococci of groups A, B, C, and G, *Streptococcus bovis, S. anginosus (milleri)*, and *S. pneumoniae*, indicate that most of the antibiotic resistance determinants are carried by the chromosome (3, 4, 12–14; this study). In fact, only 7 of 71 (10%) beta-hemolytic streptococci of groups A, B, C, and G harbor R plasmids; these plasmids code for resistance to erythromycin (Em^r) or to both erythromycin and chloramphenicol (11). However, all of the strains harboring R plasmids also carry on the chromosome a nonconjugative tetracycline resistance (Tc^r) determinant (11, 14). All of the other strains studied are plasmid-free except a few that harbor cryptic plasmids.

Chromosomal antibiotic resistance determinants that transfer by conjugation were found in only 25% of the 152 strains studied; usually, the determinants transfer en bloc and at low frequency (10^{-6} to 10^{-9} transconjugants per donor). Transfer usually occurs with recipients belonging to virtually all streptococcal and enterococcal species, more rarely with *Listeria monocytogenes*, and exceptionally with *Staphylococcus aureus* (3, 14).

All of the Tc^r streptococcal strains that we studied are also resistant to minocycline (Mn^r). Tc^r-Mn^r is considered a single marker. The Tc^r Mn^r phenotype is usually conferred by the Tet M (2) determinant and less frequently by the Tet O determinant (20).

Two types of mobile chromosomal elements carrying antibiotic resistance determinants have been described so far in streptococci and enterococci: conjugative transposons such as Tn*916* (6), which are about 16 to 20 kb, and composite conjugative elements such as Tn*3701* (16), which are usually larger than 50 kb.

The purpose of this report is to describe the molecular structure of Tn*3701* and to trace its dissemination among plasmid-free antibiotic-resistant streptococci of groups A, B, C, G, and D (*S. bovis*), *S. anginosus*, and *S. pneumoniae*.

Molecular Structure of Tn*3701*

Streptococcus pyogenes A454 (group A) carries Tn*3701*, a conjugative chromosomal element encoding Em^r and Tc^r-Mn^r. Tn*3701* transfers to the chromosome of streptococcal and of Rec$^+$ and Rec$^-$ *Enterococcus faecalis* recipients. This transfer occurs only in filter matings (18).

Different regions of the A454 chromosome corresponding to Tn*3701* were cloned by ligating *Eco*RI-digested chromosomal DNA to pUC8 (24) and by screening recombinant plasmids for homology to pIP1116 (16); pIP1116 was derived from the hemolysin plasmid pIP964 by the translocation of a 44.0-kb DNA fragment from the chromosome of A454 (18). Of the 75 recombinant plasmids hybridizing with pIP1116, four, which did not hybridize with the chromosome of antibiotic-susceptible group A streptococci, were selected as probes for further hybridization experiments. These plasmids, designated pIP1145, pIP1147, pIP1148, and pIP1149, carry inserts of 9.6, 4.9, 4.4, and 4.0 kb, respectively; the cloned fragments did not hybridize among themselves (16).

Hybridization experiments in which the A454 chromosome was probed with pAM170 (an *Escherichia coli* recombinant plasmid carrying Tn*916* [8]) revealed the presence of a structure similar to Tn*916* (22) that was designated Tn*3703* (16).

We have constructed a linear restriction map of a 50-kb region of the A454 chromosome (Fig. 1A) corresponding to the internal part of Tn*3701* (the extremities of this element have not yet been identified). We consider Tn*3701* to be a composite element, since it contains in its central region the transposon Tn*3703* (19.7 kb). Tn*3703* differs from Tn*916* by the presence of an *Eco*RI site and a 3.0-kb region on which the Em^r determinant is located. As in Tn*916*, the single *Hind*III site on Tn*3703* falls within the *tetM* gene (Fig. 1B).

Insertion of Tn*3701* in Various Streptococci and in *E. faecalis*

Molecular analysis of the elements carried by the wild-type strain A454 as well as by various transconjugants (obtained by matings with A454, as the donor) revealed how Tn*3701* inserts in the chromosome of different hosts (16). We examined 13 streptococcal transconjugants belonging to six different species and 12 *E. faecalis* transconjugants. Tn*3701* inserted without any apparent change in its structure in all streptococcal transconjugants as well as in one of the 12 *E. faecalis* transconjugants. Each of seven *E. faecalis* transconjugants carried one copy of Tn*3701* plus an additional copy of Tn*3703*. Four *E. faecalis* trans-

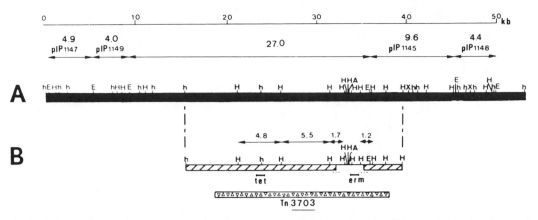

FIG. 1. Map of the A454 chromosomal region containing 50.0 kb of Tn*3701*. (A) Restriction sites for *Ava*I (A), *Eco*RI (E), *Hinc*II (H), *Hind*III (h), and *Xba*I (X). (B) Representation of the *Hind*III-*Hinc*II region of Tn*3701* homologous to Tn*916*. The sizes of some *Hinc*II fragments are indicated. Symbols: ▨, regions homologous to Tn*916*; ....., positions where homology to Tn*916* is thought to stop; *erm* and *tet*, locations of genes encoding erythromycin and tetracycline resistance genes, respectively; ▽△▽△▽, Tn*3703*. (From C. Le Bouguénec et al. [16].)

conjugants carried Tn*3703* alone, inserted at different sites of the chromosome. The independent translocation of Tn*3703* from A454 to an *E. faecalis* host, either onto the chromosome or onto a hemolysin plasmid (see below), is an additional genetic argument that Tn*3701* is a composite structure. That this transposition is limited to the *E. faecalis* recipient suggests that the host specificity of Tn*3703* is more restricted than that of Tn*3701*. Those transconjugants in which Tn*3703* was alone displayed no further conjugative transfer. In this respect, Tn*3703* differs markedly from Tn*916*. These results suggest that the genes responsible for the conjugative transfer of Tn*3701* are located outside of Tn*3703*.

Translocation of Tn*3703* onto the Hemolysin Plasmid pIP964

Translocation of Tn*3703* onto the hemolysin (Hly) plasmid pIP964 (1) gave rise to derivative plasmids that carried inserts of different sizes in various fragments of pIP964; most of these derivatives conferred an altered hemolytic phenotype to their *E. faecalis* hosts (10, 18).

Two of the pIP964 derivative plasmids, one encoding Em^r and the other encoding Tc^r-Mn^r, were used to test the ability of various Hly plasmids to coexist in the same cell. A series of plasmids, including pIP964, pAD1, pAMγ1, and pJH2, was found to constitute a distinct incompatibility group, designated IncHly (7).

Translocation of Tn*3703* onto pIP964 was exploited for the localization of genes implicated in hemolysin expression. The β-hemolysin of *E. faecalis* had been found by Granato and Jackson (9) to consist of two components that produce lytic activity only when both are present. Insertion of

Tn*3703* in either the *Eco*RI E (pIP1077) or the *Eco*RI G (pIP1117) fragment of pIP964 inactivated hemolysin expression. *E. faecalis* strains harboring either of these derivatives were nonhemolytic. However, hemolysis was restored in complementation tests (Fig. 2) when two *E. faecalis* strains, one harboring pIP1077 and the other harboring pIP1117, were plated in proximity. In this way, the genes encoding the two components necessary for hemolytic activity, designated lytic and colytic factors, were localized on the *Eco*RI G and E fragments of pIP964, respectively (19).

Dissemination of Tn*3701* among Different Streptococci

The dispersion of Tn*3701* and the relationship between Tn*3701* and other chromosomal elements were studied by DNA-DNA hybridization experiments. The probes used were the *Escherichia coli* recombinant plasmids pIP1145, pIP1147, pIP1148, pIP1149 (16), and pAM170 (8). The four former probes contain different regions of Tn*3701* lying outside of Tn*3703*; pAM170

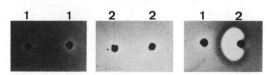

FIG. 2. Example of a hemolysis complementation test performed between bacterial pairs harboring pIP964-derived plasmids. 1, pIP1117; 2, pIP1077. Hemolytic activity of an *E. faecalis* strain harboring pIP1077 was evidenced when this strain was grown in the vicinity of another *E. faecalis* strain harboring pIP1117. (From C. Le Bouguénec et al. [19].)

contains Tn916, which was used as probe instead of Tn3703. pAT101 (21) and pUOA4 (23), which carry the Tet M and Tet O determinants, respectively, were also used as probes. The chromosomal DNA of the streptococcal strains studied here was digested with EcoRI and HincII.

Table 1 summarizes the results of hybridization experiments obtained with EcoRI-digested DNA of 124 strains. The clinical isolates tested belonged to streptococci of groups A, B, C, and G, S. bovis, S. anginosus, and S. pneumoniae (17; this study). All but seven of these strains were Tc^r Mn^r. Most of them were also resistant to other antibiotics such as erythromycin (macrolide-lincosamide-streptogramin B phenotype), chloramphenicol, and high levels of aminoglycosides. One S. pyogenes (group A), one S. anginosus, and 11 S. pneumoniae strains harbored cryptic plasmids; all other streptococci were plasmidfree. The strains were divided into several categories according to the results obtained by hybridization with the different probes used here.

Seventeen strains displayed homology with each of the four probes derived from Tn3701 as well as with Tn916 (Fig. 3). In most of these strains, the hybridizing fragments were different in size from those detected in A454, the original strain carrying Tn3701. These results indicate that some strains carry chromosomal elements with composite structures closely related to that of Tn3701 and that other strains carry regions of homology with Tn3701 having a structural organization different from that of Tn3701. Two of the 17 strains, S. agalactiae B109 (12) and S. pneumoniae BM6001 (4), carry chromosomal conjugative elements encoding multiple antibiotic resistance that were previously characterized, Tn3951 (67.0 kb) (15) and Ω(cat tet) (65.5 kb) (25), respectively. The comparison of Tn3701 with these two elements demonstrated that, like Tn3701, both Tn3951 and Ω(cat tet) are composite

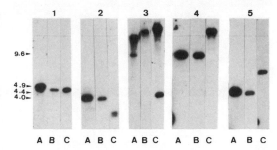

FIG. 3. Hybridization of EcoRI-digested chromosomal DNA (autoradiogram) with Tn3701. Lanes: A, A454; B, S. anginosus MG39; C, S. pneumoniae BM6047. Panels 1 to 5 represent α-^{32}P-labeled pIP1147, pIP1149, pAM170, pIP1145, and pIP1148, respectively. Molecular sizes (in kilobases) of internal Tn3701 fragments are shown on the left.

elements containing a Tn916-like structure in their central regions (17).

In 53 of the 124 strains studied, homologous sequences were revealed with two to four of the cloned Tn3701 regions, including Tn3703. In this category fell most of the S. bovis and S. pneumoniae strains examined. Homology to four of the cloned regions of Tn3701, including Tn3703, was detected in the chromosomal DNA of S. pneumoniae BM4200 (4). In a previous study of BM4200, the conjugative genetic element carried by this strain was described as a 25.3-kb conjugative transposon, designated Tn1545 (5). On the basis of mating experiments and hybridization data, we concluded that BM4200 carries a composite conjugative element resembling Tn3701; Tn1545, which has a structure similar to that of Tn916, constitutes part of this larger element (17).

In 16 of the strains, homology was found only with one to three of the cloned regions of Tn3701

TABLE 1. Presence of chromosomal elements in different streptococci

DNA-DNA homology with:	Streptococci of groups A, B, C, and G	S. bovis (group D)	S. anginosus (milleri)	S. pneumoniae
5 of the Tn3701 regions (including Tn3703)	7 (5)^a	2 (1)	5 (1)	3 (2)
2 to 4 of the Tn3701 regions (including Tn3703)	9 (6)	19 (1)	3	22 (4)
1 to 3 of the Tn3701^b regions (outside of Tn3703)	9 (4)	3	4 (1)	0
Tn916	11 (5)	4	9 (3)	2 (2)
Only Tet M or Tet O	3	0	2	1
None detected	4^c	0	2^c	0
Total strains	43	28	25	28

^{a}Numbers in parentheses indicate number of strains in which antibiotic resistance determinants were conjugative.

^{b}Except in one strain that was susceptible to tetracyclines, homology with only the Tet M or Tet O determinant was detected in all the strains.

^{c}These strains were susceptible to tetracyclines.

lying outside Tn*3703*. However, either the Tet M or, more rarely, the Tet O determinant was also detected in these strains. It is possible that Tn*3701*-like composite elements result from the insertion of Tn*916*-like structures into preexisting genetic elements like those found in these 16 strains.

In 26 of the strains, only Tn*916*-like structures were detected, without any other of the cloned Tn*3701* regions. An *Eco*RI site was found in 62% of the structures resembling Tn*916*, including those present in the composite elements; an example is given in Fig. 3 (panel 3, lanes A and C). We present in Fig. 4 (panels a) some examples of hybridization between *Hinc*II-digested chromosomal DNA and Tn*916*. In each strain, we detected several hybridizing fragments, four of which correspond in size to four (5.5, 4.8, 1.6, and 1.1 kb) of the five internal *Hinc*II fragments of Tn*916* (6). The same filters were then probed with pAT101. Homology with Tet M was revealed, in most cases, on a *Hinc*II fragment of 4.8 kb, as in Tn*916* (lanes 1 to 4, 6, 7, and 9). In a few cases, Tet M was located on a fragment of a different size, as in the example shown in panel b, lane 5. In a certain number of strains, such as those shown in lanes 8 and 10 (panel b), Tet M was located both on a 4.8-kb fragment and on a second fragment, suggesting either the presence of two copies of Tet M or, perhaps, the presence of a *Hinc*II site in this gene. Four of the 26 strains carried two elements: one similar to Tn*916* and one detectable only by its ability to transfer by conjugation. None of the additional elements hybridized to our probes, and therefore they remain unidentified.

In six strains, homology was revealed with either the Tet M or the Tet O determinant. For those strains carrying Tet M, Tn*916* hybridized only with a *Hinc*II fragment identical in size to that hybridizing with Tet M. Finally, the chromosomal DNA of the other six strains, which were susceptible to tetracycline-minocycline, did not detectably hybridize with any of the probes used here.

Concluding Remarks

The results presented here attest to the existence of several categories of chromosomal genetic elements displaying various degrees of relatedness either to Tn*3701* (56.4% of the strains) or to Tn*916* (21% of the strains); some of the strains of the latter category may carry, in addition to Tn*916*, a second independent element. About 13% of the strains carry elements having regions of homology with Tn*3701* lying outside of Tn*3703*.

In each category we found conjugative and nonconjugative elements. In fact, of the elements carried by 112 of the 124 strains studied here, only 31.5% transferred by conjugation. Some of the conjugative and nonconjugative elements had indistinguishable hybridization patterns. Consequently, under the experimental conditions presented here, we could not correlate the capacity to transfer by conjugation and the structure of the element.

In conclusion, our study reveals the diversity and complexity of the chromosomal genetic elements coding for antibiotic resistance that are found in various streptococci.

We thank Chantal Le Bouguénec and Karen Pepper for criticism of the manuscript and Odette Rouelland for secretarial

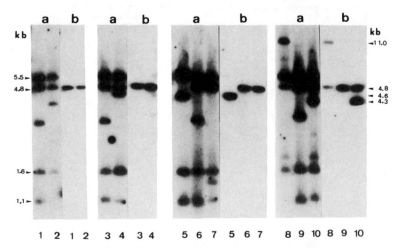

FIG. 4. Hybridization of *Hinc*II-digested chromosomal DNA (autoradiogram) with Tn*916* and Tet M. Panels a and b represent α-^{32}P-labeled pAM170 and pAT101, respectively; lanes 1 to 10 represent different *S. bovis* clinical isolates. Molecular sizes (in kilobases) of four of the five internal Tn*916* fragments are shown on the left. Molecular sizes of the fragments hybridizing with the Tet M probe are presented on the right.

assistance. We thank D. Clewell, P. Courvalin, and D. Taylor for giving us pAM170, pAT101, and pUOA4, respectively.

This work was supported by grant 873008 (Molecular Biology of Streptococci and Enterococci: Mobile Genetic Elements and Virulence Factors) from the Institut National de la Santé et de la Recherche Médicale to T.H.

LITERATURE CITED

1. **Borderon, E., G. Bieth, and T. Horodniceanu.** 1982. Genetic and physical studies of *Streptococcus faecalis* hemolysin plasmids. *FEMS Microbiol. Lett.* **14:**51–55.

2. **Burdett, V., J. Inamine, and S. Rajagopalan.** 1982. Heterogeneity of tetracycline resistance determinants in *Streptococcus. J. Bacteriol.* **149:**995–1004.

3. **Buu-Hoi, A., and T. Horaud.** 1985. Genetic basis of antibiotic resistance in group A, C and G streptococci, p. 231–232. *In* Y. Kimura, S. Kotami, and Y. Shiokawa (ed.), *Recent Advances in Streptococci and Streptococcal Diseases.* Reedbooks, Bracknell, England.

4. **Buu-Hoi, A., and T. Horodniceanu.** 1980. Conjugative transfer of multiple antibiotic resistance markers in *Streptococcus pneumoniae. J. Bacteriol.* **143:**313–320.

5. **Caillaud, F., C. Carlier, and P. Courvalin.** 1987. Physical analysis of the conjugative shuttle transposon Tn*1545. Plasmid* **7:**58–60.

6. **Clewell, D. B., and Gawron-Burke.** 1986. Conjugative transposons and the dissemination of antibiotic resistance in streptococci. *Annu. Rev. Microbiol.* **40:**635–659.

7. **Colmar, I., and T. Horaud.** 1987. *Enterococcus faecalis* hemolysin-bacteriocin plasmids belong to the same incompatibility group. *Appl. Environ. Microbiol.* **53:**567–570.

8. **Gawron-Burke, C., and D. B. Clewell.** 1984. Regeneration of insertionally inactivated streptococcal DNA fragments after excision of transposon Tn*916* in *Escherichia coli*: strategy for targeting and cloning of genes from gram-positive bacteria. *J. Bacteriol.* **159:**214–221.

9. **Granato, P. A., and R. W. Jackson.** 1969. Bicomponent nature of lysin from *Streptococcus zymogenes. J. Bacteriol.* **100:**865–868.

10. **Horaud, T., C. Le Bouguénec, and G. de Cespédès.** 1987. Genetic and molecular analysis of streptococcal and enterococcal chromosome-borne antibiotic resistance markers, p. 74–78. *In* J. J. Ferretti and R. Curtiss III (ed.), *Streptococcal Genetics.* American Society for Microbiology, Washington, D.C.

11. **Horaud, T., C. Le Bouguénec, and K. Pepper.** 1985. Molecular genetics of resistance to macrolides, lincosamides and streptogramin B (MLS) in streptococci. *J. Antimicrob. Chemother.* **16**(Suppl. A):111–135.

12. **Horodniceanu, T., L. Bougueleret, and G. Bieth.** 1981. Conjugative transfer of multiple-antibiotic resistance markers in beta-hemolytic group A, B, F, and G streptococci in the absence of extrachromosomal deoxyribonucleic acid. *Plasmid* **5:**127–187.

13. **Horodniceanu, T., A. Buu-Hoi, F. Delbos, and G. Bieth.** 1982. High-level aminoglycoside resistance in group A, B, G, D *(Streptococcus bovis)*, and viridans streptococci. *Antimicrob. Agents Chemother.* **21:**176–179.

14. **Horodniceanu, T., C. Le Bouguénec, A. Buu-Hoi, and G. Bieth.** 1982. Conjugative transfer of antibiotic resistance markers in beta-hemolytic streptococci in the presence and absence of plasmid DNA, p. 105–108. *In* D. Schlessinger (ed.), *Microbiology—1982.* American Society for Microbiology, Washington, D.C.

15. **Inamine, J., and V. Burdett.** 1985. Structural organization of a 67-kilobase streptococcal conjugative element mediating multiple antibiotic resistance. *J. Bacteriol.* **161:**620–626.

16. **Le Bouguénec, C., C. de Cespédès, and T. Horaud.** 1988. Molecular analysis of a composite chromosomal conjugative element (Tn*3701*) of *Streptococcus pyogenes. J. Bacteriol.* **170:**3930–3936.

17. **Le Bouguénec, C., G. de Cespédès, and T. Horaud.** 1990. Presence of chromosomal elements resembling the composite structure Tn*3701* in streptococci. *J. Bacteriol.* **172:**727–734.

18. **Le Bouguénec, C., T. Horaud, G. Bieth, R. Colimon, and C. Dauguet.** 1984. Translocation of antibiotic resistance markers of a plasmid-free *Streptococcus pyogenes* (group A) strain onto different streptococcal hemolysin plasmids. *Mol. Gen. Genet.* **194:**377–387.

19. **Le Bouguénec, C., T. Horaud, C. Geoffroy, and J. E. Alouf.** 1988. Insertional inactivation by Tn*3701* of pIP964 hemolysin expression in *Enterococcus faecalis. FEMS Microbiol. Lett.* **49:**455–458.

20. **Levy, S. B., L. M. McMurry, V. Burdett, P. Courvalin, W. Hillen, M. C. Roberts, and D. E. Taylor.** 1989. Nomenclature of tetracycline resistance determinants. *Antimicrob. Agents Chemother.* **33:**1373–1374.

21. **Martin, P., P. Trieu-Cuot, and P. Courvalin.** 1986. Nucleotide sequence of the *tet*M tetracycline resistance determinant of the streptococcal conjugative shuttle transposon Tn*1545. Nucleic Acids Res.* 14:7047–7058.

22. **Senghas, E., M. J. Jones, M. Yamamoto, C. Gawron-Burke, and D. B. Clewell.** 1988. Genetic organization of the bacterial conjugative transposon Tn*916. J. Bacteriol.* **170:**245–249.

23. **Taylor, D. E., K. Hiratsuka, H. Ray, and E. K. Manavathu.** 1987. Characterization and expression of a cloned tetracycline resistance determinant from *Campylobacter jejuni* plasmid pUA466. *J. Bacteriol.* **169:**2984–2989.

24. **Vieira, J., and J. Messing.** 1982. The pUC plasmids, an M13mp7-derived system for insertion mutagenesis and sequencing with synthetic universal primers. *Gene* **19:**259–268.

25. **Vijayakumar, M. N., S. D. Priebe, and W. R. Guild.** 1986. Structure of a conjugative element in *Streptococcus pneumoniae. J. Bacteriol.* **166:**978–984.

Molecular Dissection of the Transposition Mechanism of Conjugative Transposons from Gram-Positive Cocci

PATRICK TRIEU-CUOT, CLAIRE POYART-SALMERON, CECILE CARLIER,
AND PATRICE COURVALIN

Unité des Agents Antibactériens, Institut Pasteur, 75724 Paris Cedex 15, France

The conjugative transposon Tn*1545*, originally detected in the chromosome of *Streptococcus pneumoniae* BM4200, confers resistance to kanamycin and structurally related aminoglycosides *(aphA3)*, to macrolide-lincosamide-streptogramin B antibiotics *(ermAM)*, and to tetracycline *(tetM)* (8). This 25.3-kilobase (kb) element is a member of a closely related family of conjugative transposons that includes Tn*916*, Tn*918*, and Tn*920* from *Enterococcus faecalis* and Tn*919* from *Streptococcus sanguis* (25). These transposons can be conjugatively transferred, in the absence of plasmid, to a large variety of gram-positive bacteria, and they integrate at various loci of the genome of the recipient (8). Transposition, but not conjugation, also occurs in *Escherichia coli* (9). The two elements that have been the most extensively characterized are Tn*1545* and Tn*916*. The nucleotide sequences of the extremities of Tn*1545* are almost identical to those of Tn*916* for at least 250 bp (5, 7). Tn*1545* and Tn*916* are flanked by terminal imperfect (20 of 26 bp) inverted repeated sequences, but unlike most transposons, they possess variable base pairs at their extremities and do not generate a duplication of the target DNA upon insertion. Another unusual property of these elements is their ability to excise precisely from the target DNA (i.e., the targets are identical before insertion and after excision) in gram-positive and gram-negative hosts devoid of a homologous recombination system. With both transposons, however, the target after excision can differ from its original sequence by 2 bp (Tn*1545*) or 3 bp (Tn*916*) (7, 23).

Evidence for a Covalently Closed Circular Form of Tn*1545*

The current hypothesis for the mechanism of mobility (transposition and conjugation) of these transposons is that following excision, a nonreplicative circular intermediate of the element can undergo intracellular transposition to a new site, can be conjugatively transferred to a new host in which it transposes, or can be lost from the progeny during cell division (7, 12). Transposon Tn*1545* excises at a very high frequency (nearly 100%) when cloned on a multicopy vector into a *recA E. coli* strain, whereas it is more stable (less than 10% loss) after cloning on a low-copy-number replicon. We took advantage of the latter

property to characterize the nonreplicative circular intermediate of Tn*1545*. The 26.6-kb *Eco*RI fragment of plasmid pIP964 containing Tn*1545* (9) was cloned into the low-copy-number vector pSC101. Analysis, by agarose gel electrophoresis of crude lysates, of randomly selected *E. coli* transformants revealed the presence, in every clone, of both the excised and unexcised forms of the hybrid plasmid. We thus inferred (although, as expected, they were not visible on agarose gels) the presence of nonreplicative circular intermediates of Tn*1545*. Two 21-mer oligonucleotides specific for the left and right termini of Tn*1545* (5) were used to amplify a DNA fragment spanning the joined extremities of the hypothetical circular intermediate. Plasmid DNA purified from the transformants studied was used as a substrate for amplification by the polymerase chain reaction, followed by cloning and sequencing. The results obtained (Fig. 1) demonstrate that following excision, Tn*1545* is a circular structure with ends separated by either of the two hexanucleotides that were present at the transposon-target junctions and at the excision site of the corresponding replicon (23, 23a).

Localization of the Genes Necessary for Excision and Integration

We have previously shown that a 2,058-bp *Sau*3A right-junction fragment of transposon Tn*1545* (Fig. 2) specifies two gene products that are absolutely required for excision of the element (23). The DNA sequences of these genes, designated *xis-Tn* and *int-Tn*, have been determined, and the corresponding proteins, Xis-Tn and Int-Tn, have been identified in a bacterial cell-free coupled transcription-translation system. The genetic organization (Fig. 2) and the biological activities of these two proteins are reminiscent of those of lambdoid bacteriophage-encoded proteins Xis and Int, respectively. Using an in vivo *trans*-complementation assay, we have demonstrated that Int-Tn alone catalyzes, albeit at a low level, excision of a deletion derivative of Tn*1545* defective for this function and that Xis-Tn strongly stimulates the activity of Int-Tn (23). We have also shown that Int-Tn alone is able to catalyze in vivo integration of a circular intermediate of Tn*1545* defective for integration and

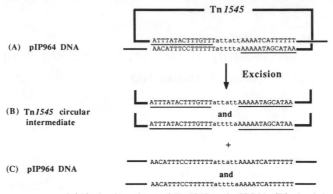

FIG. 1. Nucleotide sequences of (A) the junctions of Tn*1545* and pIP964, (B) the joined extremities of the nonreplicative circular intermediate of Tn*1545* after excision, and (C) the target site of pIP964 after excision. The 6-bp overlap sequence is indicated in lowercase. DNA of Tn*1545* is depicted by a thick line; that of pIP964 is depicted by a thin line. The sequences of the extremities of Tn*1545* are underlined. Only one strand of DNA is presented.

excision. These functional analogies are strengthened by the fact that the corresponding proteins display local similarities. Xis-Tn is a slightly basic protein (net charge of +3), 67 amino acid residues long, with an amino-terminal half similar to those of Xis excisionase of bacteriophage P22 and, to a lesser extent, of phages λ and φ80 (23). These three Xis proteins are required for normal excision of the corresponding phage, and ho-

mology among their amino acid sequences is limited to this NH$_2$-terminal region (18). Int-Tn is a highly basic (net charge, +17) 405-amino-acid protein that belongs to the Int-related family of site-specific recombinases. This group of proteins comprises phage-encoded integrases, transposon-encoded peptides (transposase and resolvase), and DNA invertases. Although these proteins display an overall sequence diversity, they can be

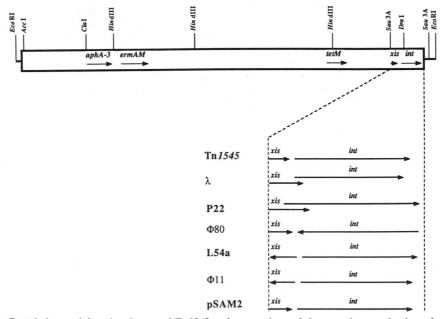

FIG. 2. Restriction and functional map of Tn*1545* and comparison of the genetic organization of *xis-Tn* and *int-Tn* of Tn*1545* with those of *xis* and *int* of bacteriophages λ, P22, φ80, L54a, and φ11 and of plasmid pSAM2. *aphA-3*, 3'-Aminoglycoside phosphotransferase type III determinant; *ermAM*, erythromycin resistance methylase determinant; *tetM*, tetracycline resistance determinant; *xis*, excisionase; *int*, integrase. Arrows indicate direction and extent of transcription. Only relevant restriction sites are shown.

Domain I   **Domain II**

```
Int (Tn1545)   NH2 -//- 213 ydeililkTGlRisEfggltipdldfe 240 - // - 342 HsLRhtfctnyanaG-mnpkalqyimGHan-iamtlnYya 380 -//- COOH 405

Int (λ)        NH2 -//- 199 rlamelavvTGqRvgdlcemkwsdivdg 226 - // - 306 HeLRslsa-rlyekq-isdkfaqhllGHks-dtmasqY-r 342 -//- COOH 356

Int (Φ80)      NH2 -//- 243 vflvkfimlTGcRtaEirlserswfrld 270 - // - 350 HdmRrtiatnlselG-cpphviekllGHqm-vgvmahYn- 386 -//- COOH 416

Int (P2)       NH2 -//- 181 kkiailclsTGaRwgEaarlkaeniihn 208 - // - 273 HaLRhsfathfminG-gsiitlqrilGHtr-ieqtmvYah 309 -//- COOH 343

Int (P4)       NH2 -//- 232 miavklsllTfvRssElrfarwdefdfd 259 - // - 346 HgfRtmargalgesGlwsddaiergslHsernnvraaYih 386 -//- COOH 438

Int (P22)      NH2 -//- 202 ksvvefalsTGlRrsniinlewqqidmq 230 - // - 312 HdlRhtwaswlvqaG-vpisvlqemgGwes-iemvrrYah 349 -//- COOH 387

Int (186)      NH2 -//- 191 etvvriclaTGaRwsEaeslrksqlaky 218 - // - 237 HvLRhtfashfmmnG-gnilvlqrvlGHtd-ikmtmrYah 273 -//- COOH 296

Cre (P1)       NH2 -//- 205 tagvekalslGvtklverwisvsgvadd 232 - // - 290 HsaRvgaardmaraG-vsipeimqagGwtn-vnivmnYir 326 -//- COOH 343

Int (Φ11)      NH2 -//- 183 rqltrllfysGlRigEalalqvwkdydki 211 - // - 295 HhLRhsyasylinnG-vdmyllmelmrHsnitetiqtYsh 332 -//- COOH 348

Int (L54)      NH2 -//- 170 agavevqalTGmRigEllalqvkdvdlk 240 - // - 298 HtLRhthisllaemn-islkaimkrvGHrdekttikvTth 336 -//- COOH 354

Int (HP1)      NH2 -//- 195 glivriclaTGaRwsEaetltqsqvmpy 210 - // - 280 HvLRhtfashfmmnGgnilvlkleilGHst-iemtmrYah 318 -//- COOH 337

Int (pSAM2)    NH2 -//- 198 ayiv-vallTGaRteElraltwdhvflk 223 - // - 327 reLRhsfvsllsdrG-vpleeisrlvGHsgtavteevYrk 365 -//- COOH 388
                                     *   *                                       *    *
```

FIG. 3. Local homology among the integrase family of site-specific recombinases from Tn*1545*, pSAM2, and lambdoid phages. Domains I and II are those described by Argos et al. (2). The amino acid sequence of Int (Tn*1545*) is from Poyart-Salmeron et al. (23); that of Int (λ) is from Hoess et al. (16); those of Int (φ80) and Int (P22) are from Leong et al. (18); those of Int (P2), Int (P4), and Int (186) are from Argos et al. (2); that of Cre (P1) is from Sternberg et al. (26); those of Int (φ11) and Int (L54a) are from Ye et al. (30); that of Int (HP1) is from Goodman and Scocca (13); that of Int (pSAM2) is from Boccard et al. (4). Numbers indicate positions of the corresponding residues in the sequence of each protein. Gap (-) were introduced to maximize homology. Positions at which a minimum of 9 of the 12 proteins contain the same amino acid are shown by underlined uppercase letters; positions at which a minimum of 9 of the 12 proteins contain homologous amino acids are shown by underlined lowercase letters. Chemically similar residues are I, L, V and M; D and E; R and K; Q and N; S and T; and F and Y. Residues identical in all sequences are marked by asterisks. Arrow indicates the tyrosine residue likely to be covalently linked to DNA during recombination.

aligned in their carboxy-terminal halves, where two domains (I and II), thought to be part of in their active sites, are well conserved (2). In particular, a histidyl, an arginyl, and a tyrosyl residue are perfectly conserved in domain II (Fig. 3) (14, 22). This invariant tyrosyl residue is likely to be covalently linked to the DNA during recombination. These two conserved regions were found in Int-Tn at the same relative position (Fig. 3).

A Model for Excision and Integration of Tn1545 and Related Elements

The regional similarity observed between Int-Tn and Int-related site-specific recombinases suggests that Int-Tn belongs to this family (2). Within this group of enzymes, the best-characterized recombination events are those catalyzed by Int (λ), Cre (P1), and Flp from the 2μm yeast plasmid for which in vitro systems have been developed. All three recombinases bind to specific sequences and generate staggered nicks with 5′ protruding ends of six (Cre), seven (Int), or eight (Flp) bases (1, 11, 15, 19). Based on the functional properties of the phage-encoded integrases and on analyses of the nucleotide sequences of (i) the transposon-target junctions, (ii) joined extremities of the circular intermediate obtained after excision, and (iii) targets before insertion and after excision, we proposed a model for Int-catalyzed excision and integration of Tn1545 and related conjugative transposons (23, 23a). The model postulates that excision of Tn1545 occurs by reciprocal site-specific recombination between nonhomologous regions of the transposon-target junctions (23). It stipulates that in the presence of Xis-Tn, Int-Tn generates 5′-hydroxyl protruding 6 (possibly 7)-bp staggered nicks at the extremities of the transposon and catalyzes strand exchange reactions leading to a nonreplicative covalently closed circular form of the transposon and to an excised form of the target replicon (Fig. 4). Demonstration that the extremities of the circularized element are separated by either of the two hexanucleotides that were present at the transposon-target junctions in the donor replicon provides further support for this model (23a). Although excisive and integrative recombinations can be viewed as a reversible reaction, they correspond in fact to two different pathways: Int-Tn alone catalyzes integration, whereas efficient excision requires Xis-Tn and Int-Tn. The λ Xis protein is a small basic polypeptide that binds cooperatively to two tandemly repeated sites in the P arm and thus stimulates the binding of Int (31). The DNA bend induced by λ Xis presumably facilitates the formation of a productive synapsis (27). One can reasonably speculate that the stimulatory effect of Xis-Tn on Int-Tn-catalyzed excision is similar to that exerted by λ Xis on λ Int.

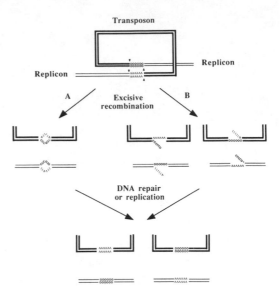

FIG. 4. Proposed model for excision of the conjugative transposons from gram-positive cocci. Following specific binding to the ends of the element in the presence of Xis-Tn, Int-Tn generates 5′-hydroxyl protruding staggered nicks at the borders of the 6 (possibly 7)-bp overlap sequences. (A) The presence of two *cis* DNA-protein complexes in a productive synapsis results in a complete strand exchange reaction leading to an excised form of the target replicon and to a nonreplicative covalently closed circular form of the transposon. (B) The top or the bottom strand at the recombination site is resealed, leaving a nick in the complementary strand. Following excision, the heteroduplexes or the unsealed DNA at the recombination sites can be removed by the mismatch repair system of the host or by replication for the target replicon. Elimination of either of the nonhomologous DNA segments results in the random production of one of the two possible excision products, independently, in the target replicon and in the circular intermediate. DNA of the transposon is depicted by a thick line; that of the target replicon is depicted by a thin line. ○○○○○○ and △△△△△△ represent two different 6-bp overlap sequences; ▼ and ▲ indicate the positions of Int-Tn cleavage sites on both strands of DNA. Because of sequence identity between the known targets and the left end of Tn1545 and Tn916, the overlap region could be 7 bp long (23).

The integration-excision cycle of bacteriophage λ occurs by two sequential reciprocal strand exchanges within a 15-bp-long homologous core region present in both bacterial (*attB*) and phage (*attP*) attachment sites (for a recent review, see reference 17). The strand exchanges are catalyzed by λ Int, which introduces 7-bp staggered nicks within the core sequence (11, 19). The position of the strand breaks defines the overlap region in which the products of recombination contain DNA strands derived from the parents (28, 29). The requirement for homology between the overlap regions is not absolute, and illegitimate recombination involving secondary *att* sites (*saf* mutants) has been described (3, 20, 21, 28). Crosses between *saf* and wild-type *att* sites produce heteroduplex recombinant sites that contain mismatches (3, 28). All characterized excisions or integrations of Tn*1545* and Tn*916* involve Int-Tn-mediated illegitimate recombination between 6 (or 7)-bp overlap regions that can differ at up to three positions (7, 23, 23a). The striking similarity between the integration-excision systems of Tn*1545* and of bacteriophage λ led us to suggest that Int-Tn catalyzes reciprocal strand exchange between the cleavage sites via the formation of a heteroduplex in the overlap region (Fig. 4A). Indirect evidence for the presence of a heteroduplex between the joined termini of circular intermediates of Tn*916* has been recently reported (6). In the two recombinant products formed, the overlap regions contain one strand from each parental site. Subsequent in vivo or in vitro (polymerase chain reaction) replication through these heteroduplexes must generate equivalent

amounts of the two types of segregants in each outcome product. This is indeed the case after excision of Tn*1545* from pIP964 (two mismatches; Fig. 1A) in *E. coli* (23, 23a). Similarly, excision of Tn*916* from pAM120 (three mismatches) generates equal amounts of the two expected hexanucleotides at the excision site of the target replicon (7). In vivo, heteroduplexes between the termini of the nonreplicative circular intermediate of the transposon can be corrected by mismatch repair, and again unbiased segregation of progeny molecules is expected: a DNA strand should not be preferentially repaired, since no strand is undermethylated relative to the other. In another pathway proposed for strand exchange (23), Int-Tn first cleaves and religates at random either the top or the bottom DNA strand, whereas the second cleavage is not followed by ligation (Fig. 4B). The formation of such DNA templates is homology independent, and following repair of the unpaired strand, equivalent amounts of both strands should always segregate

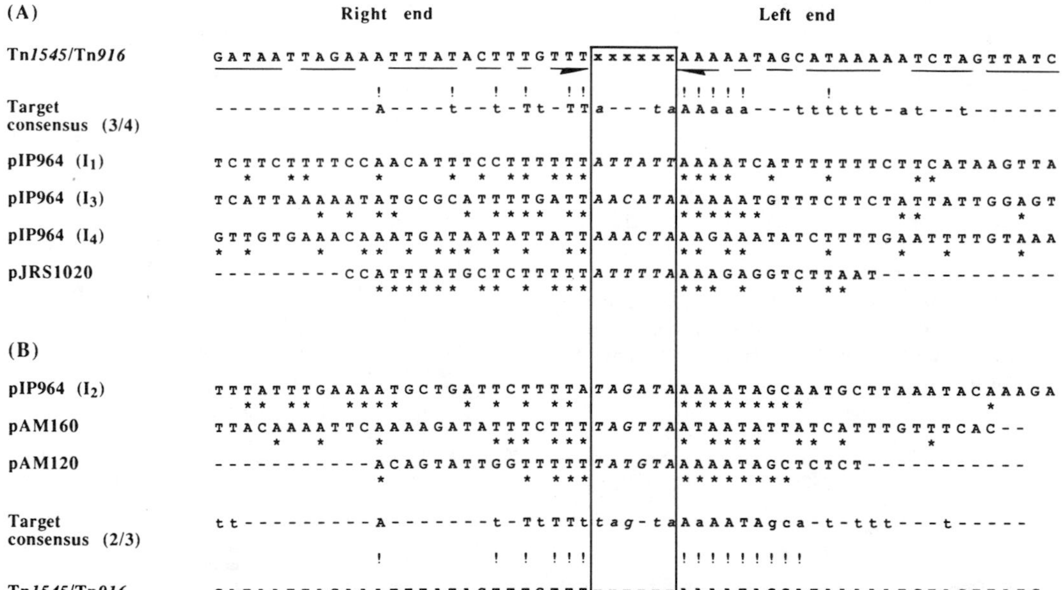

FIG. 5. Consensus sequence for integration of transposons Tn*1545* and Tn*916*. The target DNA sequences of variants of Tn*1545* and Tn*916* having five (A) or four (B) adenyl residues in the left extremity were aligned relative to the Int-Tn cleavage sites; the overlap region is written in italic and boxed. The sequence of pIP964 (I₁ and I₂) is from Caillaud and Courvalin (5), those of pAM120 and pAM160 are from Clewell et al. (7), and that of pJRS1020 is from Caparon and Scott (6). The nucleotide sequences of the joined termini of the two variants of Tn*1545* and Tn*916* are included for comparison at the top and bottom of the figure; horizontal arrows denote terminal imperfect inverted repeats, and xxxxxx represents the 6 (possibly 7)-bp variable overlap regions. Right and left ends refer to integrated transposons. Identical bases, in pairwise comparisons, between every target and the termini of the transposon are indicated by asterisks. The consensus sequences for integration are based on a minimum score of identity depending on the number of sequences considered (three of four in panel A; two of three in panel B). Identities between the termini of the transposons and the target consensus sequences are indicated by exclamation marks.

in the two recombinant products. This possibility is very unlikely, since excision of Tn916 from pAM160 (three mismatches) yields only one type of sequence at the excision site of the target (7). Preferential segregation of mismatches into one of the two recombinant products, a feature already observed in the λ system (28), suggests that both strand exchanges occur at one side of the mutant sites (3).

Target Specificity of Tn1545 and Related Elements

The Tn1545 and Tn916 integration sites described so far are structurally related and share a certain degree of similarity with the transposons' termini, in particular in the vicinity of the recombining sites (Fig. 5). The overlap regions of the circular intermediates of the elements are flanked by 7-bp sequences composed almost entirely of thymyl and adenyl residues at the right and left ends, respectively (Fig. 5). Similar or degenerated palindromes based on this motif are present in all of the Tn1545 and Tn916 integration sites, and it is likely that they constitute a structural signal for integration. By analogy with other site-specific recombination systems, these "obligatory regions" might constitute binding sites for the integrase (10, 24). Sequence homology between the overlap regions could also influence the efficiency of integration of the elements. We have demonstrated that exchange of overlapping sequences between DNA of the target and of the conjugative transposons of gram-positive cocci occurs during successive cycles of integration and excision. Thus, nonreplicative circular intermediates of the elements differing only in their overlap regions could display different efficiencies for integration at a given target site.

This work was supported by a grant (ATP "Bactéries Lactiques") from the Centre National de la Recherche Scientifique. C.P.-S. was a recipient of a fellowship from the Fondation pour la Recherche Médicale.

LITERATURE CITED

1. **Andrews, B. J., G. A. Proteau, L. G. Beatty, and P. D. Sadowski.** 1985. The FLP recombinase of the 2μ circle DNA of yeast: interaction with the target sequences. *Cell* **40**:795–803.
2. **Argos, P., A. Landy, K. Abremski, S. B. Egan, E. Haggard-Ljungquist, R. H. Hoess, M. L. Kahn, B. Kalionis, S. V. L. Narayana, L. S. Pierson III, N. Sternberg, and J. M. Leong.** 1986. The integrase family of site-specific recombinases: regional similarities and global diversity. *EMBO J.* **5**:433–440.
3. **Bauer, C., J. F. Gardner, and R. I. Gumport.** 1985. Extent of sequence homology required for bacteriophage λ site-specific recombination. *J. Mol. Biol.* **181**:187–197.
4. **Boccard, F., T. Smokvina, J. L. Pernodet, A. Friedmann, and M. Guérineau.** 1989. The integrated conjugative plasmid pSAM2 of *Streptomyces ambofaciens* is related to temperate bacteriophages. *EMBO J.* **8**:973–980.
5. **Caillaud, F., and P. Courvalin.** 1987. Nucleotide sequence of the ends of the conjugative shuttle transposon Tn1545. *Mol. Gen. Genet.* **209**:110–115.
6. **Caparon, M. G., and J. R. Scott.** 1989. Excision and insertion of the conjugative transposon Tn916 involves a novel recombination mechanism. *Cell* **59**:1027–1034.
7. **Clewell, D. B., S. E. Flannagan, Y. Ike, J. M. Jones, and C. Gawron-Burke.** 1988. Sequence analysis of termini of conjugative transposon Tn916. *J. Bacteriol.* **170**:3046–3052.
8. **Courvalin, P., and C. Carlier.** 1986. Transposable multiple antibiotic resistance in *Streptococcus pneumoniae*. *Mol. Gen. Genet.* **205**:291–297.
9. **Courvalin, P., and C. Carlier.** 1987. Tn1545: a conjugative shuttle transposon. *Mol. Gen. Genet.* **206**:259–264.
10. **Craig, N. L.** 1988. The mechanism of conservative site-specific recombination. *Annu. Rev. Genet.* **22**:77–105.
11. **Craig, N. L., and H. A. Nash.** 1983. The mechanism of phage λ site-specific recombination: site-specific breakage of DNA by Int topoisomerase. *Cell* **35**:795–803.
12. **Gawron-Burke, C., and D. B. Clewell.** 1982. A transposon in *Streptococcus faecalis* with fertility properties. *Nature* (London) **300**:281–284.
13. **Goodman, S. D., and J. Scocca.** 1989. Nucleotide sequence and expression of the gene for the site-specific integration protein from bacteriophage HP1 of *Haemophilus influenzae*. *J. Bacteriol.* **171**:4232–4240.
14. **Gronostajski, R. M., and P. D. Sadowski.** 1985. The FLP recombinase of the *Saccharomyces cerevisiae* 2μm plasmid attaches covalently to DNA via a phosphotyrosyl linkage. *Mol. Cell. Biol.* **5**:3274–3279.
15. **Hoess, R. H., and K. Abremski.** 1985. Mechanism of strand cleavage and exchange in the Cre-*lox* site-specific recombination system. *J. Mol. Biol.* **181**:351–362.
16. **Hoess, R. H., C. Foeller, K. Bidwell, and A. Landy.** 1980. Site-specific recombination functions of bacteriophage λ: DNA sequence of regulatory regions and overlapping structural genes for Int and Xis. *Proc. Natl. Acad. Sci. USA* **77**:2482–2486.
17. **Landy, A.** 1989. Dynamic, structural, and regulatory aspects of λ site-specific recombination. *Annu. Rev. Biochem.* **58**:913–949.
18. **Leong, J. M., S. E. Nunes-Duby, A. B. Oser, C. F. Lesser, P. Youderian, M. M. Susskind, and A. Landy.** 1986. Structural and regulatory diversity among site-specific recombination genes of lambdoid phages. *J. Mol. Biol.* **189**:603–616.
19. **Mizuuchi, K., R. Weisberg, L. Enquist, M. Mizuuchi, M. Burasczyska, C. Foeller, P.-L. Hsu, W. Ross, and A. Landy.** 1981. Structure and function of the *att* site: size, Int-binding sites, and location of the crossover point. *Cold Spring Harbor Symp. Quant. Biol.* **45**:429–437.
20. **Nash, H., C. E. Bauer, and J. F. Gardner.** 1987. Role of homology in site-specific recombination of bacteriophage λ: evidence against joining of cohesive ends. *Proc. Natl. Acad. Sci. USA* **84**:4049–4053.
21. **Nunes-Düby, S. E., L. Matsumoto, and A. Landy.** 1987. Site-specific recombination intermediates trapped with suicide substrates. *Cell* **50**:779–788.
22. **Pargellis, C. A., S. E. Nunes-Duby, L. Moitoso de Vargas, and A. Landy.** 1988. Suicide recombination substrates yield covalent λ integrase-DNA complexes and lead to identification of the active site tyrosine. *J. Biol. Chem.* **263**:7678–7685.
23. **Poyart-Salmeron, C., P. Trieu-Cuot, C. Carlier, and P. Courvalin.** 1989. Molecular characterization of two proteins involved in the excision of the pneumococcal transposon Tn1545: homologies with other site-specific recombinases. *EMBO J.* **8**:2425–2433.
23a. **Poyart-Salmeron, C., P. Trieu-Cuot, C. Carlier, and P. Courvalin.** 1990. The integration-excision system of the streptococcal transposon Tn1545 is structurally and functionally related to those of lambdoid phages. *Mol. Microbiol.* **4**:1513–1521.
24. **Sadowski, P.** 1986. Site-specific recombinases: changing partners and doing the twist. *J. Bacteriol.* **165**:341–347.

25. **Senghas, E., J. M. Jones, M. Yamamoto, C. Gawron-Burke, and D. B. Clewell.** 1988. Genetic organization of the bacterial conjugative transposon Tn916. *J. Bacteriol.* **170:**245–249.

26. **Sternberg, N., B. Sauer, R. H. Hoess, and K. Abremski.** 1986. Bacteriophage P1 *cre* gene and its regulatory region. Evidence for multiple promoters and for regulation by DNA methylation. *J. Mol. Biol.* **187:**197–212.

27. **Thompson, J. F., L. Moitoso de Vargas, C. Koch, R. Kahmann, and A. Landy.** 1987. Cellular factors couple recombination with growth phase: characterization of new component in the λ site-specific recombination pathway. *Cell* **50:**901–908.

28. **Weisberg, R. A., L. W. Enquist, C. Foeller, and A. Landy.** 1983. Role for DNA homology in site-specific recombination. The isolation and characterization of a site affinity mutant of coliphage λ. *J. Mol. Biol.* **170:**319–342.

29. **Weisberg, R. A., and A. Landy.** 1983. Site-specific recombination in phage lambda, p. 211–250. *In* R. Hendrix, J. Roberts, F. Stahl, and R. Weisberg (ed.), *Lambda II*. Cold Spring Harbor Laboratory, Cold Spring Harbor, N.Y.

30. **Ye, Z. H., S. A. Buranen, and C. Y. Lee.** 1990. Sequence analysis and comparison of *int* and *xis* genes from staphylococcal bacteriophages L54a and φ11. *J. Bacteriol.* **172:**2568–2575.

31. **Yin, S., W. Bushman, and A. Landy.** 1985. Interaction of the λ site-specific recombination protein Xis with attachment site DNA. *Proc. Natl. Acad. Sci. USA* **82:**1040–1044.

Mechanism of Transposition of Conjugative Transposons

JUNE R. SCOTT

*Department of Microbiology and Immunology, Emory University Health Sciences Center,
Atlanta, Georgia 30322*

Antibiotic-resistant clinical isolates of several different groups of streptococci often do not contain plasmids. Instead, elements named conjugative transposons have been identified as the carriers of resistance markers in these organisms (recently reviewed by Clewell and Gawron-Burke [9]; 15, 18).

Conjugative transposons have a broad host range, since they are capable of transposition in both gram-positive and gram-negative bacteria. So far, it appears that all conjugative transposons carry *tetM* (1), a determinant of tetracycline resistance which is expressed in all bacteria investigated, including both gram-negative and gram-positive organisms. Many conjugative transposons, which may be as large as 60 or more kb, carry additional resistance markers as well (2, 7, 21, 27). The elements may be unique (consisting of a single transposon) or composite (two regions of the element may transpose independently) (21).

Conjugative transposons differ in important ways from other classes of transposons because they do not cause duplication of the target sequence upon insertion (3, 4, 8, 17). They cause gram-positive bacteria in which they reside to become donors in a conjugative event in which the transposon is transferred to another cell (10, 11). In the recipient, the transposon is found by Southern blot analysis to be located in a different DNA region from its location in the donor (10, 13). The transposition event, therefore, occurs during or after the mating. These elements are highly promiscuous. They can be transferred by themselves or on plasmids between unrelated species and even to members of distant genera (12, 19, 20, 25; P. Mullany, M. Wilks, I. Lamb, C. Clayton, B. Wren, and S. Tabaqchali, *J. Gen. Microbiol.*, in press). Thus, it appears that these transposons are very important medically in the spread of antibiotic resistance.

In most strains, the target for insertion of conjugative transposons is not specific, although in a few strains unique insertion sites have been reported (15, 16; Mullany et al., in press). These elements have been extremely valuable as insertional mutagens because they were believed to insert at random locations. However, our recent work in *Bacillus subtilis* suggests that there are target limitations.

Several different conjugative transposons have been described. One of the first, Tn*916* (11), has served as a prototype because of its small size (16.4 kb). Although the other well-studied element, Tn*1545*, carries additional resistance determinants, its ends are almost identical in sequence to those of Tn*916* (4, 8), and an enzyme (the product of ORF2) which is encoded near one end of the element and is required for conjugative transposition (24) is functionally interchangeable between these two elements (M. J. Storrs et al., unpublished data; see below). Many of the larger elements found in clinical streptococcal isolates also show homology to the ends of Tn*916* and therefore may be considered members of the same family. The only conjugative transposon reported to differ at its ends is one identified in *Clostridium difficile* (Mullany et al., in press).

We have capitalized on the promiscuity of Tn*916* to use it as a shuttle vector to transfer cloned DNA from *Escherichia coli* to group A streptococci (22). This technique, described in more detail by M. G. Caparon and J. R. Scott (*Methods Enzymol.*, in press), should be generally applicable for bacteria that have not been found to be transformable. It requires a transformable gram-positive intermediate host as recipient of the transposon containing the cloned DNA. This host then becomes a conjugative donor for the transposon. This approach may be used either for random chromosomal insertion of a cloned gene in single copy or for allelic replacement. By such allelic replacement, we were able to construct a deletion of the structural gene for the M protein *(emm)* in a group A streptococcal strain which is otherwise isogenic with its *emm6.1* parent (22) (Fig. 1).

A Circular Transposon Molecule Can Serve as an Intermediate in Transposition

Clewell and associates proposed that transposition of Tn*916* proceeds through a mechanism more closely related to that of induction of the prophage lambda than to that used by other types of transposons (9, 13, 14). They suggested that excision of the transposon was the rate-limiting step and would produce a circular intermediate that could then integrate into a different DNA target. When we examined the cosmid clone of the *Streptococcus pyogenes* DNA containing the region including a Tn*916* insertion responsible for reduction in M-protein RNA yield (the *mry-1* mutant), we observed an unexpected extra band in agarose gel electrophoresis (5). By restriction en-

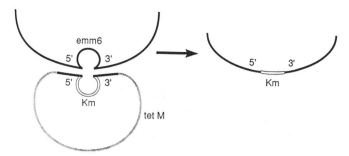

FIG. 1. Allelic replacement of *emm6* by *aphA3* (kanamycin resistance). The dark line represents chromosomal DNA of the *emm6*⁺ *S. pyogenes* strain JRS4. The gray circle represents the covalently closed circular form of Tn916 which is introduced into JRS4 by mating with a *B. subtilis* strain and contains the sandwich cloned at the *Bst*XI site. The sandwich contains the *aphA3* (kanamycin resistance [Km] gene; represented in white) cloned between the region 5′ of *emm6* and the region 3′ of *emm6*. The homologous regions 5′ and 3′ of *emm6* on the transposon and chromosome of the recipient *S. pyogenes* strain pair with each other, and recombination leads to substitution on the kanamycin resistance gene for *emm6* and to loss of the rest of the transposon, including the *tetM* gene.

donuclease analysis and Southern blot hybridization, we demonstrated that this band contained a covalently closed supercoiled form of Tn916 with no detectable associated target DNA.

We reasoned that if this were the predicted intermediate form for transposition, it should be able to carry out all of the reactions that Tn916 can perform with an efficiency at least as great as that of the cosmid containing Tn916. To test for transposition by the large (16.4-kb) supercoiled transposon molecule, we used *B. subtilis* protoplasts as recipients because no DNA rearrangements occur on entry in this system. The purified covalently closed circular Tn916 transposed into *B. subtilis* protoplasts about 150 times as efficiently as the cosmid DNA carrying the transposon from which it was derived (5). To confirm that transformation resulted from transposition, we determined that the transposon had inserted at different locations in the *B. subtilis* transformants (5). Furthermore, the transformants were able to act as conjugative donors. The *B. subtilis*::Tn916 strains transferred Tn916 at a low frequency to an *S. pyogenes* recipient, confirming that the transposon was functionally intact (5). Therefore, a covalently closed Tn916 molecule can serve as an intermediate in transposition.

Mechanism of Excision as Deduced from the Sequence of Excisant Joints

Excision of conjugative transposons was at first believed to regenerate the original target sequence that had been present prior to insertion of the element, since in most cases insertional mutants reverted to the wild-type phenotype when tetracycline resistance was lost (4, 14). However, for Tn916, comparison of the DNA sequence of the unmutagenized target with that of the same region of the excisant (the molecule from which the element has excised) indicated

discrepancies (6, 8). Therefore, when Tn916 insertion mutant genes are cloned in *E. coli*, it is important not to assume that the excisants, which arise frequently in that organism, have the wild-type sequence. It is necessary to clone the region from the unmutagenized parental strain to be sure that the true wild type is obtained (Caparon and Scott, in press).

Analysis of the regions near the ends of Tn916 in the two classes of excisants that arose spontaneously from the cloned *mry*::Tn916 mutant DNA in *E. coli* (6) indicated that the bases present at the joint corresponded either to the bases left of the inserted transposon or to the bases to its right. These flanking regions of approximately five bases have been named coupling regions. Thus, Tn916 appears to excise, taking with it one of the two coupling regions and leaving the other behind (Fig. 2). A similar type of explanation was among those proposed by Clewell et al. (8) on the basis of their data. Although the Courvalin group initially reported only one type of excisant from each insertion of Tn1545 (4), when the Tn916 results were explained to them, they found the second excisant for Tn1545 and subscribed to the model that we had proposed to them (24).

Sequence of the Joint of the Circular Transposon Molecule

If excision is a reciprocal recombination process, the circular transposon molecule that has been excised should have the coupling region that is not left behind in the excisant (Fig. 1). To determine the sequence of the joint that was spliced together during excision to form the circular transposon molecule, several approaches were possible. We avoided sequencing directly from the circles because we anticipated the presence of a mixture of two types of molecules. A mixture of sequences for this coupling region would not

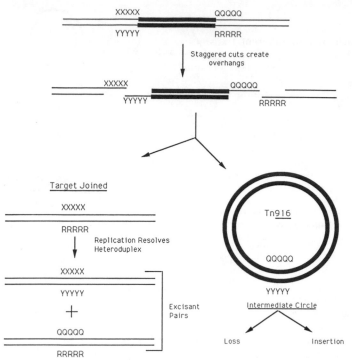

FIG. 2. Model for excision of Tn916. The thick lines represent Tn916, and the thin lines represent the DNA adjacent to the transposon. The coupling sequences are indicated by the hypothetical nucleotide pairs X-Y and Q-R. A staggered cleavage of the phosphodiester backbone on the 3' side of the coupling sequence with both strands (first line) generates molecules with 3' single-stranded ends (second line). The target and transposon sequences are joined by their 3' single-stranded regions to generate an excisant molecule and the transposon circle. Because there is no apparent requirement for homology between the two coupling sequences, both the excisant and the transposon circle contain heteroduplexes consisting of the base pairs originally present in the coupling sequences. Semiconservative replication resolves the heteroduplex present in the excisant and generates a pair of molecules (excisant pairs), one of which has the left, and the other of which has the right, coupling sequence at the site of excision. (Reproduced from reference 6 with permission.)

allow us to determine what the actual sequence in a single molecule was. Therefore, we (6) cloned a small AluI fragment that included the joint of the circular transposon. Each clone contained a different assortment of other AluI fragments that were ligated together randomly with the joint fragment, indicating that each clone resulted from transformation of E. coli by a single ligated DNA molecule.

Surprisingly, the plasmid DNA from five of the nine joint region chimeras examined contained two different nucleotides at equivalent positions for several base pairs in the joint region (6). A mixture of two kinds of joint region chimeras was present in each clone, as shown by retransformation with the DNA containing the mixed sequence. Since the original transformant resulted from a single DNA molecule, the chimeric plasmid must have contained a heteroduplex region at the joint of the circular transposon, which resolved upon subsequent replication. This conclusion was confirmed by direct restriction endonuclease analysis of the circular transposon molecules (6). An unmatched base pair within a DraI site in the heteroduplex region made the DraI site unrecognizable to the restriction enzyme.

These results showed that the joint sequence of the circular transposon did contain the coupling regions, as predicted. However, a single transposon molecule contained both coupling regions together in a heteroduplex joint. This led to the model (6) for excision of conjugative transposons presented in Fig. 2.

Mechanism of Insertion as Determined by Sequence Analysis

If insertion is the reciprocal of excision, each strand of the circular intermediate transposon would bring with it one of the two coupling regions that flanked the transposon in its previous target (Fig. 3). To test this, the strains obtained by transforming purified circular Tn916 into B. subtilis protoplasts (second-generation insertions) were used (6), since the sequence of the joints of the circular transposon had been determined. From each of four different transformants, the region containing the transposon in the *B. subtilis* chromosome was cloned into a cosmid vector and the sequence flanking the transposon was analyzed. Each had one and only one of the two coupling sequences that flanked the original transposon insertion, as predicted. On the other end, each had a new coupling sequence that had presumably come from the *B. subtilis* chromosomal target site. This work demonstrates that the model presented in Fig. 2 and 3 is correct in outline (6).

Variability in the Number of T's at the Transposon Terminus

Three different laboratories (4, 6, 8) found that five or six T's may be present at the right end of Tn*916*, and it seems likely that excision enzymes may have to interact with these. In the circular transposon molecule, in addition to the class of joint sequences described above (class I), we found a second class that may be explained if the right end of the transposon is displaced within this run of T's (6). This can occur if the factors that bind here slip and therefore use a different T for the transposon terminus. Only two of the nine plasmid chimeras that we studied had class II joints (five T's).

Mismatch Repair

Most bacteria, including *E. coli*, have efficient mechanisms to repair base mismatches in duplex DNA. The isolation of circular transposon molecules containing heteroduplex joints was therefore surprising. The four transposons of the nine investigated that did not contain mismatches may have resulted from repair, which probably occurred following introduction of the purified deproteinized DNA into *E. coli*.

Repair does not appear to occur during insertion of Tn*916*, since second-generation transformants have different coupling sequences flanking the transposon (17). We therefore proposed the possibility that conjugative transposons have a mechanism that inhibits mismatch repair. It is possible, for example, that the recombination enzymes involved in excision and insertion remain associated with the heteroduplexes and that this association prevents the corrective action of enzymes responsible for the repair of mismatches. It will be interesting to determine whether mismatched joint regions are actually required for insertion of the circular transposon molecule.

Enzymes Involved in Transposition

To begin an analysis of transposon-encoded functions involved in transposition, Clewell's group (26, 30) isolated Tn*5* insertions in Tn*916*

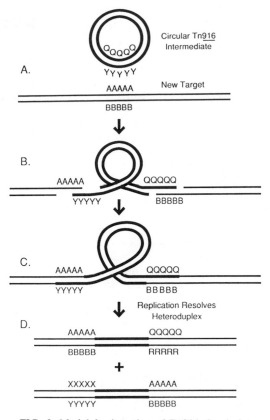

FIG. 3. Model for insertion of Tn*916*. Symbols are the same as in Fig. 1; A and B represent the complementary bases of the target. Staggered cleavage at the 3′ ends of the new target and the heteroduplex-containing joint of the transposon circle (A) generates molecules with 3′ single-stranded ends (B). The transposon circle and target are covalently joined by their single-stranded ends to create a new insertion with a heteroduplex at each end (C). Replication resolves the heteroduplexes and generates a pair of molecules in which one member is flanked by the target sequence at one end and a coupling sequence from the previous insertion at its other end. The second member of the pair would be flanked by the target sequence on the side opposite its location in the first member and would have at its other end the second of the two coupling sequences of the first insertion (D). The two forms would segregate at the first replication event. (Reproduced from reference 6 with permission.)

and classified them as to ability to transpose to the chromosome of *Streptococcus faecalis* following protoplast transformation (function 1), transpose to a conjugative plasmid as assayed by mating to a new strain (function 2), or perform complete conjugational transposition (function 3). As anticipated, a large part of the transposon is required for conjugational transfer (function 3).

A defect in function 2 was interpreted to mean that the mutant transposon was unable to transpose to a coresident conjugative plasmid. It was therefore surprising that several mutants were defective for function 2 but were proficient in function 1 (protoplast transformation). One possible interpretation of this class is that the mating-out assay of function 2 measures complementation by the plasmid of a conjugation defect in the transposon and that the two elements are transferred to the recipient as independent molecules. In this case, these transposon mutants might be in *cis*-acting conjugation functions or in conjugation functions that are specific to the transposon and thus cannot be complemented by the plasmid.

Two insertion mutants were found to be unable to perform any of the transposon functions measured (1, 2, or 3), and it was suggested that these mutants might have defects in excision (26).

In an alternative approach to define the functions needed for transposition, Courvalin's group (24) constructed a minitransposon containing a drug resistance marker between the ends of Tn*1545*. Several constructs were designed to contain additional DNA from one transposon end (the left end of Tn*916*, which is equivalent to the right end of Tn*1545*). Although cloning of the smaller constructs in *E. coli* was successful, attempts to clone the larger piece led to excision and loss of the transposon, leaving the excisant plasmid (the vector with the target region from which the transposon excised) behind. This indicated that at least for the minitransposon, a particular region at the right of Tn*1545* was required for excision. Comparison of the maps suggests that this correlates with the location of the proposed excision function of Senghas et al. (26). The analysis of the sequence of this region (24) revealed two open reading frames, the larger of which (ORF2) shows homology with the lambda integrase family of proteins and appears to be sufficient for excision. Insertional inactivation of ORF2 in Tn*916* leads to a defect in conjugational transposition of the element which can be complemented by an actively transcribed ORF2 gene (Storrs et al., unpublished data).

To investigate the possibility of repression of transposition by a resident Tn*916*, we introduced a Tn*916* element marked with an erythromycin resistance determinant (Tn*916E*) by conjugation into a *B. subtilis* strain containing a resident Tn*916* (with *tetM*). Erythromycin-resistant transconjugants occurred at about the same frequency as when Tn*916E* was introduced into a strain with no transposon (23). Furthermore, the incoming element was usually found at a site distant from the resident one, indicating that transposition into a strain with a resident conjugative transposon is much more frequent than homologous recombination between the Tn*916* elements. This shows that a resident Tn*916* element does not repress transposition following conjugation, at least when the entering transposon is chromosomally located.

Conclusion

An improved understanding of the mechanism of transposition of conjugative transposons is of value both in applied and in basic science. As suggested by Gawron-Burke and Clewell (14), these elements are extremely useful as insertional mutagens in gram-positive organisms like streptococci. After the transposon insertion mutant has been isolated, the transposon-containing chromosomal segment can be cloned into *E. coli*. However, now that it is clear that there are two types of excisants, both of which may appear to be phenotypically wild type (6, 8), it is important to be aware that only one of the two will have the wild-type sequence. To obtain this, the unmutagenized strain must be used (either for cloning or for polymerase chain reaction sequence analysis) (Caparon and Scott, in press).

When we have an understanding of how the excision enzyme recognizes the transposon ends and of what constitutes a recognizable target, this system may be useful for recombination in vitro. More immediately, though, conjugative transposons can be used as shuttle vectors between *E. coli* and nontransformable species of bacteria (22). For many strains of streptococci, this now permits the introduction of a gene of interest or the replacement of the wild-type allele of any gene with one that has been specifically altered by in vitro techniques.

Most exciting to me, perhaps, is the unique mechanism of recombination employed by conjugative transposons (6). Although in general outline they move by a mechanism that sounds like lambda prophage excision and insertion, lambda is unable to deal with mismatched bases in the overlap region (28, 29). Tn*916*, on the other hand, may actually require such mismatches. I look forward to learning more about the molecular details and enzymology of this type of recombination event.

Most of the work in my laboratory on conjugative transposons was performed by Mike Caparon, and I am very grateful to him for discussions which helped formulate our ideas about these elements.

Our work was supported in part by grant AI20723 from the National Institutes of Health.

LITERATURE CITED

1. **Burdett, V., J. Inamine, and S. A. Rajagopalan.** 1982. Heterogeneity of tetracycline resistance determinants in *Streptococcus. J. Bacteriol.* **149:**995–1004.
2. **Buu-Hoi, A., and T. Horodniceanu.** 1980. Conjugative transfer of multiple antibiotic resistance markers of *Streptococcus pneumoniae. J. Bacteriol.* **143:**313–320.
3. **Caillaud, F., C. Carlier, and P. Courvalin.** 1987. Physical analysis of the conjugative shuttle transposon Tn*1545. Plasmid* **17:**58–60.
4. **Caillaud, F., and P. Courvalin.** 1987. Nucleotide sequence

of the ends of the conjugative shuttle transposon Tn*1545*. *Mol. Gen. Genet.* **209:**110–115.

5. **Caparon, M. G., and J. R. Scott.** 1987. Identification of a gene that regulates expression of M protein, the major virulence determinant of group A streptococci. *Proc. Natl. Acad. Sci. USA* **84:**8677–8681.

6. **Caparon, M G., and J. R. Scott.** 1989. Excision and insertion of the conjugative transposon Tn*916* involves a novel recombination mechanism. *Cell* **59:**1027–1034.

7. **Carlier, C., and P. Courvalin.** 1982. Resistance of streptococci to aminoglycoside-aminocyclitol antibiotics, p. 162–166. *In* D. Schlessinger (ed.), *Microbiology—1982*. American Society for Microbiology, Washington, D.C.

8. **Clewell, D. B., S. E. Flannagan, Y. Ike, J. M. Jones, and C. Gawron-Burke.** 1988. Sequence analysis of termini of conjugative transposon Tn*916*. *J. Bacteriol.* **170:**3046–3052.

9. **Clewell, D. B., and C. Gawron-Burke.** 1986. Conjugative transposons and the dissemination of antibiotic resistance in streptococci. *Annu. Rev. Microbiol.* **40:**635–659.

10. **Courvalin, P., and C. Carlier.** 1986. Transposable multiple antibiotic resistance in *Streptococcus pneumoniae*. *Mol. Gen. Genet.* **205:**291–297.

11. **Franke, A. E., and D. B. Clewell.** 1981. Evidence for a chromosome-borne resistance transposon (Tn*916*) in *Streptococcus faecalis* that is capable of conjugal transfer in the absence of a conjugative plasmid. *J. Bacteriol.* **145:**494–502.

12. **Gaillard, J. L., P. Berche, and P. J. Sansonnetti.** 1986. Transposon as a tool to study the role of hemolysin in the virulence of *Listeria monocytogenes*. *Infect. Immun.* **52:**50–55.

13. **Gawron-Burke, C., and D. B. Clewell.** 1982. A transposon in *Streptococcus faecalis* with fertility properties. *Nature* (London) **300:**281–284.

14. **Gawron-Burke, C., and D. B. Clewell.** 1984. Regeneration of insertionally inactivated streptococcal DNA fragments after excision of transposon Tn*916* in *Escherichia coli*: strategy for targeting and cloning genes from gram-positive bacteria. *J. Bacteriol.* **159:**214–221.

15. **Gawron-Burke, C., R. Wirth, M. Yamamoto, S. Flannagan, G. Fitzgerald, F. An, and D. B. Clewell.** 1986. Properties of the streptococcal transposon Tn*916*, p. 191–197. *In* S. Hamada, S. Michalek, H. Kiyono, L. Menaker, and J. McGhee (eds.), *Molecular Microbiology and Immunobiology of Streptococcus mutans*. Elsevier, New York.

16. **Hill, C., C. Daly, and G. F. Fitzgerald.** 1985. Conjugative transfer of the transposon Tn*919* to lactic acid bacteria. *FEMS Microbiol Lett.* **30:**115–119.

17. **Ike, Y., R. A. Craig, B. A. White, Y. Yagi, and D. B. Clewell.** 1983. Modification of *Streptococcus faecalis* sex pheromones after acquisition of plasmid DNA. *Proc. Natl. Acad. Sci. USA* **80:**5369–5373.

18. **Jones, J. M., C. Gawron-Burke, S. E. Flannagan, M. Yamamoto, E. Senghas, and D. B. Clewell.** 1987. Structural and genetic studies of the conjugative transposon Tn*916*, p. 54–60. *In* J. J. Ferretti and R. Curtiss III (ed.), *Streptococcal Genetics*. American Society for Microbiology, Washington, D.C.

19. **Kathariou, S., D. S. Stephens, P. Spellman, and S. A. Morse.** 1990. Transposition of Tn*916* to different sites in the chromosome of *Neisseria meningitidis*: a genetic tool for meningococcal mutagenesis. *Mol. Microbiol.* **4:**729–735.

20. **Kauc, L., and S. H. Goodgal.** 1989. Introduction of transposon Tn*916* DNA into *Haemophilus influenzae* and *Haemophilus parainfluenzae*. *J. Bacteriol.* **171:**6625–6628.

21. **Le Bouguénec, C., G. de Cespédès, and T. Horaud.** 1988. Molecular analyis of a composite chromosomal conjugative element (Tn*3701*) of *Streptococcus pyogenes*. *J. Bacteriol.* **170:**3930–3936.

22. **Norgren, M., M. G. Caparon, and J. R. Scott.** 1989. A method for allelic replacement that uses the conjugative transposon Tn*916*: deletion of the *emm*6.1 allele in *Streptococcus pyogenes* JRS4. *Infect. Immun.* **57:**3846–3850.

23. **Norgren, M., and J. R. Scott.** 1991. Presence of conjugative transposon Tn*916* in the recipient strain does not impede transfer of a second copy of the element. *J. Bacteriol.* **173:**319–324.

24. **Poyart-Salmeron, C., P. Trieu-Cuot, C. Carlier, and P. Courvalin.** 1989. Molecular characterization of two proteins involved in the excision of the conjugative transposon Tn*1545*: homologies with other site-specific recombinases. *EMBO J.* **8:**2425–2433.

25. **Scott, J. R., P. A. Kirchman, and M. G. Caparon.** 1988. An intermediate in transposition of the conjugative transposon Tn*916*. *Proc. Natl. Acad. Sci. USA* **85:**4809–4813.

26. **Senghas, E., J. M. Jones, M. Yamamoto, C. Gawron-Burke, and D. B. Clewell.** 1988. Genetic organization of the bacterial conjugative transposon Tn*916*. *J. Bacteriol.* **170:**245–249.

27. **Shoemaker, N. B., M. D. Smith, and W. R. Guild.** 1980. DNase-resistant transfer of chromosomal *cat* and *tet* insertions by filter mating in pneumococcus. *Plasmid* **3:**80–87.

28. **Weisberg, R., and A. Landy.** 1983. Site-specific recombination in phage lambda, p. 211–250. *In* F. W. Stahl, J. Roberts, and R. A. Weisberg (ed.), *Lambda II*. Cold Spring Harbor Laboratory, Cold Spring Harbor, N.Y.

29. **Weisberg, R. A., L. W. Enquist, C. Foeller, and A. Landy.** 1983. A role for DNA homology in site-specific recombination: the isolation and characterization of a site-affinity mutant of coliphage lambda. *J. Mol. Biol.* **181:**189–197.

30. **Yamamoto, M., J. M. Jones, E. Senghas, C. Gawron-Burke, and D. B. Clewell.** 1987. Generation of Tn*5* insertions in streptococcal conjugative transposon Tn*916*. *Appl. Environ. Microbiol.* **53:**1069–1072.

Comparative Analysis of cAD1- and cCF10-Induced Aggregation Substances of *Enterococcus faecalis*

REINHARD WIRTH,[1] STEPHEN OLMSTED,[2] DOMINIQUE GALLI,[1] AND GARY DUNNY[2]

Lehrstuhl für Mikrobiologie, Universität München, D-8000 Munich 19, Federal Republic of Germany,[1] and Bioprocess Technology Institute, University of Minnesota, St. Paul, Minnesota 55108[2]

Pheromone-induced conjugal transfer of DNA in *Enterococcus faecalis* involves the secretion of sex pheromones by recipient cells and the subsequent recognition of these pheromones by the donor cell (5). Each pheromone is specific for a particular plasmid or group of related plasmids. In response to these pheromones, a cascade of complex events occurs; probably all structural and regulatory genes necessary for this cascade are plasmid encoded. Phenotypically, cells can be observed to form clumps (aggregates) in broth within 45 min after induction by sex pheromone. This is accompanied by the expression of plasmid-encoded conjugal transfer genes by the donor cells and mating (3).

Aggregation of donor and recipient cells is mediated by a surface protein, aggregation substance (5, 15). This protein is plasmid encoded, and its expression is induced by the specific pheromone. It was shown recently that aggregation substance appears as "hairlike" structures on the cell surface of induced donor cells (11). The combination of immunogold labeling and high-resolution scanning electron microscopy allowed studies on the distribution of aggregation substance. It could be demonstrated that the adhesin is not evenly distributed over the cell surface but appears to be localized in regions not close to the septum (16); i.e., aggregation substance may not be incorporated into newly formed cell wall. The genes encoding aggregation substance from two different *E. faecalis* plasmids, pAD1 (10) and pCF10 (S. Olmsted and G. Dunny, unpublished data), have now been identified and sequenced. This chapter focuses on the identification of these genes, their positions relative to other transfer genes on the respective plasmids, and the similarity in the protein structures for which they encode.

Gene Identification

pAD1-encoded aggregation substance. The structural gene for pAD1-encoded aggregation substance called Asa1 (aggregation substance of pAD1) was identified in the Munich laboratory during a comparative study on all known sex pheromone plasmids. It was found that all sex pheromone plasmids (except pAM373) contain one homologous DNA region; in pAD1, some Tn917 insertions in this region led to a nonclumping phenotype (8).

Since pAD1-encoded aggregation substance was purified in the Munich laboratory as a 78-kDa protein (but see below for the size of the mature protein) by a detergent extraction, the N-terminal amino acid sequence of the adhesin could be established. Oligonucleotides deduced therefrom hybridized with the pAD1 region homologous to the other sex pheromone plasmids. Therefore, this DNA region was sequenced (10); a total of four open reading frames (ORFs) were detected, three of which run counterclockwise on the pAD1 restriction map; ORF4 would be transcribed clockwise. The proteins deduced from the sequences of ORF1, ORF3, and ORF4 would be 13.0, 12.3, and 13.1 kDa in size, with isoelectric points of 4.12, 10.19, and 8.53, respectively. However, no data are available indicating whether these ORFs code for proteins.

ORF2 codes for a protein of 142.3 kDa, a value which does not agree with the size of the originally purified form of aggregation substance. It was shown subsequently that mature (i.e., lacking the signal peptide) pAD1-encoded aggregation substance (hereafter referred to as Asa1 protein) has a molecular weight of 137 kDa and can be extracted, together with the 78-kDa form, from induced cells by a lysozyme treatment (10). In addition, it was shown that the 78- and 137-kDa forms of aggregation substance have the same N-terminal amino acid sequence, that the two proteins are immunologically related, and that limited proteolysis resulted in a partially identical cleavage pattern (10). An antiserum against the purified 78-kDa protein inhibited the clumping response in an up to 800-fold dilution and was used in the above-mentioned scanning electron microscopic studies.

pCF10-encoded aggregation substance. Transfer regions of plasmid pCF10, which encoded tetracycline resistance (6), were originally defined by Tn917 insertional mutagenesis and called tra 1 to 9 (1). The identification of these regions, varying in size from 1 to 5.2 kb, was based on phenotypes and physical locations of the inserts. Negative and positive regulatory regions were shown to abolish the inducible phenotype and to increase the transfer frequency (in some cases resulting in a constitutive clumping phenotype), respectively.

A more thorough analysis of the tra region was

initiated by cloning the various regions into *Escherichia coli*-*E. faecalis* shuttle vectors pWM401 and pWM402 (18). Further mutagenesis of the cloned fragments with Tn5 was carried out, and expression of the cloned fragments in both *E. coli* and *E. faecalis* was investigated. Subsequently, the entire *Eco*RI c and e fragments were sequenced (S.-M. Kao, Ph.D. thesis, Cornell University, Ithaca, N.Y., 1990; see also Dunny et al., this volume). Specific genes that are involved in the positive regulation of induction were identified; these were called *prgR*, -*S*, and -*X* (for pheromone-responsive gene; 4). Two genes, *prgA*, and *prgB*, were also identified and found to encode proteins expressed on the surface of donor cells (2). The first encodes a product that has been shown to be involved in surface exclusion (7). The product of *prgB* is a protein with a molecular size of about 150 kDa, as determined by sodium dodecyl sulfate-polyacrylamide gel electrophoresis (SDS-PAGE). This protein is named Asc10 (aggregation substance of pCF10) and has been demonstrated to mediate the aggregation of donor and recipient cells. Three lines of evidence support this notion: (i) induction by sex pheromone results in the simultaneous expression of Asc10 and the clumpy phenotype, (ii) clones containing the positive regulatory region and the structural gene for Asc10 express the protein constitutively and show constitutive clumping in broth culture, and (iii) monoclonal antibodies to Asc10 inhibit clumping and subsequent mating, as does a specific polyclonal serum against Asc10 or Fab fragments therefrom (4). Furthermore, a very high degree of homology exists between the Asa1 and Asc10 proteins (see below).

Organization of the Genes

Nomenclature. In this chapter, we wish to invoke the use of a proposed nomenclature (10) for proteins and genes involved in the sex pheromone system of *E. faecalis* (there is agreement between the three groups of D. B. Clewell, G. Dunny, and R. Wirth on this nomenclature). Proteins will be named according to their function and also with respect to the plasmid by which they are encoded. This means that all aggregation substances are designated by one letter and a number which refers to the corresponding sex pheromone plasmid; Asc10 and Asa1, therefore, stand for aggregation substances coded by pCF10 and pAD1, respectively. Surface exclusion proteins are named accordingly: Sec10 stands for surface exclusion protein coded by pCF10. Such a nomenclature will lead to an unambiguous naming of sex pheromone-induced proteins. If possible, the structural genes should also be designated according to this nomenclature. Therefore, *asa1* encodes pAD1-specific aggregation substance. At least for the moment, genes named and published under other names will keep their designations.

Comparison of pAD1 and pCF10 regions coding for aggregation substance. A comparison of the regions of pCF10 and pAD1 coding for the aggregation substances is given in Fig. 1. On the physical map of pCF10, the *prgB* gene is located around the 35-kb region, with transcription in the clockwise direction (1; Dunny et al., this volume).

The solid lines represent the corresponding DNA regions of pCF10 and pAD1; boxes indicate genes that have been shown to code for *E. faecalis* surface proteins. Transcription for these three genes, as shown in Fig. 1, is from left to right; i.e., with respect to the restriction maps (1, 10), the given DNA regions are drawn in opposite directions. The solid arrows indicate sequenced ORFs; there is no direct proof, however, that these ORFs indeed code for proteins. For pCF10, these ORFs refer (from left to right) to *prgX*, -*R*, and -*S*, respectively (4); the ORF at the right end of the pCF10 map is highly homologous to a corresponding ORF of pAD1. For pAD1, the identities and locations of the first three ORFs (indicated by broken arrows) are only indirectly deduced from Tn917 insertions (8, 14, 17). As is the case for the first three pCF10-encoded ORFs, they might compose the key regulatory region for the sex pheromone-induced cascade of complex events. The two solid arrows 5' and 3' to the *asa1* gene indicate sequenced ORFs. The above-men-

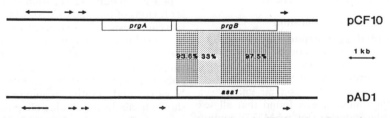

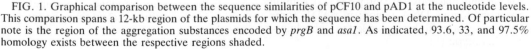

FIG. 1. Graphical comparison between the sequence similarities of pCF10 and pAD1 at the nucleotide levels. This comparison spans a 12-kb region of the plasmids for which the sequence has been determined. Of particular note is the region of the aggregation substances encoded by *prgB* and *asa1*. As indicated, 93.6, 33, and 97.5% homology exists between the respective regions shaded.

tioned ORF4 (see discussion of pAD1-encoded aggregation substance) reading outward from the *asa1* 5′ region is not indicated in Fig. 1.

It is evident from Fig. 1 that there is a very high degree of homology for the regions covering the structural genes for pCF10- and pAD1-encoded aggregation substances. The homology starts at an identical ribosome binding site and extends to amino acid position 265, with an identity of 93.6%. The region from amino acids 266 to 559 shows only 33% identity, followed by a region of 97.5% identity which covers also the small ORF following the structural gene for aggregation substance. These values refer to amino acid identity; on the DNA level, identity values are only 1 to 4% lower. A dot matrix comparison of the Asc10 and Asa1 proteins is shown in Fig. 2.

No homology is observed between pCF10 and pAD1 5′ to the structural gene of pAD1-encoded aggregation substance; the small ORF 5′ to the *asa1* gene shows no homology to available pCF10 sequences. Since no sequence data are available for the pAD1 region between the putative regulatory region and ORF1 (located 5′ to *asa1*), it is not proven that a gene analogous to *prgA*, which codes for the pCF10-specific surface exclusion protein, might be located here. Interestingly, strains harboring a plasmid with Tn*917* insertions in this region lack an inducible surface protein similar in size to Sec10.

At the moment, it seems that at least the structural genes for pCF10- and pAD1-encoded aggregation substances (*prgB* and *asa1*) are highly homologous to each other. It well might be, however, that regulatory functions for the surface proteins are not located at the corresponding regions. Data from the Munich laboratory show the existence of a common transcript covering ORF1 and the *asa1* gene. On the other hand, an ORF1 analog was not found in pCF10; preliminary RNA data indicate that the pheromone-

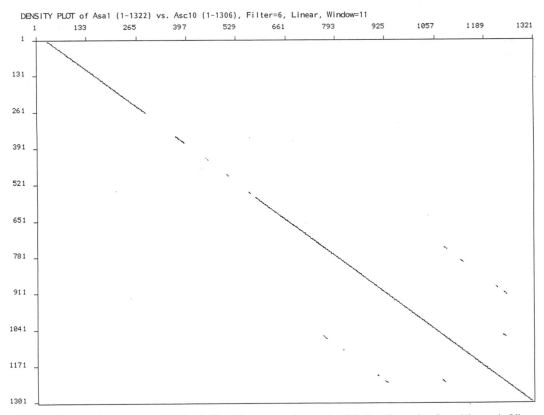

FIG. 2. Dot matrix plot using DDMatrix (Intelligenetics suite version 5.3; Intelligenetics, Inc., Mountain View, Calif.), created to compare the nucleotide similarities between Asa1 and Asc10. Using a window size (*w*) of 11 centered around each coordinate (*i, j*), a density plot is made for the number of nucleotide matches (*c*). The program then calculates the formula $c(i, j)/w$. If the fraction is less than 1, it is multiplied by 10 and rounded to the nearest whole integer. Scores of 100% are denoted by an asterisk. To create the actual graph, a cutoff point or filter of 6 has been used such that all scores over 6 are used. The results reflect those illustrated in Fig. 1. A strong similarity exists in the regions between 1 and 280 and again between 600 and the C terminus, with some variations between 280 and 600.

inducible transcript of the *prgB* region is approximately the same size as the *prgB* gene.

Protein Structure

Since pAD1- and pCF10-encoded aggregation substances are (except for a region from amino acids 266 to 559) nearly identical, their structures will not be described separately.

The Asa1 and Asc10 proteins should be 1,296 and 1,305 amino acids in length, respectively; however, a signal peptide of 43 amino acids is cleaved from the protein during transport through the membrane, as shown for Asa1 by N-terminal amino acid sequencing (10). The mature proteins migrate in SDS-PAGE with an apparent molecular size of ca. 150 kDa; for Asa1, it has been demonstrated that various denaturing conditions lead to altered (reversible) migrating forms in SDS-PAGE. In this connection, it should be mentioned that aggregation substance lacks any cysteine residue. Secondary structure predictions according to Gascuel and Golmard (12) indicate the existence of a short coil structure (positions 60 to 120) and an α-helical region at positions 150 to 220. No other significant secondary structures could be identified; the idea that the protein is of rather globular structure is supported by electron microscopic data for the Asa1 protein (11, 16). The protein extends only 18 nm over the cell surface, which is in marked contrast to the M6 protein of *Streptococcus pyogenes*, which is less than half the size of aggregation substance but extends at least 55 nm over the cell surface.

Predictions on hydrophobic regions indicate that the proteins have a C-terminal transmembrane region; this region shows high homology to other membrane anchors (see appendix 1, this volume) and is preceded by a region which, according to others (9, 13), might constitute a cell wall-associated region. The region around positions 700 to 720 might be membrane associated; aggregation substance itself is classified by computer predictions as an integral membrane protein.

The two proteins show a typical modular structure for surface proteins of gram-positive bacteria. The N terminus consists of a signal peptide, cleaved from the mature protein during transport through the membrane. The C terminus is composed of a cell wall-associated region followed by a membrane anchor at the very C-terminal end. However, in contrast to many other surface proteins from gram-positive bacteria (see appendix 1, this volume), there are no internal repeats observed for aggregation substance (neither on the protein nor on the DNA level). The adhesin shows another unusual feature for surface proteins of gram-positive bacteria: the amino acid motifs Arg-Gly-Asp-Ser and Arg-Gly-Asp-Val are present around positions 600 and 930. These motifs were identified as recognition sequences for eukaryotic membrane receptors, the so-called integrins (13a). This opens the very interesting possibility that *E. faecalis* aggregation substances mediate binding to eukaryotic cells, indicating a possible role of the adhesin as a virulence factor.

This work was supported by grant SFB 145/B8 to R.W. from the Deutsche Forschungsgemeinschaft and by Public Health Service grant AI19310 to G.D. from the National Institutes of Health.

LITERATURE CITED

1. **Christie, P. J., and G. M. Dunny.** 1986. Identification of regions of the *Streptococcus faecalis* plasmid pCF-10 that encode antibiotic resistance and pheromone response functions. *Plasmid* 15:230–241.

2. **Christie, P. J., S.-M. Kao, J. C. Adsit, and G. M. Dunny.** 1988. Cloning and expression of genes encoding pheromone-inducible antigens of *Enterococcus (Streptococcus) faecalis*. *J. Bacteriol.* 170:5161–5168.

3. **Clewell, D. B., and K. E. Weaver.** 1989. Sex pheromones and plasmid transfer in *Enterococcus faecalis*. *Plasmid* 21:175–184.

4. **Dunny, G. M.** 1990. Genetic functions and cell-cell interactions in the pheromone-inducible plasmid transfer system of *Enterococcus faecalis*. *Mol. Microbiol.* 4:689–696.

5. **Dunny, G. M., B. L. Brown, and D. B. Clewell.** 1978. Induced cell aggregation and mating in *Streptococcus faecalis*: evidence for a bacterial sex pheromone. *Proc. Natl. Acad. Sci. USA* 75:3479–3484.

6. **Dunny, G. M., C. Funk, and J. C. Adsit.** 1981. Direct stimulation of antibiotic resistance by sex pheromones in *Streptococcus faecalis*. *Plasmid* 16:270–278.

7. **Dunny, G. M., D. L. Zimmerman, and M. L. Tortorello.** 1985. Induction of surface exclusion (entry exclusion) by *Streptococcus faecalis* sex pheromones: use of monoclonal antibodies to identify an inducible surface antigen involved in the exclusion process. *Proc. Natl. Acad. Sci. USA* 82:8582–8586.

8. **Ehrenfeld, E. E., and D. B. Clewell.** 1987. Transfer functions of the *Streptococcus faecalis* plasmid pAD1: organization of plasmid DNA encoding response to sex pheromone. *J. Bacteriol.* 169:3473–3481.

9. **Fahnestock, S. R., P. Alexander, J. Nagle, and D. Filpula.** 1986. Gene for an immunoglobulin-binding protein from a group G streptococcus. *J. Bacteriol.* 167:870–880.

10. **Galli, D., F. Lottspeich, and R. Wirth.** 1990. Sequence analysis of *Enterococcus faecalis* aggregation substance encoded by the sex pheromone plasmid pAD1. *Mol. Microbiol.* 4:895–904.

11. **Galli, D., R. Wirth, and G. Wanner.** 1989. Identification of aggregation substances of *Enterococcus faecalis* cells after induction by sex pheromones—an immunological and ultrastructural investigation. *Arch. Microbiol.* 151:486–490.

12. **Gascuel, O., and J. L. Golmard.** 1988. Description of the methods implemented in GGBSM. *CABIOS* 4:357–365.

13. **Hollingshead, S. K., V. A. Fischetti, and J. R. Scott.** 1986. Complete nucleotide sequence of type 6M protein of the group A *Streptococcus*—repetitive structure and membrane anchor. *J. Biol. Chem.* 261:1677–1686.

13a. **Hynes, R. O.** 1987. Integrins: a family of cell surface receptors. *Cell* 48:549–554.

14. **Ike, Y., and D. B. Clewell.** 1984. Genetic analysis of the pAD1 pheromone response in *Streptococcus faecalis*, using transposon Tn917 insertional mutagenesis and cloning. *J. Bacteriol.* 158:777–783.

15. **Kessler, R. E., and Y. Yagi.** 1983. Identification and partial characterization of a pheromone-induced adhesive surface antigen of *Streptococcus faecalis*. *J. Bacteriol.* 155:714–721.

16. **Wanner, G., H. Formanek, D. Galli, and R. Wirth.** 1989.

Localization of aggregation substances of *Enterococcus faecalis* after induction by sex pheromones—an ultrastructural comparison using immuno labelling, transmission and high resolution scanning electron microscopic techniques. *Arch. Microbiol.* **151:**831–836.

17. **Weaver, K. E., and D. B. Clewell.** 1988. Regulation of the pAD1 sex pheromone response in *Enterococcus faecalis*:

construction and characterization of *lacZ* transcriptional fusions in a key control region of the plasmid. *J. Bacteriol.* **170:**4343–4352.

18. **Wirth, R., F. Y. An, and D. B. Clewell.** 1986. Highly efficient protoplast transformation system for *Streptococcus faecalis* and a new *Escherichia coli-S. faecalis* shuttle vector. *J. Bacteriol.* **165:**831–836.

Properties of Conjugative Transposon Tn916

DON B. CLEWELL,[1,2] SUSAN E. FLANNAGAN,[1] LOIS A. ZITZOW,[2] YAN A. SU,[1] PING HE,[1]
ELISABETH SENGHAS,[1†] AND KEITH E. WEAVER[1‡]

*Department of Biologic and Materials Sciences, School of Dentistry,[1] and Department of Microbiology and
Immunology, School of Medicine,[2] The University of Michigan, Ann Arbor, Michigan 48109*

The conjugative transposon Tn916 [16.4 kb; carries *tet*(M)] is a member of a large and widely disseminated family of mobile elements exhibiting a broad bacterial host range (3). Movement is believed to involve an excision event (8, 9) that is rate limiting and gives rise to a circular intermediate incapable of autonomous replication; biochemical evidence for such structures has been reported (14; see below). The nature of intercellular transfer is unknown but might occur much like that of a plasmid, with reestablishment of the circle in the recipient followed by insertion into the chromosome. Poyart-Salmeron et al. (11) and Caparon and Scott (1) have proposed that excision involves staggered nicks at the termini giving rise to 5- or 6-base overlapping ends that are not necessarily complementary but become ligated with each other. (See chapters by Scott and by Trieu-Cuot et al. in this volume for additional details.)

Tn916 has been mapped and studied genetically by using Tn5 as an insertional mutagen (15). Tn5 insertions near the left end of Tn916 eliminated the ability to excise in *Escherichia coli* and, at the same time, the ability to insert into the *Enterococcus faecalis* chromosome upon transformation of protoplasts. However, mutations in the remainder of Tn916 did not prevent insertion into the chromosome. A majority of the latter were unable to act as donors in subsequent mating experiments, thereby defining a large contiguous region essential for conjugative functions. With regard to those mutants that would not excise or transpose, it was found that excision in *E. coli* could be complemented in strains carrying the left end of Tn916 on a coresident plasmid. Poyart-Salmeron et al. (11) have reported a similar result involving the equivalent end of the highly similar Tn1545. In addition, they identified two open reading frames whose products, ORF1 and ORF2, exhibit local homology with the proteins Xis and Int of bacteriophage lambda (see Trieu-Cuot et al., this volume). Essentially identical open reading frames have been observed in the case of Tn916 (Fig. 1). The deduced ORF1 pro-

tein consists of 67 amino acid residues (molecular weight, 8,110), whereas ORF2 corresponds to 405 residues (molecular weight, 47,046). Both reading frames would be transcribed toward the end of the transposon. A recognizable transcription terminator is not evident between the reading frames; however, the presence of a possible promoter between the two reading frames suggests that ORF2 might be capable of independent expression.

Tn916 Does Not Prevent Uptake of Additional Homologous Elements

In overnight matings on filter membranes, Tn916 transfers at frequencies of 10^{-5} to 10^{-9} per donor, depending on the donor strain (8). This variability probably reflects the position of the donor transposon and the effect or involvement of adjacent DNA. Transconjugants frequently acquire multiple copies of the element (up to five or six), and it is not uncommon for less than 50% to obtain only one copy. It would therefore appear that upon entrance, the transposon may undergo several transposition events before settling into a more stable state. Since the transposition event itself is viewed as a conservative process, the accumulation of more than one copy might result from replication of chromosomal DNA through the initially inserted element followed by transposition of only one of the daughters.

It was of interest to examine whether the transposon requires the accumulation of a negatively acting repressor before stabilizing. If a repressor is eventually expressed at significant levels and in *trans*, this process might be expected to result in an observable immunity, that is, a lowering of the uptake rate of Tn916 by conjugation if a similar element were already established in the recipient. Experiments to test this, or the possibility that entry exclusion functions are active, have utilized the homologous transposon derivative Tn916ΔE (13), which carries a gene encoding erythromycin resistance *(erm)* in place of *tet*. *E. faecalis* CG110 (8; a JH2-2 host harboring six copies of Tn916) and RH120 (13; an OG1SS host [chromosomal resistance mutations to streptomycin and spectinomycin] harboring one copy of Tn916ΔE) were mated with each other. Transfer of *tet* from CG110 to RH120 in overnight filter matings (2)

†Present address: Gesellschaft für Biotechnologische Forschung, D-3300 Braunschweig, Federal Republic of Germany.
‡Present address: Department of Microbiology, School of Medicine, University of South Dakota, Vermillion, SD 57069.

CGA<u>TTGCTGG</u> TAAAACAACT TTTA<u>TGAAAT</u> CCAAATAAGT GATTTGGAAA <u>GGAGG</u>ATTTT ATG AAG CAG ACT GAC ATT CCT   81
   -35   -10   RBS   M   K   Q   T   D   I   P
                                                                    ORF1

ATT TGG GAA CGT TAT ACC CTA ACC ATT GAA GAA GCG TCA AAA TAT TTT CGT ATT GGC GAA AAC AAG CTA   150
 I   W   E   R   Y   T   L   T   I   E   E   A   S   K   Y   F   R   I   G   E   N   K   L

CGA CGC TTG GCA GAG GAA AAT AAA AAT GCA AAT TGG CTG ATT ATG AAT GGC AAT CGT ATT CAG ATT AAA   219
 R   R   L   A   E   E   N   K   N   A   N   W   L   I   M   N   G   N   R   I   Q   I   K

CGA AAA CAA TTT GAA AAA ATT ATA GAT ACA TTG GAC GCA ATC TAG CG<u>TCGC C</u>AAAGGGTCT TGTATATGA<u>T</u>   290
 R   K   Q   F   E   K   I   I   D   T   L   D   A   I   *   -35

<u>AAAAT</u>AGTAT TAAGTCGTAT CAAGGCTCTT TCCATAAAGG AAA<u>GGAG</u>CAA ATGCC ATG TCA GAA AAA AGA CGT GAC AAT   369
 -10   RBS   M   S   E   K   R   R   D   N
                                                                     ORF2

AGA GGT CGA ATC TTA AAG ACT GGA GAG AGC CAA CGA AAA GAC GGA AGA TAC TTA TAC AAA TAT ATA GAT   438
 R   G   R   I   L   K   T   G   E   S   Q   R   K   D   G   R   Y   L   Y   K   Y   I   D

TCA TTT GGA GAA CCG CAA TTT GTT TAC TCG TGG AAA CTT GTG GCT ACA GAC CGA GTA CCA GCA GGA AAG   507
 S   F   G   E   P   Q   F   V   Y   S   W   K   L   V   A   T   D   R   V   P   A   G   K

CGT GAT TGT ATC TCA CTT AGA GAG AAA ATC GCA GAG TTA CAG AAA GAC ATT CAT GAT GGT ATT GAT GTT   576
 R   D   C   I   S   L   R   E   K   I   A   E   L   Q   K   D   I   H   D   G   I   D   V

GTA GGA AAG AAA ATG ACA CTC TGC CAG CTT TAC GCA AAA CAG AAC GCT CAA AGA CCA AAG GTT AGA AAA   645
 V   G   K   K   M   T   L   C   Q   L   Y   A   K   Q   N   A   Q   R   P   K   V   R   K

AAC ACT GAA ACT GGA CGC AAA TAT CTT ATG GAT ATT TTG AAG AAA GAC AAG TTA GGT GTA AGA AGT ATT   714
 N   T   E   T   G   R   K   Y   L   M   D   I   L   K   K   D   K   L   G   V   R   S   I

GAC AGT ATT AAG CCA TCA GAC GCT AAA GAA TGG GCT ATT AGA ATG AGT GAA AAT GGT TAT GCT TAT CAA   783
 D   S   I   K   P   S   D   A   K   E   W   A   I   R   M   S   E   N   G   Y   A   Y   Q

ACC ATC AAT AAC TAC AAA CGT TCT TTA AAG GCT TCA TTC TAT ATT GCT ATA CAA GAT GAT TGT GTT CGG   852
 T   I   N   N   Y   K   R   S   L   K   A   S   F   Y   I   A   I   Q   D   D   C   V   R

AAG AAT CCA TTT GAC TTT CAA CTG AAA GCA GTT CTT GAT GAT GAT ACT GTC CCT AAG ACC GTA CTA ACA   921
 K   N   P   F   D   F   Q   L   K   A   V   L   D   D   D   T   V   P   K   T   V   L   T

GAA GAA CAG GAA GAA AAA CTG TTA GCC TTT GCA AAA GCT GAT AAA ACC TAC AGC AAA AAT TAT GAT GAA   990
 E   E   Q   E   E   K   L   L   A   F   A   K   A   D   K   T   Y   S   K   N   Y   D   E

ATT CTG ATA CTC TTA AAA ACA GGT CTT CGT ATT TCA GAG TTT GGT GGT TTG ACA CTT CCA GAT TTA GAT 1059
 I   L   I   L   L   K   T   G   L   R   I   S   E   F   G   G   L   T   L   P   D   L   D

TTT GAG AAT CGT CTT GTC AAT ATA GAC CAT CAG CTA TTG AGA GAT ACT GAA ATT GGG TAC TAC ATT GAA 1128
 F   E   N   R   L   V   N   I   D   H   Q   L   L   R   D   T   E   I   G. Y   Y   I   E

ACA CCA AAG ACC AAA AGT GGC GAA CGT CAA GTT CCT ATG GTT GAA GAA GCC TAT CAA GCA TTT AAG CGA 1197
 T   P   K   T   K   S   G   E   R   Q   V   P   M   V   E   E   A   Y   Q   A   F   K   R

GTG TTA GCG AAT CGA AAG AAT GAT AAG CGT GTT GAG ATT GAT GGA TAT AGT GAT TTC CTC TTT CTT AAT 1266
 V   L   A   N   R   K   N   D   K   R   V   E   I   D   G   Y   S   D   F   L   F   L   N

AGA AAG AAC TAT CCA AAA GTG GCA AGT GAT TAC AAC GGC ATG ATG AAA GGT CTT GTT AAG AAA TAC AAT 1335
 R   K   N   Y   P   K   V   A   S   D   Y   N   G   M   M   K   G   L   V   K   K   Y   N

AAG TAT AAC GAG GAT AAA TTG CCA CAC ATC ACT CCA CAT AGT TTG CGA CAT ACA TTC TGT ACC AAC TAT 1404
 K   Y   N   E   D   K   L   P   H   I   T   P   H   S   L   R   H   T   F   C   T   N   Y

GCA AAT GCA GGA ATG AAT CCA AAG GCA TTA CAG TAC ATT ATG GGA CAT GCT AAT ATA GCC ATG ACG CTG 1473
 A   N   A   G   M   N   P   K   A   L   Q   Y   I   M   G   H   A   N   I   A   M   T   L

AAC TAT TAC GCA CAT GCA ACA TTC GAT TCT GCA ATG GCA GAA ATG AAA CGC TTG AAT AAA GAG AAG CAA 1542
 N   Y   Y   A   H   A   T   F   D   S   A   M   A   E   M   K   R   L   N   K   E   K   Q

CAG GAG CGT CTT GTT GCT TAG TAGTACA AATGAATTTA CTACTTATTT ACCACTTCTG ACAGCTAAGA CATGAGGAAA 1620
 Q   E   R   L   V   A   *

TATGCAAAGA AACGTGAAGT ATCTTCCTAC AGTAAAAATA CTCGAAAGCA CATAGAATAA GGCTTTACGA GCATTTAAGA 1700

AAATATAAAA AGATAATTAG AAATTTATAC TTTGTTTAGA 1740

FIG. 1. Nucleotide sequence of the left end of Tn*916*. Nucleotide 1737 represents the left end of the transposon. Two open reading frames designated ORF1 and ORF2 are indicated. Ribosome binding sites (RBS) and possible promoter sequences (−10 and −35) are noted.

occurred at essentially the same frequency (about 10^{-6} per donor) as that observed with use of a transposon-free recipient (OG1SS). The recipient *erm* gene was lost in only 3 of 250 transconjugants screened, implying that homologous recombination between transposons was not a predominant occurrence. The reciprocal experiment, involving transfer of *erm* from RH120 to CG110 and a JH2-2 control, gave similar results except that in this case the mating frequency was about 10^{-8} per donor. As expected, no loss of tetracycline resistance was observed in this case, since all six *tet* elements would have to have been displaced. Thus, the presence of an element in the recipient does not appear to prevent or even reduce the uptake and establishment of a new element.

Southern blot analyses of transconjugants demonstrated the nontandem presence of both Tn916 and Tn916ΔE (data not shown; see Table 1). In addition, transconjugants were able to donate *tet* or *erm* independently. However, depending on the strain and which gene (*erm* or *tet*) was selected initially among transconjugants, the second marker cotransferred 29 to 100% of the time (Table 1). It is evident, therefore, that coresident elements may sometimes transfer in a linked fashion or that the initiation of transfer of one element triggers the independent excision and transfer of another (perhaps by *trans*-acting factors).

Transposon Residing in Recipient Does Not Prevent Zygotic Induction

It was earlier reported that "zygotic induction" occurred when a conjugative plasmid carrying Tn916 entered a plasmid-free, transposon-free recipient (8). Using plasmid pAM180 (a pAM81::Tn916 derivative), selection for a plasmid marker *(erm)* resulted in a significant percentage of transconjugants that had lost Tn916 (segregated) or had acquired the transposon on the chromosome. The plasmid appeared intact with the transposon excised.

A similar experiment using the highly conjugative hemolysin/bacteriocin plasmid pAD1 (4) carrying Tn916 was conducted. The donor strain FA2-2(pAM250) (4) was mated (overnight on filter membranes) separately with both OG1SS (transposon-free) and the isogenic RH120 (harbors chromosomal Tn916ΔE). The only drug selection was for a recipient chromosomal marker (streptomycin resistance on horse blood THB agar [4]); however, the bacteriocin determined by the plasmid causes a strong self-selection for transconjugants that acquire bacteriocin immunity due to uptake of the plasmid. Thus, hemolytic transconjugants arose at frequencies of about 5×10^{-2} per donor. In both cases about 50% of the hemolytic transconjugants were sensitive to tetracycline (i.e., excision and loss of the transposon occurred upon transfer). The

TABLE 1. Transfer of Tn916 and Tn916ΔE from the same host[a]

Background	Donor no.	Copy number of: Tn916 (Tc^r)[b]	Copy number of: Tn916ΔE (Em^r)[c]	Transfer/donor with selection for Tc^r: Frequency	Transfer/donor with selection for Tc^r: % That also received Em^r	Transfer/donor with selection for Em^r: Frequency	Transfer/donor with selection for Em^r: % That also received Tc^r	Frequency of transfer per donor with selection for Tc^r + Em^r
CG110	2	6	2	2.4×10^{-5}	45	4.4×10^{-6}	58	7.1×10^{-7}
	6	6	1	3.3×10^{-5}	33	5.2×10^{-6}	70	9.4×10^{-7}
	7	5	2	5.9×10^{-5}	29	1.7×10^{-5}	63	3.3×10^{-6}
RH120	64	2	1	9.1×10^{-7}	56	9.1×10^{-7}	89	5.2×10^{-7}
	177	1	1	4.7×10^{-7}	100	8.7×10^{-7}	75	$<2.0 \times 10^{-7}$
	248	2	2	1.2×10^{-6}	70	1.7×10^{-6}	94	7.4×10^{-7}

[a] Transfer in all cases was to transposon-free recipients (OG1SS for CG110-derived donors or JH2-2 for RH120-derived donors), with antibiotic selection for the recipient chromosomal markers plus the transposon marker(s) as indicated.
[b] Determined by Southern blot hybridization analyses as previously described (9), using pAM120 as a probe.
[c] Determined by using pVA891 (which contains the appropriate *erm* gene) as a probe.

data imply that an already present transposon does not prevent zygotically induced excision upon uptake.

The basis of zygotically induced excision and transposition remains unknown; however, there is some evidence that it may be affected by the nature in which the DNA enters the recipient cell. When pAM180 (pAM81::Tn916) was transformed into protoplasts (19) of *E. faecalis*, with selection for *erm*, all 35 transformants examined carried an intact plasmid still carrying Tn916. As noted above, the same plasmid yielded a high frequency of zygotically induced excision and transposition when introduced into cells by conjugation. The difference may be due to whether the transposon enters via a single strand (as is the likely case when the transposon is transferred on a plasmid) or a double strand (as is likely in protoplast transformation). In the cases in which *E. coli* plasmid chimeras were introduced into *E. faecalis* (15), transposition into the chromosome may have been due primarily to intermediates already present in the plasmid preparations (see above). It is not clear, however, whether this was the case in our earlier study in which *Streptococcus sanguis* (9) was transformed by such chimeras. Since the latter utilized the natural transformation system believed to involve uptake of only single strands, it is conceivable that zygotic induction could have played a role.

Inability of Tn916 To Mobilize a Nearby Plasmid Marker Located in cis

Previous studies showed that when Tn916 resided on the chromosome of *E. faecalis*, it transferred conjugatively with no detectable transfer of chromosomal markers (7). When the nonconjugative plasmid pAD2 (which encodes resistance to erythromycin, streptomycin, and kanamycin) was present, no transfer of plasmid markers could be detected (3). These observations were consistent with the model of excision and independent transfer of Tn916. Further examination of this point has involved efforts to detect mobilization of a closely positioned marker located in *cis* outside the transposon. The plasmid designated pAM2524KT (17) is a nonconjugative deletion derivative of pAD1 carrying Tn916 and an *erm* gene located about 1 kb outside the transposon. Filter matings (overnight) were performed, using as a donor strain OG1X(pAM2524KT) and as a recipient strain FA2-2 (plasmid-free with chromosomal resistances to rifampin and fusidic acid). Transconjugants that had acquired *tet* (i.e., Tn916) arose at a frequency of 4.2×10^{-6} (average of three experiments). None of the transconjugants acquired *erm*. In fact, when erythromycin-resistant transconjugants were selected (in the absence of tetracycline), no transconjugants arose ($<2.7 \times 10^{-8}$). These results

are consistent with the view that excision occurs prior to intercellular transfer.

Tn916 Transposition in *E. coli*

Some degree of transposition has been noted for Tn916 and Tn1545 in *E. coli*, but neither appears able to transfer intercellularly between *E. coli* strains (5, 9; unpublished data). The following more recent study provides further information on intracellular transposition of Tn916 in *E. coli*.

The plasmid chimera pAM120 (9) corresponds to an *Eco*RI restriction fragment of pAD1 containing Tn916 cloned in pGL101 (a derivative of pBR322). When *E. coli* DH1(pAM120) was grown in the absence of selection for *tet* or the plasmid marker *bla*, curing of the entire plasmid occurred at a relatively high frequency (50% loss after two overnight subcultures in broth). However, about 4% of the plasmid-free cells were found to carry Tn916 on the chromosome. Of those that maintained plasmid DNA, as many as 15% were found to have Tn916 excised and established on the chromosome. There were frequently multiple inserts on the chromosome (Southern blot hybridization revealed one to five copies, with about 30% having more than one), and there was a rough (but not absolute) positive correlation between the level of tetracycline resistance and the number of transposon copies. In cases where only one copy was found, resistance was barely detectable (growth at tetracycline concentrations of no more than 1 to 2 µg/ml). It is clear from these observations that Tn916 is fully capable of transposing from pAM120 to the *E. coli* chromosome; indeed, it is likely that there is preference for existence on the chromosome rather than the multicopy plasmid.

Tn916ΔE could also be established on the *E. coli* chromosome by simply transforming with the suicide (enterococcal) plasmid derivative pAD1::Tn916ΔE. (Transformants arose at frequencies of 1 to 13 per µg of DNA.) In this case the majority of the inserts examined (44 of 46) were present as single copies, and pAD1 sequences were not detectable. The highly conjugative *E. coli* plasmid pOX38 (a derivative of F [16]) was introduced into plasmid-free strains harboring either Tn916 or Tn916ΔE on the chromosome. To examine the ability to transpose from the chromosome to pOX38, these strains were mated (overnight in L broth) with the *E. coli* plasmid-free recipient HB101 (10), with selection (using the recipient streptomycin resistance allele) for transconjugants harboring the transposon marker (*erm* or *tet*, depending on the donor used) or for pOX38 (which encodes kanamycin resistance) alone. The results showed that transposition occurred at a frequency of about 10^{-6} per cell per generation and that the transposon could be found at different sites in the plasmid.

Concluding Remarks

The fact that a resident transposon does not significantly prevent the uptake and establishment of an exogenous homologous element implies an absence of immunity and entry exclusion functions. Even the zygotically induced excision that occurs when Tn916 enters passively via a conjugative plasmid was not affected by an already present transposon in the recipient. This suggests the absence of a high level of strongly inhibiting substance acting in *trans* in cells harboring the transposon. The fact that protoplast transformation of a pAM81::Tn916 derivative into *E. faecalis* cells did not result in a high level of transposition while the same plasmid derivative readily gave rise to excision and transposition when introduce via mating suggests that the difference may be related to having entered the cell as a single strand (assuming that only a single strand of the plasmid transfers during conjugation, something that has not been proven in this case but is known for certain plasmids in gram-negative systems [18]). Perhaps a transient hemi-methylated site(s) occurring after the complementary strand is replicated in the recipient is an important activation factor. Methylation is known to play a significant role in the regulation of transposable elements such as IS10 (12) and IS50 (20).

It is noted that the uptake of an intermediate of Tn916 from the donor is probably significantly different from the uptake of a replicon carrying the transposon. In the former case, the system is already activated and probably committed to events necessary to complete transposition. From the perspective of the recipient, the process is beyond that which would normally be under any negative control.

It is conceivable that a particular gene product(s) which might normally act as a negative control factor acts only in *cis*. It is also possible that regulation depends significantly on factors outside the transposon. This would be consistent with transposition frequencies that vary widely (over a 10,000-fold range), depending on the location of the insert. That is, events such as stimulation by read-in from outside promoters or the level of homology between the ends of the transposon and adjacent DNA (or both) may be important with respect to how frequently transposition is initiated. This could also explain the variability in the accumulation of multiple copies in transconjugants. Secondary transpositions may reflect a low stability at the site of the initial insert—a site that may have been especially attractive for related reasons (e.g., a high level of homology with the end junction region of the

intermediate). Subseqent transpositions, occurring for the same reasons, could lead to the accumulation of a number of transposons. Those transconjugants in which only one copy appears to establish may represent a case in which the initial insertion was reasonably stable. It is noteworthy that some degree of accumulation of multiple copies may be of evolutionary importance to ensure survival of the element. Indeed, other types of transposons (i.e., nonconjugative) are known to move by a replicative process or by a process that indirectly results in an increase in copy number (6).

This study was supported by Public Health Service grants AI10318 and DE02731 from the National Institutes of Health.

LITERATURE CITED

1. **Caparon, M. G., and J. R. Scott.** 1989. Excision and insertion of the conjugative transposon Tn916 involves a novel recombination mechanism. *Cell* **59:**1027–1034.
2. **Clewell, D. B., F. Y. An, B. A. White, and C. Gawron-Burke.** 1985. *Streptococcus faecalis* sex pheromone (cAM373) also produced by *Staphylococcus aureus* and identification of a conjugative transposon (Tn918). *J. Bacteriol.* **162:**1212–1220.
3. **Clewell, D. B., and C. Gawron-Burke.** 1986. Conjugative transposons and the dissemination of antibiotic resistance in streptococci. *Annu. Rev. Microbiol.* **40:**635–659.
4. **Clewell, D. B., P. K. Tomich, M C. Gawron-Burke, A. E. Franke, Y. Yagi, and F. Y. An.** 1982. Mapping of *Streptococcus faecalis* plasmids pAD1 and pAD2 and studies relating to transposition of Tn917. *J. Bacteriol.* **152:**1220–1230.
5. **Courvalin, P., and C. Carlier.** 1987. Tn1545: a conjugative shuttle transposon. *Mol. Gen. Genet.* **206:**259–264.
6. **Craig, N. L., and N. Kleckner.** 1987. Transposition and site-specific recombination, p. 1054–1070. *In* F. C. Neidhardt, J. L. Ingraham, K. B. Low, B. Magasanik, M. Schaechter, and H. E. Umbarger (ed.), *Escherichia coli and Salmonella typhimurium: Cellular and Molecular Biology.* American Society for Microbiology, Washington, D.C.
7. **Franke, A. E., and D. B. Clewell.** 1981. Evidence for a chromosome-borne resistance transposon (Tn916) in *Streptococcus faecalis* that is capable of "conjugal" transfer in the absence of a conjugative plasmid. *J. Bacteriol.* **145:**494–502.
8. **Gawron-Burke, C., and D. B. Clewell.** 1982. A transposon in *Streptococcus faecalis* with fertility properties. *Nature* (London) **300:**281–284.
9. **Gawron-Burke, C., and D. B. Clewell.** 1984. Regeneration of insertionally inactivated streptococcal DNA fragments after excision of transposon Tn916 in *Escherichia coli*: strategy for targeting and cloning of genes from gram-positive bacteria. *J. Bacteriol.* **159:**214–221.
10. **Maniatis, T., E. F. Fritsch, and J. Sambrook.** 1982. *Molecular Cloning: a Laboratory Manual*, p. 504. Cold Spring Harbor Laboratory, Cold Spring Harbor, N.Y.
11. **Poyart-Salmeron, C., P. Trieu-Cuot, C. Carlier, and P. Courvalin.** 1989. Molecular characterization of two proteins involved in the excision of the conjugative transposon Tn1545: homologies with other site-specific recombinases. *EMBO J.* **8:**2425–2433.
12. **Roberts, D. E., B. C. Hoopes, W. R. McClure, and N. Kleckner.** 1985. IS10 transposition is regulated by DNA adenine methylation. *Cell* **43:**117–130.
13. **Rubens, C. E., and I. M. Heggen.** 1988. Tn916ΔE: a Tn916 transposon derivative expressing erythromycin resistance. *Plasmid* **20:**137–142.

14. **Scott, J. R., P. A. Kirchman, and M. G. Caparon.** 1988. An intermediate in transposition of the conjugative transposon Tn*916*. *Proc. Natl. Acad. Sci. USA* **85:**4809–4813.

15. **Senghas, E., J. M. Jones, M. Yamamoto, C. Gawron-Burke, and D. B. Clewell.** 1988. Genetic organization of the bacterial conjugative transposon Tn*916*. *J. Bacteriol.* **170:**245–249.

16. **Way, J. C., M. A. Davis, D. Morisato, D. E. Roberts, and N. Kleckner.** 1984. New Tn*10* derivatives for transposon mutagenesis for construction of *lacZ* operon fusions by transposition. *Gene* **32:**369–379.

17. **Weaver, K. E., and D. B. Clewell.** 1989. Construction of *Enterococcus faecalis* pAD1 miniplasmids: identification of a minimal pheromone response regulatory region and evaluation of a novel pheromone-dependent growth inhibition. *Plasmid* **22:**106–119.

18. **Willetts, N. S., and B. Wilkins.** 1984. Processing of plasmid DNA during bacterial conjugation. *Microbiol. Rev.* **48:**24–41.

19. **Wirth, R., F. Y. An, and D. B. Clewell.** 1986. A highly efficient protoplast transformation system for *Streptococcus faecalis* and a new *Escherichia coli-S. faecalis* shuttle vector. *J. Bacteriol.* **165:**831–836.

20. **Yin, J. C. P., M. P. Krebs, and W. S. Reznikoff.** 1988. Effect of *dam* methylation on Tn*5* transposition. *J. Mol. Biol.* **199:**35–45.

Molecular Analysis of the Gene Encoding Alkaline Phosphatase in *Streptococcus faecalis* 10C1

C. B. ROTHSCHILD, R. P. ROSS, AND A. CLAIBORNE

Department of Biochemistry, Wake Forest University Medical Center, Winston-Salem, North Carolina 27103

Alkaline phosphatase (EC 3.1.3.1) is a metalloenzyme containing both zinc and magnesium ions that catalyzes the nonspecific hydrolyses of a wide variety of phosphomonoesters (for a review, see reference 2). The enzyme itself appears to be ubiquitous and has been studied from various biological sources, including humans and *Escherichia coli*. The *E. coli* enzyme has been studied most extensively and exists as a dimer of 94 kDa with a pH optimum of 8.0. This alkaline phosphatase is encoded by the *phoA* gene, part of the phosphate utilization (*pho*) regulon which is induced by the products of *phoB* and *phoR* during phosphate starvation (11). Normally, alkaline phosphatase is synthesized as an inactive precursor with a signal peptide attached and is transported extracellularly (or to the periplasmic space for *E. coli*), where the active dimeric form appears with the signal peptide removed. The three-dimensional structure of *E. coli* alkaline phosphatase is now well understood, the crystal structure being recently refined to a resolution of 2.0 Å (0.2 nm) (6).

An important application of alkaline phosphatase is its use in gene fusion experiments using *E. coli phoA* as a reporter gene (9). Since the protein is enzymatically active only when outside the plasma membrane (i.e., it is inactive in the cytoplasm), the gene can be used to study such diverse matters as protein localization and membrane topology. These experiments utilize fusions in which the *phoA* gene minus its promoter and signal sequence are joined to a gene of interest. Enzymatic activity can be detected by simple spectrophotometric assay, and there exist a number of selective and indicator media for genetic screenings and selections.

We have identified, cloned, and determined the DNA sequence of the gene for *Streptococcus faecalis* alkaline phosphatase (*phoS*). Previous studies had identified a phosphatase activity from *S. mutans* that hydrolyzed *p*-nitrophenyl phosphate with a pH optimum of 8.0; chemical and kinetic studies, however, identified no active-site serine residues or SH groups important for enzyme activity (7). Although alkaline phosphatases are found in organisms from humans to bacteria, their exact physiological role (except perhaps for *E. coli*) is as yet unclear. Like the *E. coli* gene, *phoS* has applications in the study of membrane proteins, specifically those from *Streptococcus* species. These could include streptococcal K$^+$-ATPase (10) or other clinically important membrane proteins such as the penicillin-binding proteins in pneumococci (4) and glucosyltransferase genes from *S. mutans* (8). In addition, it is hoped that the information presented below will make a significant contribution to our overall knowledge of the structure and function of alkaline phosphatase.

Cloning and Analysis of the *phoS* Gene

Cloning. An *S. faecalis* genomic library was initially constructed in λgt11 to isolate a number of streptococcal chromosomal genes by immunological screening. Briefly, genomic DNA isolated from *S. faecalis* 10C1 was mechanically sheared, size fractionated (3 to 8 kb), and following the addition of *Eco*RI linkers, ligated directly to *Eco*RI-digested λgt11. After in vitro packaging, recombinants were isolated at a frequency of 8×10^5/μg of streptococcal DNA. Subsequent immunological screening of this library by using alkaline phosphatase-conjugated secondary antisera led to the identification of positive recombinant plaques at a frequency of 10^{-4}. Controls, however, revealed that these plaques expressed an alkaline phosphatase activity even in the absence of secondary antisera. Several positive transductants were isolated and plaque purified by serial dilution and replating. Restriction analysis of DNA isolated from three such plaques revealed that each contained an apparently identical insert of 3.2 kb, which indicated that the library had in fact been screened to saturation.

To determine the size of the *phoS* gene product, lysates prepared from recombinants were analyzed for alkaline phosphatase activity following polyacrylamide gel electrophoresis. These clones were first grown as lysogens (in *E. coli* Y1089) in both the absence and presence of isopropyl-1-thio-β-D-galactopyranoside (IPTG). Following heat induction, the resultant lysates were resolved on 10% polyacrylamide denaturing gels and blotted onto nitrocellulose. These blots were blocked with bovine serum albumin (1%) overnight and then exposed to the chromogenic alkaline phosphatase substrate mixture nitroblue tetrazolium (NBT)–5-bromo-4-chloro-3-indolyl phosphate (BCIP).

The alkaline phosphatase-positive (AP$^+$) recombinants produced a 47-kDa protein that retained alkaline phosphatase activity even after electrophoresis and electroblotting (Fig. 1, lanes 3 and 4). This protein was not detected in uninfected *E. coli* Y1089 (lanes 1 and 2), nor was it present in lysogens derived from AP$^-$ recombinant plaques (lanes 5 and 6). Proteins from AP$^+$ lysogens were also probed with monoclonal mouse anti-β-galactosidase followed by anti-mouse immunoglobulin G-conjugated alkaline phosphatase. After exposure to BCIP and NBT, a high-molecular-mass protein (116 kDa) corresponding to *E. coli* β-galactosidase was detected in the lysate derived from the IPTG-containing culture but not in the uninduced culture (lanes 8 and 7, respectively). Again the 47-kDa protein was present in both induced and uninduced AP$^+$ clones. Thus, IPTG strongly increases the expression of β-galactosidase but has little or no effect on expression of *phoS*. These results demonstrate that *phoS* is not cloned as a β-galactosidase fusion protein and that its expression is not subject to regulation of the IPTG-inducible vector promoter.

Genetic analysis. The 3.2-kb insert encoding *phoS* was subsequently excised from λgt11 by restriction with *Eco*RI and subcloned into pUC13, yielding plasmids pAP01 (Fig. 2) and pAP02, which contained the insert in opposite orientations. Restriction endonuclease mapping of the insert in pAP01 revealed three *Hin*dIII, two *Pst*I, and single *Sal*I and *Xba*I sites; these sites were first determined experimentally and later verified by sequencing. Sequencing was performed according to the method of Bankier et al. (1), in which the insert DNA was sonicated to produce random fragments of approximately 200 bp, which were subcloned into M13, sequenced, and melded by computer analysis. Sequencing the entire 3.2-kb fragment encoding *phoS* revealed an open reading frame encoding a protein of 471 amino acids with an ATG start codon and internal *Hin*dIII and *Pst*I sites. Nine base pairs immediately upstream from this start codon, a possible ribosome binding site (GGAGG) was identified. A putative −10 (TATAGT) sequence was also evident 36 bp upstream from the ATG codon; however, a corresponding −35 sequence is not included in the 3.2-kb insert DNA.

Sequence comparisons. A search of the NBRF protein sequence data base with the 471-amino-acid open reading frame revealed other alkaline phosphatases as those sequences sharing the greatest homology with the streptococcal sequence. In all, the *phoS* gene product is approximately 31 and 37% identical to human and *E. coli* sequences, respectively, which themselves are 31% identical. The amino acid sequence of the *S. faecalis* alkaline phosphatase was then compared with that of other known alkaline phosphatases by using the alignment program CLUSTAL (5). This program aligns multiple sequences, allowing deletions and insertions to yield the maximum number of identities. The streptococcal alkaline phosphatase residues included in these identities are Asp-105–Ser-106–Ala-107 (Fig. 3) and Arg-170, which with the

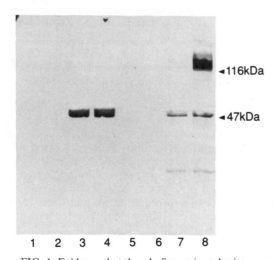

FIG. 1. Evidence that the *phoS* gene is under its own transcriptional control in *E. coli* and codes for a 47-kDa protein. Samples were resolved on a 10% polyacrylamide gel and electroblotted onto nitrocellulose. Lanes 3 and 7 (minus-IPTG cultures) and 4 and 8 (plus-IPTG cultures) contain λgt11 lysates expressing *phoS*. Negative controls are included in lanes 1, 2, 5, and 6. Lanes 7 and 8 were screened with mouse β-galactosidase antisera by using alkaline phosphatase-conjugated secondary antisera; lanes 1 to 6 were treated with alkaline phosphatase substrates NBT and BCIP directly. Protein bands of 47 kDa (lanes 3, 4, 7, and 8) and 116 kDa (lane 8), corresponding to the *phoS* gene product and *E. coli* β-galactosidase, respectively, are clearly visible.

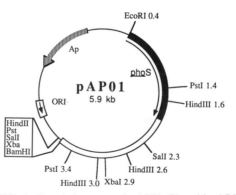

FIG. 2. Restriction map of pAP01. Plasmid pAP01 was generated by cloning the 3.2-kb *Eco*RI insert of a λgt11 clone expressing the *phoS* gene into pUC13.

```
E. coli      PDYVTDSAASATAWSTGVKTYNGALGVDIHEK---------DHPTILEMAKAAGLATGN
human        DRQVPDSAATATAYLCGVKANFQTIGLSAAARFNQCNTTRGNEVISVMNRAKQAGKSVGV
yeast        DSLVTDSAAGATAFACALKSYNGAIGVDPHHR---------PCGTVLEAAKLAGYLTGL
rat          NAQVPDSAGTATAYLCGVKANEGTVGVSAATERTRCNTTQGNEVTSILRWAKDAGKSVGI
S. faecalis  EENVTDSASAATAMAAGVKTYNNAIALDNDKS---------KTETVLERAKKVGKSTGL
          100      *.***..***     ..*.   .....              ...  ** .*   .*  149
```

FIG. 3. CLUSTAL alignment of amino acids 100 to 149 of streptococcal alkaline phosphatase with homologous regions of *E. coli*, human, yeast, and rat alkaline phosphatases. To optimize alignments, gaps (-) have been introduced. Identities (*) and conservative changes (·) are indicated.

three metal ions (two Zn^+ and Mg^{2+}) and their ligands can be regarded as the active site of the enzyme. Eight amino acids reported to be involved in metal binding are conserved in the streptococcal sequence: His-289, His-421, and Asp-285, which bind Zn1; His-328, Asp-57, and Asp-327, which bind Zn2; and Asp-57, Glu-280, and Thr-159, which bind Mg^{2+}. In contrast to the *E. coli* and mammalian enzymes, however, which contain four and six half-cystines, respectively, the streptococcal enzyme has only one such residue (located close to the putative N terminus at position 20). In the *E. coli* enzyme, these four residues are involved in forming two disulfide bridges which are located distal to the active site. Interestingly, using CLUSTAL, the *E. coli* and streptococcal enzymes were assigned C-terminal deletions of some 40 and 14 amino acids, respectively, when aligned with the human enzyme.

The streptococcal and *E. coli* sequences were also compared by using the program GAP in the Genetics Computer Group sequence analysis software package (3). Overall, seven deletions had to be introduced into the mature *E. coli* alkaline phosphatase sequence to accommodate the streptococcal sequence. These deletions occur at positions 17 to 30, 89 to 92, 174 to 175, 189 to 199, 213 to 219, 245 to 248, and 268 to 289 of the mature *E. coli* sequence. A predicted three-dimensional model of the streptococcal alkaline phosphatase based on the now well-defined structure of the *E. coli* enzyme was then constructed (Fig. 4). In this model, all seven deletions are found to be located away from the active site of the enzyme. In addition, the largest deletion occurring at 268 to 289 consists of a loop in the *E. coli* sequence which, if deleted, might not be expected to adversely disrupt the overall enzyme structure (Fig. 4, deletion 7).

Putative *phoS* signal sequence. All alkaline phosphatases studied to date contain a signal peptide sequence that is removed during transport of

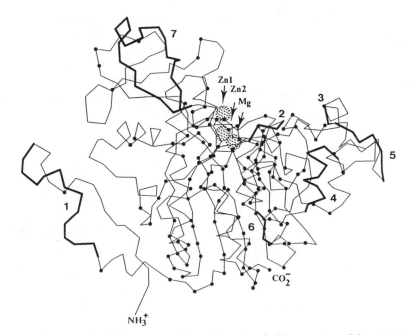

FIG. 4. C_{α} trace of a monomer of alkaline phosphatase from *E. coli*. The sequence of the streptococcal enzyme is superimposed such that black dots and thick black lines (1 to 7) correspond to conservative residues and deletions, respectively. The three metals are also indicated. (Courtesy of Eunice E. Kim and Harold W. Wyckoff, Yale University.)

the enzyme either to the periplasmic space in *E. coli* or en route to the cytoplasmic membrane for eukaryotic enzymes. A search for such a sequence in the streptococcal enzyme immediately revealed a putative leader peptide containing 14 amino acids (Ala-Leu-Leu-Gly-Val-Thr-Leu-Leu-Thr-Phe-Thr-Thr-Leu-Ala) preceded by three basic residues (Lys-Lys-Arg) located immediately behind the initiator methionine. This sequence is most likely cleaved between Ala-18 and Gly-19 during extracellular transport of the protein.

Characterization and Uses of *phoS*

The physiological function of the streptococcal alkaline phosphatase is most likely the scavenging of phosphate from the extracellular environment. Like other alkaline phosphatases, the streptococcal enzyme is probably active only after cleavage of the signal peptide and transport from the cytoplasm. The gene encoding this enzyme, *phoS*, is strongly expressed in *E. coli* under its own transcriptional control. This expression probably initiates from the almost consensus -10 sequence identified upstream from the structural gene even though a -35 region is not included in our clone. Like *E. coli* alkaline phosphatase, *phoS* codes for a protein of 471 amino acids. The streptococcal and *E. coli* alkaline phosphatases are also equally homologous to the mammalian enzymes. A multiple sequence alignment of *E. coli*, human, yeast, rat, and *Streptococcus* alkaline phosphatases identified 53 invariant amino acids present in all sequences. These most likely represent residues important to the structure and function of the enzyme and are expected to be found also in other alkaline phosphatases. Included among these residues is the tripeptide Asp-Ser-Ala, which has been found to be highly conserved in all serine hydrolases; this Ser has been shown to be phosphorylated during the catalytic phosphatase reaction. X-ray crystallographic studies with the mature *E. coli* enzyme suggest that the interaction of Arg-166 with Asp-101 may constrain this active-site serine in the proper position for nucleophilic attack on the incoming phosphate. Since these residues are conserved in alignments of streptococcal and *E. coli* enzymes, a similar interaction probably occurs in the streptococcal alkaline phosphatase. In addition, amino acids predicted to be involved in metal ligand binding are also conserved. The multiple sequence alignments also introduced deletions or insertions in some of the sequences to maximize homology. Notably, the mammalian enzymes contain a 10-amino-acid insertion at position 128 in the mature *E. coli* phosphatase, and the *E. coli* enzyme was assigned a long C-terminal deletion. Although the streptococcal enzyme shares only 29 to 37% identity with other alkaline

phosphatases tested, it does appear to retain the core three-dimensional structure of the *E. coli* enzyme (the seven deletions that were introduced in the *E. coli* structure to accommodate the streptococcal sequence were all located distal to the active site of the enzyme).

As a reporter gene, the streptococcal *phoS* gene should find widespread uses and acceptability in the study of streptococcal gene expression and membrane topology. Such applications would necessitate alteration of the promoter region and signal peptide sequence, respectively. Advantages of alkaline phosphatase as a reporter gene include the sensitivity and simplicity of the assay as well as the availability of inexpensive chromogenic substrates that can be used in enzyme assays or plate media.

This work was supported in part by Public Health Service grant GM35394 from the National Institutes of Health and by a grant-in-aid from the American Heart Association. Al Claiborne is an Established Investigator of the American Heart Association.

Eunice Kim and Harold Wyckoff (Yale University) are gratefully acknowledged for providing Fig. 4 and for valuable discussions.

LITERATURE CITED

1. **Bankier, A. T., K. M. Weston, and B. G. Barrell.** 1987. Random cloning and sequencing by the M13/dideoxynucleotide chain termination method. *Methods Enzymol.* **155:**51–93.
2. **Coleman, J. E., and P. Gettins.** 1983. Alkaline phosphatase, solution structure, and mechanism. *Adv. Enzymol.* **55:**381–452.
3. **Devereux, J., P. Haeberli, and O. Smithies.** 1984. A comprehensive set of sequence analysis programs for the VAX. *Nucleic Acids Res.* **12:**387–395.
4. **Handwerger, S., and A. Tomasz.** 1986. Alterations in penicillin binding proteins of clinical and laboratory isolates of pathogenic pneumococci with low levels of penicillin resistance. *J. Infect. Dis.* **153:**83–89.
5. **Higgins, D. G., and P. M. Sharp.** 1988. CLUSTAL: a package for performing multiple sequence alignment on a microcomputer. *Gene* **73:**237–244.
6. **Kim, E. E., and H. W. Wyckoff.** 1989. Structure of alkaline phosphatases. *Clin. Chim. Acta* **186:**175–188.
7. **Knuttila, M. L. E., and K. K. Makinen.** 1972. Purification and characterization of a phosphatase specifically hydrolyzing *p*-nitrophenyl phosphate from an oral strain of *Streptococcus mutans*. *Arch. Biochem. Biophys.* **152:**685–701.
8. **Kuramitsu, H. K., T. Shiroza, S. Sato, and M. Hayakawa.** 1987. Genetic analysis of *Streptococcus mutans* glycosyltransferases, p. 209–211. *In* J. J. Ferretti and R. Curtiss III (ed.), *Streptococcal Genetics*. American Society for Microbiology, Washington, D.C.
9. **Manoil, C., J. J. Mekalanos, and J. Beckwith.** 1990. Alkaline phosphatase fusions: sensors of subcellular location. *J. Bacteriol.* **172:**515–518.
10. **Solioz, M., S. Mathews, and P. Furst.** 1987. Cloning of the K⁺-ATPase of *Streptococcus faecalis*. Structural and evolutionary implications of its homology to the KdpB-protein of *Escherichia coli*. *J. Biol. Chem.* **262:**7358–7362.
11. **Wanner, B. L.** 1987. Bacterial alkaline phosphatase gene regulation and the phosphate response in *Escherichia coli*, p. 12–19. *In* A. Torriani-Gorini, F. G. Rothman, S. Silver, A. Wright, and E. Yagil (ed.), *Phosphate Metabolism and Cellular Regulation in Microorganisms*, 1st ed. American Society for Microbiology, Washington, D.C.

Organization of Tn5253, the Pneumococcal Ω(cat tet) BM6001 Element

MOSES N. VIJAYAKUMAR, PATRICIA AYOUBI, AND ALI O. KILIC

Department of Botany and Microbiology, Oklahoma State University, Stillwater, Oklahoma 74078

Tn5253, formerly called the Ω(cat tet) element, was originally detected in the chromosome of the plasmid-free clinical isolate *Streptococcus pneumoniae* BM6001 (2, 8). The transposon encodes functions for its transfer en bloc among and within several streptococcal species via a process requiring cell-to-cell contact (26). A variety of conjugative transposons have been identified among streptococci so far. Besides transfer functions, each transposon may also carry one or more antibiotic resistance determinants, and the common feature among most if not all of these is the presence of a homologous tetracycline resistance (Tc^r) determinant (28) of the type Tet M (1). Other antibiotic resistance determinants include chloramphenicol (Cm^r), kanamycin (Km^r), erythromycin (Em^r), and streptomycin (Sm^r). Some of the other conjugative transposons include the Tn916 (Tc) from *Enterococcus faecalis* DS16 (3, 9), Tn1545 (Tc Em Km) from *S. pneumoniae* BM4200 (5, 6, 11), Tn3701 (Tc) from *S. pyogenes* A454 (13, 17), and Tn3951 (Tc Em Cm) from *S. agalactiae* B109 (14, 15, 27). On the basis of size, these conjugative elements could be grouped into two types, one ranging from 16 to 25 kb and the other around 60 kb.

We have been studying the structural and genetic organization of Tn5253 as carried by the pneumococcal laboratory strain DP1322 (25). By inserting the *Escherichia coli* vector plasmid pVA891 (20) at many sites specifically within Tn5253, we were able to clone and to recover parts of the element in *E. coli* (31). Physical analysis of the passenger DNAs from these plasmids made it possible to construct a detailed restriction map of this 65.5-kb element, to localize the drug resistance determinants, and to identify its junction and target regions in the pneumococcal chromosome (30; Fig. 1). From an arbitrarily chosen left end of the element, the *cat* gene is 14 kb inside and is flanked by direct repeats of about 3 kb, a copy of which is also present in the wild-type Rx1 chromosome. These were thought to be possibly related to the frequent spontaneous curing of *cat* (25; S. D. Priebe, Ph.D. thesis, Duke University, Durham, N.C., 1986). Another internal region of 2.5 kb is also present in Rx1 in two copies. It remains to be seen whether these represented insertion sequence elements or their remains. The *tet* gene is located about 44 kb from the left end.

In Rx1, Tn5253 prefers to insert at a specific site (30), and this is the same spot to which Tn3951 transfers when Rx1 is mated with *S. agalactiae* B109 (12). However, unlike these larger conjugative transposons, Tn916 inserts at several sites in the chromosome when introduced into Rx1 (Priebe, Ph.D. thesis, 1986). With the cloned fragments derived from Tn5253, we sought to undertake a comparative analysis of related conjugative transposons. Here we report some of the results which indicate that Tn5253 is a composite element of two independent conjugative transposons, Tn5251 and Tn5252.

The *tet* in Tn5253 Resides on an Independent Conjugative Transposon, Tn5251

The prevalence of a Tet determinant of type M and extensive homology between regions of DNA surrounding this gene among most of the conjugative transposons suggested that a smaller element such as the *tet*-carrying Tn916 (16 kb) could have served as a progenitor in the evolution of the larger elements (4, 12). Autoaccumulation of other heterologous elements was speculated to have resulted in the observed increase in size. If so, the termini of all of the conjugative transposons would be expected to carry some degree of homology. However, two experimental observations were not consistent with this speculation. First, we had observed that plasmid pAM118 (10), containing the entire Tn916, failed to hybridize to either of the termini of Tn5253 when used as a probe in blot hybridization experiments. The homology within Tn5253 to Tn916 was confined only to the region containing the *tet* gene in one contiguous segment (data not shown).

Further, when a 23-kb *Xba*I fragment containing the *tet* region from Tn5253 was cloned into pVA891 to create pVJ403, the plasmid was stable in *E. coli* if tetracycline selection was maintained. In the absence of selective pressure, an 18-kb segment containing the *tet* gene was excised and lost. The restriction maps of several of the deletion plasmids (Fig. 2B) were all similar, suggesting the excision of a defined segment of DNA from pVJ403, and one of the deletion derivatives, pVJ403Δtet, was kept for further studies. Under similar conditions, the excision of a related 25-kb conjugative transposon, Tn1545, from a plasmid replicon and transposition into the *E. coli*

ΩBM6001 ELEMENT (Tn 5253)

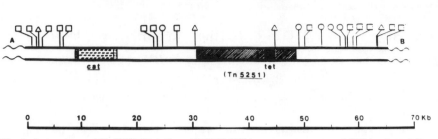

FIG. 1. Physical structure of Tn*5253*. Notation: Tn*5253* DNA; ~, pneumococcal DNA; A and B, *cat* and *tet* ends of the element, respectively; ▰▰▰, *cat* region; ⬚, direct repeats flanking the *cat* segment; ▰▰, *tet* region that transposes when removed from Tn*5253*; ☐, *Xba*I; ⬦, *Kpn*I; ☐, *Bam*HI.

chromosome had been shown to occur (7), suggesting that the excision of *tet* from pVJ403 also could have been related to its transposition. To determine whether *tet* resided on an independent transposon even though it was recovered from a part of the larger Tn*5253*, pVJ403 was introduced into competent pneumococcal Rx1 cells. As the entry of donor DNA during pneumococcal transformation is in single-stranded form (16), plasmid establishment would require entry of two overlapping complementary molecules and subsequent generation of an intact circle (22). Since the vector portion, pVA891, was incapable of autonomous replication in streptococci (20), we did not expect any Em[r] transformants, and none was found. As the wild-type Rx1 genome did not carry homology to any portion of the passenger DNA in pVJ403, genetic recombination by the normal pathway was not expected. However, about 50 Tc[r] transformants per 2 × 10[7] CFU per 10 μg of plasmid DNA resulted. All were sensitive to erythromycin, indicating that the vector was lost and that insertion of the heterologous Tet marker did not involve the homology-dependent insertion-duplication pathway (29). To determine whether the transposition of *tet* during transformation involved unique or multiple target sites, chromosomal DNAs from several Tc[r] transformants were analyzed in blot hybridization experiments using pVJ403 and pVJ403Δ*tet* as probes. That the transposition did not involve any sequences beyond the 18-kb *tet* segment in pVJ403 was evident, since pVJ403Δ*tet* did not react with any of the samples (data not shown). On the other hand, pVJ403 strongly hybridized to at least two fragments representing chromosome-*tet* element junction regions in each case (not shown). The differences in sizes of the junction fragments in different clones indicated random insertion of the element and ruled out the possibility of any plasmid forms. Further, three Tc[r] clones were used as donors in filter mating experiments with Rx1 recipients to test whether *tet* could be conjugally transferred from these transformants. Two were able to transfer *tet* at a frequency of 3 × 10[-5] per donor under conditions in which transfer of the chromosomal marker, *str*, could not be detected. From these results, it was clear that the Tet determinant was within a conjugative transposon that we now termed Tn*5251*. The restriction map of this 18-kb transposon showed significant similarities to the 16-kb transposon, Tn*916* (24), suggesting possible common ancestry.

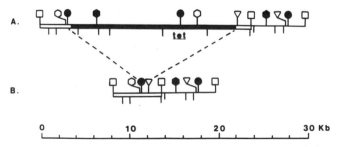

FIG. 2. Restriction endonuclease map of the *E. coli* plasmid pVJ403 (A) and the deletion derivative pVJ403Δ*tet* (B). Symbols: ☐, *Xba*I; ⬦, *Kpn*I; ▽, *Bam*HI; ●, *Hin*dIII; ●, *Eco*RI; ⊤, *Hin*cII; ▬, 18-kb Tn*5251* that excises in *E. coli* in the absence of tetracycline selection. The ends of Tn*5251* were determined from DNA sequencing studies (not shown). The location of the Tet determinant is indicated.

Tn5253 is a Composite Structure of Two Independent Conjugative Transposons, Tn5251 and Tn5252

Since the properties of Tn5251 (in particular, the target selection following conjugal transfer) were different from those of Tn5253, we sought to determine the role of Tn5251 in the transfer of the larger element. If transfer of the entire Tn5253 during conjugation was due to the presence of Tn5251, Tn5253 devoid of Tn5251 would be transfer deficient. To induce the deletion of Tn5251 from within Tn5253, using plasmid pVJ403Δtet, we employed the strategy shown in Fig. 3. As mentioned earlier, Tn5251 was excised and lost from pVJ403 when propagated in E. coli without tetracycline selection, giving rise to the deletion derivative pVJ403Δtet. This derivative was digested with XbaI, and the 5-kb fusion fragment was isolated and fed to competent DP1324 cells carrying the entire Tn5253. Because of the homology provided by the flanking regions in the fusion fragment, Tn5251 was expected to be deleted during transformation. It is perhaps worthwhile pointing out that in pneumococcal transformation, a single-stranded donor DNA molecule could introduce either deletion or insertion of a heterologous DNA segment into the chromosome as long as the region of alteration is flanked on either side by homologous DNA of sufficient length to permit efficient register. After 2 h was allowed for phenotypic expression in liquid broth, the transformants were plated on nonselective medium. The following day, 4,000 colonies were replica plated prior to screening for Tcs and Cmr transformants. Thirty Tcs clones were found.

Physical analysis of the chromosomal DNAs from four of these clones in blot hybridization experiments using pVJ403 as a probe confirmed the deletion of the 18-kb Tn5251 from within Tn5253 in each case. One of these Cmr Tcs clones was designated SP1000. To determine whether SP1000 was capable of conjugal transfer of the sequences beyond Tn5251 within Tn5253, SP1000 cells were used as donors in filter mating experiments with Rx1 recipients. Interestingly, Cm transferred at a frequency of 10^{-6} to 10^{-7} per donor, which was comparable to that of the intact parental Tn5253. Since the DNA beyond Tn5251 within Tn5253, as in SP1000, was found to be capable of conjugal transfer by itself, this segment of DNA was termed Tn5252.

Concluding Remarks

The results presented above showed that the streptococcal conjugative transposon Tn5253 was a composite of at least two mobile elements. In parallel to our findings, Le Bouguénec et al. have localized transposon Tn3703 within the >50-kb conjugative transposon Tn3701, carried by S. pyogenes A454 (17–19). Hybridization studies done in other laboratories (2) and ours indicate that Tn5251 may be closely related to the Tn916, Tn1545, and Tn3703 class of transposons and could be distinct from the Tn5252 class of elements. Besides, the observed differences in the target selection following conjugation between the two types of elements may be due to different modes of transfer (30; Priebe, Ph.D. thesis, 1986). Convincing evidence of a circular intermediate in the transposition of Tn916 has been presented (23), and it is likely that the other tran-

Tn5253

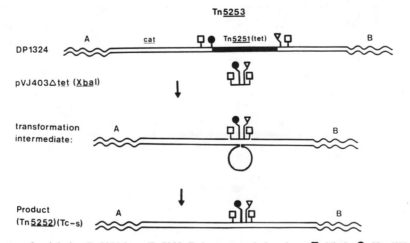

FIG. 3. Strategy for deleting Tn5251 from Tn5253. Relevant restriction sites: ⬠, XbaI; ●, HindIII; ▼, BamHI. The passenger DNA, the 5-kb fusion fragment carrying sequences flanking Tn5251 in Tn5253, from pVJ403Δtet was used as donor DNA. The donor molecule taken up as a single strand was expected to displace the resident strand and pair with the complementary strand, inducing the intervening segment containing Tn5251 to loop out. After one round of replication and segregation of markers, Tcs transformants arose.

sposons of this type function similarly. However, while simultaneously entering plasmids were being restricted, the lack of restriction of Tn5253 during conjugal transfer into recipients carrying the DpnII system suggests an alternate transfer pathway for the larger conjugative transposons (12). Even though Tn5251 and Tn5252 are both capable of independent conjugal transfer, the separation of these elements has not been observed when they are associated together as Tn5253.

The origin and composition of the larger conjugative transposons may turn out to be somewhat complicated, as indicated by the presence of Tn5251 and the two insertion sequence-like elements within Tn5253. The cat region flanked by direct repeats in Tn5253 has been shown to be homologous to the staphylococcal plasmid pC194 (21). Further work may reveal whether a propensity exists for the autoaccumulation of various genetic units into prototype elements such as Tn5252 and Tn3701 to form larger conjugative structures. Without the 18-kb Tn5251 and 7.5-kb cat segments, the remaining portion of DNA of Tn5253 constituting Tn5252 is about 40 kb. Localization of transfer-related and other genes within Tn5252 may provide further insight into the nature of this interesting element.

This work was supported in part by Health Research grant HN9-003 from the Oklahoma Center for the Advancement of Science and Technology to M.N.V.

LITERATURE CITED

1. **Burdett, V., J. Inamine, and S. Rajagopalan.** 1982. Heterogeneity of tetracycline resistance determinants in Streptococcus. J. Bacteriol. **149:**995–1004.
2. **Buu-Hoi, A., and T. Horodniceanu.** 1980. Conjugative transfer of multiple antibiotic resistance markers in Streptococcus pneumoniae. J. Bacteriol. **143:**313–320.
3. **Clewell, D. B., G. F. Fitzgerald, L. Dempsey, L. E. Pearce, A. White, Y. Yagi, and C. Gawron-Burke.** 1985. Streptococcal conjugation: plasmids, sex pheromones, and conjugative transposons, p. 194–203. In S. A. Mergenhagen and B. Rosan (ed.), Molecular Basis of Oral Microbial Adhesion. American Society for Microbiology, Washington, D.C.
4. **Clewell, D. B., and C. Gawron-Burke.** 1986. Conjugative transposons and the dissemination of antibiotic resistance in streptococci. Annu. Rev. Microbiol. **40:**635–659.
5. **Courvalin, P., and C. Carlier.** 1986. Transposable multiple antibiotic resistance in Streptococcus pneumoniae. Mol. Gen. Genet. **205:**291–297.
6. **Courvalin, P., and C. Carlier.** 1987. Tn1545: a conjugative shuttle transposon. Mol. Gen. Genet. **206:**259–264.
7. **Courvalin, P., C. Carlier, and F. Caillaud.** 1987. Functional anatomy of the conjugative shuttle transposon Tn1545, p. 61–64. In J. J. Ferretti and R. Curtiss III (ed.), Streptococcal Genetics. American Society for Microbiology, Washington, D.C.
8. **Dang-Van, A., G. Tiraby, J. F. Acar, W. V. Shaw, and D. H. Bouanchaud.** 1978. Chloramphenicol resistance in Streptococcus pneumoniae: enzymatic acetylation and possible plasmid linkage. Antimicrob. Agents Chemother. **13:**577–582.
9. **Franke, A., and D. B. Clewell.** 1981. Evidence for a chromosome-borne resistance transposon (Tn916) in Strepto-

coccus faecalis that is capable of transfer in the absence of a conjugative plasmid. J. Bacteriol. **145:**494–502.
10. **Gawron-Burke, C., and D. B. Clewell.** 1984. Regeneration of insertionally inactivated streptococcal fragments after excision of transposon Tn916 in Escherichia coli: strategy for targeting and cloning genes from gram-positive bacteria. J. Bacteriol. **159:**214–221.
11. **Guild, W. R., S. Hazum, and M. D. Smith.** 1981. Chromosomal location of conjugative R determinants in strain BM4200 of Streptococcus pneumoniae, p. 610. In S. B. Levy, R. C. Clowes, and E. L. Koenig (ed.), Molecular Biology, Pathogenicity, and Ecology of Bacterial Plasmids. Plenum Publishing Corp., New York.
12. **Guild, W. R., M. D. Smith, and N. B. Shoemaker.** 1982. Conjugative transfer of chromosomal R determinants in Streptococcus pneumoniae, p. 88–92. In D. Schlessinger (ed.), Microbiology—1982. American Society for Microbiology, Washington, D.C.
13. **Horaud, T., C. Le Bouguénec, and G. de Cespédès.** 1987. Genetic and molecular analysis of streptococcal and enterococcal chromosome-borne antibiotic resistance markers, p. 74–78. In J. J. Ferretti and R. Curtiss III (ed.), Streptococcal Genetics. American Society for Microbiology, Washington, D.C.
14. **Horodniceanu, T., L. Bougueleret, and G. Bieth.** 1981. Conjugative transfer of multiple-antibiotic resistance markers in beta hemolytic group A, B, F, and G streptococci in the absence of extrachromosomal deoxyribonucleic acid. Plasmid **5:**127–137.
15. **Inamine, J. M., and V. Burdett.** 1985. Structural organization of a 67-kilobase streptococcal conjugative element mediating multiple antibiotic resistance. J. Bacteriol. **161:**620–626.
16. **Lacks, S. A.** 1977. Binding and entry of DNA in pneumococcal transformation, p. 179–232. In J. Ressig (ed.), Microbial Interactions. Chapman and Hall, London.
17. **Le Bouguénec, C., G. de Cespédès, and T. Horaud.** 1988. Molecular analysis of a composite chromosomal conjugative element (Tn3701) of Streptococcus pyogenes. J. Bacteriol. **170:**3930–3936.
18. **Le Bouguénec, C., G. de Cespédès, and T. Horaud.** 1990. Presence of chromosomal elements resembling the composite structure Tn3701 in streptococci. J. Bacteriol. **172:**727–734.
19. **Le Bouguénec, C., T. Horaud, G. Bieth, R. Colimon, and C. Dauguet.** 1984. Translocation of antibiotic resistance markers of a plasmid-free Streptococcus pyogenes (group A) strain into different streptococcal hemolysin plasmids. Mol. Gen. Genet. **194:**377–387.
20. **Macrina, F. L., R. P. Evans, J. A. Tobian, D. L. Hartley, D. B. Clewell, and K. R. Jones.** 1983. Novel shuttle plasmid vehicles for Escherichia-Streptococcus transgeneric cloning. Gene **25:**145–150.
21. **Pepper, K., G. de Cespédès, and T. Horaud.** 1988. Heterogeneity of chromosomal genes encoding chloramphenicol resistance in streptococci. Plasmid **19:**71–74.
22. **Saunders, C. W., and W. R. Guild.** 1981. Pathway of plasmid transformation in pneumococcus: open circular and linear molecules are active. J. Bacteriol. **146:**517–526.
23. **Scott, J. R., P. A. Kirchman, and M. G. Caparon.** 1988. An intermediate in transposition of the conjugative transposon Tn916. Proc. Natl. Acad. Sci. USA **85:**4809–4813.
24. **Senghas, E., J. M. Jones, M. Tamamoto, C. Gawron-Burke, and D. B. Clewell.** 1988. Genetic organization of the bacterial conjugative transposon Tn916. J. Bacteriol. **170:**245–249.
25. **Shoemaker, N. B., M. D. Smith, and W. R. Guild.** 1979. Organization and transfer of heterologous chloramphenicol and tetracycline resistance genes in pneumococcus. J. Bacteriol. **139:**432–441.
26. **Shoemaker, N. B., M. D. Smith, and W. R. Guild.** 1980. DNase-resistant transfer of chromosomal cat and tet insertions by filter mating in pneumococcus. Plasmid **3:**80–87.
27. **Smith, M. D., and W. R. Guild.** 1982. Evidence for transposition of the conjugative R determinants of Streptococcus

agalactiae B109, p. 109–111. *In* D. Schlessinger (ed.), *Microbiology—1982*. American Society for Microbiology, Washington, D.C.

28. **Smith, M. D., S. Hazum, and W. R. Guild.** 1981. Homology among *tet* determinants in conjugative elements of streptococci. *J. Bacteriol.* **148:**232–240.

29. **Vasseghi, H., J. P. Claverys, and A. M. Sicard.** 1981. Mechanism of integrating foreign DNA during transformation of *Streptococcus pneumoniae*, p. 137–153. *In* M. Polsinelli and G. Mazza (ed.), *Transformation—1980*. Cotswold Press, Oxford.

30. **Vijayakumar, M. N., S. D. Priebe, and W. R. Guild.** 1986. Structure of a conjugative element in *Streptococcus pneumoniae*. *J. Bacteriol.* **166:**978–984.

31. **Vijayakumar, M. N., S. D. Priebe, G. Pozzi, J. M. Hageman, and W. R. Guild.** 1986. Cloning and physical characterization of chromosomal conjugative elements in streptococci. *J. Bacteriol.* **166:**972–977.

Comparison of an Enterococcal Gentamicin Resistance Transposon and β-Lactamase Gene with Those of Staphylococcal Origin

SUSAN L. HODEL-CHRISTIAN,[1,2] MONICA C. SMITH,[1] KAREN K. ZSCHECK,[1]
AND BARBARA E. MURRAY[1,3]

*Center for Infectious Diseases,[1] Department of Microbiology,[2] and Department of Internal Medicine,[3]
University of Texas Medical School, Houston, Texas 77030*

Enterococci have developed resistance to virtually all useful antibiotics and presumably as a result are playing a larger role as a causative agent of nosocomial infections (11). Enterococci have both intrinsic or inherent resistances (e.g., tolerance to β-lactams and low-level resistance to aminoglycosides) and acquired resistances (e.g., β-lactamase and vancomycin resistance and high-level resistance to aminoglycosides). As a result of the latter resistance, the otherwise synergistic treatment of enterococcal infections with penicillin and an aminoglycoside has lost its effectiveness. In our laboratory, we have been studying the acquisition of high-level resistance to gentamicin and the presence of β-lactamase in *Enterococcus faecalis*. Two strains studied (one from Houston, Tex. [HH22], and one from Pennsylvania [PA]) carry the genetic determinants for both aminoglycoside resistance and β-lactamase on related plasmids.

Gentamicin Resistance Transposon in *E. faecalis*

High-level gentamicin resistance (Gm^r) (MIC, $\geq$2,000 µg/ml) in *E. faecalis* was first observed in 1979 in France, and nine *E. faecalis* isolates from the United States were reported to have high-level resistance to all commercially available aminoglycosides in 1983 (7, 10). One of these original isolates, HH22, was later shown to contain a conjugative plasmid, pBEM10, that encoded Gm^r and β-lactamase (Bla$^+$). Gm^r in both enterococci and staphylococci is the result of a bifunctional enzyme having 6'-acetyltransferase and 2"-phosphotransferase activities. The gene encoding this bifunctional enzyme is transposon borne in some Australian isolates of *Staphylococcus aureus* (Tn4001) and U.S. isolates of *Staphylococcus epidermidis* (Tn4031) (8, 9, 17). Because identical nucleotide sequences have been obtained for the region encoding 6'-acetyltransferase and 2"-phosphotransferase in Tn4001 and the Gm^r *E. faecalis* plasmid pIP800 (1, 5, 16), we investigated the possibility that the Gm^r gene resides on a transposon on pBEM10.

To determine whether the Gm^r gene of enterococci could transpose, we first generated a trans-fer-deficient (Tra$^-$) mutant of pBEM10, using plasmid pTV1-ts as a temperature-sensitive delivery system for the erythromycin resistance (Em^r) transposon Tn917, as detailed elsewhere (6). Briefly, plasmid pBEM10 was mated into *E. faecalis* OG1X(pTV1-ts), and following growth at the nonpermissive temperature, Gm^r Em^r Bla$^+$ Cm^s colonies [OG1X(pBEM10::Tn917)] were selected and screened for loss of the ability to transfer resistances by conjugation. The conjugative, tetracycline resistance (Tet^r) plasmid pCF10 was then mated into OG1X(pBEM10::Tn917, Tra$^-$) (2, 3). Cross-streak mating was used to detect transposition of the Gm^r gene from pBEM10::Tn917 into the coresiding pCF10. Transconjugants were selected that had cotransfer of Tet^r and Gm^r in the absence of Em^r and Bla$^+$ transfer, as this pattern suggested that the Gm^r gene had inserted into pCF10 and had been carried into the recipient as an integral part of the Tet^r plasmid. Insertion was verified by restriction endonuclease (RE) analysis and DNA-DNA hybridization. Using an *E. faecalis* strain with an intact homologous recombination system, three transconjugants were obtained that had the Gm^r gene inserted into different locations in pCF10; these derivatives are referred to as A, B, and C. Two of the insertions (A and C) were within 100 to 200 bp of each other, and the third (B) was located over 20 kb away. To verify that the Gm^r gene in pBEM10 was carried on a transposon having the ability to move independently of homologous recombination, we cotransferred pCF10 and pBEM10::Tn917 to *E. faecalis* UV202 (Rif^r), a recombination-deficient strain (20), by taking advantage of the ability of pCF10 to cotransfer pBEM10::Tn917 at a very low frequency. Then UV202(pBEM10::Tn917, pCF10) was cross-streak mated with OG1SSp, a standard *E. faecalis* recipient strain, and Gm^r Tet^r Em^s Bla$^-$ transconjugants were selected. Presence of the Gm^r gene in pCF10 was verified again by RE analysis and DNA-DNA hybridization using the Gm^r gene probe. Two different insertion sites in pCF10 were observed following transposition in UV202. One insertion site was near Tn925, and the other location resulted in colonies that were constitutive clumpers. The Gm^r transposon in *E. faecalis* has been termed Tn*5281*.

RE analysis and DNA-DNA hybridization were used to compare Tn*5281* with the staphylococcal Gm^r transposons Tn*4001* and Tn*4031*. Our data suggest that Tn*5281* is identical to Tn*4001* and Tn*4031*, as all three transposons have symmetrically located sites for *Hin*dIII (2.5 kb apart), *Cla*I (slightly more than 2.5 kb apart), and *Hae*III (3.9 kb apart) that hybridize with a Gm^r gene probe (containing the *Alu*I fragment of pIP800 cloned into pSF815A) (5, 6). All of these restriction sites are located in the terminal inverted repeats, termed IS*256* in Tn*4001* (1). An additional *Hae*III site is located 0.3 kb outside of the forementioned *Hae*III site in each terminus. Digestion of plasmid DNA containing Tn*4031* and Tn*5281* revealed that the transposons have the same RE pattern when digested with *Hinc*II, *Sca*I, and *Alu*I. Figure 1 contains a restriction map comparison of Tn*4001*, Tn*4031*, and Tn*5281*.

To further verify that Tn*5281* was related to the staphylococcal transposons and that the termini of Tn*5281* were composed of inverted repeats similar to IS*256*, we analyzed plasmid DNA by using a probe containing a portion of the terminal inverted repeats of Tn*4031*. Plasmids pBEM10,

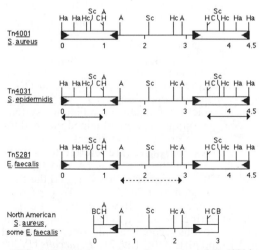

FIG. 1. Restriction endonuclease maps of Tn*4001* in Australian isolates of Gm^r *S. aureus* (1, 8, 16), Tn*4031* located in U.S. isolates of Gm^r *S. epidermidis* (17), the Gm^r transposon in *E. faecalis*, and the region of the Gm^r gene in North American isolates of *S. aureus* and *E. faecalis* X-PA and X-H197. Boxes with arrows indicate the positions of terminal inverted repeats (IS*246* of Tn*4001*). The dashed-line arrow denotes the region contained in the Gm^r gene probe; solid-line arrows indicate areas used for the IS*256*-like probe. Restriction endonuclease abbreviations: Ha, *Hae*III; H, *Hin*dIII; C, *Cla*I; B, *Bgl*II; Hc, *Hinc*II; Sc, *Sca*I; A, *Alu*I. For clarity, only relevant *Cla*I, *Hinc*II, and *Alu*I sites are shown.

pCF10 and its Tn*5281*-bearing derivatives A, B, and C, and pGO121 (containing Tn*4031* from *S. epidermidis*) were digested with *Hae*III and *Hae*III-*Hin*dIII and hybridized with a probe specific for the terminal inverted repeats of Tn*4031* (IS*256*-like probe) (Fig. 2). Since Tn*5281* is similar to Tn*4001* and Tn*4031*, hybridization was expected to be observed between the probe and two *Hae*III fragments and two *Hae*III-*Hin*dIII fragments. Both A and C yielded the expected results (Fig. 2b, lanes 3' and 5'); however, both pBEM10 and B contain three copies of the IS*256*-like element (Fig. 2). Additional results obtained following hybridization of the IS*256*-like probe with *Hin*dIII-digested pBEM10 and B indicated that two of the three IS*256*-like elements are located side by side as a tandem repeat and are positioned in the same orientation with respect to each other.

We also studied plasmid DNA from other Gm^r *E. faecalis* isolates from various geographic locations (Pennsylvania; Houston, Tex; Thailand; and Chile) to determine whether Tn*5281*-like elements are present in other *E. faecalis* strains. Table 1 summarizes the results obtained following hybridization with the Gm^r gene probe. All Gm^r plasmids tested contained a 2.5-kb *Hin*dIII fragment that hybridized with the Gm^r gene probe. All isolates tested also showed hybridization of a 1.5-kb *Alu*I fragment, which would be expected since nucleotide sequences of both Tn*4001* and pIP800 are known to have two *Alu*I sites located 1.5 kb apart in the *aacA-aphD* coding region. In addition to pBEM10, two isolates from Chile and another Houston plasmid containing the 3.9-kb *Hae*III fragment known to be present in Tn*5281*, Tn*4001*, and Tn*4031*. These isolates also had patterns identical to those seen in pBEM10 (containing Tn*5281*) following digestion with both *Hinc*II and *Sca*I and hybridization to the Gm^r gene probe. Other isolates, however, had different patterns. Three Gm^r *E. faecalis* isolates (from Pennsylvania, Houston, and Thailand) lacked the 3.9-kb *Hae*III fragment of Tn*5281*, Tn*4001*, and Tn*4031*. Two of these (Pennsylvania and Houston) had *Bgl*II sites located 3.0 kb apart, similar to the nonmobile Gm^r determinants found in North American isolates of *S. aureus* that are known to have *Bgl*II sites symmetrically located 3.15 kb apart (8, 17). These strains also had hybridization patterns different from those observed for Tn*5281* following digestion with *Hinc*II and *Sca*I. All but one of these various isolates (from Thailand) contained a 2.5-kb *Cla*I fragment that hybridized with the Gm^r gene probe.

The restriction maps of various Gm^r determinants in *E. faecalis* are shown in Fig. 1. Both PA (Pennsylvania) and X-H197 (Houston) have RE and DNA-DNA hybridization patterns similar to those observed from North American isolates of *S. aureus* that have nonmobile Gm^r determinants,

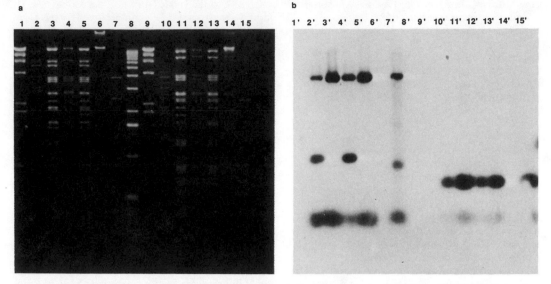

FIG. 2. (a) Agarose (0.7%) gel electrophoresis of restriction endonuclease digestions with *Hae*III (lanes 2 to 7) and *Hae*III-*Hin*dIII (lanes 10 to 15) of pBEM10 (lanes 2 and 10); plasmid DNA from derivatives A (lanes 3 and 11), B (lanes 4 and 12), and C (lanes 5 and 13); pCF10 (lanes 6 and 14); and pGO121 (lanes 7 and 15). Molecular weight standards: lambda DNA digested with *Hin*dIII (lanes 1 and 9) and 1-kb ladder (lane 8). Digestion of pCF10 was incomplete. (b) Autoradiograph of a filter of the agarose gel in panel a following hybridization with the ^{32}P-labeled IS*256*-like probe.

whereas X-H181 (Houston) and two strains from Chile appear to contain Tn*5281*. The strain from Thailand is not similar to either Tn*5281* or the other enterococci. Hybridization data obtained from another Thailand isolate of Gm^r *E. faecalis* also show divergence from both the Tn*5281* and North American patterns. These data indicate that there are both similarities and differences present in the regions encoding Gm^r in *E. faecalis* and that the Gm^r transposon, Tn*5281*, is present within the enterococci from at least two distinct geographic locations.

In conclusion, the Gm^r determinant located on *E. faecalis* plasmid pBEM10 is carried on a transposon, Tn*5281*, that is similar, if not identical, to the staphylococcal transposons Tn*4001* and Tn*4031*. Our data suggest that the original copy of Tn*5281* in pBEM10 has a tandem repeat of the

IS*256*-like element at one end and that Tn*5281* apparently can jump with or without the double insertion sequence element. Some Gm^r *E. faecalis* isolates from diverse geographic locations share these similarities, but others are divergent, including some isolates that are more similar to the North American isolates of *S. aureus*. The isolates from Thailand appear to be extremely divergent in the region surrounding the Gm^r gene compared with the other two patterns.

β-Lactamase in *E. faecalis*

β-Lactamase was first reported in enterococci in the Gm^r isolate HH22 and subsequently shown to be encoded on the large, conjugative pheromone-responsive plasmid pBEM10 described above (12–14). A 1.4-kb β-lactamase-encoding region of pBEM10 was found to have an RE

TABLE 1. Summary of hybridization with the Gm^r gene probe

Origin	Fragment hybridizing (kb)				Conclusion
	*Hin*dIII	*Hae*III	*Bgl*II	*Cla*I	
Houston	2.5	3.9	>15	2.5	Tn*5281*
Pennsylvania	2.5	7.5	3.0	2.5	Like nonmobile *S. aureus*
Houston	2.5	3.9	>15	2.5	Tn*5281*-like
Houston	2.5	9–10	3.0	2.5	Like nonmobile *S. aureus*
Thailand	2.5	9–10	6	3.0	Neither
Chile	2.5	3.9	>15	2.5	Tn*5281*-like
Chile	2.5	3.9	>15	2.5	Tn*5281*-like

digestion and hybridization pattern identical to that of an equal-size region encompassing the *bla* gene from the staphylococcal plasmid pI258 (18). A second β-lactamase-encoding fragment from *E. faecalis* PA also has been cloned into *Escherichia coli* and compared by RE digestion analysis and hybridization with the β-lactamase-encoding regions from pBEM10, the conjugative staphylococcal plasmid pCRG1600, and pI258 (19, 21). Hybridization was observed between a *bla* probe from pI258 and an equal-size *Hin*dIII-*Xba*I fragment present in all four plasmids. The *bla* structural gene in each plasmid is encompassed by two adjacent *Eco*RV fragments. The larger *Eco*RV fragment contains the upstream region of the gene and is of an identical size in all four plasmids; in the plasmid from PA, the smaller *Eco*RV fragment is ~50 bp smaller than the corresponding fragment in the other plasmids, indicating that the downstream *Eco*RV site is located closer to the *bla* gene than in the other plasmids. Proximal to the upstream *Eco*RV site, the enterococcal regions have RE digestion patterns that are markedly different from each other and from those of the staphylococcal upstream regions published for a number of *bla* plasmids (21). Hybridization studies using the region downstream of *bla* from pCRG1600 as a probe showed strong homology between this probe and the corresponding region of PA but not of pBEM10. It therefore appears that although the enterococcal *bla* gene originated from a staphylococcal strain, the entire region, particularly the upstream region, did not emanate from *S. aureus* or has undergone significant rearrangements.

The sequence of the *bla* gene from pBEM10 has been determined and compared with the published nucleotide sequences of several staphylococcal *bla* genes (K. Zscheck and B. E. Murray, *Program Abstr. 29th Intersci. Conf. Antimicrob. Agents Chemother.*, abstr. 1121, p. 72, 1989). The coding region for the enterococcal *bla* gene is identical to those of two type A staphylococcal *bla* genes but differs from those of two other genes by 2 nucleotides (both changes are silent) and 12 nucleotides (4). This gene also is highly homologous to a type D *bla* gene that has been sequenced and the partial sequence available from two type C genes.

Despite the similarity of the structural genes, the enterococcal and staphylococcal β-lactamases differ in regulation, as there is constitutive expression of the enzyme in *E. faecalis* HH22 as well as the other enterococcal Bla+ strains studied, whereas the enzyme is inducible in most *S. aureus* strains. An explanation for the difference in regulation is not evident in the nucleotide sequences, as the *bla* genes have identical promoter and operator regions. Staphylococcal *bla* is reported to be under the control of a repressor, and

studies are currently under way to determine whether constitutive expression of enterococcal β-lactamase is due to the absence of this repressor or to other factors. The staphylococcal and enterococcal β-lactamases also differ in their cellular location. In *S. aureus* the β-lactamase is released into the surrounding environment, whereas it is cell bound in *E. faecalis*. The difference in location appears to be a host-related phenomenon, since when the enterococcal *bla* was put into a staphylococcal background, the enterococcal β-lactamase was released into the surrounding medium; when the staphylococcal *bla* was put in an enterococcal host, the enzyme became cell bound.

In addition to the geographic areas of the two β-lactamase-producing enterococci described above, there are at least 10 other locations in which such organisms have been found: eight cities in the United States; Buenos Aires, Argentina; and Beirut, Lebanon. Most of these isolates also have resistance to high levels of gentamicin. We have recently compared 12 Bla+ enterococcal isolates from 10 locations by using pulsed-field gel electrophoresis of *Sma*I-digested total genomic DNA; this enzyme typically generates from 15 to 20 fragments varying from 20 to 500 kb from enterococcal chromosomes (15). Bla+ isolates from Connecticut, Massachusetts, Lebanon, and Argentina had marked restriction fragment length polymorphism and thus very different RE digestion patterns; these restriction patterns differed from those of other Bla+ enterococci and of non-β-lactamase-producing strains (Fig. 3). However, isolates from Houston, Philadelphia, Pittsburgh, Delaware, Florida, and Virginia appeared to be derivatives of a single strain,

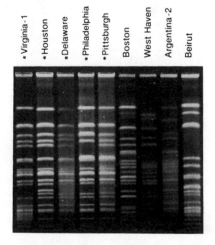

FIG. 3. Pulsed-field gel electrophoresis of nine Bla+ enterococci. The fragments on this gel range from ≈20 to ≈450 kb. The isolates marked with asterisks represent the common strain.

(B. E. Murray, K. V. Singh, S. M. Markowitz, H. A. Lopardo, J. E. Patterson, M. J. Zervos, E. Rubeglio, G. M. Eliopoulos, L. B. Rice, F. W. Goldstein, G. Caputo, R. Nasnas, L. S. Moore, E. S. Wong, and G. Weinstock, *J. Infect. Dis.*, in press). These isolates had identical or highly similar RE digestion patterns, with most differing from others in this group by two to three fragments. The relatedness of fragments of the same size also was shown by isolating two different fragments of ~171 and ~27 kb of one isolate and hybridizing these fragments to digests of the other β-lactamase producers. For the related isolates, a single fragment the same size as the probe hybridized, but with the dissimilar strains, one or more fragments of different sizes hybridized. Thus, it appears that a single clone of β-lactamase-producing *E. faecalis* spread between Houston, Florida, and the four mid-Atlantic cities. Pulsed-field gel electrophoresis also was used to show that this same strain spread to a number of patients in a hospital in Virginia.

Conclusion

In these studies, we have shown that in some enterococci, as in staphylococci, Gm[r] is encoded on a transposon and that the transposons in these two genera are similar if not identical. However, in other enterococci, either Gm[r] is on a different element or the transposon has been altered extensively. The *bla* gene of enterococci also is identical to ones from staphylococci. This gene is on a conjugative plasmid that transfers at a high frequency, which should lead to its dissemination into other strains. However, using pulsed-field gel electrophoresis, we have shown that in six of the eight cities in the United States in which Bla[+] enterococci are found, the isolates are colonized by a single strain. This technique appears to be a powerful new method for epidemiologic analysis of enterococci as well as a number of other organisms.

LITERATURE CITED

1. **Byrne, M. E., D. A. Rouch, and R. A. Skurray.** 1989. Nucleotide sequence analysis of IS*256* from the *Staphylococcus aureus* gentamicin-tobramycin-kanamycin-resistance transposon Tn*4001*. *Gene* **81**:361–367.
2. **Dunny, G. M., J. C. Adsit, and C. Funk.** 1981. Direct stimulation of the transfer of antibiotic resistance by sex pheromones in *Streptococcus faecalis*. *Plasmid* **6**:270–278.
3. **Dunny, G. M., B. L. Brown, and D. B. Clewell.** 1978. Induced cell aggregation and mating in *Streptococcus faecalis*: evidence for a bacterial sex pheromone. *Proc. Natl. Acad. Sci. USA* **75**:3479–3483.
4. **East, A. K., and K. G. H. Dyke.** 1989. Cloning and se-

quence determination of six *Staphylococcus aureus* beta-lactamases and their expression in *Escherichia coli* and *Staphylococcus aureus*. *J. Gen. Microbiol.* **135**:1001–1015.
5. **Ferretti, J. J., K. S. Gilmore, and P. Courvalin.** 1986. Nucleotide sequence analysis of the gene specifying the bifunctional 6′-aminoglycoside acetyltransferase 2″-aminoglycoside phosphotransferase enzyme in *Streptococcus faecalis* and identification and cloning of gene regions specifying the two activities. *J. Bacteriol.* **167**:631–638.
6. **Hodel-Christian, S. L., and B. E. Murray.** 1990. Mobilization of the gentamicin resistance gene in *Enterococcus faecalis*. *Antimicrob. Agents Chemother.* **34**:1278–1280.
7. **Horodniceanu, T., L. Bougueleret, N. El-Solh, G. Bieth, and F. Delbos.** 1979. High-level, plasmid-borne resistance to gentamicin in *Streptococcus faecalis* subsp. *zymogenes*. *Antimicrob. Agents Chemother.* **16**:686–689.
8. **Lyon, B. R., M. T. Gillespie, M. E. Byrne, J. W. May, and R. A. Skurray.** 1987. Plasmid-mediated resistance to gentamicin in *Staphylococcus aureus*: the involvement of a transposon. *J. Med. Microbiol.* **23**:101–110.
9. **Lyon, B. R., J. W. May, and R. A. Skurray.** 1984. Tn*4001*: a gentamicin and kanamycin resistance transposon in *Staphylococcus aureus*. *Mol. Gen. Genet.* **193**:554–556.
10. **Mederski-Samoraj, B. D., and B. E. Murray.** 1983. High-level resistance to gentamicin in clinical isolates of enterococci. *J. Infect. Dis.* **147**:751–757.
11. **Murray, B. E.** 1990. The life and times of the enterococcus. *Clin. Microbiol. Rev.* **3**:46–65.
12. **Murray, B. E., F. Y. An, and D. B. Clewell.** 1988. Plasmids and pheromone response of the β-lactamase producer *Streptococcus (Enterococcus) faecalis* HH22. *Antimicrob. Agents Chemother.* **32**:547–551.
13. **Murray, B. E., D. A. Church, A. Wanger, K. Zscheck, M. E. Levison, M. J. Ingerman, E. Abrutyn, and B. Mederski-Samoraj.** 1986. Comparison of two β-lactamase-producing strains of *Streptococcus faecalis*. *Antimicrob. Agents Chemother.* **30**:861–864.
14. **Murray, B. E., B. Mederski-Samoraj, S. K. Foster, J. L. Brunton, and P. Hartford.** 1986. *In vitro* studies of plasmid-mediated penicillinase from *Streptococcus faecalis* suggest a staphylococcal origin. *J. Clin. Invest.* **77**:289–293.
15. **Murray, B. E., K. V. Singh, J. D. Heath, B. R. Sharma, and G. M. Weinstock.** 1990. Comparison of genomic DNA of different enterococcal isolates using restriction endonucleases with infrequent recognition sites. *J. Clin. Microbiol.* **28**:2059–2063.
16. **Rouch, D. A., M. E. Byrne, Y. C. Kong, and R. A. Skurray.** 1987. The *aacA-aphD* gentamicin and kanamycin resistance determinant of Tn*4001* from *Staphylococcus aureus*: expression and nucleotide sequence analysis. *J. Gen. Microbiol.* **133**:3039–3052.
17. **Thomas, W. D., and G. L. Archer.** 1989. Mobility of gentamicin resistance genes from staphylococci isolated in the United States: identification of Tn*4031*, a gentamicin resistance transposon from *Staphylococcus epidermidis*. *Antimicrob. Agents Chemother.* **33**:1335–1341.
18. **Wanger, A. R., and B. E. Murray.** 1990. Comparison of enterococcal and staphylococcal β-lactamase plasmids. *J. Infect. Dis.* **161**:54–58.
19. **Weber, D. A., and R. V. Goering.** 1988. Tn*4201*, a β-lactamase transposon in *Staphylococcus aureus*. *Antimicrob. Agents Chermother.* **32**:1164–1169.
20. **Yagi, Y., and D. B. Clewell.** 1980. Recombination-deficient mutant of *Streptococcus faecalis*. *J. Bacteriol.* **143**:966–970.
21. **Zscheck, K. K., and B. E. Murray.** 1988. Restriction mapping and hybridization studies of a β-lactamase-encoding fragment from *Streptococcus (Enterococcus) faecalis*. *Antimicrob. Agents Chemother.* **32**:768–769.

Genetic Manipulation of Streptococci by Chromosomal Integration of Recombinant DNA

GIANNI POZZI,[†] MARCO R. OGGIONI,[†] RICCARDO MANGANELLI,[†] AND
PIERFAUSTO PLEVANI

Istituto di Microbiologia, Università di Verona, 37134 Verona, Italy

As for any microorganism, the possibility of genetically engineering streptococci is important for studying their biology. Genetic manipulation of streptococci has many applications, especially in the study of pathogenicity and in biotechnology. In particular, we are interested in constructing stable recombinant strains of nonpathogenic streptococci that express on the surface heterologous antigens, to be used as live vectors for vaccines.

Since recombinant DNA technology has been available, the use of self-replicating plasmids has been the most common approach to genetic manipulation of bacteria. Recombinant plasmids can be introduced into bacterial cells by a variety of genetic techniques: natural transformation, artificial transformation, protoplast transformation, transduction, conjugative mobilization, and electroporation. The major limitations of this approach are due to the fact that recombinant plasmids are often unstable. Also, fine studies on the physiology of gene expression cannot be performed when genes are carried on multicopy plasmids. An alternative approach is to integrate recombinant DNA molecules into the bacterial chromosome. Chromosomal integration of recombinant DNA has two major advantages: it is possible to (i) construct stable recombinant strains and (ii) create stable merodiploids in which a single additional copy of a chromosomal gene is integrated in the chromosome, away from its locus.

In naturally transformable streptococci, such as *Streptococcus pneumoniae* and *Streptococcus gordonii* Challis (formerly *Streptococcus sanguis*), heterologous DNA ligated to chromosomal sequences can be integrated into the chromosome during transformation (5, 6, 8, 12, 13), whereas in nontransformable streptococci, the same result can be achieved by using conjugative transposons as vectors of recombinant DNA molecules (10).

Chromosomal Integration and Gene Expression in *S. gordonii*

To integrate recombinant DNA molecules into the chromosome of *S. pneumoniae* or *S. gordonii*

Challis, it is possible to exploit the natural competence of these bacteria for genetic transformation. Chromosomal DNA ligated to the DNA to be integrated provides the homology that drives the integration process. Depending on how the constructs are made, there are two modes by which heterologous DNA can be integrated into the chromosome of naturally transformable streptococci: by insertion duplication or by flanking homology. When a segment of chromosomal DNA is simply ligated to the heterologous DNA, the integration during transformation leads to duplication of the homologous segment, with formation of two direct repeats flanking one copy of the heterologous sequences (6, 8). On the other hand, when homology is present on both sides, i.e., the heterologous DNA is cloned within a fragment of chromosomal DNA, integration occurs by flanking homology, without creating repeats but simply interrupting the chromosomal sequences (8).

If heterologous DNA is ligated to a chromosomal digest and used for transformation, it is possible to obtain random integration of the heterologous sequences into the streptococcal chromosome. We used the latter approach to integrate promoterless genes into the chromosome of *S. gordonii*, in order to obtain strains expressing these genes after in vivo transcriptional fusion with chromosomal promoters. The method was based on the presence of a restriction site a few base pairs upstream of the translation initiation codon, so that cleavage with the restriction enzyme would leave the gene promoterless but with an intact ribosome binding site (Fig. 1). This restriction site was used to ligate in vitro the promoterless gene with random fragments of chromosomal DNA. The ligation mixture was used to transform the naturally competent *S. gordonii* Challis, and several hundred transformants were selected. Among them, it was possible to select strains with different levels of expression and with different degrees of stability, since each transformant was the product of an integration event involving a different fragment of chromosomal DNA. A promoterless M6 protein gene (*emm6.1*) and a promoterless *cat* gene were used in these experiments. For selection purposes, an *erm* gene was cloned adjacent to *emm6.1*, within

[†]Present address: Dipartimento di Biologia Moleculare, Sezione di Microbiologia, Università di Siena, 53100 Siena, Italy.

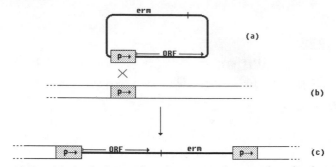

FIG. 1. In vivo transcriptional fusion by integration of a promoterless gene into the chromosome of a naturally transformable streptococcus. A promoterless gene linked to the selectable marker *(erm)* is ligated in vitro to random chromosomal fragments, using a restriction site a few base pairs upstream of the translation initiation codon of its open reading frame (ORF) (a). The ligation mixture is used to transform the recipient strain (b), and the homologous sequences allow chromosomal integration of the promoterless gene and of the *erm* marker (c). By this process, the heterologous DNA is integrated in the chromosome between two direct repeats of the homologous chromosomal fragment ( ). Since some of the chromosomal fragments contain promoters (P→), it is possible to obtain expression of the promoterless gene after in vivo transcriptional fusion with chromosomal promoters (c).

the same restriction fragment (Fig. 1), whereas the integration of *cat* was detected by directly selecting for chloramphenicol-resistant transformants.

Results concerning the integration of *emm6.1* and *cat* into the chromosome of *S. gordonii* that are of general interest with respect to the construction of recombinant strains are as follows: (i) for 13 of 13 transformants analyzed by Southern blot, it was shown that both restriction sites at the junction between heterologous DNA and chromosome were intact; (ii) 10 of the 13 transformants analyzed had only one copy of the heterologous gene integrated into the chromosome; (iii) the stability of the phenotypes of the transformants varied considerably among different transformants, but it was always possible to recover stable transformants (Fig. 2); and (iv) 196 of 700 transformants (28%) selected for the *erm* marker also expressed the M6 protein.

Using the approach described above, we were able to clone and characterize a promoter from the chromosome of *S. gordonii* that is stable in *Escherichia coli* and can promote heterologous gene expression in *S. gordonii* and in *S. pneumoniae*, both when integrated into the chromosome and when carried on an autonomous plasmid (P. Plevani et al., unpublished data). It was also possible to isolate strains that expressed on the surface large amounts of a heterologous protein, the M6 protein of *Streptococcus pyogenes* (G. Pozzi et al., unpublished data).

Conjugative Transposons as Vectors of Recombinant DNA

As discussed above, heterologous DNA can be easily integrated into the chromosomes of natu-

rally transformable streptococci. Genetic manipulation of nontransformable streptococci is more difficult and relies mainly on electroporation and conjugative mobilization of plasmid DNA. We devised a genetic system that exploits the properties of conjugative transposons to transfer recombinant DNA molecules to the chromosomes of nontransformable streptococci (7, 9, 10). By transformation, a heterologous gene is first integrated into a conjugative transposon carried on the chromosome of a naturally competent streptococcus (*S. pneumoniae* or *S. gordonii*), and

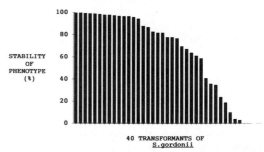

FIG. 2. Stability of the phenotypes of 40 transformants of *S. gordonii*, in which a promoterless heterologous gene was integrated into the chromosome and expressed after transcriptional fusion with resident promoters. After growth in liquid culture without selection for about 50 generations, each transformant was plated, and the phenotype of 200 CFU was analyzed. Results were expressed as percentage of CFU that retained the phenotype encoded by the integrated heterologous DNA. The data represent cumulative results of two different sets of experiments, in which 22 transformants expressed M6 protein (after chromosomal integration of *emm6.1*) and 18 expressed chloramphenicol resistance (after integration of *cat*).

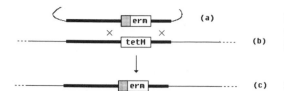

FIG. 3. Transfer of recombinant DNA among streptococci by a conjugative transposon (DNA sequences of the conjugative transposon are represented with a thick line). A heterologous gene ( ) is cloned in the insertion vector adjacent to *erm* (a), and the recombinant plasmid is used as donor in transformation of an intermediate host (*S. pneumoniae* or *S. gordonii*) carrying a conjugative transposon (b). The heterologous gene and *erm* are integrated into the conjugative transposon, inactivating the *tetM* marker (c), and then they are transferred to the chromosomes of other streptococci by conjugation.

then it is transferred by conjugation to the chromosomes of other bacteria (Fig. 3).

Streptococcal conjugative transposons are mobile chromosomal elements that carry the tetracycline resistance gene *tetM* and are capable of conjugative transfer in intra- and interspecific matings (1–4, 11). Our host-vector system for gene transfer to the chromosome of nontransformable streptococci is based on the conjugative transposon Ω(*cat tet*)6001 (now called Tn*5253*). Tn*5253* is 65.5 kb in size and has one specific insertion site in the streptococcal chromosome (14, 15). The insertion vector is pDP36, an *E. coli* plasmid that contains DNA sequences of the *tetM* gene of Tn*5253* interrupted by an *erm* gene (10). The strategy for gene transfer with the Tn*5253*-pDP36 host-vector system is as follows: (i) the gene to be transferred is cloned in pDP36 adjacent to *erm*, and (ii) the recombinant plasmid is used as donor in transformation of a naturally competent streptococcus carrying Tn*5253* integrated into the chromosome; transformants are selected in which integration of *erm* and of the adjacent gene leads to inactivation of the *tetM* gene; these transformants are stable and (iii) can be used as donors in conjugation experiments (Fig. 3).

Tn*5253* has a limited host range: it was found on the chromosome of a clinical strain of *S. pneumoniae*, and its conjugal transfer has been shown only to *S. pyogenes*, *S. agalactiae*, *S. gordonii* Challis, and *Enterococcus faecalis* (4, 7, 10, 11). Conjugative transposons with a host range broader than that of Tn*5253* could also be used as vectors of recombinant DNA, according to the scheme described above. For instance, we have had encouraging preliminary results with the broad-host-range Tn*916* (1, 2). It should be noted that insertion vector pDP36 and its derivative can be used as insertion vectors not only for Tn*5253*

but also for other conjugative transposons, since the target for integration is the *tetM* gene, which is common to all of these genetic elements.

This work was supported in part by grants from the Consiglio Nazionale delle Ricerche (Progetto Finalizzato Ingegneria Genetica) and the North Atlantic Treaty Organization (0840/88).

LITERATURE CITED

1. **Clewell, D. B.** 1981. Plasmid, drug resistance, and gene transfer in the genus *Streptococcus*. *Microbiol. Rev.* **45:** 409–436.
2. **Clewell, D. B., and C. Gawron-Burke.** 1986. Conjugative transposons and the dissemination of antibiotic resistance in streptococci. *Annu. Rev. Microbiol.* **40:**635–659.
3. **Courvalin, P., and C. Carlier.** 1986. Transposable multiple antibiotic resistance in *Streptococcus pneumoniae*. *Mol. Gen. Genet.* **205:**291–297.
4. **Guild, W. R., M. D. Smith, and N. B. Shoemaker.** 1982. Conjugative transfer of chromosomal R determinants in *Streptococcus pneumoniae*, p. 88–92. *In* D. Schlessinger (ed.), *Microbiology—1982*. American Society for Microbiology, Washington, D.C.
5. **Mannarelli, B. M., and S. A. Lacks.** 1984. Ectopic integration of chromosomal genes in *Streptococcus pneumoniae*. *J. Bacteriol.* **106:**867–873.
6. **Morrison, D. A., M.-C. Trombe, M. K. Hayden, G. A. Waszak, and J.-D. Chen.** 1984. Isolation of transformation-deficient *Streptococcus pneumoniae* mutants defective in control of competence, using insertion-duplication mutagenesis with the erythromycin resistance determinant of pAMβ1. *J. Bacteriol.* **159:**870–876.
7. **Oggioni, M. R., and G. Pozzi.** 1989. Ω6001-mediated conjugative mobilization of the cloned M6 protein gene to the chromosomes of different strains of *Streptococcus pyogenes*, p. 189–193. *In* L. O. Butler, C. Harwood, and B. E. B. Moseley (ed.), *Genetic Transformation and Expression*. Intercept, Andover, Hants, United Kingdom.
8. **Pozzi, G., and W. R. Guild.** 1985. Modes of integration of heterologous plasmid DNA into the chromosome of *Streptococcus pneumoniae*. *J. Bacteriol.* **161:**909–912.
9. **Pozzi, G., R. A. Musmanno, and M. R. Oggioni.** 1989. Gene exchange in streptococci: the conjugative chromosomal element Ω6001 as the vector of recombinant DNA molecules, p. 183–188. *In* L. O. Butler, C. Harwood, and B. E. B. Moseley (ed.), *Genetic Transformation and Expression*. Intercept, Andover, Hants, United Kingdom.
10. **Pozzi, G., R. A. Musmanno, E. A. Renzoni, M. R. Oggioni, and M. G. Cusi.** 1988. Host-vector system for integration of recombinant DNA into chromosomes of transformable and nontransformable streptococci. *J. Bacteriol.* **170:**1969–1972.
11. **Shoemaker, N. B., M. D. Smith, and W. R. Guild.** 1979. Organization and transfer of heterologous chloramphenicol and tetracycline resistance genes in pneumococcus. *J. Bacteriol.* **139:**432–441.
12. **Vasseghi, H., and J. P. Claverys.** 1983. Amplification of a chimeric plasmid carrying an erythromycin resistance determinant introduced into the genome of *Streptococcus pneumoniae*. *Gene* **21:**285–292.
13. **Vasseghi, H., J. P. Claverys, and A. M. Sicard.** 1981. Mechanisms of integrating foreign DNA during transformation of *Streptococcus pneumoniae*, p. 137–153. *In* M. Polsinelli and G. Mazza (ed.), *Transformation—1980*. Cotswold Press, Oxford.
14. **Vijayakumar, M. N., S. D. Priebe, and W. R. Guild.** 1986. Structure of a conjugative element in *Streptococcus pneumoniae*. *J. Bacteriol.* **166:**978–984.
15. **Vijayakumar, M. N., S. D. Priebe, G. Pozzi, J. H. Hageman, and W. R. Guild.** 1986. Cloning and physical characterization of chromosomal conjugative elements in streptococci. *J. Bacteriol.* **166:**972–977.

II. Molecular and Genetic Analysis of Pneumococci

comA Locus of *Streptococcus pneumoniae*: Sequence and Predicted Structures Identify a New Member of the Bacterial ATP-Dependent Transport Protein Family

FRANCIS M. HUI, SUSAN M. LANDOWSKI, AND DONALD A. MORRISON

Laboratory for Molecular Biology, Department of Biological Sciences, University of Illinois at Chicago, Chicago, Illinois 60680

Competence for genetic transformation in *Streptococcus pneumoniae* and *Streptococcus sanguis* is regulated in response to population density through the mediation of an extracellular protein, called competence factor (CF) (18, 19). CF has been characterized as a small, basic protein (20), but details of its biochemistry, regulation, and mode of secretion from the cell have not been determined. Most known mutations affecting transformation in this system do not interfere with this regulatory circuit but rather affect some step of the pathway for DNA uptake and recombination in competent cells (14). However, a few mutations do alter the regulation of competence. For example, *trt* mutants (11) are constitutively competent, even in trypsin-containing media, perhaps because they circumvent the CF portion of the cycle. In contrast, *com* mutants block competence induction. Some of the *com* mutants are conditional mutants, capable of wild-type levels of competence if induced by CF supplied in the supernatant from a wild-type competent culture.

Conditional *com* mutations include those at the *comAB* locus of *S. pneumoniae* (15). DNA spanning 6 kb at this locus was cloned, and insertion mutations were used to identify a region of 4 to 5 kb required for normal competence induction (1). The cloned DNA is expressed in the heterologous *Escherichia coli* system, revealing adjacent genes for two proteins (77 and 49 kDa), designated ComA and ComB (Fig. 1) (2). Three possible functions for this locus were proposed: (i) synthesis or secretion of CF, (ii) a CF receptor and signaling pathway, or (iii) release of an inhibitory factor by the Com⁻ insertion mutants.

We have obtained the first sequence of a gene at this locus, *comA*. Details of the sequence will be presented elsewhere. An open reading frame coincided with the location deduced earlier for ComA, both in position and in size (Fig. 1). The amino acid sequence deduced for ComA is shown in Fig. 2. The 717-amino-acid sequence contained one cysteine, would have a pI of 5.8 and a molecular weight of 80,290, and consisted of two moieties, one hydrophilic and the other hydrophobic.

A search for related sequences in the PIR data base (Table 1) revealed several proteins with a high degree of similarity to ComA. The highest similarity scores were for the *E. coli* HlyB protein, the murine and human multidrug resistance proteins, and finally other members of a bacterial ATP-dependent active transport protein family (6). As the alignment of ComA with HlyB in Fig. 2 shows, the basis for the high similarity score with HlyB lay in a large conserved region surrounding the ATP binding site motifs of HlyB and extending throughout the entire C-terminal half of the protein. The strongest sequence similarities to the other members of the family were also in this region. In addition to the large region of sequence similarity, ComA also displayed a structural analogy to the other members of this transport protein family in possessing six putative membrane-spanning segments distributed throughout its N-terminal half. This portion of the sequence contained as well several 20-amino-acid regions of strong similarity between ComA and HlyB.

The *E. coli* HlyB protein belongs to a family of homologous bacterial membrane proteins, each of which is required for the export of a specific extracellular protein. The exported proteins include hemolysins of *Proteus* species (9, 10), hemolysins of *Actinobacillus actinomycetemcomitans* (12) and *Actinobacillus pleuropneumoniae* (4), the *Pasteurella haemolytica* leukotoxin (7, 17), and the bifunctional toxin of *Bordetella pertussis* (3). In each of these cases, the *hlyB* homolog is part of a gene cluster that includes the

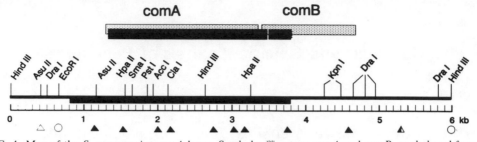

FIG. 1. Map of the *S. pneumoniae comA* locus. Symbols: ▨, genes *comA* and *comB*, as deduced from gene truncation studies (2); ■, genes deduced from the DNA sequence; ▬, extent of sequence determined in both strands for identification of ComA. Triangles and circles indicate the positions of insertion mutations and the outer ends of insertion-duplication mutations, respectively. Filled symbols, Com⁻; open symbols, Com⁺.

```
HLYB    -    MDS--------CHKIDYGLYALEILAQYHN---VSVNPEEIKHRFDTD   -37
                       .: :. .: .. :            :.
COMA    -    MKFGKRHYRPQVDQMDCGVASLAMVFGYYGSYYFLAHLRELAKTTMDG   -48


HLYB    -    GTGLGLTSWLLAAKSLELKVKQVK---KTIDRLNFIF-LPALVWREDGRH   -83
               : :::    . :  .  ..:    :      :    : :.:
COMA    -    TTALGL---VKVAEEIGFETRAIKADMTLFDLPDLTFPFVAHVLKEGKLL   -95


HLYB    -    FILTKISKEVNRYLIFDLEQR-NPRVLEQSEFEALYQGHIILITSRSSV-  -131
                      .  ::.       :     :   .:   .:
COMA    -    HYYVVTGQDKDSIHIADPDPGVKLTKLPRERFEEEWTG-VTLFMAPSPDY  -144


                                  ------------------
HLYB    -    -TGKLAKFDFTWFIPAIIKYRRIFIETLVVSVFLQLFALITPLFFQVVMD  -180
               :  :    ::: ..: :. .. ...... .. . ..:
COMA    -    KPHKEQKNGLLSFIPILVKQRGLIANIVLATLLVTVINIVGSYYLQSIID  -194
                                  ------------------


            ---------------
HLYB    -    KVLVHRGFSTLNVITVALSVVVVFEIILSGLRTYIFAHSTSRIDVELGAK  -230
              .   :::  .:.. :  .:  ::: .   . ...
COMA    -    TYVPDQMRSTLGIISIGLVIVYILQQILSYAQEYLLLVLGQPLSIDVILS  -244
                                  ------------------ ---------------


                                          -------------
HLYB    -    LFRHLLALPISYFESRRVGDTVARVRELDQIRNFLTGQALTSVLDLLFSL  -280
              .:: ::.:.: .:: :. :.:   .  :  :.   :. ::. ..
COMA    -    YIKHVFHLPMSFFATRRTGEIVSRFTDANSIIDALASTILSIFLDV-STV  -293
              --                                 ------- ---


            ------  ------------------
HLYB    -    IFFAVMWYY-SPKLTLVILFSLPCYAAWSVFISPILRRRLDDKFSRNADN  -329
              . ...  . .: .:: :.          .  :    :.
COMA    -    VIISLVLFSQNTNLFFMTLLLALPIYTVIIFAFMKPFEKMNRDTMEANAVL -343
              ------  ------------------


                                    ------------------
HLYB    -    QSFLVESVTAINTIKAMAVSPQMTNIWDKQLAGYVAAGFKVTVLATIGQQ  -379
              : ..:. . : :::... :  .  :: :. :   . .:
COMA    -    SSSIIEDINGIETIKSLTSESQRYQKIDKEFVDYLKKSFTYSRAES-QQK  -392


            ------ -----------  ------------------
HLYB    -    GIQLIQKTVMIIN-LWLGAHLVISGDLSIGQLIAFNMLAGQIVAPVIRLA  -428
              . . ..  :: .:: ::. :. ::.:::::..: :     :  :
COMA    -    ALKKVAHLLLNVGILWMGAVLVMDGKMSLGQLITYNTLLVYFTNPLENII  -442
              ------------------
```

FIG. 2. Alignment of *S. pneumoniae* ComA with *E. coli* HlyB. The ComA sequence deduced from the DNA sequence as in Fig. 1 was aligned with the *E. coli* hemolysin A secretion protein, HlyB, using the method of Myers and Miller (16), with open gap cost of 5 and unit gap cost of 2. NB-1 and NB-2 indicate nucleotide binding sites (5). Symbols: :, identical amino acid residues; ., similar residues (AST, DE, NQ, RK, ILMV, and FYW). A short sequence shared only by all members of the ATP-dependent transport family is indicated (Transport), as are positions in HlyB that are conserved in most members of the transport superfamily defined by Higgins et al. (5) (*). ------, Membrane-spanning segment (8).

```
HLYB   -  QIWQDFQQVGISVTRLGDVLNSPTESYHGKLTLPEIN---GDITFRNIRF  -475
             ..    :   .. ::  .:    .:  : :  ..   ..   ::..:... .
COMA   -  NLQTKLQTAQVANNRLNEVYLVASE-FEEKKTVEDLSLMKGDMTFKQVHY  -491


                    ***  **  **  *****  ********  *  *       *  *
HLYB   -  RYKPDSPVILDNINLSIKQGEVIGIVGRSGSGKSTLTKL,₂ₙFYIPENGQ  -525
            .:        .:   :::..  ::   .   ::  :::::::.::.:..   ::  :  .:
COMA   -  KY-GYGRDVLSDINLTVPQGSKVAFVGISGSGKTTLAKMMVNFYDPSQGE  -540
                                     |_____|
                                            NB-1


            *    *  **      *   **   ****  **   *    **  ***
HLYB   -  VLIDGHDLALADPNWLRRQVGVVLQDNVLLNRSIIDNISL-ANPGMSVEK  -574
             .  .  :    :    :      ::    .   .  :     .  :  .:..:.  :  :  :.  :
COMA   -  ISLGGVNLNQIDKKALRQYINYLPQQPYVFNGTILENLLLGAKEGTTQED  -590


               *  *           *          *        ****************    *
HLYB   -  VIYAAKLAGAHDFISELREGYNTIVGEQGAGLSGGQRQRIAIARALVNNP  -624
             .. :   ::     .  :     .:.    :::.::::::::::::.:::::.
COMA   -  ILRAVELAEIREDIERMPLNYQTELTSDGAGISGGQRQRIALARALLTDA  -640
                                         |_____|  |_____|...
                                            Transport


             *******  *****        *  *         ***  *  *  *    *   ****
HLYB   -  KILIFDEATSALDYESEHVIMRNMHKICKGRTVIIIAHRLSTVKNADRII  -674
             .::  ::::::.::   .:   :.  :.   :    .:.:  :::::.     .....
COMA   -  PVLILDEATSSLDILTEKRIVDNL--IALDKTLIFIAHRLTIAERTEKVV  -690
         ..._____|
                 NB-2


             **   ****    *    **    *
HLYB   -  VMEKGKIVEQGKHKELLSEPESLYSYLYQLQSD  -707
             :..  :::::  :::  .::.       :  .  :
COMA   -  VLDQGKIVEEGKHADLLAQ-GGFYAHLV---NS  -717
```

FIG. 2—*Continued*.

structural gene for the exported protein as well as one or more additional genes for its modification or export.

While the similarity to HlyB clearly suggests that ComA has a transport function, the role of ComA in CF production could be indirect. However, a simple hypothesis accommodating both the strong similarity of ComA to HlyB and the CF-deficient phenotype of *comA* mutants is that the ComA protein functions in secretion of CF

TABLE 1. Protein sequences similar to deduced sequence of *S. pneumoniae* ComA protein

Data base sequence matched[a]		Homology score[b]	
Gene designation	Protein	I	O
HlyB	Hemolysin secretion protein	129	857
HMDR	Human multidrug resistance protein	166	604
MMDR	Mouse multidrug resistance protein	174	581
MalK	Inner membrane maltose permease	96	236
HisP	Histidine permease protein	89	224
ChlD	Molybdenum transport protein	100	213
NodI	Nodulation protein I	87	213
OPPF	Oligopeptide permease protein F	74	203
MbpX	Membrane protein, liverwort	87	196
RbsA	Ribose transport protein	79	190
OPPD	Oligopeptide permease protein D	103	153
NUO	NADH-ubiquinone oxidoreductase	54	91

[a] Gene designations are from original literature; descriptions are as in the PIR data base. Proteins with the 12 highest optimized scores (13) are listed.

[b] I, Initial score; O, optimized score.

from the cell. Indeed, the commonly observed structure of related bacterial toxin loci, involving an ATP-dependent secretion protein and other modification or secretion proteins linked to the toxin structural gene, suggests that *comA* may also be part of a complex locus responsible for synthesis, modification, and secretion of CF.

This work was supported in part by a grant from the National Science Foundation Genetic Biology Program.

LITERATURE CITED

1. **Chandler, M. S., and D. A. Morrison.** 1987. Competence for genetic transformation in *Streptococcus pneumoniae*: molecular cloning of *com*, a competence control locus. *J. Bacteriol.* **169:**2005–2011.
2. **Chandler, M. S., and D. A. Morrison.** 1988. Identification of two proteins encoded by *com*, a competence control locus of *Streptococcus pneumoniae*. *J. Bacteriol.* **170:**3136–3141.
3. **Glaser, P., H. Sakamoto, J. Bellalou, A. Ullmann, and A. Danchin.** 1988. Secretion of cyclolysin, the calmodulin-sensitive adenylate cyclase-haemolysin bifunctional protein of *Bordetella pertussis*. *EMBO J.* **7:**3997–4004.
4. **Gygi, D., J. Nicolet, J. Frey, and M. Cross.** 1990. Isolation of the *Actinobacillus pleuropneumoniae* haemolysin gene and the activation and secretion of the prohaemolysin by the HlyC, HlyB and HlyD proteins of *Escherichia coli*. *Mol. Microbiol.* **4:**123–128.
5. **Higgins, C. F., M. P., Gallagher, M. L. Mimmack, and S. R. Pearce.** 1988. A family of closely related ATP-binding subunits from prokaryotic and eukaryotic cells. *Bioessays* **8:**111–116.
6. **Higgins, C. F., I. D. Hiles, G. P. C. Salmond, D. R. Gill, J. A. Downie, I. J. Evans, I. B. Holland, L. Gray, S. D. Buckel, and A. W. Bell.** 1986. A family of related ATP-binding subunits coupled to many distinct biological processes in bacteria. *Nature* (London) **323:**448–450.
7. **Highlander, S. K., M. Chidambaram, M. J. Engler, and G. M. Weinstock.** 1989. DNA sequence of the *Pasteurella haemolytica* leukotoxin gene cluster. *DNA* **8:**15–28.
8. **Klein, P., M. Kanehisa, and C. DeLisi.** 1985. The detection and classification of membrane-spanning proteins. *Biochim. Biophys. Acta* **815:**468–476.
9. **Koronakis, V., M. Cross, B. Senior, E. Koronakis, and C. Hughes.** 1987. The secreted hemolysins of *Proteus mira-* *bilis*, *Proteus vulgaris*, and *Morganella morganii* are genetically related to each other and to the alpha-hemolysin of *Escherichia coli*. *J. Bacteriol.* **169:**1509–1515.
10. **Koronakis, V., E. Koronakis, and C. Hughes.** 1988. Comparison of the haemolysin secretion protein HlyB from *Proteus vulgaris* and *Escherichia coil* site-directed mutagenesis causing impairment of export function. *Mol. Gen. Genet.* **213:**551–555.
11. **Lacks, S. A., and B. Greenberg.** 1973. Competence for DNA uptake and deoxyribonuclease action external to cells in the genetic transformation of *Diplococcus pneumoniae*. *J. Bacteriol.* **114:**152–163.
12. **Lally, E. T., E. E. Golub, I. R. Kieba, N. S. Taichman, J. Rosenbloom, J. C. Rosenbloom, C. W. Gibson, and D. R. Demuth.** 1989. Analysis of the *Actinobacillus actinomycetemcomitans* leukotoxin gene. *J. Biol. Chem.* **264:**15451–15456.
13. **Lipman, D. J., and W. R. Pearson.** 1985. Rapid and sensitive protein similarity searches. *Science* **227:**1435–1441.
14. **Morrison, D. A., S. A. Lacks, W. R. Guild, and J. M. Hageman.** 1983. Isolation and characterization of three new classes of transformation-deficient mutants of *Streptococcus pneumoniae* that are defective in DNA transport and genetic recombination. *J. Bacteriol.* **156:**281–290.
15. **Morrison, D. A., M. C. Trombe, M. K. Hayden, G. A. Waszak, and J. D. Chen.** 1984. Isolation of transformation-deficient *Streptococcus pneumoniae* mutants defective in control of competence, using insertion-duplication mutagenesis with the erythromycin resistance determinant of pAMβ1. *J. Bacteriol.* **159:**870–876.
16. **Myers, E. W., and W. Miller.** 1988. Optimal alignments in linear space. *CABIOS* **4:**11–17.
17. **Strathdee, C. A., and R. Y. C. Lo.** 1989. Cloning, nucleotide sequence, and characterization of genes encoding the secretion function of the *Pasteurella haemolytica* leukotoxin determinant. *J. Bacteriol.* **171:**916–928.
18. **Tomasz, A.** 1966. Model for the mechanism controlling the expression of the competent state in pneumococcus cultures. *J. Bacteriol.* **91:**1050–1061.
19. **Tomasz, A., and R. D. Hotchkiss.** 1964. Regulation of the transformability of pneumococcal cultures by macromolecular cell products. *Proc. Natl. Acad. Sci. USA* **51:**480–486.
20. **Tomasz, A., and J. L. Mosser.** 1966. On the nature of the pneumococcal activator substance. *Proc. Natl. Acad. Sci. USA* **55:**58–66.

Generalized Mismatch Repair in *Streptococcus pneumoniae*

MARC PRUDHOMME, VINCENT MÉJEAN, BERNARD MARTIN, ODILE HUMBERT, AND
JEAN-PIERRE CLAVERYS

*Centre de Recherche de Biochimie et de Génétique Cellulaires du Centre National de la Recherche Scientifique,
Université Paul Sabatier, 31062 Toulouse Cedex, France*

Investigations of variation in marker transformation efficiencies in *Streptococcus pneumoniae* have led to the detailed characterization of the Hex mismatch repair system (reviewed in reference 5). Further studies on *hex* mutants revealed an increase in spontaneous mutation rates (34, 35), which was taken to indicate that the Hex system corrects potentially mutagenic mismatches resulting from DNA replication errors. Since then, the Mut system of *Escherichia coli* has been characterized as another generalized mismatch repair system similarly involved in genetic stability. Evidence is now accumulating that such systems are widespread from bacteria to fungi and probably to higher eukaryotes. Here we survey results obtained in a study of the Hex system and discuss several lines of evidence which suggest that generalized mismatch repair systems have been conserved during evolution.

Specificity of Mismatch Repair

The Hex system can process different base-base mismatches with different efficiencies (reviewed in reference 10). Tentative mismatch ranking as a function of decreasing repair efficiency is G/T = A/C = G/G > C/T > A/A > T/T > A/G (10). No evidence for repair of C/C has been obtained so far. The Hex system also functions in the correction of short insertions and deletions (5, 9). Similar specificity of repair has been reported for the Mut system of *E. coli* (5, 14, 19, 32) and for the PMS (postmeiotic segregation) system of *Saccharomyces cerevisiae* (2–4, 16, 37). Differential repair of the same mismatch within different sequence contexts has been observed for both Hex (5, 10) and Mut (14) systems. However, contrary to what has been suggested for the Mut system (14), the efficiency of repair of a given mismatch by the Hex system is not related in a simple way to G + C content in the neighboring sequence (10).

The Hex System and Its Counterparts in Other Organisms

Two *hex* genes, *hexA* and *hexB*, have been identified (6) and cloned in a pneumococcal host-vector system (23, 25). Mutation in either of these genes abolishes mismatch repair and confers a mutator phenotype. The homology existing between HexA and MutS (12, 26) and between HexB and MutL (22, 27) gives strong support to the hypothesis that the Hex system and the Mut system of *E. coli* (and of the related bacterium *Salmonella typhimurium*) evolved from a common ancestor (5). The finding that PMS1 of *S. cerevisiae* is also homologous to HexB and MutL (17, 22, 27) suggests that evolutionarily related mismatch repair systems exist in widely diverged organisms. The recent finding of the human (8) and mouse (21) gene products exhibiting homology with HexA-MutS reinforces this idea.

Long-Tract Mismatch Correction

Integrating DNA may contain mismatches of the type recognized and corrected by Hex (G/T, A/C, etc.; see above) or poorly recognized (e.g., C/C). The transformation frequency is 20 times lower in the former case than in the latter. Both genetic evidence (reviewed in reference 5) and physical evidence (24) for long-tract excision have been obtained. We have reinvestigated the mechanism of mismatch repair in transformation, using as the donor a cloned pneumococcal fragment radioactively labeled on one strand only. This fragment either was completely homologous to the recipient chromosome or carried a transition mutation (approximately in the middle of the donor fragment; Fig. 1). The fate of donor label was analyzed by lysis of the transformed cells and separation by agarose gel electrophoresis of chromosomal fragments generated by restriction endonucleases (Fig. 1). In the absence of mismatch at the donor-recipient heteroduplex stage in transformation, homology-dependent integration of donor label occurred. DNA sequences corresponding to the center of the donor fragment were integrated approximately two times more efficiently than were flanking regions. When a transition mismatch between donor and recipient DNA was present, Hex-mediated mismatch repair resulted in the loss of nearly all donor label (Fig. 1). Results are consistent with a strand-specific repair process in which strand discrimination is based on the recognition of single-strand breaks (11). Such a process would lead to excision of the entire donor (invading) strand (Fig. 2). This differs from results obtained in vitro for Mut-directed mismatch repair, which suggest that correction is bordered by a break on one side and by the mismatch on the other (31). The difference

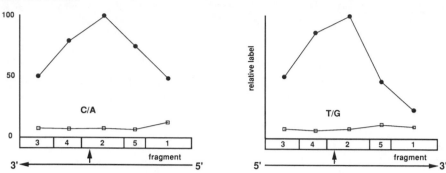

FIG. 1. Mechanism of mismatch repair in transformation. Donor DNA uniformly labeled with ^{32}P on one strand only was prepared in vitro from M13 recombinant clones carrying a 5,445-bp fragment from the *ami* locus (1) (pm1 and pm2, wild-type sequence in either orientation; pm3 and pm4, fragment with the *amiC9* mutation in either orientation). T7 DNA polymerase was used to synthesize complementary strand on these recombinant M13 single-stranded DNA templates in the presence of 5'-[α-^{32}P]dATP, using as the primer a 20-mer complementary to M13 sequences located around the *Xmn*I site (position 357). Wild-type *S. pneumoniae* cells (R800) were exposed for 5 min to DNA. After exposure to DNA, cells were incubated for 20 min at 32°C in the presence of DNase I (to remove unabsorbed DNA) and 6-(p-hydroxyphenylazo)-uracil (a specific inhibitor of replication in gram-positive bacteria [24]) before extraction of chromosomal DNA. Extracted DNA was digested with *Hind*III, *Bam*HI, and *Cla*I, which generated five fragments (numbered 1 to 5 according to decreasing size), and subjected to agarose gel electrophoresis. Autoradiography and densitometer tracing (LKB Ultroscan) were used to quantitate the labeling of each fragment. Relative label was calculated by correcting for fragment size. The position of the *amiC9* mutation (which corresponds to a transition mutation [10]), and hence of the mismatch, within fragment 2 is indicated by an arrow. Symbols: ●, wild-type *ami* fragment; □, *amiC9* mutation. (Left) pm1 and pm3; (right) pm2 and pm4. The relative orientation of the labeled donor strand is indicated at the bottom. The mismatch present at the donor-recipient heteroduplex level is C/A (A mutant; left panel) or T/G (T mutant; right panel).

between the two repair systems may be only circumstantial and simply reflect the availability of two breaks in mismatch repair reactions occurring during transformation.

Thus, both Hex- and Mut-directed repair processes involve long excision tracts, and both systems are very likely to use the same strand discrimination signal. Indeed, although the *E. coli* Mut system is directed by the state of methylation of dGATC sequences (reviewed in reference 5), results obtained both in vivo (20) and in vitro (31) show that, as hypothesized by Lacks and co-workers (18), the ultimate signal for strand discrimination is a single-strand break. This break is introduced by the *E. coli* MutH

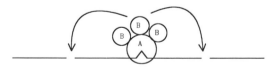

FIG. 2. Model for Hex-mediated mismatch repair in transformation. HexA protein (A) binds to the mismatch (ˆ) present at the donor-recipient heteroduplex stage. HexB protein (B) interaction with this DNA-protein complex triggers the search for a DNA break. Such a break could preferentially be found at one or both extremities of the donor (invading) strand. The ensuing correction would result in removal of the entire donor strand, thus aborting recombination.

protein immediately 5' to the dG of the dGATC sequence (36).

In Vivo Expression of Pneumococcal *hex* Genes in *E. coli*

The homology existing between Hex and Mut proteins, together with the similarities in specificity and mechanism of mismatch repair (see above) has led us to investigate the effect of expressing *hex* genes in *E. coli*. The *hexA* and *hexB* genes were placed under the control of an inducible promoter on a multicopy plasmid and introduced either individually or together in *E. coli* *mutS* and *mutL* mutant strains. We analyzed the effect of *hex* gene expression on forward mutation rates (resistance to rifampin). Indeed, since *mutS* and *mutL* strains exhibit a mutator phenotype (reviewed in reference 5), complementation of a *mut* mutation was expected to result in a decrease in mutation rate. No such effect was observed in the *mut* strains complemented by *hexA*, *hexB*, or both. However, we observed that expression of *hexA* in wild-type *E. coli* cells conferred a mutator phenotype similar to that conferred by *mut* mutations (Table 1). No effect of *hexB* gene expression was observed either individually or in conjunction with *hexA* (data not shown). Since expression of *hexA* did not affect mutation rate in *mutS* cells (data not shown), we conclude that the mutator effect induced in wild-type *E. coli* cells by HexA results from inhibition of the Mut

TABLE 1. Effect of *hexA* gene expression on
mutation rate in *E. coli* GM147 cells[a]

Genotype	Mutation rate (10^{10})
Wild type	44
Wild type (*hexA*[+])	4,086
mutS::Tn*10*	4,016

[a]Independent cultures (at least 35 for each determination)
were inoculated from single colonies and grown to stationary
phase, and clones resistant to rifampin (100 μg/ml) were scored.
Mutation rates, expressed as mutations per cell per generation,
were calculated from mutation frequencies (defined as the number of rifampin-resistant mutants per sensitive cell).

system. Given the homology between HexA and
MutS (see above) and the demonstration that purified MutS protein binds in vitro to DNA substrates containing mismatches (33), we suggest
that HexA binds to mismatches resulting from
replication errors. This binding may protect the
mismatches from repair by the Mut system. The
HexA protein could thus either be unable to interact with Mut proteins to carry out mismatch
repair or produce nonfunctional complexes.

Concluding Remarks

Reduction of yield of recombinants due to repair of mismatched base pairs present within heteroduplex joints, as observed in transformation
(see above), is not limited to *S. pneumoniae*. Indeed, several instances of reduction of the yield
of recombinants by the Mut system of *E. coli* have
been reported, during conjugation (7) and in
phage-phage (13, 15) and plasmid-phage (30) recombination. This antirecombinogenic action has
been suggested to reflect a functional role of generalized mismatch repair systems in controlling
the fidelity of genetic exchange (28). The drastic
(Mut-mediated) reduction of genetic exchange
observed between two bacterial species, *E. coli*
and *Salmonella typhimurium*, that are approximately 20% divergent (29) has led to the attractive hypothesis that generalized mismatch repair
systems play fundamental roles in preventing in-
ter- and intrachromosomal rearrangements (by
suppressing recombination between members of
a gene family) and in maintaining a genetic barrier between species (29). However, preliminary
experiments investigating the effect of the Hex
system on recombination between partially divergent sequences (in the range of 5 to 15% divergence) show that mismatch repair can be
saturated by a single fragment harboring several mismatches (unpublished observations). This
finding suggests that the Hex system is not a major
barrier to genetic exchange between species. Further work with other members of the generalized
mismatch repair family is thus necessary to assess
the generality of this hypothesis.

We are grateful to Alexandra Gruss for editing the manuscript.

This work was supported in part by Contrat de Recherche
Externe 861005 from the Institut National de la Recherche
Médicale.

LITERATURE CITED

1. **Alloing, G., M. C. Trombe, and J. P. Claverys.** 1990. The
ami locus of the Gram-positive bacterium *Streptococcus
pneumoniae* is similar to binding protein-dependent transport operons of Gram-negative bacteria. *Mol. Microbiol.*
4:633–644.
2. **Bishop, D. K., J. Andersen, and R. D. Kolodner.** 1989.
Specificity of mismatch repair following transformation of
Saccharomyces cerevisiae with heteroduplex plasmid DNA.
Proc. Natl. Acad. Sci. USA **86:**3713–3717.
3. **Bishop, D. K., and R. D. Kolodner.** 1986. Repair of heteroduplex plasmid DNA after transformation into *Saccharomyces cerevisiae. Mol. Cell. Biol.* **8:**3401–3409.
4. **Bishop, D. K., M. S. Williamson, S. Fogel, and R. D.
Kolodner.** 1987. The role of heteroduplex correction in
gene conversion in *Saccharomyces cerevisiae. Nature* (London) **328:**362–364.
5. **Claverys, J. P., and S. A. Lacks.** 1986. Heteroduplex deoxyribonucleic acid base mismatch repair in bacteria. *Microbiol. Rev.* **50:**133–165.
6. **Claverys, J. P., H. Prats, H. Vasseghi, and M. Gherardi.**
1984. Identification of *Streptococcus pneumoniae* mismatch
repair genes by an additive transformation approach. *Mol.
Gen. Genet.* **196:**91–96.
7. **Feinstein, S. I., and K. B. Low.** 1986. Hyper-recombining
recipient strains in bacterial conjugation. *Genetics* **113:**13–
33.
8. **Fujii, H, and T. Shimada.** 1989. Isolation and characterization of cDNA clones derived from the divergently transcribed gene in the region upstream from the human
dihydrofolate reductase gene. *J. Biol. Chem.* **264:**10057–
10064.
9. **Gasc, A. M., P. Garcia, D. Baty, and A. M. Sicard.** 1987.
Mismatch repair during pneumococcal transformation of
small deletions produced by site-directed mutagenesis.
Mol. Gen. Genet. **210:**369–372.
10. **Gasc, A. M., A. M. Sicard, and J. P. Claverys.** 1989. Repair
of single- and multiple-substitution mismatches during
recombination in *Streptococcus pneumoniae. Genetics*
121:29–36.
11. **Guild, W. R., and N. B. Shoemaker.** 1976. Mismatch correction in pneumococcal transformation: donor length and
hex-dependent marker efficiency. *J. Bacteriol.* **125:**125–
135.
12. **Haber, L. T., P. P. Pang, D. I. Sobell, J. A. Mankovich,
and G. C. Walker.** 1988. Nucleotide sequence of the *Salmonella typhimurium mutS* gene required for mismatch
repair: homology of MutS and HexA of *Streptococcus pneumoniae. J. Bacteriol.* **170:**197–202.
13. **Huisman, O., and M. S. Fox.** 1986. A genetic analysis of
primary products of bacteriophage lambda recombination.
Genetics **112:**409–420.
14. **Jones, M., R. Wagner, and M. Radman.** 1987. Repair of
a mismatch is influenced by the base composition of the
surrounding nucleotide sequence. *Genetics* **115:**605–610.
15. **Karimova, G. A., P. S. Grigoriev, and V. N. Rybchin.** 1985.
Participation of genes for the system of mismatched bases
correction in genetic recombination of *Escherichia coli.
Mol. Genet. Mikrobiol. Virusol.* **10:**29–34.
16. **Kramer, B., W. Kramer, M. S. Williamson, and S. Fogel.**
1989. Heteroduplex DNA correction in *Saccharomyces
cerevisiae* is mismatch specific and requires functional *PMS*
genes. *Mol. Cell. Biol.* **9:**4432–4440.
17. **Kramer, W., B. Kramer, M. S. Williamson, and S. Fogel.**
1989. Cloning and nucleotide sequence of DNA mismatch
repair gene of *PMS1* from *Saccharomyces cerevisiae*: homology of PMS1 to procaryotic MutL and HexB. *J. Bacteriol.* **171:**5339–5346.
18. **Lacks, S. A., J. J. Dunn, and B. Greenberg.** 1982. Identification of base mismatches recognized by the heterodu-

plex-DNA-repair system of *Streptococcus pneumoniae*. *Cell* **31**:327–336.

19. **Lahue, R. S., K. G. Au, and P. Modrich.** 1989. DNA mismatch correction in a defined system. *Science* **245**:160–164.

20. **Längle-Rouault, F., G. Maenhaut-Michel, and M. Radman.** 1987. GATC sequences, DNA nicks and the MutH function in *Escherichia coli* mismatch repair. *EMBO J.* **6**:1121–1127.

21. **Linton, J. P., J.-Y. J. Yen, E. Selby, Z. Chen, J. M. Chinski, K. Liu, R. E. Kellems, and G. F. Crouse.** 1989. Dual bidirectional promoters at the mouse *dhfr* locus: cloning and characterization of two mRNA classes of the divergently transcribed *rep-1* gene. *Mol. Cell. Biol.* **9**:3058–3072.

22. **Mankovich, J. A., C. A. McIntyre, and G. C. Walker.** 1989. Nucleotide sequence of the *Salmonella typhimurium* *mutL* gene required for mismatch repair: homology of MutL to HexB of *Streptococcus pneumoniae* and to Pms1 of the yeast *Saccharomyces cerevisiae*. *J. Bacteriol.* **171**:5325–5331.

23. **Martin, B., H. Prats, and J. P. Claverys.** 1985. Cloning of the *hexA* mismatch repair of *Streptococcus pneumoniae* and identification of the product. *Gene* **34**:293–303.

24. **Méjean, V., and J. P. Claverys.** 1984. Effect of mismatched base pairs on the fate of donor DNA in transformation of *Streptococcus pneumoniae*. *Mol. Gen. Genet.* **197**:467–471.

25. **Prats, H., B. Martin, and J. P. Claverys.** 1985. The *hexB* mismatch repair gene of *Streptococcus pneumoniae*: characterization, cloning and identification of the product. *Mol. Gen. Genet.* **200**:482–489.

26. **Priebe, S., S. Hadi, B. Greenberg, and S. A. Lacks.** 1988. Nucleotide sequence of the *hexA* gene for DNA mismatch repair in *Streptococcus pneumoniae* and homology of HexA to MutS of *Escherichia coli* and *Salmonella typhimurium*. *J. Bacteriol.* **170**:190–196.

27. **Prudhomme, M., B. Martin, V. Méjean, and J. P. Claverys.** 1989. Nucleotide sequence of the *Streptococcus pneumo-niae hexB* mismatch repair gene: homology of HexB to MutL of *Salmonella typhimurium* and to PMS1 of *Saccharomyces cerevisiae*. *J. Bacteriol.* **171**:5332–5338.

28. **Radman, M.** 1988. Mismatch repair and genetic recombination, p. 169–192. *In* R. Kucherlapati and G. R. Smith (ed.), *Genetic Recombination*. American Society for Microbiology, Washington, D.C.

29. **Rayssiguier, C., D. S. Thaler, and M. Radman.** 1989. The barrier to recombination between *Escherichia coli* and *Salmonella typhimurium* is disrupted in mismatch-repair mutants. *Nature* (London) **342**:396–401.

30. **Shen, P., and H. V. Huang.** 1989. Effect of base pair mismatches on recombination via the RecBCD pathway. *Mol. Gen. Genet.* **218**:358–360.

31. **Su, S.-S., M. Grilley, R. Thresher, J. Griffith, and P. Modrich.** 1989. Gap formation is associated with methyl-directed mismatch correction under conditions of restricted DNA synthesis. *Genome* **104**:104–111.

32. **Su, S.-S., R. S. Lahue, K. G. Au, and P. Modrich.** 1988. Mispair specificity of methyl-directed DNA mismatch correction *in vitro*. *J. Biol. Chem.* **263**:6829–6835.

33. **Su, S.-S., and P. Modrich.** 1986. *Escherichia coli mutS*-encoded protein binds to mismatched DNA base pairs. *Proc. Natl. Acad. Sci. USA* **83**:5057–5061.

34. **Tiraby, G., and M. S. Fox.** 1973. Marker discrimination in transformation and mutation of pneumococcus. *Proc. Natl. Acad. Sci. USA* **70**:3541–3545.

35. **Tiraby, G., and A. M. Sicard.** 1973. Integration efficiencies of spontaneous mutant alleles of *amiA* locus in pneumococcal transformation. *J. Bacteriol.* **116**:1130–1135.

36. **Welsh, K. M., A.-L. Lu, S. Clark, and P. Modrich.** 1987. Isolation and characterization of the *Escherichia coli mutH* gene product. *J. Biol. Chem.* **262**:15624–15629.

37. **White, J. H., S. Fogel, and K. Lusnak.** 1985. Mismatch-specific postmeiotic segregation frequency in yeast suggests a heteroduplex recombination intermediate. *Nature* (London) **315**:350–352.

Restriction/Modification Systems of Pneumococci: Why Two Methylases in the *Dpn*II System?

SANFORD A. LACKS, SYLVIA S. SPRINGHORN, AND SUSANA CERRITELLI

Biology Department, Brookhaven National Laboratory, Upton, New York 11973

The systems for DNA restriction and modification in *Streptococcus pneumoniae* are unusual in three respects. First, the *Dpn*I restriction endonuclease, which is present in some strains of *S. pneumoniae* (formerly called *Diplococcus pneumoniae*), only acts on a methylated site in DNA (8). Typically, bacterial restriction endonucleases act on unmethylated sites in DNA, and the cells that produce them protect their own DNA by methylating the recognition site. Since *Dpn*I attacks only methylated DNA, cells that make it do not require modification of their DNA. Second, strains of *S. pneumoniae* that do not harbor *Dpn*I contain a complementary restriction system, that of *Dpn*II (16). The *Dpn*I and *Dpn*II systems are mutually exclusive because DNA from a *Dpn*I strain, which is not methylated at 5′-GATC-3′ sites, is susceptible to *Dpn*II, and DNA from a *Dpn*II strain, which contains 5′-GmeATC-3′, is susceptible to *Dpn*I (9). Third, genetic and biochemical analysis of the *Dpn*I and *Dpn*II systems (4, 5, 10) revealed an additional, unexpected protein in each of the restriction systems. In the *Dpn*I system an 18-kDa protein of unknown function, as well as the endonuclease, is produced. In the *Dpn*II system three proteins are produced, whereas in a typical restriction system, which like *Dpn*II cleaves at unmethylated sites, only two proteins are generally found: the restriction endonuclease and a modification methylase (21). The function and significance of the additional protein in the *Dpn*II system will be a major focus of this chapter.

Genetic Basis of the *Dpn*I and *Dpn*II Systems

The *Dpn*I and *Dpn*II systems are encoded by genetic cassettes alternatively located at the same position in the pneumococcal chromosome (10). The cassettes are bordered by the same genes, *orfL* and *orfR*, on either side (Fig. 1). This cassette form of chromosomal localization ensures the presence of *Dpn*I genes only in *Dpn*I cells and *Dpn*II genes only in *Dpn*II cells, an arrangement required by the mutually exclusive nature of the two systems. It also allows the exchange of one system for another by bacterial transformation, which is the natural mode of genetic exchange in *S. pneumoniae*, by providing homologous regions adjacent to the cassettes.

The *Dpn*I cassette contains two genes. The first, *dpnC*, encodes a 30-kDa polypeptide corresponding to the *Dpn*I endonuclease (5). The second, *dpnD*, produces a polypeptide of 18 kDa. When hyperexpressed in *Escherichia coli*, the latter product is found in a structure 10 times this size. Nothing is known about the aggregation state of the protein in *S. pneumoniae* or what its function may be in the restriction system.

A laboratory strain of *S. pneumoniae*, called Rx, contains neither restriction endonuclease. This null strain apparently was derived from a *Dpn*I strain by a single mutation in the *dpnC* gene. It is missing a single T·A base pair in a run of eight T·A base pairs at positions 565 to 572 (as numbered in reference 10). This deletion terminates the DpnC polypeptide at one-third its normal length.

The *Dpn*II cassette contains three genes (Fig. 1). The first (*dpnM*) and last (*dpnB*) genes encode the expected modification methylase (DpnM) and restriction endonuclease (DpnB or *Dpn*II) that normally compose such a typical type II restriction system. However, a middle gene, *dpnA*, was found to encode another methylase, DpnA, which like DpnM methylates adenine at 5′-GATC-3′ sequences in DNA, but which differs from DpnM in several properties (1).

Examination of the nucleotide sequences of the *Dpn* cassettes indicates the presence of a single promoter at the start of each cassette and a transcription terminator at each end (5, 10). However, such a simple mode of transcription could not allow the regulation required to establish a *Dpn*II system in a null cell, which was shown to occur readily (16). It is possible either that the initial transcripts are processed or that other transcripts are synthesized. Translational control mechanisms may also intervene. Although, as indicated in Fig. 1, some of the genes display normal ribosome binding sites of the Shine-Dalgarno type (18), the DpnM and DpnA methylases appear to be translated from atypical ribosome binding sites (4). These sites have been tentatively identified in the DNA as 5′-AATTTCT-(4 or 5 nucleotides)-TATA-(9 or 10 nucleotides)-ATG-3′ (4, 14). Their use may be related to translational control of *Dpn*II gene expression. The slight overlapping of the open reading frames in

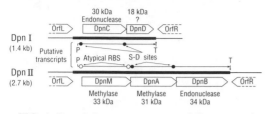

FIG. 1. Restriction gene cassettes of *S. pneumoniae* and their products. Symbols: _____, *S. pneumoniae* chromosome; _____, *dpn* cassette; ▷, coding regions of open reading frames; →, direction of transcription from putative promoters (P) to terminators (T); ●, Shine-Dalgarno (S-D) sites; ○, putative atypical ribosome binding sites (RBS).

both cassettes, which is evident in Fig. 1, may provide translational coupling in synthesis of the proteins.

Properties of *Dpn*I and *Dpn*II Endonucleases

Physically, the two methylases and the endonuclease of the *Dpn*II system appear to be dimers of the corresponding polypeptides (4). Although the evidence for dimeric structure, from gel filtration and sedimentation behavior, is unequivocal for DpnA and DpnB, DpnM gave values intermediate between monomer and dimer. Conceivably, the DpnM protein is in equilibrium between the two forms. Interestingly, the symmetry shown by crystals of DpnM is indicative of a monomeric protein in the crystal structure. When sedimented in 0.5 M NaCl, the *Dpn*I endonuclease behaved as a monomer. However, the *Dpn*I protein was not examined under physiological salt conditions, in which it may exist as dimer. The enzymatic properties of both the *Dpn*I and *Dpn*II endonucleases would appear to require a dimeric structure.

The enzymatic properties of *Dpn* system endonucleases are crucial for their apparent roles in nature. Neither enzyme can act on single-stranded DNA, whether it is methylated or not (20). When the substrate is double stranded, both *Dpn*I and *Dpn*II will cleave at the palindromic 5′-GATC-3′ sites only when both DNA strands are methylated and unmethylated, respectively, on the N6 of adenine (20). Hemimethylated DNA, in which one strand is methylated and the other is not, is not cleaved by either enzyme on either strand. Simultaneous recognition of two unmethylated or two methylated sites would require a double recognition site or a dimeric enzyme.

Restriction of Viral Infection

The primary biological purpose of the *Dpn* systems appears to be the restriction of viral infection. Bacteriophage grown on a *Dpn*I strain is reduced almost 10^6-fold in its infectivity for a *Dpn*II cell, and the converse is also true (16).

The unmethylated phage DNA in the first case and the methylated DNA in the second case, being double stranded when injected into the cell, are readily degraded by the *Dpn*II and *Dpn*I endonucleases, respectively.

The presence of complementary restriction systems in *S. pneumoniae* also supports a major role for the systems in protecting the bacteria from viruses (11). It is assumed that natural populations of pneumococci contain cells of both restriction phenotypes. Initiation of a viral epidemic in such a population would perforce begin with a single infecting phage particle. If that phage had grown on a *Dpn*I host, it could attack a *Dpn*I cell, giving rise to more phage with unmethylated DNA, which would wipe out the *Dpn*I portion of the population. However, the *Dpn*II-containing cells in the population would restrict the phage DNA and survive. If the initial phage had grown on a *Dpn*II host, the *Dpn*I-containing cells would survive. Thus, the presence of complementary restriction systems serves the species by allowing a remnant of a population to survive a viral epidemic.

Effect of Restriction Systems on Bacterial Transformation

Chromosomal transformation is not affected by *Dpn* restriction enzymes in the recipient cell. Cells are transformed with respect to a chromosomal marker at the same frequency, whether or not the donor DNA is methylated, in both *Dpn*I- and *Dpn*II-containing recipients (8, 12). This lack of restriction effect presumably reflects the molecular fate of DNA in transformation. In the transformation of *S. pneumoniae* and other gram-positive bacterial species, donor DNA is processed during its uptake by the recipient cell (7). The double-stranded donor DNA suffers single-strand breaks when it is bound to the cell surface, and one strand is degraded to oligonucleotides as the complementary strand segment enters the cell. The single-strand segment is subsequently integrated into the recipient cell chromosome. Neither the single strands that enter the cell nor the hemimethylated heteroduplex DNA that results from integration into the chromosome is susceptible to the *Dpn* restriction enzymes.

Because a linear single strand of a plasmid that enters a cell of *S. pneumoniae* via the transformation pathway cannot by itself circularize and replicate, two complementary plasmid strands that enter separately must interact to establish a plasmid (17). Since both donor strands that reconstitute the plasmid would be methylated or not, as the case may be, greater susceptibility of plasmid transfer than of chromosomal transformation to *Dpn* restriction would be expected. However, plasmid establishment is only mildly restricted by the *Dpn*I and *Dpn*II systems, with transfer reduced to approximately 40% in the cross-transformation (12). In the case of methylated donor plasmids transferred to DpnI-containing cells, this result can be explained by the considerable new synthesis needed to repair gaps in the reconstituted structure (17). Those reconstituted plasmids in which the 5'-GATC-3' sites were located only in regions of repair synthesis would show no susceptible sites (12). However, in the case of transfer of an unmethylated plasmid to a DpnII-containing recipient, repair synthesis would create susceptible sites. Therefore, it was puzzling that the restriction effect observed on unmethylated plasmid transfer to a DpnII-containing strain was so slight.

Behavior of Mutants in the dpnA Gene

A deletion mutation within the dpnA gene, called dpnA275, was constructed in a recombinant plasmid by removal of a DraI segment, and the mutation was introduced into the chromosome of a DpnII strain (1). An isogenic pair of strains carrying the wild-type and dpnA275 mutant DpnII gene cassettes in their chromosomes was compared with respect to restriction of plasmid establishment and of chromosomal transformation (1). The plasmids used, pJS3, pMV158, and pLS70, contained 2, 8, and 11 5'-GATC-3' sites, respectively. Comparison with results for transformation of the null restriction strain as a control afforded a measure of possible variation in quality or quantity of the donor DNA.

The results shown in Table 1 confirm earlier work with a wild-type recipient (12) in showing no restriction of chromosomal transformation and only limited restriction, approximately 50%, of plasmid establishment. Results for the dpnA275 recipient strain were markedly different

(Table 1). Establishment of unmethylated plasmids was restricted to a level between 0.2 and 11% of methylated plasmid establishment inversely depending on the number of DpnII restriction sites in the plasmids. Chromosomal transformation was minimally affected in the dpnA275 mutant.

A plausible explanation for the severe restriction of plasmid establishment in dpnA mutant but not in wild-type recipients is that the methylation of incoming single-stranded DNA by the DpnA methylase protects the plasmid sites from later attack by the DpnII endonuclease (Fig. 2). In the absence of such methylation, attack by DpnII would follow conversion of the plasmid DNA to a double-stranded unmethylated form by either association with a complementary donor plasmid strand or repair synthesis of unpaired segments after partial plasmid reconstitution. Thus, the biological function of the DpnA methylase may be to allow the transfer of plasmids from DpnI strains of S. pneumoniae or from other bacterial species to DpnII strains.

Properties of DpnM and DpnA Methylases

To test the possible role of the DpnA methylase in protecting incoming plasmid DNA, the methylating activities of purified DpnM and DpnA were compared with single-stranded and double-stranded DNA substrates unmethylated at 5'-GATC-3' sites (1). Both methylases were active on double-stranded DNA, although DpnM gave a higher specific activity. DpnA was highly active in methylating the single-stranded DNA of phage M13, whereas DpnM showed no detectable activity on M13 DNA. Thus, the protection for incoming plasmid DNA can be provided only by DpnA and not by DpnM. DpnM is presumably

TABLE 1. Effect of dpnA mutation on restriction of plasmid transfer[a]

Recipient DpnII genotype	Donor DNA[b]	Marker	DpnI strain (unmodified)	DpnII strain (modified)	Ratio, I/II
			Transformants/ml with donor DNA from:		
Null	pJS3	Cm^r	1,300	800	1.62
	pMV158	Tc^r	2,800	2,600	1.08
	pLS70	Tc^r	6,100	17,000	0.36
	Chromosome	Str^r	90,000	100,000	0.90
Wild	pJS3	Cm^r	570	740	0.77
	pMV158	Tc^r	4,900	10,000	0.49
	pLS70	Tc^r	59,000	160,000	0.37
	Chromosome	Str^r	250,000	260,000	0.96
A275	pJS3	Cm^r	2,400	28,000	0.086
	pMV158	Tc^r	290	35,000	0.0083
	pLS70	Tc^r	230	380,000	0.0006
	Chromosome	Str^r	320,000	760,000	0.42

[a]Adapted from data in reference 1.
[b]Plasmids pJS3, pMV158, and pLS70 carry 2, 8, and 11 DpnI/II sites, respectively.

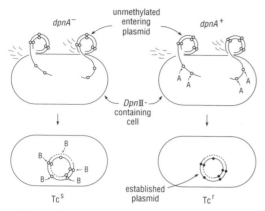

FIG. 2. Role of DpnA methylase in enabling un-methylated plasmid transfer into cells containing the *Dpn*II restriction system. Because of the degradative processing of DNA entering the cell by the transformation pathway, plasmid establishment requires the reconstitution of a plasmid from complementary strands that separately enter the cell. In a *dpnA* mutant host, unmethylated plasmid DNA, upon reconstitution to a double-stranded form, is cleaved by the *Dpn*II endonuclease. In a *dpnA*+ host, the DpnA methylase methylates the single strands upon entry, so that the reconstituted plasmid is protected from the *Dpn*II endonuclease. O, Unmethylated GATC site; ●, methylated GATC site; A, DpnA methylase; B, *Dpn*II endonuclease.

better suited for the maintenance of cellular DNA methylation.

In addition to acting on single-stranded DNA, DpnA differs from DpnM in its less stringent requirement for the 5′-GATC-3′ target sequence, as first shown by its in vitro ability to methylate DNA already methylated at 5′-GATC-3′ sites (4). By varying single-stranded oligonucleotides in a single base of the recognition sequence, it was found that recognition of ATC in the sequence was critical for methylation, but variation of the first base, G, was tolerated. The significance, if any, of this degeneracy is unknown.

The amino acid sequence of DpnA is distinct from that of DpnM (10); only three small boxes of similarity, common to a variety of DNA adenine methylases, are evident. DpnM was found to be homologous to the Dam methylase of *E. coli* (14) (Fig. 3). DpnA is similar in sequence to the *Hin*fI methylase of *Haemophilus influenzae* (3), which methylates adenine in 5′-GANTC-3′ (Fig. 4); 40% of the DpnA residues are identical to the 255 amino-terminal residues of the *Hin*fI methylase (1). Despite their adjacent location in the *Dpn*II gene cassette, *dpnA* and *dpnM* arose not by gene duplication but rather by derivation from different ancestral prototypes of DNA adenine methylase genes.

Structural Studies on *Dpn* Restriction System Enzymes

There is virtually no significant amino acid sequence similarity between any pair of the *Dpn*I or *Dpn*II system proteins. This is particularly interesting because all three of the *Dpn*II system proteins recognize the same 5′-GATC-3′ sequence in DNA, and the *Dpn*I endonuclease recognizes a slight modification of that sequence. The basis for *Dpn*I binding to its recognition site may be a zinc finger structure (5). A zinc finger motif, CPNCGNNPLNHFENNRPVAD-FYCNHC, is present between amino acid residues 34 and 59 in the protein. Multiple fingers with such motifs have been found in eukaryotic regulatory proteins that bind to DNA at specific sites (15, 19). *Dpn*I may be an example of how

```
DpnM: MKIKEIKKVTLQPFTKWTGGKRQLLPVIRELIPKTYNRYFEPFVGGGALFFDLAPKDAVINDFNAELINCYQQIKDNPQE  80
      .::      : :: ::: :   :. .::  . ..:::::.: .:..   .. :.: .:: :. .:  .:
Dam:  MKKNRA--FLKWAGGKYPLLDDIKRHLPKG-ECLVEPFVGAGSVFLNTDFSRYILADINSDLISLYNIVKMRTDE  72

DpnM: LIEILKVHQEYNSKEYYLDLRSADRDERIDMMSEVQRAARILYMLRVNFNGLYRVNSKNQFNVPYGRYKNPKIVDEELIS 160
      ...    ..   .. .  .::     .:    . :: .::. :. .::: :.: .:::::::: :  . ::.
Dam:  YVQAARELFVPETNCAEVYYQF--REEFNKSQDPFRRAVLFLYLNRYGYNGLCRYNLRGEFNVPFGRYKKPYFPEAELYH 150

DpnM: AISVYINNNQLEIKVGDFEKAIVDVRTGDFVYFDPPYIPLSETSAFTSYTHEGFSFADQVRLRDAFKRLSDTGAYVMLSN 240
      :   .    ..:  .:-:  :  :: :::: ::: :  . :.. .: .:  . : ...::
Dam:  FAEKAQNAFFYCESYADS-MARADDASV--VYCDPPYAPLSATANFTAYHTNSFTLEQQAHLAEIAEGLVERHIPVLISN 227

DpnM: SSSALVEELYKDFNIHYVEATRTNGAKSSSRGKISEIIVTNYEK 284
      . : ::: .:  ..:. :.   .: :. :...
Dam:  HDTMLTREWYQRAKLHVVKVRRSISSNGGTRKKVDELLALYKPGVVSPAKK 278
```

FIG. 3. Comparison of DpnM methylase of *S. pneumoniae* and Dam methylase of *E. coli*. Amino acid sequences are shown from the amino termini (upper left) to the carboxyl termini (lower right). They are aligned to give maximum correspondence. Dashes indicate gaps produced by this alignment in one or the other sequence. Numbers at the right correspond to positions in the individual polypeptide sequences. Dots over the Dpn sequence indicate every 10th residue. Two dots indicate identical residues, and one dot indicates amino acid residues with similar properties.

```
DpnA:  MKNNEYKYGGVLMTKPYYNKNKMILVHSDTFKFLSKMKPESMDMIFADPPYFLSNGGISNSG-GQVVSVDKGDWDKISSF  79
       :  :  ....:  :  .  :.:.::::::::.  :       :..:   .:::.  :
HinF:         MMKENINDFLNTILKGDCIEKLKTIPNESIDLIFADPPYFMQTEGKLLRTNGDEFSGVDDEWDKFNDF  68


DpnA:  EEKHEFNRKWIRLAKEVLKPNGTVWISGSLHNIYSVGMALEQEGFKILNNITWQKTNPAPNLSCRYFTHSTETILWARKN  159
       :     :..  . .::  ..:. ::. :::  .:. ...   : :::. :.:::: ::..  :   .  ::.::   :
HinF:  VEYDSFCELWLKECKRILKSTGSIWVIGSFQNIYRIGYIMQNLDFWILNDVIWNKTNPVPNFGGTRFCNAHETMLWCSKC  148


DpnA:  DKKARHYYNYDLMKELNDGKQMKDVWTGSLTK-----KVEKWAGKHPTQKPEYLLERIILASTKEGDYILDPFVGSGTTG  234
       ::  .::  ::.  :. ::. ::      :  :::::  :.:.:   :..::::.::::
HinF:  -KKNKFTFNYKTMKHLNQEKQERSVWSLSLCTGKERIKDEEGKKAHSTQKPESLLYKVILSSSKPNDVVLDPFFGTGTTG  227


DpnA:  VVAKRLGRRFIGIDAEKEYLKIARKRLEAENETN  268
       :::  :::  .:::. :    :.  .:  :::
HinF:  AVAKALGRNYIGIEREQKYIDVAEKRLREIKPNPNDIELLSLEIKPPKVPMKTLIEADFLRVGQTLFDKNENAICIVTQD  307


DpnA:  GNVKDNEETLSIHKMSAKYLNKTNNNGWDYFYLFRNNNFITLDSLRYEYTNQ  359
```

FIG. 4. Comparison of DpnA methylase of *S. pneumoniae* and *Hin*fI methylase of *H. influenzae*. See legend to Fig. 3 for details.

zinc fingers were used individually in ancestral bacteria for DNA recognition and binding.

Aside from a few small boxes of similar amino acid residues, which are common to DNA adenine methylases in general (4, 13), the two *Dpn*II system methylases exhibit no homology. As indicated above, they even stem from different ancestral methylase genes. The fact that both the DpnM and DpnA methylases and also the *Dpn*II endonuclease differ in primary structure makes it more intriguing that they nevertheless recognize the same sequence, 5'-GATC-3'. They must achieve this recognition by different three-dimensional structures. We hope to determine these structures by X-ray diffraction. To this end, we have succeeded in crystallizing the DpnM methylase (2) and the *Dpn*II endonuclease (unpublished results). The methylase crystals exhibited very good diffraction. Their unit cell dimensions and space group, $P2_12_12_1$, indicate that the methylase is present as a monomer in the crystal (2).

Conclusion

The DpnA methylase is unusual in its ability to methylate single-stranded DNA. It appears to have evolved for this particular function inasmuch as the *Dpn*II restriction system, of which it is a part, already contains a potent methylase for double-stranded DNA, DpnM. The primary biological role of DpnA may be to enable plasmid transmission to *Dpn*II-containing cells via the transformation pathway of DNA entry, which introduces DNA in single-stranded form (6). Methylation of the incoming strand would protect the subsequently reconstituted plasmid from *Dpn*II cleavage. The mechanism of chromosomal transformation itself ensures its resistance to restriction (20). Thus, the systems of *S. pneumoniae* that allow the species to benefit from genetic exchange and plasmid transfer are largely immune

to restriction, which is presumably directed at the prevention of bacterial virus infection. The fact that the *dpnA* gene was apparently incorporated and maintained by the *Dpn*II restriction system solely for the purpose of allowing plasmid transfer points up the considerable importance of systems for genetic exchange in the survival of living species.

We appreciate the assistance of Bill Greenberg in determining the *dpnC* mutation in the Rx strain.

This research was conducted at Brookhaven National Laboratory under the auspices of the U.S. Department of Energy Office of Health and Environmental Research. It was supported by Public Health Service grant GM29721.

LITERATURE CITED

1. **Cerritelli, S., S. S. Springhorn, and S. A. Lacks.** 1989. DpnA, a methylase for single-strand DNA in the *Dpn*II restriction system, and its biological function. *Proc. Natl. Acad. Sci. USA* **86:**9223–9227.
2. **Cerritelli, S., S. W. White, and S. A. Lacks.** 1989. Crystallization of the DpnM methylase from the *Dpn*II restriction system of *Streptococcus pneumoniae*. *J. Mol. Biol.* **208:**841–842.
3. **Chandrasegaran, S., K. D. Lunnen, H. O. Smith, and G. G. Wilson.** 1988. Cloning and sequencing the *Hin*fI restriction and modification genes. *Gene* **70:**387–392.
4. **de la Campa, A. G., P. Kale, S. S. Springhorn, and S. A. Lacks.** 1987. Proteins encoded by the *Dpn*II restriction gene cassette: two methylases and an endonuclease. *J. Mol. Biol.* **196:**457–469.
5. **de la Campa, A. G., S. S. Springhorn, P. Kale, and S. A. Lacks.** 1988. Proteins encoded by the *Dpn*I restriction gene cassette: hyperproduction and characterization of the *Dpn*I endonuclease. *J. Biol. Chem.* **263:**14696–14702.
6. **Lacks, S.** 1962. Molecular fate of DNA in genetic transformation of pneumococcus. *J. Mol. Biol.* **5:**119–131.
7. **Lacks, S. A.** 1988. Mechanisms of genetic recombination in gram-positive bacteria, p. 43–85. *In* R. Kucherlapati and G. Smith (ed.), *Genetic Recombination*. American Society for Microbiology, Washington, D.C.
8. **Lacks, S., and B. Greenberg.** 1975. A deoxyribonuclease of *Diplococcus pneumoniae* specific for methylated DNA. *J. Biol. Chem.* **250:**4060–4066.
9. **Lacks, S., and B. Greenberg.** 1977. Complementary specificity of restriction endonucleases of *Diplococcus pneumoniae* with respect to DNA methylation. *J. Mol. Biol.* **114:**153–168.

10. **Lacks, S. A., B. M. Mannarelli, S. S. Springhorn, and B. Greenberg.** 1986. Genetic basis of the complementary DpnI and DpnII restriction systems of S. pneumoniae: an intercellular cassette mechanism. *Cell* **46:**993–1000.

11. **Lacks, S. A., B. M. Mannarelli, S. S. Springhorn, B. Greenberg, and A. G. de la Campa.** 1987. Genetics of the complementary restriction systems *Dpn*I and *Dpn*II revealed by cloning and recombination in *Streptococcus pneumoniae*, p. 31–41. *In* J. J. Ferretti and R. Curtiss III (ed.), *Streptococcal Genetics*. American Society for Microbiology, Washington, D.C.

12. **Lacks, S. A., and S. S. Springhorn.** 1984. Transfer of recombinant plasmids containing the gene for *Dpn*II DNA methylase into strains of *Streptococcus pneumoniae* that produce *Dpn*I or *Dpn*II restriction endonucleases. *J. Bacteriol.* **158:**905–909.

13. **Lauster, R.** 1989. Evolution of type II DNA methyltransferases: a gene duplication model. *J. Mol. Biol.* **206:**313–321.

14. **Mannarelli, B. M., T. S. Balganesh, B. Greenberg, S. S. Springhorn, and S. A. Lacks.** 1985. Nucleotide sequence of the *Dpn*II DNA methylase gene of *Streptococcus pneumoniae* and its relationship to the *dam* gene of *E. coli*. *Proc. Natl. Acad. Sci. USA* **82:**4468–4472.

15. **Miller, J., A. D. McLachlan, and A. Klug.** 1985. Repetitive zinc-binding domains in the protein transcription factor IIIA from *Xenopus* oocytes. *EMBO J.* **4:**1609–1614.

16. **Muckerman, C. C., S. S. Springhorn, B. Greenberg, and S. A. Lacks.** 1982. Transformation of restriction endonuclease phenotype in *Streptococcus pneumoniae*. *J. Bacteriol.* **152:**183–190.

17. **Saunders, C. W., and W. R. Guild.** 1980. Monomer plasmid DNA transforms *Streptococcus pneumoniae*. *Mol. Gen. Genet.* **181:**57–62.

18. **Shine, J., and L. Dalgarno.** 1975. Determinant of cistron specificity in bacterial ribosomes. *Nature* (London) **254:**34–38.

19. **Vincent, A.** 1986. TFIIIA and homologous genes. The 'finger' proteins. *Nucleic Acids Res.* **14:**4385–4391.

20. **Vovis, G. F., and S. Lacks.** 1977. Complementary action of restriction enzymes Endo R.*Dpn*I and Endo R.*Dpn*II on bacteriophage fl DNA. *J. Mol. Biol.* **115:**525–538.

21. **Wilson, G. G.** 1988. Type II restriction-modification systems. *Trends Genet.* **4:**314–318.

Lytic Enzymes of *Streptococcus pneumoniae* and Its Bacteriophages: a Model of Molecular Evolution

RUBENS LÓPEZ, ERNESTO GARCÍA, JOSÉ L. GARCÍA, EDUARDO DÍAZ, ALICIA ROMERO, JESÚS M. SANZ, JOSÉ M. SÁNCHEZ-PUELLES, PEDRO GARCÍA, AND CONCEPCIÓN RONDA

Centro de Investigaciones Biológicas, Consejo Superior de Investigaciones Científicas, Velázquez 144, Madrid 28006, Spain

Autolysins are enzymes that hydrolyze covalent bonds of cell walls. Most microorganisms contain one or more autolysins, but the search for the physiological roles of these enzymes has been a matter of continuous debate because of the difficulties in obtaining deletion mutations in the genes encoding these enzymes. *Streptococcus pneumoniae* contains a primary autolytic enzyme, *N*-acetylmuramoyl-L-alanine amidase (15), as well as *N*-acetylglucosaminidase (11). The development of a technique for the rapid detection of amidase-deficient mutants of *S. pneumoniae* (7) has allowed us to clone and sequence the amidase-encoding gene *lytA* (10). We have demonstrated that the primary translation product of this gene is a low-activity form of the enzyme (E form) (5). The fully active form of the amidase (C form) is found only in pneumococci that contain choline in the teichoic acids of the cell wall. Although the E form can be converted in vitro to the C form by 2% choline, the catalytic activity of the pneumococcal amidase is inhibited by high concentrations of choline (20 mM or higher) (2, 13). Our experimental approach has provided a way to characterize a series of mutants, including a strain completely deleted in the *lytA* gene (22). It has been established that the amidase is one of the agents responsible for the separation of the daughter cells at the end of cell division (20), and it also seems to contribute to the virulence of *S. pneumoniae* (1). Furthermore, this enzyme participates in the liberation of the progeny bacteriophage into the medium, a process that appears to be shared with the lytic enzymes of the pneumococcal bacteriophages (21).

Modular Organization of the Cell Wall Lytic Enzymes of Pneumococci

We have demonstrated that Cp-1, a small virulent phage infecting *S. pneumoniae*, encodes its own lytic enzyme (CPL1). A fragment of Cp-1 DNA containing the gene *cpl1*, coding for CPL1, has been cloned and expressed in high amounts in *Escherichia coli*. CPL1 was purified to electrophoretic homogeneity by using affinity chromatography, and the enzyme was characterized as a muramidase (lysozyme). This lysozyme required the presence of choline in the teichoic acids of

the pneumococcal cell walls for activity, but free choline at concentrations of 3 mM or higher inhibited noncompetitively the activity of CPL1 (8).

The biochemical similarities found between the host amidase and the CPL1 lysozyme suggested that the genes coding for these enzymes also should show homology at the nucleotide level. The pneumococcal *lytA* gene hybridized with 7.9-kb *Hin*dIII fragments of DNA from phages Cp-1 and Cp-5 and with a 6.1-kb fragment of Cp-9 DNA (Fig. 1A). Remarkably, homology was not detectable with *Hin*dIII fragments of Cp-7 DNA. Using appropriate overlapping restriction fragments, we established the strategy for determining the nucleotide sequences of the *cpl1* and *cpl9* genes, coding for the CPL1 and CPL9 lytic enzymes, respectively (6, 12). The DNA sequences of *cpl1* and *cpl9* showed only one open reading frame of significant length, 1,071 nucleotides, in both genes. In *cpl9*, 52 of the first 600 nucleotides (8.7%) differed from the *cpl1* gene, whereas only 20 changes were found in the remaining nucleotides (12). Only 10 amino acids were different between CPL1 and CPL9 muramidases (Fig. 2).

When the nucleotide sequences of *cpl1* and *cpl9*, two genes virtually identical, were compared with that of *lytA*, no significant similarity was found in the 5′ regions of the genes, whereas a high similarity was found in the 3′ regions. Comparisons between CPL1 and the pneumococcal amidase showed that 73 of the carboxy-terminal 142 amino acid residues were identical and that 55 of the 69 substitutions were conservative (Fig. 2).

More recently, we have cloned from the DNA of the pneumococcal temperate phage HB-3 a lysin gene (*hbl*) that codes for an *N*-acetylmuramoyl-L-alanine amidase (HBL) (18). We demonstrated that different restriction fragments of HB-3 DNA containing the *hbl* gene hybridized with *lytA*, which facilitated the determination of the nucleotide sequence of the *hbl* gene. A remarkable nucleotide similarity (87.1%) between the *lytA* and *hbl* genes has been found (19).

One of the most interesting features of CPL1, CPL9, HBL, and the host amidase is the presence, in the C-terminal half of each enzyme, of a set of six repeated sequences (P1 to P6), each

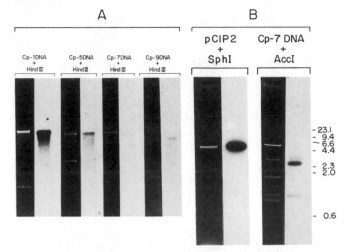

FIG. 1. Ethidium bromide-stained gels and Southern blot hybridization analyses of restriction fragments obtained from Cp-1, Cp-5, Cp-7, and Cp-9 DNAs. Ethidium bromide-stained 0.7% agarose gels show the restriction patterns obtained from digestion of the different DNAs with HindIII (A) and of Cp-7 DNA with AccI (B). The autoradiograph to the right of each gel shows the fragments blotted to Hybond membranes and hybridized at 50°C for 5 h with ^{32}P-labeled pGL80 (A) or a ^{32}P-labeled 1.4-kb AccI-SphI fragment of Cp-1 DNA (B) that contains the 5'-flanking region of the cpl1 gene and the sequence coding only for the N-terminal half of CPL1 (12).

about 20 amino acids long (Fig. 2). When the amino-terminal sequence of the muramidase of the fungus *Chalaropsis* sp., an enzyme that degrades either choline- or ethanolamine (EA)-containing pneumococcal cell walls, was compared with those of CPL1 and CPL9, similarity was found (4, 6, 12); most important, the two amino acids that had been reported to be at least partially responsible for the activity of the *Chalaropsis* lysozyme (Asp-10 and Glu-37) are located the same distance apart in the N-terminal moiety of the CPL lysozymes (Fig. 2). Analysis of the results reported above suggested that the lytic enzymes of *S. pneumoniae* and its bacteriophages have a modular organization. We have proposed that the C-terminal domains of these enzymes may be responsible for the specific recognition of choline-containing cell walls as well as for the noncompetitive inhibition of the catalytic activity of these enzymes by high concentrations of choline, whereas the active centers of these enzymes may be located in their N-terminal domains.

Cp-7 Codes for a Peculiar Cell Wall Lysin

As pointed out above, we found no hybridization band when Cp-7 DNA was probed against the *lytA* gene. To define the location of the *cpl7* gene within the genome of Cp-7, we digested Cp-7 DNA with appropriate restriction enzymes, and the fragments separated by electrophoresis were blotted and hybridized with the ^{32}P-labeled 1.4-kb AccI-SphI fragment of Cp-1 DNA that represents the 5'-flanking region of the *cpl1* gene

encoding the N-terminal half of CPL1. We found that this fragment hybridized with two AccI fragments that map at the right end of the genome of Cp-7 (Fig. 1B). An appropriate fragment from Cp-7 DNA was cloned and sequenced. Only one open reading frame of significant length (1,026 nucleotides) was found, and the protein encoded has been biochemically characterized as a muramidase (12).

In a comparison of the *cpl1* and *cpl7* genes, we found that only 17.3% of the first 600 nucleotides were different (12). The most striking differences between the *cpl1* and *cpl9* genes and the *cpl7* gene were found within the 400 bp preceding the termination codon of *cpl7*, where the sequences diverge completely. Most remarkably, the set of six repeated sequences (P1 to P6) found in the C-terminal domains of *lytA*, *hbl*, *cpl1*, and *cpl9*, each about 60 nucleotides long, was replaced by a 2.8 perfect tandem repeat of 144 nucleotides (M1 to M3) in *cpl7*. From the alignments of the amino acid sequences of the muramidases CPL1 and CPL9 and of the host and the HBL amidases (Fig. 2), we have suggested above that the C-terminal domains of all of these enzymes may be responsible for recognition of the choline residues present in the pneumococcal cell wall. If this hypothesis is correct, the biological and biochemical properties of the CPL7 muramidase should dramatically differ from those reported above for the other lytic enzymes of *S. pneumoniae* and its bacteriophages. Extracts obtained from *E. coli* cells harboring pCP70, a plasmid that contains the *cpl7*

```
                                                                        *
CPL-1                                                           MVKKNDLFVDV
CPL-9                                                           MVKKNDLFIDV
CPL-7                                                           MVKKNDLFVDV
```

```
                              *
LYTA           MEINVSKLRTDLPQVGVQPYRQVHAHSTGNPHSTVQNEADYHWRKDPELG
HBL            MDIDRNRLRTGLPQVGVQPYRQVHAHSTGNRNSTVQNEADYHWRKDPELG
CPL-1   SSHNGYDITGILEQMGTTNTIIKISESTTYLNPCLSAQVEQSNPIGFYHFARFGGDVAEAERE
CPL-9   SSHNGYDITGILEQMGTTNTIVKISESTTYLNPCLSAQVEQSTPIGFYHFARFGGDVAEAERE
CPL-7   ASHQGYDISGILEEAGTTNTIIKVSESTSYLNPCLSAQVSQSNPIGFYHFAWFGGNEEEAEAE
```

```
LYTA    FFSHIVGNGCIMQVGPVDNGAWDVGGGWNAETYAAVELIESHSTKEEFMTDYRLYIELLRNLA
HBL     FFSHVVGNFRIMQVGPVNNGSWDVGGGWNAETYAAVELIESHSTKEEFMADYRLYIELLRNLA
CPL-1   AQFFLDNVPMQVKYLVLDYEDDPSGDAQANTNACLRFMQMIADAGYKPIYYSYKPFTHDNVDY
CPL-9   AQFFLDNVPTQVKYLVLDYEDDPSGNAQANTNACLRFMQMIADAGYTPIYYSYKPFTLDNVDY
CPL-7   ARYFLDNVPTQVKYLVLDYEDHASASVQRNTTACLRFMQIIAEAGYTPIYYSYKPFTLDNVDY
```

```
LYTA    DEAGLPKTLDTGSLAGIKTHEYCTNNQPNNHSDHVDPYPYLAKWGISREQFKHDIENGLTIET
HBL     DEAGLPKTLDTDDLAGIKTHEYCTNNQPNNHSDHVDPYPYLASWGISREQFKQDIENGLSAAT
CPL-1   QQILAQFPNSLWIAGYGLNDGTANFEYFPSMDGIRWWQYSSNPFDKNIVLLDDEEDDKPKTAG
CPL-7   QQILAQFPNSLWIAGYGLNDGNADFEYFPSMDGIRWWQYSSNPFDKNIVLLDDEEDEKPKTAG
CPL-9   QQILAQFPNSLWIAGYGLNDGTANFEYFPSMDGIRWWQYSSNPFDKNIVLLDDEKEDNINNEN
```

```
        |---------P1---------|----------P2--------|--------P3---------|
LYTA    GWQKNDTGYWYVHSDGSYPKDKFEKINGTWYYFDSSGYMLADRWRKHTDGNWYWFDNSGEMAT
HBL     GWQKNGTGYWYVHSDGSYSKDKFEKINGTWYYFDGSGYMLSDRWKKHTDGNWYYFDQSGEMAT
CPL-1   TWKQDSKGWWFRRNNGSFPYNKWEKIGGVWYYFDSKGYCLTSEWLKDNE-KWYYLKDNGAMAT
CPL-9   TWKQDSKGWWFRRNNGSFPYNKWEKIGGVWYYFDSKGYCLTSEWLKDNE-KWYYLKDNGAMVT
CPL-7   TLKSLTTVANEVIQGLWGNGQERYDSLANRGYDPQAVQDKVNEILNAREIADLTTVANEVIQG
        |------------------M1-----------------------|----------
```

```
        |---------P4---------|----------P5----------|--------P6---------|
LYTA    GWKKIADKWYYFNEEGAMKTGWVKYKDTWYYLDAKEGAMVSNAFIQSADGTGWYYLKPDGTLA
HBL     GWKKIADKWYYFDVEGAMKTGWVKYKDTWYYLDAKEGAMVSNAFIQSADGTGWYYLKPDGTLA
CPL-1   GWVLVGSEWYYMDDSGAMVTGWVKYKNNWYYMTNERGNMVSNEFIKS--GKGWYFMNTNGELA
CPL-9   GWVLVGSEWYYMDDSGAMVTGWVKYKNNWYYMTNERGNMVSNEFIKS--GKGWYFMNTNGELA
CPL-7   LWGNGQERYDSLANRGYDPQAVQDKVNEILNAREIADLTTVANEVIQGLWGNGQERYDSLANR
        ----------------M2-------------------|-------------M3---------
```

```
LYTA    DRPEFTVEPDGLITVK
HBL     DKPEFTVEPDGLITVK
CPL-1   DNPSFTKEPDGLITVA
CPL-9   DNPSFTKEPDGLITVA
CPL-7   GYDPQAVQDKVNELLS
        -----------|
```

FIG. 2. Alignment of the deduced amino acid sequences of the bacterial amidase (LYTA), the bacteriophage HB-3 amidase (HBL), and three bacteriophage muramidases (CPL1, CPL7, and CPL9). Amino acids residues are shown in one-letter code. Asterisks indicate the two putative active-site amino acids of the muramidases. P1 to P6 are sets of repeated sequences of LYTA, HBL, CPL1, and CPL9, and M1 to M3 are the perfect tandem repetitions of CPL7.

gene, were able to degrade in vitro pneumococcal cell walls containing either choline or EA in the teichoic acids. On the other hand, pneumococcal cell walls containing EA instead of choline have been proven to be resistant to the lytic activity of the pneumococcal amidase (14). Furthermore, we had found that cultures of EA-grown cells required the presence of choline in the cell wall for liberation of the phage progeny (9). In contrast, when EA-grown pneumococci were infected with Cp-7, the culture lysed, liberating productive progeny of phages (12).

The fact that in the CPL7 lysozyme, which is no longer dependent on the presence of choline in the pneumococcal cell wall for catalytic activity, the C-terminal domain is totally different but the N terminus is conserved in comparison with CPL7 and CPL9 (Fig. 3) reinforced our hypothesis as to the modular organization of these enzymes and suggested that the phages may have interchanged some regions of their genomes with that of the host cell.

Construction of Chimeric Phage-Bacterial Enzymes

The availability of various cloned genes coding for the lytic enzymes discussed above represents an important tool for use in the construction of chimeric proteins between host and phage lytic enzymes, which should provide additional valuable information about domain organization and target recognition. We propose that the domains may be interchangeable and that their recombination should create active chimeric enzymes having novel properties. On the basis of the known properties of the *lytA* and *cpl1* genes, we have constructed chimeric proteins. The new proteins were generated through in vitro recombination at equivalent positions of both DNAs, as defined by the alignment at sites approximating the junction zone of the N- and C-terminal domains of the enzymes (Fig. 3). Since the *lytA* gene contains a *Sna*BI site in this zone whereas the *cpl1* gene does not, we introduced, by site-directed mutagenesis, this restriction site at a position comparable to that of the *lytA* gene (3a). The introduction of this novel site in *cpl1* was neutral with respect to the reading frame and to enzymatic activity. Plasmids pCL and pLC contain the chimeric genes coding for a pair of reciprocal chimeric proteins between the pneumococcal amidase and the phage lysozyme (Fig. 3). The correctness of the construction was confirmed by restriction enzyme analysis and by sequencing the region surrounding the *Sna*BI restriction site.

Extracts obtained from *E. coli* HB101(pCL) and *E. coli* HB101(pLC) demonstrated the presence of active lytic enzymes (named CLL and LCA, respectively) that degraded pneumococcal cell walls. The CLL enzyme, which consisted of the N-terminal domain of the CPL1 lysozyme and the C-terminal domain of the pneumococcal amidase (Fig. 3), acts as a lysozyme and requires conversion to achieve full enzymatic activity, a peculiar property of the host pneumococcal amidase and of the HB-3 amidase. In contrast, the chimeric protein LCA, consisting of the N-terminal domain of the host amidase and the C-terminal domain of the CPL1, was characterized as an amidase that did not require conversion. As pointed out above, choline inhibited the activity of the CPL1 lysozyme at a concentration (2 mM) lower than that required by the host amidase (20 mM). This property was interchanged between the chimeric proteins. Finally, new chimeric enzymes showing novel properties have been created by recombination between the *lytA* and *cpl7* genes (*lc7* gene) (E. Díaz, R. López, and J. L. García, *J. Biol. Chem.*, in press).

Concluding Remarks

Comparison of the amino acid sequences of the lytic enzymes of *S. pneumoniae* and its bacteri-

Gene	Enzymatic activity	Choline dependence for activity	Conversion	Amino acid number
lytA	amidase	Yes	Yes	
cpl-1	muramidase	Yes	No	
cpl-9	muramidase	Yes	No	
cpl-7	muramidase	No	No	
hbl	amidase	Yes	Yes	
lca	amidase	Yes	No	
cll	muramidase	Yes	Yes	
lc7	amidase	No	No	
c-cpl-1	—	Yes	—	
c-lytA	—	Yes	—	

FIG. 3. Comparison of the most relevant properties of the lytic muramidases of phages Cp-1, Cp-7, Cp-9, the phage amidase HBL, the pneumococcal amidase, the chimeric enzymes LCA, CLL, and LC7, and the truncated polypeptides C-CPL1 and C-LytA. Symbols: ■, N-terminal portion of the amidases; ◨, N-terminal portion of the CPL1, CPL7, CPL9, and CLL muramidasas; ▨, C-terminal homologous repetition of the host amidase, HBL amidase, CPL1, CPL9 muramidases, the chimeric enzymes LCA and CLL, and the truncated proteins; □, C-terminal modular repetition of the CPL7 muramidase and the LC7 chimeric protein; ▮, interlinking region between the N- and C-terminal regions; ▤, nonhomologous region. The amino acid sequences showing a z value higher than 4.7 according to Lipman and Pearson (17) are represented by identical shading.

ophages has provided valuable information about the relationship between structure and function in this system. We suggest that these cell wall lytic enzymes originated from the fusion of two functionally different modules. The formation of chimeric enzymes by genetic manipulation of some of the genes coding for these enzymes demonstrated, in a direct experimental way, our previous proposal that the active center of each enzyme is located in the N-terminal domain, whereas the C-terminal domain should be responsible for recognition of the choline-containing cell wall substrate and also for the inhibitory effect of choline on these enzymes. In addition, we have found that the two domains are interchangeable and independently functional. We have recently cloned the 3' moieties of the *lytA* and *cpl1* genes and expressed the C-terminal domains of the pneumococcal amidase and of the CPL1 lysozyme without the N-terminal domain (23). We observed that the truncated proteins conserved their biochemical properties (e.g., binding to choline), a result that reinforces our conclusions as to the organization of domains (Fig. 3).

Although the theory of a modular evolution of proteins is supported by an increasing number of experimental data, there are few examples in which the most relevant aspects of this theory can be considered as a whole. In addition, despite the peculiar relatedness between bacteria and phages, homologies between bacterial and phage genes have been described only in a limited number of cases (3). The lytic genes of *S. pneumoniae* and its bacteriophages provide a model of modular evolution that is distinctive in two respects: (i) the need for a group of enzymes capable of degrading the cell wall, a property shared with most bacterium-phage systems, and (ii) the uncommon presence of choline in the teichoic acids of the pneumococcal cell wall, which seems to act as an element of selective pressure preserving some regions of the lytic enzymes. These assumptions are consistent with the concept that changes in a precise sequence are closely related to functionality, which ensures that those sequences that play a critical role in functionality remain more stable (16). The fact that the lysozyme from pneumococcal phage Cp-7, which is not regulated by choline, contains an N-terminal domain practically identical to that of the CPL1 or CPL9 lysozymes but a completely different C-terminal domain also supports the idea that the lytic enzymes may be the result of an interchange of modules; in this sense, our chimeric constructions should provide experimental models for the formation of ancestral genes by fusion of distinct genetic modules.

We thank E. Cano and M. Carrasco for excellent technical assistance.

This work was supported by a grant from Programa Sectorial de Promoción General del Conocimiento (PB87-0214). E.D., A.R., J.M.S.-P., and J.M.S. were the recipients of fellowships from the Fondo de Investigaciones Sanitarias and from the Ministerio de Educación y Ciencia.

LITERATURE CITED

1. **Berry, A. M., R. A. Lock, D. Hansman, and J. C. Patton.** 1989. Contribution of autolysin to virulence of *Streptococcus pneumoniae. Infect. Immun.* **57:**2324–2330.
2. **Briese T., and R. Hakenbeck.** 1985. Interaction of the pneumococcal amidase with lipoteichoic acid and choline. *Eur. J. Biochem.* **146:**417–427.
3. **Campbell, A.** 1988. Phage evolution and speciation, p. 1–14. *In* R. Calendar (ed.), *The Bacteriophages*, vol. 1. Plenum Press, New York.
3a.**Díaz, E., R. López, and J. L. García.** 1990. Chimeric phage-bacterial enzymes: a clue to the molecular evolution of genes. *Proc. Natl. Acad. Sci. USA* **87:**8125–8129.
4. **Fouche, P. B., and J. H. Hash.** 1978. The *N,O*-diacetylmuramidase of *Chalaropsis* species. *J. Biol. Chem.* **253:**6787–6793.
5. **García, E., J. L. García, C. Ronda, P. García, and R. López.** 1985. Cloning and expression of the pneumococcal autolysin gene in *Escherichia coli. Mol. Gen. Genet.* **201:**225–230.
6. **García, E., J. L. García, P. García, A. Arraras, J. M. Sánchez-Iuelles, and R. López.** 1988. Molecular evolution of lytic enzymes of *Streptococcus pneumoniae* and its bacteriophages. *Proc. Natl. Acad. Sci. USA* **85:**914–918.
7. **García, E., C. Ronda, J. L. García, and R. López.** 1985. A rapid procedure to detect the autolysin phenotype in *Streptococcus pneumoniae. FEMS Microbiol. Lett.* **29:**77–81.
8. **García, J. L., E. García, A. Arraras, P. García, C. Ronda, and R. López.** 1987. Cloning, purification, and biochemical characterization of the pneumococcal bacteriophage Cp-1 lysin. *J. Virol.* **61:**2573–2580.
9. **García, P., E. García, C. Ronda, A. Tomasz, and R. López.** 1983. Inhibition of lysis by antibody against phage-associated lysin and requirement of choline residues in the cell wall for progeny phage release in *Streptococcus pneumoniae. Curr. Microbiol.* **8:**137–140.
10. **García, P., J. L. García, E. García, and R. López.** 1986. Nucleotide sequence and expression of the pneumococcal autolysin gene from its own promoter in *Escherichia coli. Gene* **43:**265–272.
11. **García, P., J. L. García, E. García, and R. López.** 1989. Purification and characterization of the autolytic glycosidase of *Streptococcus pneumoniae. Biochem. Biophys. Res. Commun.* **158:**251–256.
12. **García, P., J. L. García, E. García, J. M. Sánchez-Puelles, and R. López.** 1990. Modular organization of the lytic enzymes of *Streptococcus pneumoniae* and its bacteriophages. *Gene* **86:**137–142.
13. **Giudicelli, S., and A. Tomasz.** 1984. Attachment of pneumococcal autolysin to wall teichoic acids, an essential step in enzymatic wall degradation. *J. Bacteriol.* **158:**1188–1190.
14. **Höltje, J.-V., and A. Tomasz.** 1975. Specific recognition of choline residues in the cell wall teichoic acids by the *N*-acetylmuramyl-L-alanine amidase of pneumococcus. *J. Biol. Chem.* **250:**6072–6076.
15. **Howard, L. V., and H. Gooder.** 1974. Specificity of the autolysin of *Streptococcus (Diplococcus) pneumoniae. J. Bacteriol.* **117:**796–804.
16. **Jollès, P., and J. Jollès.** 1984. What's new in lysozyme research? *Mol. Cell. Biochem.* **63:**165–189.
17. **Lipman, D. J., and W. R. Pearson.** 1985. Rapid and sensitive protein similarities searches. *Science* **227:**1435–1441.
18. **Romero, A., R. López, and P. García.** 1990. Characterization of the pneumococcal bacteriophage HB-3 amidase: cloning and expression in *Escherichia coli. J. Virol.* **64:**137–142.
19. **Romero, A., R. López, and P. García.** 1990. Sequence of

the *Streptococcus pneumoniae* bacteriophage HB-3 amidase reveals high homology with the major host autolysin. *J. Bacteriol*. **172**:5064–5070.

20. **Ronda, C., J. L. García, E. García, J. M. Sánchez-Puelles, and R. López.** 1987. Biological role of the pneumococcal amidase. Cloning of the *lytA* gene in *Streptococcus pneumoniae*. *Eur. J. Biochem*. **164**:621–624.

21. **Ronda, C., R. López, A. Tapia, and A. Tomasz.** 1977. Role of the pneumococcal autolysin (murein hydrolyase) in the release of the progeny bacteriophage and in the bacteriophage-induced lysis of the host cell. *J. Virol*. **21**:366–374.

22. **Sánchez-Puelles, J. M., C. Ronda, J. L. García, P. García, R. López, and E. García.** 1986. Searching for autolysin functions. Characterization of a pneumococcal mutant deleted in the *lytA* gene. *Eur. J. Biochem*. **158**:289–293.

23. **Sánchez-Puelles, J. M., J. M. Sanz, J. L. García, and E. García.** 1990. Cloning and expression of gene fragments encoding the choline-binding domain of the pneumococcal murein hydrolases. *Gene* **89**:69–75.

Analysis of Some Putative Protein Virulence Factors of *Streptococcus pneumoniae*

G. J. BOULNOIS,[1] T. J. MITCHELL,[1] K. SAUNDERS,[1] R. OWEN,[1] J. CANVIN,[1]
A. SHEPHERD,[1] M. CAMARA,[1] R. WILSON,[2] C. FELDMAN,[2] C. STEINFORT,[2] L. BASHFORD,[3]
C. PASTERNAK,[3] AND P. W. ANDREW[1]

Department of Microbiology, University of Leicester, Leicester, LE1 9HN,[1] Host Defence Unit, Brompton Hospital, London SW3 6LR,[2] and Department of Biochemistry, St. Georges Hospital Medical School, London SW17 0RE,[3] England

Streptococcus pneumoniae, or the pneumococcus, remains an important agent of disease in both developing and Western countries. The pneumococcus is the most important cause of community-acquired pneumonia and is an important agent in septicemia and meningitis (9, 27).

The pneumococcus is found asymptomatically in the nasopharynx. The events involved in the transition from this carrier state to an invasive state are unclear. However, the immune status of the individual is critical. An important nonspecific mechanism of defense against the pneumococcus is the normal functioning of the ciliated epithelial cells of the respiratory tract (6).

It is well documented that particular bacterial components also contribute directly to the ability of the organism to cause disease. The polysaccharide capsule is clearly essential but not sufficient for virulence, and pneumococcal cell wall components have been shown to be potent inducers of inflammation in the respiratory tract (36) and brain (35). There is emerging evidence that certain pneumococcal protein products also contribute to the disease process. For example, a surface protein (PspA) is found on clinical isolates of the pneumococcus, and insertional inactivation of the *pspA* gene reduced or abolished the virulence of certain pneumococcal strains (4).

There are a number of other pneumococcal products with activities suggestive of a role in disease. These include a membrane-damaging toxin (pneumolysin), an immunoglobulin A1 (IgA1) protease, a neuraminidase, and hyaluronidase.

Pneumolysin: a Structure-Function Study

Pneumolysin is a thiol-activated toxin and is a member of a family of similar toxins produced by diverse gram-positive bacteria (30). The predicted amino acid sequences of four of these toxins (pneumolysin, streptolysin O, listeriolysin O, and perfringolysin O) reveal them to be highly homologous (13, 18, 37, 38). These toxins are cytolytic, and it is generally assumed that the toxins bind to target cells via cholesterol and insert into the lipid bilayer, where they form oligomers

that act as transmembrane pores (30). A free sulfhydryl group was thought to be essential for the cytolytic activity of alveolysin and was thought to play a role in facilitating binding to target cells (8), implying that the sulfhydryl group mediates an interaction with cholesterol. Thiol activation has been postulated to involve intramolecular disulfide bond formation (reviewed in reference 30), but this view was rendered untenable by the observation that pneumolysin (38) and three other thiol-activated toxins contained only one cysteine residue (13, 18, 37). However, this sequence analysis reinforced the notion that the single cysteine was important because it lies in the largest contiguous stretch of sequence identity between the toxins. This conserved sequence, called the cysteine motif, has the sequence –ECTGLAWEWWR–. However, replacement of the single cysteine in pneumolysin (28) and streptolysin O (25) by alanine indicated no essential role of thiol groups for in vitro activity of these toxins: the modified toxins had full cytolytic and cytotoxic activities.

We have used site-directed mutagenesis to explore in more detail the role of the cysteine motif in pneumolysin activity. Several changes made in this motif varied in their effects on the extent of cytolytic activity, although all modified toxins had detectable activity. This finding suggests that the overall structure of the motif, rather than specific residues with an essential role, is important for activity. Interestingly, each modified toxin, regardless of cytolytic activity, was able to bind erythrocytes and free cholesterol and to form oligomers in membranes as normal (28; unpublished data). Thus, the cysteine motif probably functions in the generation of functional lesions rather than in receptor binding and oligomerization.

We have shown that wild-type pneumolysin causes leakage of labeled phosphorylcholine from Lettre cells and that this leakage is partly inhibited by Ca^{2+} and Zn^{2+}. In preliminary experiments, toxins modified in the cysteine motif had an altered ability to promote leakage and the lesions formed were less susceptible to inhibition by cations (unpublished data). We tentatively suggest that the cysteine motif is involved, either directly or indirectly, in the generation of active lesions in target membranes.

This observation implies that the receptor and cholesterol binding regions of pneumolysin and those involved in interactions required for oligomerization are distinct from the cysteine motif. Our preliminary experiments indicate that a conserved histidine residue in the thiol-activated toxins plays a role in oligomerization. A His-367→Arg modified pneumolysin is devoid of cytolytic activity but binds target cells normally, yet it fails to oligomerize in membranes.

Pneumolysin also activates the classical pathway of complement (24). In vivo, this process may consume complement components and thereby compromise opsonization of invading bacteria. All of the mutants described above which affect cytolytic activity had no effect on complement activation, suggesting that regions of toxin required for cytolysis and complement activation are different. We were led to one region of pneumolysin involved in complement activation by the observation that pneumolysin shared limited sequence homology with the human acute-phase protein, C-reactive protein (CRP). Both pneumolysin and CRP activate complement, and CRP has been shown to play a role in host defense against pneumococcal infection (19, 39). A continuous stretch of amino acids in CRP (residues 121 to 187) had homology with two noncontiguous regions within pneumolysin (residues 257 to 297 and 368 to 397, termed regions 1 and 2, respectively). We have made a number of amino acid changes in region 1 by site-directed mutagenesis, but none of these affected complement activation. In contrast, some changes in region 2 decreased complement activation but to different extents. This was probably not a consequence of gross structural changes in the toxin, because each retained full cytolytic activity and all changes made were of a conservative nature. These findings suggested that the CRP-like region 2 of pneumolysin is involved in complement activation.

Pneumolysin-mediated complement activation was abolished by addition of ethylene glycol-bis(β-aminoethyl ether)-N,N,N',N'-tetraacetic acid (EGTA) and Mg^{2+} and by prior removal of IgG from serum. Thus, complement activation was by the classical pathway and IgG dependent. We have observed that pneumolysin binds IgG and that purified Fc but not Fab fragments are bound. A Tyr-384→Phe modified pneumolysin, as well as showing reduced complement activation, also bound IgG less well. This finding suggests that complement activation by pneumolysin involves IgG binding via the Fc domain of the antibody.

Pneumolysin: Potential In Vivo Roles

Several lines of evidence point to a role for pneumolysin in disease. First, pneumolysin is produced during human infection, since anti-pneumolysin antibody titers increase during infection (12). Second, immunization of mice with native pneumolysin confers some protection against nasal challenge with virulent pneumococci (23), and a defined pneumolysin-negative mutant of the pneumococcus exhibits reduced virulence in the same murine model (3). Pneumolysin-negative pneumococci were cleared more rapidly from the bloodstream than was the wild-type parent (3). To what extent this reflects the ability of pneumolysin to interfere with the functioning of polymorphonuclear leukocytes (22) and immune cells (7) is unclear.

We have shown that pneumococcal culture filtrates can cause slowing of cilia beating in organ cultures of human respiratory epithelium and that this was caused by pneumolysin alone (32). Slowing of cilia beating, in the absence of obvious damage to the epithelium, was apparent when toxin was used in nanogram quantities (C. Feldman et al., submitted for publication). Thus, pneumolysin may facilitate bacterial access to the lower respiratory tract. In this location, pneumolysin may have a number of effects.

We have shown that pneumolysin alone induces all of the salient histological features of pneumococcal infection in the rat lung (Feldman et al., submitted). Pneumococci or purified recombinant pneumolysin (0.8 μg) was injected through the apical lobe bronchus (in some cases partially ligated to narrow but not occlude the bronchus) into the apical lobe of the rat lung. After 1 and 7 days, the animals were sacrificed and lung sections were processed for histological assessment. Infection with the pneumococcus was frequently fatal, although no animals died from injection of toxin. Both pneumococci and pneumolysin induced a pneumonia localized to the apical lobe, and the pathology was most severe if the bronchus had been previously ligated. The histological features of the apical lobe following inoculation with bacteria or toxin were identical. The cytolytic and complement-activating activities of pneumolysin both contribute to the pathology, since injection of toxin modified by site-directed mutagenesis in either the cysteine motif or region 2 of the CRP-like domain caused significantly less inflammation. Thus, if pneumolysin is delivered to the lower respiratory tract, it seems likely that it will contribute to the inflammatory process and supplement the inflammatory properties of cell wall components. In addition, in the lower respiratory tract it may facilitate bacterial growth by interfering with phagocyte function and complement-mediated opsonization.

Immunization of mice with native pneumolysin confers limited protection against nasal challenge with virulent pneumococci (23). In collaboration with J. Paton, we have demonstrated that recombinant pneumolysin (and its modified variants) also confers protection. Typically, immunized animals survive longer than nonimmunized controls (23). We have attempted to define the basis of this protection by using a murine model of pneumococcal pneumonia and septicemia. In this model, 10^6 to 10^7 virulent pneumococci are instilled into the nose of an anesthetized animal. Shortly after challenge, and after recovery of the animal from the anesthetic, about 20% of the inoculum is found in the lung. Bacterial growth in the lung over the first 24 h of infection is accompanied by the appearance of pneumococci in the blood, although the numbers in the blood do not increase over this period. Between 24 and 48 h there is an explosive growth of bacteria in the blood, which is reflected in high bacterial counts in the spleen, liver, and brain, leading to death of the animal. After 48 h, pneumolysin is readily detected by Western immunoblotting in homogenates of the lung. The ease of detection of pneumolysin, given the number of bacteria present, has led us to speculate that production of pneumolysin may be enhanced in the in vivo environment.

Using this model, we have monitored the organ distribution of bacteria in control and pneumolysin-immunized mice following nasal challenge with pneumococci. Bacterial growth in the lung of immunized animals is similar to that in the nonimmunized controls. Bacterial growth in the blood of immunized animals is slightly delayed, although the final levels reached are similar to those in control animals. Thus, a high bacterial load in the blood and lungs, in the presence of antipneumolysin antibodies, is not sufficient to cause death. The major difference between the control and immunized animals was the extent to which bacteria were found in the brain. Although low numbers of bacteria were found in the brain of immunized animals somewhat earlier than in their nonimmunized counterparts, the levels reached in the brain of the immunized animals were several orders of magnitude lower than in the controls. Thus, pneumolysin may facilitate pneumococcal access to (and/or growth in) the brain.

On the basis of these experiments, it is clear that pneumolysin is an important virulence factor of the pneumococcus. Many questions remain regarding the mode of action and in vivo role of the toxin and the regulation of pneumolysin gene expression.

Neuraminidase

The pneumococcus is a prolific producer of glycosidases, and one has neuraminidase activity (10). Neuraminidase cleaves terminal N-acetylneuraminic (NeuNAc) acid from glycoproteins and gangliosides. Some studies have indicated the existence of multiple forms of these enzymes which may be isoenzymes (31, 33, 34), although others have suggested that they are proteolytic fragments of a single enzyme (16). There is circumstantial evidence to suggest that neuraminidase may contribute to pneumococcal virulence. For example, intracerebral injection of neuraminidase into mice releases NeuNAc from tissues and results in death (14). In patients with pneumococcal meningitis, neuraminidase activity and elevated levels of NeuNAc have been detected in cerebrospinal fluid (21), and the latter correlate with poor prognosis (20). Furthermore, neuraminidase has been detected in patients with persistent middle ear effusion, and this correlates with the presence of the pneumococcus (15). Purified pneumococcal neuraminidase (16) has been used to immunize mice, and these animals showed a small increase in survival times following nasal challenge with pneumococci (17).

The precise role (if any) that neuraminidase plays is not clear. It may act to decrease the viscosity of mucus and compromise its protective effect at the mucosal surface (29). Alternatively, it may aid attachment by removing terminal NeuNAc from oligosaccharides to expose putative receptors for pneumococcal attachment (1). It has been proposed (26) that neuraminidase protentiates the activity of IgA1 protease.

Berry et al. (2) reported the cloning and analysis of a pneumococcal neuraminidase gene. We constructed a library of pneumococcal DNA in lambda EMBL301 and analyzed lysates for neuraminidase activity by using the fluorogenic substrate methylumbelliferyl-N-acetylneuraminic acid (2). DNA from a neuraminidase-expressing recombinant was subcloned, using partial Sau3A digests, into pJDC9 (5) to yield pMC2150. The restriction map of the insert in pMC2150 was completely different from that of the neuraminidase gene cloned by Berry et al. The insert in pMC2150 hybridized to DNA from laboratory and clinical isolates of pneumococci but not to the DNA fragment cloned by Berry et al. DNA from laboratory strains and a clinical isolate of the pneumococcus hybridized to both versions of the neuraminidase gene. Thus, individual isolates have the capacity to express both versions of the enzyme, which may conceivably have different substrate specificities.

IgA1 Protease and Hyaluronidase

In common with many bacteria that colonize and infect at mucosal surfaces, the pneumococcus cleaves human IgA1 in the hinge region of the heavy chain (26). The significance of this activity is not clear, but it is reasonable to suppose that it might compromise immune protection at the mucosal surface. However, no definitive evidence exists to suggest that any of these enzymes function as a virulence factor.

We have recently cloned the gene for pneumococcal IgA1 protease. The library described above was screened for IgA1 protease activity by incubating radiolabeled human IgA1 with pooled lysates of the library. Sodium dodecyl sulfate-polyacrylamide gel electrophoresis and autoradiography were used to monitor cleavage of antibody. One recombinant that expressed IgA1 protease stably is now being analyzed.

Many pneumococcal isolates produce hyaluronidase (11). This enzyme, which is produced by many pathogens, has been referred to as spreading factor and postulated to promote spread of bacteria through tissues. We have recently cloned the gene for pneumococcal hyaluronidase. Plaques of a pneumococcal gene library in lambda gt11 were transferred to a nitrocellulose filter, which was then placed on a hyaluronic acid-containing agar plate. The hyaluronic acid within the plate was subsequently precipitated with detergent, and hyaluronidase activity was observed as a clearing in an otherwise opaque gel. Several recombinants were identified in this way and are being investigated.

In a survey of clinical isolates of the pneumococcus from diverse anatomical locations, we observed that the vast majority expressed hyaluronidase. One strain, which failed to produce hyaluronidase naturally, was used to infect mice via the nose. This isolate proved to be highly virulent, and bacteria isolated from the lungs of infected animals produced hyaluronidase. However, this ability was lost upon subculture. Thus, hyaluronidase expression in some strains may be modulated by the in vivo environment.

G.J.B. is a Lister-Jenner Research Fellow. Work in Leicester was supported by grants from the Medical Research Council and the Wellcome Trust. R.O. and J.C. thank the Wellcome Trust and the Medical Research Council, respectively, for support.

LITERATURE CITED

1. **Anderson, B., J. Dahmen, T. Frejd, H. Leffler, G. Magnusson, G. Noori, and C. Svanborg-Eden.** 1983. Identification of an active disaccharide unit of glycoconjugate receptor for pneumococcal attachment to human pharyngeal epithelial cells. *J. Exp. Med.* **158**:559–570.

2. **Berry, A. M., J. C. Paton, E. M. Glare, D. Hansman, and D. E. A. Catcheside.** 1988. Cloning and expression of the pneumococcal neuraminidase gene in *Escherichia coli*. *Gene* **71**:299–303.

3. **Berry, A. M., J. Yother, D. E. Briles, D. Hansman, and J. C. Paton.** 1989. Reduced virulence of a defined pneumolysin-negative mutant of *Streptococcus pneumoniae*. *Infect. Immun.* **57**:2037–2042.

4. **Briles, D. E., J. Yother, and L. S. McDaniel.** 1988. Role of pneumococcal surface protein A in the virulence of *Streptococcus pneumoniae*. *Rev. Infect. Dis.* **10**(2):S372–S374.

5. **Chen, J.-D., and D. A. Morrison.** 1988. Construction and properties of a new insertion vector pJDC9, that is protected by transcriptional terminators and useful for cloning of DNA from *Streptococcus pneumoniae*. *Gene* **64**:155–164.

6. **Donowitz, G. R., and G. L. Mandell.** 1985. Acute pneumonia, p. 394–404. *In* G. L. Mandell, R. G. Douglas, and J. E. Bennett (ed.) *Principles and Practice of Infectious Disease*. Churchill Livingstone, New York.

7. **Ferrante, A., B. Rowan-Kelly, and J. C. Paton.** 1984. Inhibition of *in vitro* human lymphocyte response by the pneumococcal toxin, pneumolysin. *Infect. Immun.* **46**:585–589.

8. **Geoffroy, C., A.-M. Gilles, and J. E. Alouf.** 1981. The sulfhydryl groups of the thiol-dependent cytolytic toxin from *Bacillus alvei*: evidence for one essential sulfhydryl group. *Biochem. Biophys. Res. Commun.* **99**:781–788.

9. **Gillespie, S. H.** 1989. Aspects of pneumococcal infection including bacterial virulence, host response and vaccination. *J. Med. Microbiol.* **28**:237–248.

10. **Glasgow, L. R., J. C. Paulson, and R. L. Hill.** 1977. Systematic purification of five glycosidases from *Streptococcus (Diplococcus) pneumoniae*. *J. Biol. Chem.* **252**:8615–8623.

11. **Humphrey, J. H.** 1948. Hyaluronidase production by pneumococci. *J. Pathol.* **55**:273–275.

12. **Jalonen, E., J C. Paton, M. Kosekela, Y. Kerttula, and M. Leinonen.** 1987. Measurement of antibody responses to pneumolysin—a promising method for the presumptive aetiological diagnosis of pneumococcal pneumonia. *J. Infect.* **19**:127–134.

13. **Kehoe, M. A., L. Miller, J. A. Walker, and G. J. Boulnois.** 1987. Nucleotide sequence of the streptolysin O (SLO) gene: structural homologies between SLO and other membrane-damaging thiol-activated toxins. *Infect. Immun.* **55**:3228–3232.

14. **Kelly, R., and D. Greiff.** 1970. Toxicity of pneumococcal neuraminidase. *Infect. Immun.* **2**:115–117.

15. **LaMarco, K. L., W. F. Diven, R. H. Glew, W. J. Doyle, and E. I. Cantekin.** 1984. Neuraminidase in middle ear effusions. *Ann. Otol. Rhinol. Laryngol.* **93**:76–84.

16. **Lock, R. A., J. C. Paton, and D. Hansman.** 1988. Purification and immunological characterisation of neuraminidase produced by *Streptococcus pneumoniae*. *Microb. Pathog.* **4**:33–43.

17. **Lock, R. A., J. C. Paton, and D. Hansman.** 1988. Comparative efficacy of pneumococcal neuraminidase and pneumolysin as immunogens protective against *Streptococcus pneumoniae*. *Microb. Pathog.* **5**:461–467.

18. **Mengaud, J., M.-F. Vincente, J. Chenevert, J. M. Pereira, C. Geoffrey, B. Gicquel-Sanzey, F. Baquero, J.-C. Perez-Diaz, and P. Cossart.** 1988. Expression in *Escherichia coli* and sequence analysis of the listeriolysin O determinant of *Listeria monocytogenes*. *Infect. Immun.* **56**:766–772.

19. **Mold, C., S. Nakayama, T. J. Holzer, H. Gewurz, and T. W. DuClos.** 1981. C-reactive protein is protective against *Streptococcus pneumoniae* in mice. *J. Exp. Med.* **154**:1703–1708.

20. **O'Toole, R. D., L. Goode, and C. Howe.** 1971. Neuraminidase activity in bacterial meningitis. *J. Clin. Invest.* **50**:979–985.

21. **O'Toole, R. D., and W. L. Stahl.** 1975. Experimental pneumococcal meningitis. *J. Neurol. Sci.* **26**:167–178.

22. **Paton, J. C., and A. Ferrante.** 1983. Inhibition of human polymorphonuclear leukocyte respiratory burst, bactericidal activity, and migration by pneumolysin. *Infect. Immun.* **41**:1212–1216.

23. **Paton, J. C., R. A. Lock, and D. J. Hansman.** 1983. Effect of immunization with pneumolysin on the survival time of mice challenged with *Streptococcus pneumoniae*. *Infect. Immun.* **40**:548–552.

24. **Paton, J. C., B. Rowan-Kelly, and A. Ferrante.** 1984. Activation of human complement by the pneumococcal toxin pneumolysin. *Infect. Immun.* **43**:1085–1087.

25. **Pinkney, M., E. Beachey, and M. Kehoe.** 1989. The thiol-activated toxin streptolysin O does not require a thiol group for cytolytic activity. *Infect. Immun.* **57**:2553–2558.

26. **Reinholdt, J., M. Tomana, S. B. Mortensen, and M. Kilian.** 1990. Molecular aspects of immunoglobulin A1 degradation by oral streptococci. *Infect. Immun.* **58:**1186–1194.

27. **Roberts, R. B.** 1985. *Streptococcus pneumoniae*, p. 1142–1152. *In* G. L. Mandell, R. G. Douglas, and J. E. Bennett (ed.), *Principles and Practice of Infectious Disease*. Churchill Livingstone, New York.

28. **Saunders, F. K., T. J. Mitchell, J. A. Walker, P. W. Andrew, and G. J. Boulnois.** 1989. Pneumolysin, the thiol-activated toxin of *Streptococcus pneumoniae*, does not require a thiol group for in vitro activity. *Infect. Immun.* **57:**2547–2552.

29. **Scanlon, K. L., W. F. Diven, and R. H. Glew.** 1989. Purification and properties of *Streptococcus pneumoniae* neuraminidase. *Enzyme* **41:**143–150.

30. **Smyth, C. J., and J. L. Duncan.** 1978. Thiol-activated (oxygen-labile) cytolysins, p. 129–183. *In* J. Jeljaszewicz and T. Wadstrom (ed.), *Bacterial Toxins and Cell Membranes*. Academic Press, Inc., London.

31. **Stahl, W. L., and R. D. O'Toole.** 1972. Pneumococcal neuraminidase: purification and properties. *Biochim. Biophys. Acta* **268:**480–487.

32. **Steinfort, C., R. Wilson, T. Mitchell, C. Feldman, A. Rutman, H. Todd, D. Sykes, J. Walker, K. Saunders, P. W. Andrew, G. J. Boulnois, and P. J. Cole.** 1989. Effect of *Streptococcus pneumoniae* on human respiratory epithelium in vitro. *Infect. Immun.* **57:**2006–2013.

33. **Tanenbaum, S. W., J. Gulbinsky, M. Katz, and S.-C. Sun.** 1970. Separation, purification and some properties of pneumococcal neuraminidase isoenzymes. *Biochim. Biophys. Acta* **198:**242–254.

34. **Tanenbaum, S. W., and S.-C. Sun.** 1971. Some molecular properties of pneumococcal neuraminidase isoenzymes. *Biochim. Biophys. Acta* **229:**824–828.

35. **Tuomanen, E., H. Liu, B. Hengstler, O. Zak, and A. Tomasz.** 1985. The induction of meningeal inflammation by components of the pneumococcal cell wall. *J. Infect. Dis.* **151:**859–868.

36. **Tuomanen, E., R. Rich, and O. Zak.** 1987. Induction of pulmonary inflammation by components of the pneumococcal cell surface. *Am. Rev. Respir. Dis.* **135:**869–874.

37. **Tweten, R. K.** 1988. Nucleotide sequence of the gene for perfringolysin O (theta toxin) from *Clostridium perfringens*: significant homology with the genes for streptolysin O and pneumolysin. *Infect. Immun.* **56:**3235–3240.

38. **Walker, J. A., R. L. Allen, P. Falmagne, M. K. Johnson, and G. J. Boulnois.** 1987. Molecular cloning, characterization, and complete nucleotide sequence of the gene for pneumolysin, the sulfhydryl-activated toxin of *Streptococcus pneumoniae*. *Infect. Immun.* **55:**1184–1189.

39. **Yother, J., J. E. Volanakis, and D. E. Briles.** 1982. Human C-reactive protein is protective against fatal *Streptococcus pneumoniae* infection in mice. *J. Immunol.* **128:**2374–2376.

Pneumococcal Surface Protein A: Structural Analysis and Biological Significance

JANET YOTHER,[1] LARRY S. McDANIEL,[1] MARILYN J. CRAIN,[2]
DEBORAH F. TALKINGTON,[1†] AND DAVID E. BRILES[1,2,3]

Departments of Microbiology,[1] Pediatrics,[2] and Comparative Medicine,[3] University of Alabama at Birmingham, Birmingham, Alabama 35294

Streptococcus pneumoniae is an important cause of otitis media, meningitis, bacteremia, and pneumonia. Despite the use of antibiotics and vaccines, the prevalence of pneumococcal infections has declined little over the last 25 years (2). Classic studies demonstrated an essential role for the polysaccharide capsule in pneumococcal virulence and showed that antiserum to the type-specific capsule could protect against infection (17, 19, 20, 28, 32, 33). More recent studies in our laboratory and others have shown that antibodies to determinants of noncapsular antigens, including proteins and polysaccharides, can protect against pneumococcal infections in mouse model systems (1, 3, 5, 6, 18, 21, 23, 26, 29, 34). Antibodies against these components have been identified in human sera, suggesting that they may be important in human immunity (7, 14; M. J. Crain et al., unpublished data). Many of these antigens have been hypothesized to be important in the pathogenesis of *S. pneumoniae*. By using genetic and molecular techniques, full virulence of pneumococci has been shown to require autolysin, pneumolysin, and pneumococcal surface protein A (PspA) (3, 4, 24).

To better understand the pathogenic process of *S. pneumoniae* and to develop more effective treatments and vaccines, it is essential to further explore the nature of these and other potentially important antigens. PspA is a surface-exposed, highly immunogenic protein that is found on all isolates of *S. pneumoniae* (8, 21, 22, 24). Our laboratory has been investigating the role of PspA as both a virulence factor and a potential vaccine candidate. Here we describe our biological and molecular studies of PspA.

Occurrence and Variability of PspA

PspA has been found on all strains of *S. pneumoniae* examined. Using a rabbit antiserum, we detected PspA in all of 95 clinical and laboratory isolates examined. When reacted with seven monoclonal antibodies, 57 *S. pneumoniae* isolates exhibited 31 different patterns of reactivity. We

have proposed a protein typing system for *S. pneumoniae* based on these reactivities (8, 30). The PspA protein type is clearly independent of capsular type. These results suggest that genetic exchange in the environment has allowed for the development of a large pool of strains that are highly diverse with respect to capsule, PspA, and possibly other molecules with variable structures. Variability of PspAs from different strains is also evident in molecular sizes, which range from 67 to 99 kDa. The observed differences are stably inherited and are not the result of protein degradation. Like the PspA protein type, the molecular weight of PspA is not correlated with capsular type (31).

Immunogenic Potential of PspA

We have used several approaches to investigate the potential usefulness of PspA as a vaccine. In all cases, we have found PspA to be capable of inducing protective immune responses to pneumococcal infections in mice.

When immunized with the PspA[+] strain Rx1 (a nonencapsulated derivative of the type 2 strain D39), mice are protected against subsequent challenge with a virulent type 3 *S. pneumoniae* strain. Immunization with an otherwise isogenic PspA-deletion mutant of Rx1 extends survival by a few days but does not protect against fatal infection with the type 3 strain (24). Protection is also induced when mice are immunized with a partial PspA product representing only the amino-terminal half of the molecule (29a). This portion of PspA is expected to be surface exposed and possibly the most important in terms of eliciting an immune response. Indeed, all of our monoclonal antibodies to PspA produced against intact *S. pneumoniae* react with this partial product (8). Several of these monoclonal antibodies are also protective (21; L. S. McDaniel et al., unpublished data). Thus, PspA appears to be a dominant protection-eliciting molecule on the surface of *S. pneumoniae*.

Using partially purified PspA from a recombinant λgt11 clone, we have observed protection against challenge with several *S. pneumoniae* strains representing different capsular and PspA types (22a). This result suggests that despite the

---

†Present address: Department of Health and Human Services, Centers for Disease Control, Atlanta, GA 30333.

variability observed in PspAs from different isolates, sufficient common epitopes may be present to allow a single or at least a small number of PspAs to elicit protection against most *S. pneumoniae* strains.

PspA as a Virulence Factor

To determine whether *S. pneumoniae* requires PspA for full virulence, we constructed mutants altered in the production of PspA. These mutants either produce no PspA (PspA⁻) as the result of a deletion in the 5' end of *pspA* or produce a truncated PspA (PspA^{tr}) approximately one-half the size of the parental PspA (this is the same partial PspA product described under Immunogenic Potential of PspA). The truncated PspA is not attached to the cell but is released as a result of loss of the attachment region located in the carboxy end of the molecule (24; J. Yother, G. L. Handsome, and D. E. Briles, unpublished data; see below, PspA Structure and Cell Attachment). Introduction of either of these mutations into the capsular type 2 strain D39 results in reduced virulence in mice, as determined by 50% lethal dose, time-to-death, and blood clearance studies (24; J. Yother, W. H. Benjamin, and D. E. Briles, unpublished data). Thus, the presence of this released PspA is not sufficient for normal virulence, suggesting that either PspA must be cell associated or this truncated version of the molecule lacks features necessary for virulence. Support for the former hypothesis comes from experiments in which mice were coinfected with isogenic PspA⁺ and PspA^{tr} strains. The results were identical to those obtained when single infections were performed; i.e., the presence of the parental PspA did not alter the survival of the PspA^{tr} mutant. Both the PspA^{tr} and PspA⁻ mutants of D39 remain partially virulent, however. If sufficient bacteria are injected, following an initial decrease in number of bacteria in the blood, they are capable of normal growth and ultimately kill the animal (24; Yother, Benjamin, and Briles, unpublished data).

We have also introduced the PspA⁻ mutation into capsule type 3 and type 5 strains. Unlike the PspA mutants of the type 2 strain, these mutants are essentially avirulent: 50% lethal doses are increased by greater than 10⁴, and all bacteria are cleared from the blood within 24 h (Yother, Benjamin, and Briles, unpublished data). Thus, the effect of a *pspA* mutation on virulence is related to the strain in which the mutation occurs, and strains like the type 2 D39 apparently have a mechanism allowing them to compensate for the loss of PspA.

PspA Structure and Cell Attachment

We have cloned and sequenced the region of *pspA* coding for the carboxy-terminal two-thirds of PspA (Yother, Handsome and Briles, unpublished data). The deduced amino acid sequence reveals four distinct regions of the molecule (summarized in Fig. 1). The amino end of the sequenced DNA is highly charged and α-helical in nature. This region has homology with tropomyosin at the amino acid level (approximately 27% identity and 47% similarity). This homology is due largely to a repeating seven-residue periodicity in which the first and fourth amino acids are hydrophobic, the intervening amino acids are helix promoting, and the seventh amino acid is charged. This pattern is consistent with that of an α-helical coiled-coil molecule (9, 25, 27). N-terminal sequencing of isolated PspA indicates that the N terminus, like the internal region, exhibits potential α-helical coiled-coil structure (29a).

Following the charged helical region is a proline-rich region in which 23 of 82 amino acids are prolines. Immediately carboxy to the proline region is the first of 10 highly homologous 20-amino-acid repeats. The only significantly hydrophobic region in the sequenced portion of the molecule begins in the last repeat. This potential membrane-spanning region is followed by a short region containing several charged amino acids preceding the translational stop codon.

Several surface proteins from other gram-positive organisms have been postulated to be bound to the peptidoglycan of the cell wall via a proline-rich domain that immediately precedes an anchor region (10, 11, 15, 16). The anchor consists of a hydrophobic membrane-spanning region followed by a short charged tail that extends into the cytoplasm. PspA differs significantly from such a model in that (i) the proline region of PspA

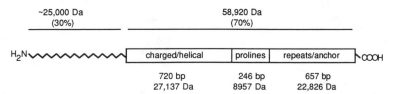

FIG. 1. Schematic of the domains of PspA. The amino acid sequence was deduced from the nucleotide sequence obtained from *pspA* cloned from *S. pneumoniae* Rx1 (Yother, Handsome, and Briles, unpublished data). Boxed areas indicate the distinct domains in the molecule. The zigzag represents unsequenced DNA.

is more proline-rich than that found in other gram-positive proteins, (ii) the proline region of PspA does not immediately precede the anchor region near the stop codon, (iii) the putative membrane-spanning region of PspA is not as hydrophobic as that of other proteins and may not be sufficiently long to traverse the membrane, and (iv) the number of charged residues in the tail of PspA is approximately one-half that observed in other proteins and only 25% of the amino acids preceding the stop codon.

Taken together, these observations suggest that PspA may be attached to the cell by a mechanism different from that of other gram-positive surface proteins. It is tempting to speculate that the repeat region of PspA is involved in attachment. Not only is it located in a position anticipated to lie close to the cell membrane, but it has a distinctly hydrophobic periodicity. In the absence of a distinct membrane anchor, repeated contacts with the membrane via this region may serve to anchor or further stabilize PspA on the cell. This type of interaction of the repeats with the membrane could bring the proline region within the cell wall. Actual attachment to the peptidoglycan may then occur via the nonhydrophobic residues of the repeat region or via the proline domain, as has been postulated for other proteins.

Using insertion-duplication mutagenesis, we have constructed mutants in which transcription through *pspA* terminates at various points in the repeat/anchor region (Yother, Handsome and Briles, unpublished data). We have found that loss of the last five repeats, the hydrophobic region, and the tail region is sufficient to abolish attachment of PspA to the cell. However, when only the hydrophobic and tail regions are eliminated, partial attachment is observed; i.e., approximately half of the mutant protein is cell associated and half is released from the cell. This observation suggests that the presence of the complete repeat region can aid in anchoring PspA but that for complete attachment the hydrophobic/tail region is necessary.

The repeat region that occurs in PspA has also been identified in the pneumococcal autolysin and the pneumococcal phages Cp-1 and Cp-9 (12, 13). Homologies of PspA with other regions of these molecules are not observed. Clearly these four proteins share a region of common ancestry. The occurrence of the repeat region in these cell wall-associated proteins suggests that this may represent a general mechanism by which several pneumococcal proteins may be attached to the cell.

Conclusions

Our studies show that PspA is a complex molecule consisting of at least four distinct domains. PspA appears to have several properties that make it distinctive not only among *S. pneumoniae* proteins but also among gram-positive proteins in general. Many of the features of PspA, such as its serological variability, its potential coiled-coil structure, and its clearly defined domains, are reminiscent of the M proteins of the group A streptococci. However, apart from the helical/charged region, PspA appears to be unique in both the types and positioning of the additional domains. Our data suggest that these domains are important in a novel type of protein-cell interaction. In addition, they are also likely to be responsible for part of the size variability observed in PspAs isolated from different *S. pneumoniae* strains (Yother, Handsome, and Briles, unpublished data).

Although the function of PspA has not yet been determined, its widespread occurrence, variability, and requirement for full virulence in mice indicate that it is likely to be important in the pathogenesis of *S. pneumoniae* infections in humans. Because of these factors and the ability to induce protective immune responses, PspA may prove useful as part of a human pneumococcal vaccine. Studies are in progress to further address this point and to determine functions for both the molecule as a whole and its domains.

This work was supported by Public Health Service grants AI21548, AI28457, and AI27201 from the National Institute of Allergy and Infectious Diseases and HD17812 from the National Institute of Child Health and Human Development.

LITERATURE CITED

1. **Au, C. C., and T. Eisenstein.** 1981. Nature of the cross-protective antigen in subcellular vaccines of *Streptococcus pneumoniae*. *Infect. Immun.* **31:**160–168.
2. **Austrian, R. A.** 1984. Pneumococcal infections, p. 257–288. *In* R. Germanier (ed.), *Bacterial Vaccines*. Academic Press, Inc., New York.
3. **Berry, A. M., R. A. Lock, D. Hansman, and J. C. Paton.** 1989. Contribution of autolysin to virulence of *Streptococcus pneumoniae*. *Infect. Immun.* **57:**2324–2330.
4. **Berry, A. M., J. Yother, D. E. Briles, D. Hansman, and J. C. Paton.** 1989. Reduced virulence of a defined pneumolysin-negative mutant of *Streptococcus pneumoniae*. *Infect. Immun.* **57:**2037–2042.
5. **Briles, D. E., C. Forman, J. C. Horowitz, J. E. Volanakis, W. H. Benjamin, Jr., L. S. McDaniel, J. Eldridge, and J. Brooks.** 1989. Measurement of antipneumococcal effects of C-reactive protein and monoclonal antibodies to pneumococcal cell wall and capsular antigens. *Infect. Immun.* **57:**1457–1464.
6. **Briles, D. E., M. Nahm, K. Schoer, J. Davie, P. Baker, J. F. Kearney, and R. Barletta.** 1981. Anti-phosphocholine antibodies found in normal mouse serum are protective against intravenous infection with type 3 *S. pneumoniae*. *J. Exp. Med.* **153:**694–705.
7. **Briles, D. E., G. Scott, B. Gray, M. J. Crain, M. Blaese, M. Nahm, P. Haber, V. Scott, and C. Commer.** 1987. Naturally occurring antibodies to phosphocholine as a potential index of antibody responsiveness to polysaccharides. *J. Infect. Dis.* **155:**1307–1313.
8. **Crain, M. J., W. D. Waltman II, J. S. Turner, J. Yother, D. F. Talkington, L. S. McDaniel, B. M. Gray, and D. E. Briles.** 1990. Pneumococcal surface protein A (PspA) is serologically highly variable and is expressed by clinically

important capsular serotypes of *Streptococcus pneumoniae*. *Infect. Immun.* **58:**3293–3299.

9. **Crick, F. H. C.** 1953. The packing of alpha-helices: simple coiled-coils. *Acta Crystallogr.* **6:**689–697.

10. **Fahnestock, S. R., P. Alexander, J. Nagle, and D. Filpula.** 1986. Gene for an immunoglobulin-binding protein from a group G streptococcus. *Infect. Immun.* **167:**870–880.

11. **Frithz, E., L. O. Heden, and G. Lindahl.** 1989. Extensive sequence homology between IgA receptor and M proteins in *Streptococcus pyogenes*. *Mol. Microbiol.* **3:**1111–1119.

12. **Garcia, E., J. L. Garcia, P. Garcia, A. Arraras, J. M. Sanchez-Puelles, and R. Lopez.** 1988. Molecular evolution of lytic enzymes of *Streptococcus pneumoniae* and its bacteriophages. *Proc. Natl. Acad. Sci. USA* **85:**914–918.

13. **Garcia, P., J. L. Garcia, J. M. Sanchez-Puelles, and R. Lopez.** 1990. Modular organization of the lytic enzymes of *Streptococcus pneumoniae* and its bacteriophages. *Gene* **86:**81–88.

14. **Gray, B. M., H. C. Dillon, and D. E. Briles.** 1983. Epidemiological studies of *Streptococcus pneumoniae* in infants: development of antibody to phosphocholine. *J. Clin. Microbiol.* **18:**1102–1107.

15. **Guss, B., M. Uhlen, B. Nilsson, M. Lindberg, J. Sjoquist, and J. Sjodahl.** 1984. Region X, the cell-wall-attachment part of staphylococcal protein A. *Eur. J. Biochem.* **138:**413–420. (Authors' correction, **143:**685.)

16. **Hollingshead, S. K., V. A. Fischetti, and J. R. Scott.** 1986. Complete nucleotide sequence of type 6 M protein of the group A *Streptococcus*. *J. Biol. Chem.* **261:**1677–1686.

17. **Kruse, W., and S. Pansini.** 1891. Untersuchungen uber den *Diplococcus pneumoniae* und verwandte Streptokokken. *Z. Infektionskr.* **11:**279–280.

18. **Lock, R. A., J. C. Paton, and D. Hansman.** 1988. Comparative efficacy of pneumococcal neuraminidase and pneumolysin as immunogens protective against *Streptococcus pneumoniae*. *Microb. Pathog.* **5:**461–467.

19. **MacLeod, C. M., R. G. Hodges, M. Heildeberger, and W. G. Bernhard.** 1945. Prevention of pneumococcal pneumonia by immunization with specific capsular polysaccharides. *J. Exp. Med.* **82:**445–465.

20. **MacLeod, C. M., and M. R. Krauss.** 1950. Relation of virulence of pneumococcal strains for mice to the quantity of capsular polysaccharide formed *in vitro*. *J. Exp. Med.* **92:**1–9.

21. **McDaniel, L. S., G. Scott, J. F. Kearney, and D. E. Briles.** 1984. Monoclonal antibodies against protease sensitive pneumococcal antigens can protect mice from fatal infection with *Streptococcus pneumoniae*. *J. Exp. Med.* **160:**386–397.

22. **McDaniel, L. S., G. Scott, K. Widenhofer, J. Carroll, and D. E. Briles.** 1986. Analysis of a surface protein of *Streptococcus pneumoniae* recognized by protective monoclonal antibodies. *Microb. Pathog.* **1:**519–531.

22a.**McDaniel, L. S., J. S. Sheffield, P. Delucchi, and D. E.** **Briles.** 1991. PspA, a surface protein of *Streptococcus pneumoniae*, is capable of eliciting protection against pneumococci of more than one capsular type. *Infect. Immun.* **59:**222–228.

23. **McDaniel, L. S., W. D. Waltman II, B. Gray, and D. E. Briles.** 1987. A protective monoclonal antibody that reacts with a novel antigen of pneumococcal teichoic acid. *Microb. Pathog.* **3:**249–260.

24. **McDaniel, L. S., J. Yother, M. Vijayakumar, L. McGarry, W. R. Guild, and D. E. Briles.** 1987. Use of insertional inactivation to facilitate studies of biological properties of pneumococcal surface protein A (PspA). *J. Exp. Med.* **165:**381–394.

25. **McLachlan, A. D., M. Stewart, and L. B. Smillie.** 1975. Sequence repeats in α-tropomyosin. *J. Mol. Biol.* **98:**281–291.

26. **Paton, J. C., R. A. Lock, and D. J. Hansman.** 1983. Effect of immunization with pneumolysin on survival time of mice challenged with *Streptococcus pneumoniae*. *Infect. Immun.* **40:**548–552.

27. **Sodek, J., R. S. Hodges, L. B. Smillie, and L. Jurasek.** 1972. Amino-acid sequence of rabbit skeletal tropomyosin and its coiled-coil structure. *Proc. Natl. Acad. Sci. USA* **69:**3800–3804.

28. **Stryker, L. M.** 1916. Variations in the pneumococcus induced by growth in immune serum. *J. Exp. Med.* **24:**49–68.

29. **Szu, S. C., S. Clarke, and J. B. Robbins.** 1983. Protection against pneumococcal infection in mice conferred by phosphocholine-binding antibodies: specificity of the phosphocholine binding and relation to several types. *Infect. Immun.* **39:**993–999.

29a.**Talkington, D. F., D. L. Crimmins, D. C. Voellinger, J. Yother, and D. E. Briles.** 1991. A 43-kilodalton pneumococcal surface protein, PspA: isolation, protective abilities, and structural analysis of the amino-terminal sequence. *Infect. Immun.* **59:**1285–1289.

30. **Waltman, W. D., II, L. S. McDaniel, B. Andersson, L. Bland, B. M. Gray, C. Svanborg-Eden, and D. E. Briles.** 1988. Protein serotyping of *Streptococcus pneumoniae* based on reactivity to six monoclonal antibodies. *Microb. Pathog.* **5:**159–167.

31. **Waltman, W. D., II, L. S. McDaniel, and D. E. Briles.** 1990. Variation in the molecular weight of PspA (pneumococcal surface protein A) among *Streptococcus pneumoniae*. *Microb. Pathog.* **8:**61–69.

32. **White, B.** 1938. *The Biology of Pneumococcus.* The Commonwealth Fund, New York.

33. **Wood, W. B., Jr., and M. R. Smith.** 1949. The inhibition of surface phagocytosis by the capsular "slime layer" of pneumococcus type III. *J. Exp. Med.* **90:**85–96.

34. **Yother, J., C. Forman, B. M. Gray, and D. E. Briles.** 1982. Protection of mice from infections with *Streptococcus pneumoniae* by anti-phosphocholine antibody. *Infect. Immun.* **36:**184–188.

Highly Variable Penicillin-Binding Proteins in Penicillin-Resistant Strains of *Streptococcus pneumoniae*

REGINE HAKENBECK, GÖTZ LAIBLE, THOMAS BRIESE,[†] CHRISTIANE MARTIN,
LYNDA CHALKLEY,[‡] CRISTINA LATORRE[§] RAILI KALLIOKOSKI,[‖] AND MAIJA LEINONEN[‖]

Max-Planck-Institut für Molekulare Genetik, D-1000 Berlin 33, Federal Republic of Germany

The intrinsic penicillin resistance of *Streptococcus pneumoniae* is due to alterations in several penicillin-binding proteins (PBPs), the target enzymes for β-lactam antibiotics. In the late 1960s, before penicillin-resistant pneumococci were recognized in many parts of the world as a clinical problem (8), important features of the stepwise development of resistance had already been described; higher resistance levels could not be transformed in a single step into a sensitive background, and consequently the participation of several genes in the development of resistance has been proposed (13). Involvement of genes other than PBP genes has not yet been demonstrated in pneumococci but cannot (and probably should not) be excluded.

In penicillin-resistant strains of *S. pneumoniae*, essential PBPs have decreased affinity for β-lactams, which prevents the inhibitory effects of these antibiotics at low concentrations. The penicillin-sensitive laboratory strain R6, which has been used as a reference strain in many investigations, contains six PBPs (1a, 1b, 2x, 2a, 2b, and 3) with molecular sizes ranging from 92 to 43 kDa (3). The high-molecular-weight PBPs 1a, 2x, and 2b are altered in resistant clinical strains (3, 6, 12, 14). In laboratory mutants, selection with piperacillin leads to low-affinity PBPs 1a and 2b; in cefotaxime-resistant mutants, PBPs 2x and 2a are primarily affected (9).

The DNA sequences of the genes from R6 coding for PBPs 2x (10) and 2b (1) are known, and those of PBP 1a (T. Briese and C. Martin, unpublished data) and of the low-molecular-weight PBP 3 (12a) have been partially sequenced. Comparison of the deduced amino acid sequences with those of *Escherichia coli* PBPs revealed that *S. pneumoniae* contains in principle the same groups of PBPs: (i) the large PBP 1a, homologous to the bifunctional *E. coli* PBPs 1a and 1b, which act as transglycosylases as well as penicillin-sensitive transpeptidases (11); (ii) the medium-size PBPs 2x and 2b, related to the *E. coli* penicillin-sensitive transpeptidases PBPs 2 and 3; and (iii) the small D,D-carboxypeptidase PBP 3, which is not involved in penicillin resistance (the only pneumococcal PBP with known enzymatic function) and shows homology to the D,D-carboxypeptidases PBPs 5 and 6 of *E. coli* and to *Bacillus subtilis* PBP 5.

PBPs of resistant clinical strains are unusual in that they have highly variable apparent M_rs, as revealed on sodium dodecyl sulfate (SDS) gels; in addition, they often cannot be easily detected as PBPs after labeling with radioactive penicillin, SDS-polyacrylamide gel electrophoresis (PAGE), and subsequent fluorography because of an extremely low affinity for β-lactams. The relationship of such variants to PBPs 1a and 2b, respectively, has been demonstrated immunologically, and peptide patterns after partial proteolysis revealed alterations in the PBP 1a-related proteins (4).

The variation seen at the protein level parallels recent data obtained on the DNA sequences of genes from resistant strains. PBP 2b genes from resistant clinical strains contained a large number of base changes, including many silent mutations (1). This finding strongly suggested that horizontal gene transfer via transformation of the naturally competent pneumococci plays an essential role, rather than the accumulation of point mutations as described in the PBP 2x gene of cefotaxime-resistant laboratory mutants (9). However, neither the origin of the resistant genes nor the epidemiology of the resistant strains is known.

To determine the relatedness of a large number of clinical isolates originating from different parts of the world, we have investigated the antigenic variation of PBPs 1a and 2b by using specific antibodies raised against the respective purified proteins from the laboratory strain R6 and have analyzed the PBP profiles. Provided that the antibodies are directed against different epitopes, such a study should allow testing of (i) the relationship between alteration of the proteins and the resistance level and (ii) the similarity or dissimilarity between resistant strains isolated at the same or different places.

[†] Present address: Institut für Virologie der Freien Universität im Robert-Koch-Institut, D-1000 Berlin 65, Federal Republic of Germany.

[‡] Present address: Microbiology Medical School, University of Witwatersrand, Johannesburg, 2193 South Africa.

[§] Present address: Service of Microbiology, Hopital San Juan de Dios, 08034 Barcelona, Spain.

[‖] Present address: National Public Health Institute, SF 00280 Helsinki, Finland.

TABLE 1. Clinical strains of penicillin-sensitive and -resistant *S. pneumoniae* investigated

Origin	No. of isolates		
	Sensitive (0.007–0.06)[a]	Relatively resistant (0.125–0.5)	Resistant (1–4)
Spain	14	3 (19, 6)[b]	26 (3, 6, 9, 14, 19, 23)
South Africa	17	6 (6A, 14, 19F)	6 (6A, 19F)
Finland	27	5 (19, 23)	4 (6)
Germany	3	5	0

[a]Penicillin MIC (micrograms per milliliter).
[b]Main serogroups associated with resistant strains.

PBP Profiles Are Variable in Both Penicillin-Sensitive and -Resistant Clinical Isolates

The clinical strains used in this study (61 penicillin-sensitive and 55 penicillin-resistant strains) and the serogroups associated with the resistant strains are summarized in Table 1. Whereas the sensitive Finnish strains were selected for serogroups 6, 19, and 23 only, the other sensitive strains were very heterogeneous, comprising a total of 16 different serogroups (not shown). Resistance was preferentially associated with serogroups 6, 19, and 23, and the Spanish collection also included six resistant type 9 strains. All other serotypes were found in single strains only.

In penicillin-sensitive strains, an unexpected variation in electrophoretic mobilities of the PBPs was observed immediately. It was most pronounced in PBP 1a and also comprised the triad of PBPs 2x, 2a, and 2b (Fig. 1A). The variation was noticed even within one serogroup, such as the serogroup 19 strains isolated in Finland (Fig. 1A). A few strains had an apparently larger PBP 3 (which has been described before in other strains [4]), and one strain had a smaller PBP 1b. As expected, the electrophoretic variation was much higher in the resistant strains (Fig. 1B). The peptide pattern of PBPs 1a and 2x obtained after partial proteolysis with V8 protease revealed extensive alterations within these proteins even in low-level-resistant strains (Fig. 2).

The PBP profiles of most resistant strains were unique, and a total of at least 34 distinct PBP profiles could be distinguished among the 55 resistant strains. Six PBP profiles were shared by several strains; six groups of Finnish or Spanish strains with three to four members each could thus be identified. This variation appears to be much higher than previously reported in a similar investigation (7), probably because of the higher resolution between individual PBPs in our study.

PBPs 1A and 2B Exist in Many Antigenic Variants

Two antisera and three monoclonal antibodies (MAbs) specific for PBP 1a and five MAbs specific for PBP 2b of the penicillin-sensitive laboratory strain R6 were used to investigate variations of the respective epitopes in the strain collection held in this laboratory.

Unexpectedly, only 14 of the 61 sensitive strains contained a reactivity pattern identical to that of the R6 strains (all antibodies reacted positively), whereas the remaining 47 strains showed another six different reactivity patterns. Most of these patterns were discriminated by the anti-PBP 2b MAb III15 (which reacted only weakly in three strains), anti-PBP 2b MAb 502 (which was nonreactive in 27 strains), or both (17 strains). Since the deduced amino acid sequences covering the penicillin-binding domain of PBP 2b from six sensitive strains did not reveal any major changes compared with the R6 PBP 2b (1), the difference between the antigenically different PBP 2b might reside within the further N-terminal sequences or might involve secondary modification. Three strains also showed a different PBP 1a, being discriminated by either the rabbit antiserum 1220 or MAb 215 (not shown).

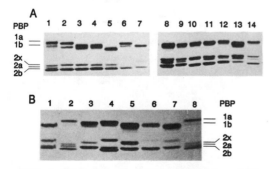

FIG. 1. High-molecular-weight PBPs of penicillin-sensitive and -resistant *S. pneumoniae*. Cell lysates were incubated with [³H]propionylampicillin (3) for 20 min at 37°C. PBPs were visualized after SDS-PAGE and fluorography. PBPs of the laboratory reference strain R6 are indicated. (A) Penicillin-sensitive strains except 6 and 7. Strains (origin, MIC [micrograms per milliliter]) are as indicated. Lanes: 1, R6; 2 to 7 (all type F, also type 19), 23077, 23558, 23276, 23556, 43359 (0.25), and 43356 (0.25); 8 to 14, 63914 (South America), 64651 (South America), 631 (Spain), 63065 (South America), 99184 (South America), 636 (Spain), and 624 (Germany). (B) Penicillin-resistant strains. Lanes: 1, 677 (Spain 2); 2 and 8, R6; 3, SpR (Germany, 0.25); 4, S1 (Germany, 0.125); 5, 343 (Germany, 0.25); 6, 8250 (South America, 0.5); 7, 60785 (South America, 4).

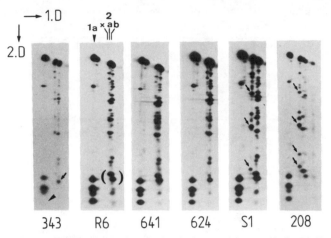

FIG. 2. Penicilloyl peptides from different strains obtained after partial proteolysis. Radioactively labeled PBPs of cell lysates were separated by SDS-PAGE. Lanes were cut and layered on the horizontal surface of a second acrylamide slab gel containing V8 protease in the stacking gel (2). Radioactive penicilloyl peptides were revealed after fluorography. The directions of first- and second-dimension electrophoresis are indicated. The positions of R6 PBPs in the first-dimension gel are shown schematically. The brackets encompass penicilloyl peptides of PBPs 2a and 2b, which are positioned differently in strains 343 and 624; the arrows point to variations in peptides derived from PBP 1a (strain 343) and PBP 2x (strains S1 and 208). All clinical strains were isolated in Germany; strains 641 and 624 are penicillin sensitive.

With the exception of four low-level-resistant strains, all resistant isolates had reactivity patterns distinct from those of the sensitive isolates. Strikingly, MAbs III15 (anti-PBP 2b) and 215 (anti-PBP 1a) discriminated all resistant strains above MICs of 0.125 and 0.25 μg/ml, respectively (Fig. 3A). The reactivities of the other antibodies did not correlate with resistance; i.e., they discriminated PBPs in a variable percentage of strains from each resistance level. Each of the antibodies investigated reacted with a different group of strains, indicating that all of them recognized epitopes distinct from each other. A total of 24 reactivity patterns were observed, 3 sensitive and 12 resistant strains showing unique reactivities. Reactivity patterns XVIII and XII, which were shared by several strains, were found exclusively associated with certain Spanish or Finnish strains, whereas reactivity pattern XIII was not found in the Spanish or Finnish strains but was detected in three South African strains and one German strain. Strains with a common reactivity pattern did not necessarily have identical PBP profiles, whereas all strains with an apparent common PBP profile always contained antigenically identical PBPs.

The results clearly indicate that during the development of resistance, PBPs 1a and 2b can be altered in several different ways and that the variation observed is independent of the resistance level. The data also confirmed the results of the PBP profile analyses in that the four groups of Spanish strains and the two groups of Finnish strains with common PBP patterns in each group

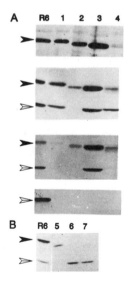

FIG. 3. Reactivities of anti-PBP antibodies. PBPs in cell lysates were visualized after SDS-PAGE and Western blotting with the help of different anti-PBP 1a antibodies (closed arrowheads) or anti-PBP 2b antibodies (open arrowheads). (A) Strains (serotype, MIC [micrograms per milliliter]) are as indicated. Lanes 1 to 4 (isolates from Finland): 43359 (19, 0.25), 43356 (19, 0.25), 43352 (6, 1), and 43349 (23, 0.5). Anti-PBP 1a antibodies (top to bottom): 1220 (rabbit antiserum); 301 (MAb); 221 (MAb); anti-PBP 2b MAbs I6, 502, and III15. (B) Cross-reactivity of S. oralis PBPs with antipneumococcal PBP antibodies. Lanes 5 to 7, S. oralis DSM 20066, 20379, and 20395. MAbs 301 (anti-PBP 1a) and 410 (anti-PBP 2b) were used for immunostaining. S. pneumoniae R6 is included as a control.

consistently contained identical antibody reactivity patterns and thus represent most likely six pneumococcal clones. Most important, the data suggest that resistant strains have originated independently on several occasions in different countries.

The antigenic variation seen in PBPs 1a and 2b is reminiscent of the nucleotide alterations recognized in DNA sequences of PBP 2b genes (1). We have also investigated PBP 2x genes of five resistant clinical isolates, which reveal an even higher degree of nucleotide variability (G. Laible, B. G. Spratt, and R. Hakenbeck, unpublished data). Only two strains, one isolated in South Africa and one isolated in Great Britain, were similar enough to suggest a common origin of the resistant genes.

S. oralis Contains PBPs Homologous to Pneumococcal PBP

Since homologous genes are proposed to participate in the transformation, we probed other streptococcal species for PBPs cross-reacting with anti-pneumococcal PBP antibodies. Group A streptococci are known to contain a PBP cross-reacting with anti-PBP 1a antiserum (5), but no reaction occurred with any of the MAbs used in this study (unpublished results). However, Streptococcus oralis strains gave strong reactions with several of the antibodies (Fig. 3B). Interestingly, each of the three strains used contained a slightly different set of PBPs that could also be antigenically differentiated (Fig. 3B).

Chromosomal DNA of S. oralis could even be used in a transformation experiment with S. pneumoniae R6 as an acceptor strain for isolating cefotaxime-resistant S. pneumoniae transformants (G. Laible, unpublished results), and those transformants contained a low-affinity PBP 2x. This experiment demonstrates not only that PBP 2x in addition to PBPs 1a and 2b are very similar in the two streptococcal species but also that DNA from a foreign species can transform pneumococci into resistant variants.

Several questions remain to be resolved. The first concerns the population structure of pneumococci; clonal analysis will certainly help to clarify whether only a few clones are involved in the emergence of resistance and to what degree resistant strains from different locations are related to each other. Since transformation of homologous genes of related species is most likely responsible for the emergence of intrinsically resistant strains, one would expect to find pneumococcal clones with members that have different PBP profiles. The reason is simply that in one transformation event involving one bacterial clone, the newly introduced gene (fragment) will recombine at different positions within the acceptor gene as long as the sequences necessary for altering the respective PBP into a functional low-affinity form remain intact.

The second question concerning the origin of the resistant DNA is more difficult to answer if one does not want to live with a hen-or-egg problem. Obtaining the answer will depend largely on access to very early resistant isolates, before extensive gene transfer between already resistant streptococcal species has occurred.

LITERATURE CITED

1. **Dowson, C. G., A. Hutchison, J. A. Brannigan, R. C. George, D. Hansman, J. Linares, A. Tomasz, J. M. Smith, and B. G. Spratt.** 1989. Horizontal transfer of penicillin-binding protein genes in penicillin-resistant clinical isolates of Streptococcus pneumoniae. Proc. Natl. Acad. Sci. USA 86:8842–8846.

2. **Ellerbrok, H., and R. Hakenbeck.** 1983. Characterization of penicillin binding proteins from Streptococcus pneumoniae by proteolysis, p. 432–436. In R. Hakenbeck, J.-V. Höltje, and H. Labischinski (ed.), The Target of Penicillin. Walter de Gruyter & Co., Berlin.

3. **Hakenbeck, R., T. Briese, H. Ellerbrok, G. Laible, C. Martin, C. Metelmann, H.-M. Schier, and S. Tornette.** 1987. Targets of β-lactams in Streptococcus pneumoniae, p. 390–399. In P. Actor, L. Daneo-Moore, M. L. Higgins, M. R. J. Salton, and G. D. Shockman (ed.), Antibiotic Inhibition of Bacterial Cell Surface Assembly and Function. American Society for Microbiology, Washington, D.C.

4. **Hakenbeck, R., H. Ellerbrok, T. Briese, S. Handwerger, and A. Tomasz.** 1986. Penicillin-binding proteins of penicillin-susceptible and -resistant pneumococci: immunological relatedness of altered proteins and changes in peptides carrying the β-lactam binding site. Antimicrob. Agents Chemother. 30:553–558.

5. **Hakenbeck, R., H. Ellerbrok, S. Tornette, and I. van de Rijn.** 1987. Common antigenic determinants of pneumococcal penicillin-binding protein (PBP) 1a and Streptococcus pyogenes PBP 2. FEMS Microbiol. Lett. 48:171–174.

6. **Hakenbeck, R., M. Tarpay, and A. Tomasz.** 1980. Multiple changes of penicillin-binding proteins in penicillin-resistant clinical isolates of Streptococcus pneumoniae. Antimicrob. Agents Chemother. 17:364–371.

7. **Jabes, D., S. Nachman, and A. Tomasz.** 1989. Penicillin-binding protein families: evidence for the clonal nature of penicillin-resistance in clinical isolates of pneumococci. J. Infect. Dis. 159:16–25.

8. **Klugman, K. P.** 1990. Pneumococcal resistance to antibiotics. Clin. Microbiol. Rev. 3:171–196.

9. **Laible, G., and R. Hakenbeck.** 1987. Penicillin-binding proteins in β-lactam-resistant laboratory mutants of Streptococcus pneumoniae. Mol. Microbiol. 1:355–363.

10. **Laible, G., R. Hakenbeck, M. A. Sicard, B. Joris, and J.-M. Ghuysen.** 1989. Nucleotide sequences of the pbpX genes encoding the penicillin-binding proteins 2x from Streptococcus pneumoniae R6 and a cefotaxime-resistant mutant, C506. Mol. Microbiol. 3:1337–1348.

11. **Nakagawa, J., S. Tamaki, S. Tomioka, and M. Matsuhashi.** 1984. Functional biosynthesis of cell wall peptidoglycan by polymorphic polypeptides. J. Biol. Chem. 259:13937–13946.

12. **Percheson, P. B., and L. E. Bryan.** 1980. Penicillin-binding components of penicillin-susceptible and -resistant strains of Streptococcus pneumoniae. Antimicrob. Agents Chemother. 18:390–396.

12a.**Schuster, C., B. Dobrinski, and R. Hakenbeck.** 1990. Unusual septum formation in Streptococcus pneumoniae mutants with an alteration in the D,D-carboxypeptidase penicillin-binding protein 3. J. Bacteriol. 172:6499–6505.

13. **Shockley, T. E., and R. D. Hotchkiss.** 1970. Stepwise introduction of transformable penicillin resistance in Pneumococcus. Genetics 64:397–408.

14. **Zighelboim, S., and A. Tomasz.** 1980. Penicillin-binding proteins of multiply antibiotic-resistant South African strains of Streptococcus pneumoniae. Antimicrob. Agents Chemother. 17:434–442.

III. Lactococci: Molecular Biology and Biotechnology

Special-Purpose Vectors for Lactococci

JAN KOK

Department of Molecular Genetics, University of Groningen, 9751 NN Haren, The Netherlands

Recent developments have rendered the lactococci (formerly designated lactic acid streptococci; 22) accessible to genetic manipulation. Traditionally, these organisms (divided into the two subspecies, *lactis* and *cremoris*) are used in the production of a variety of fermented dairy products. Their main function is the rapid acidification of milk, and the resulting low pH prevents the growth of spoilage bacteria in the fermented product. These bacteria are also involved in the development of flavor in the final product. Since some essential metabolic functions of lactococci are encoded by (unstable) plasmids, a considerable body of research has been devoted to the plasmid biology of lactococci during the last decade (for reviews, see references 5, 9, and 20). This study of lactococcal plasmids has been important for gene cloning in lactococci for two reasons. First, some of the plasmids were used to construct vector systems for cloning; second, they were the source of industrially important genes that have now been cloned (for general reviews, see references 6, 7, and 15). Researchers now have the choice of a number of general cloning vectors, each with specific advantages and disadvantages. Moreover, vectors have been developed for specific purposes, such as the isolation of promoters, terminators, and export signals and the expression of heterologous genes. In addition, vectors are available for stably integrating genes into the chromosomes of lactococci. A survey of available vectors is given below. Because several of these vectors not only replicate and are expressed in the genus *Lactococcus* but also in the genera *Streptococcus* and *Enterococcus*, this group of organisms has become more accessible to genetic analyses.

General Cloning Vectors

Although a multitude of plasmids are available in lactococci, only a few carry known functions, and most of these cannot be used to rapidly assess the parameters of an efficient gene cloning system. Therefore, in the early 1980s, a number of laboratories tried to find plasmid vectors which were easy to manipulate and select in lactococci.

Vectors already available for other microorganisms were examined, and cloning vehicles were constructed from cryptic plasmids found in lactococcal strains (Table 1). Plasmids in the first category include derivatives of the broad-host-range gram-positive conjugative plasmids pAMβ1, encoding resistance to erythromycin (Em^r), and pIP501, conferring Em^r and chloramphenicol resistance (Cm^r). The most extensively used vectors are pIL252 and pIL253, derived from pAMβ1 (23). Both plasmids have been completely sequenced. pIL252 has a low copy number (approximately five per chromosome equivalent), while the copy number of pIL253 is approximately 10 times higher. Both contain a multiple cloning site (MCS) and the easily selectable constitutive macrolides-lincosamides-streptogramidin B resistance marker conferring a high level of Em^r. Large DNA inserts (30 kb or more) can be stably maintained in these and other pAMβ1 derivatives (23). Plasmid pGB301 is a Cm^r Em^r derivative of pIP501 also used for cloning in lactococci as well as in other streptococcal species (2, 13, 14). The *Escherichia coli*-streptococcus shuttle vectors pVA838 and pSA3 share the property of being mobilizable by the pIP501-derived conjugative plasmid pVA797 and can thus be introduced into strains for which transformation protocols are not available (4, 19, 21). *Staphylococcus aureus* plasmids such as pUB110, pC194, and pE194, which are used extensively in *Bacillus subtilis* and also seem to replicate in lactobacilli, do not function in lactococci.

The second strategy has led to the development of cloning vectors derived from either the cryptic *Lactococcus lactis* subsp. *cremoris* plasmid pWV01 or from the related *L. lactis* subsp. *lactis* plasmid pSH71 (6, 10, 12). The complete nucleotide sequences of these cryptic plasmids place the plasmids in the category of small gram-positive plasmids that replicate via the rolling-circle mechanism (K. J. Leenhouts and W. M. de Vos, personal communication). Vectors constructed from these plasmids are versatile in that they rep-

TABLE 1. General cloning vectors

Vector	Replicon	Size (kb)	Antibiotic resistance gene(s)	Copy no. in *L. lactis*	MCS	Marker inactivation	Reference
Nonlactococcal							
pGB301	pIP501	9.8	Em Cm	Low	−	Yes	2
pSA3	pACYC184 pIP501	10.2	Em Cm Tc	Low	−	Yes	4
pVA838	pVA380-1 pACYC184	9.2	Em Cm	Low	+	Yes	19
pIL252	pAMβ1	4.7	Em	Low	+	No	23
pIL253	pAMβ1	5.0	Em	High	+	No	23
Lactococcal							
pGK12	pWV01	4.4	Em Cm	Low	−	Yes	12
pGL3	pWV01	5.0	Cm Km	Low	+	α complementation	D. van der Lelie, Ph.D. thesis, University of Groningen, Groningen, The Netherlands, 1989
pGKV21	pWV01	4.9	Em Cm	Low	+	Yes	30
pCTC3	pWV01	5.2	Em Cm	Low	+	Yes	34
pNZ12	pSH71	4.3	Cm Km	High	−	Yes	6
pNZ123	pSH71	2.8	Cm Km	High	+	Yes	G. Simons, personal communication
pCK1	pSH71	5.5	Cm Km	High	−	Yes	10

licate in a large number of gram-positive bacteria, (including all species of lactic acid bacteria, a number of bacilli, several streptococcal species, clostridia, and *S. aureus*) and in *E. coli*. This property has greatly facilitated the genetic research of lactococci, since it directly coupled this research to the advanced genetic manipulation protocols available for *B. subtilis* and *E. coli*. Several derivatives of pWV01 and pSH71 are available (Table 1); a map of the commonly used pWV01-derived plasmid pGK12 is given in Fig. 1. Genes conferring Cm^r, Em^r, resistance to kanamycin (Km^r), and resistance to tetracycline (Tc^r) have been used to genetically mark these plasmids. Combinations of these genes offer the possibility of cloning pieces of DNA by using marker inactivation to screen for recombinant plasmids. Various MCS sequences have also been introduced to expand vector potential in cloning experiments. Plasmid replication in *E. coli* was exploited by introducing the mp10 or mp18 α-*lacZ* fragment in pWV01 derivatives, thus enabling the selection of inserts in *E. coli* by α complementation. A recent development is the construction of a pWV01 derivative that contains the *oriT* sequence of the broad-host-range gram-negative conjugal plasmid RK2 from *E. coli* (33). This pWV01 derivative can be mobilized from an *E. coli* donor strain which provides the necessary conjugation functions in *trans* to both gram-neg-

ative and gram-positive recipients. In this way, the number of strains into which the vector can be introduced can possibly be extended since the necessity of a DNA uptake mechanism can be bypassed.

Special-Purpose Vectors

Promoter- and terminator-screening vectors. The successful exploitation of recombinant DNA techniques in future strain improvement programs will depend on knowledge of signals that promote gene expression. To study these sequences in lactococci, vectors that enable their random isolation have been developed (Table 2). In most of these vectors, the promoterless chloramphenicol acetyltransferase (*cat*) gene of either *Bacillus pumilus* (*cat*-86) or *S. aureus* (*cat*-194) was used as the reporter gene (1, 2a, 6, 30, 31). Fragments of DNA with promoter activity cloned upstream of these reporters confer Cm^r and can easily be selected. In pNZ336, a pSH71 derivative, the promoterless gene for phospho-β-galactosidase (*lacG*) of *L. lactis* NCDO 712 was used (24a). This highly expressed homologous gene may prove more useful than the heterologous *cat* genes, which express only low levels of Cm^r in lactococci. Nevertheless, the *cat* plasmids have proved useful, and a number of chromosomal and phage promoters have been analyzed. In general,

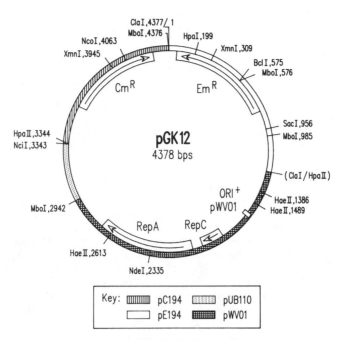

FIG. 1. Detailed restriction enzyme map of pGK12 (12). The map is based on the known sequence and shows the origin of the DNA used in construction of this plasmid.

TABLE 2. Special-purpose vectors

Vector	Replicon	Size (kb)	Probe gene[a]	Antibiotic resistance gene(s)	Reference
Promoter/terminator screening					
pGKV210	pWV01	4.5	cat-86	Em	30
pBV5030	pWV01	4.3	*cat-86	Em	2a
pMU1327	pVA380-1 pBR322	7.5	cat-194	Em	1
pNZ336	pSH71	6.9	lacG	Cm Km	24a
pGKV259	pWV01	5.0	cat-86	Em Cm	30
Signal sequence selection					
pGA14	pWV01	5.6	α-Amy	Em	26
pGB14	pWV01	5.9	β-Lac	Em	26
Integration					
pHV60A/B	pBR322	6.8		Cm	17
pKL10A/B	pBR322	3.6		Em	18
pKL203A/B	pUB110	6.5		Em Cm	18
pKL400B	pTB19	7.7		Em	18
pKL301B	pSC101	9.6		Em Cm Km	18
pE194/φ	pE194	−		Em	3
Gene expression					
pMG36	pWV01	3.7		Km	29
pMG36e	pWV01	3.6		Em	29
pNZ337	pSH71	7.3		Cm Km	Simons et al., in press

[a]cat-86 and cat-194, Chloramphenicol acetyltransferase genes from *B. pumilus* and *S. aureus*, respectively; *cat-86, constitutively expressed version of cat-86; lacG, phospho-β-galactosidase gene from *L. lactis*; α-Amy, α-amylase gene from *B. licheniformis*; β-Lac, TEM β-lactamase gene from *E. coli*.

they conformed to the consensus vegetative promoter sequences determined for *E. coli* and *B. subtilis*, with a −35 sequence TTGACA, a −10 sequence TATAAT, and a spacing of 15 to 18 bp (7, 16, 31). Some of the promoter-bearing fragments contained the start of an open reading frame. The functionality of the putative ribosome binding sites present upstream of these open reading frames was verified by making in-frame fusions to the eighth codon of the *lacZ* gene of *E. coli* (7, 31). Derivatives of the pWV01-based promoter screening vectors in which promoters were introduced were used to show that the terminator of the *B. licheniformis* penicillinase gene and that of the proteinase gene of *L. lactis* Wg2 functioned in *B. subtilis* and *L. lactis* (31; J. M. B. M. van der Vossen, Ph.D. thesis, University of Groningen, Groningen, The Netherlands, 1988).

Gene expression vectors. Knowledge of the sequences promoting transcription and translation of genes in lactococci led to the construction of vectors for expression of homologous and heterologous genes in lactococci. The expression signals and the first nine codons of the proteinase genes of both *L. lactis* SK11 and Wg2 were used to make in-frame fusions with the *E. coli lacZ* gene. The gene was functionally expressed in *L. lactis*, and the recombinant strains were able to grow on lactose as the sole energy source (8). The

same promoter has been used to express bovine prochymosin in *L. lactis* (25). Two expression vectors, pMG36 and pMG36e, were constructed by using the origin of replication of pWV01 and the strong lactococcal promoter P32 (29; Fig. 2). The MCS from pUC18 was placed between this promoter and the transcription terminator derived from the *L. lactis* Wg2 proteinase gene. Both in-frame and out-of-frame fusions can be made with the open reading frame present in the P32 sequence. With these plasmids, genes of prokaryotic as well as eukaryotic origin were expressed in lactococci, *B. subtilis*, and *E. coli*: the lysozyme genes from hen egg white and from the *E. coli* bacteriophages lambda and T4, the *lacZ* gene from *E. coli*, and the *B. subtilis* neutral proteinase gene (28, 29; M. van de Guchte, personal communication). The presence of two different antibiotic resistance genes on the vectors has proven useful in further cloning experiments. P32 is flanked by unique restriction sites and can easily be replaced by other promoter fragments. P32 has also been used to overexpress the two genes involved in the production of active proteinase in lactococci (van der Vossen, Ph.D. thesis, 1988). Promoter P59 from the same lactococcal strain has been used to increase the expression of bacteriocin genes of *L. lactis* subsp. *cremoris* 9B4 in *L. lactis* (M. van Belkum, personal communication).

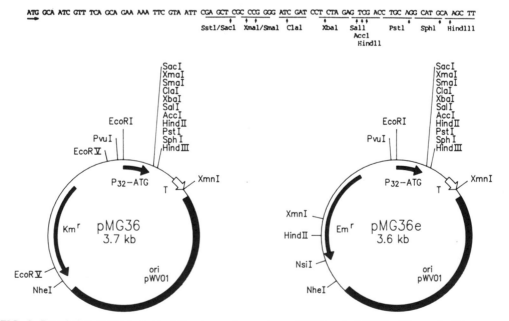

FIG. 2. Restriction enzyme maps of the expression vectors pMG36 and pMG36e. The nucleotide sequence around the MCS is given to show the start of the open reading frame present on the lactococcal promoter fragment P32 (31). Abbreviations: T, terminator of the proteinase gene of *L. lactis* Wg2 (11); P, P32; Km^r and Em^r, genes conferring resistance to kanamycin (from pPJ1) and erythromycin (from pE194), respectively. (Adapted from reference 29.)

Export signal selection vectors. The availability of large-scale fermentation technology and the status of lactococci as GRAS ("generally regarded as safe") organisms give the lactococci potential for the future synthesis of foreign gene products on an industrial scale. To examine the possibilities of protein secretion by lactococci, two approaches have been used. By looking for lactococcal secreted proteins, signal sequences effecting secretion were identified (11, 27, 32). The signal sequence present at the extreme N terminus of the *L. lactis* SK11 proteinase was used to effect bovine prochymosin secretion in *L. lactis* (25). A different approach used a pWV01-based plasmid carrying either the α-amylase gene from *Bacillus licheniformis* or the β-lactamase gene from *E. coli*. The secretion of both proteins was prevented by replacing the promoter and signal sequence codons with DNA constituting an MCS (26; Table 2). Cloning in the correct reading frame of DNA fragments carrying a promoter, ribosome binding site, start codon, and signal sequence was expected to result in synthesis and secretion of these easily assayable enzymes. Several fragments from the chromosome of *L. lactis* effecting the secretion of these reporter proteins in *E. coli* were selected and subsequently transferred to both *B. subtilis* and *L. lactis*. A selection of the isolated signals was subjected to nucleotide sequence analysis, primer extension to locate the start site of transcription, in vitro transcription and translation, and in vivo labeling to determine the size of the precursor proteins (G. Pérez Martínez, personal communication). A heterologous signal sequence has also been shown to function in *L. lactis*; the neutral proteinase from *B. subtilis* appeared to be correctly processed to the active extracellular enzyme in *L. lactis* with the aid of the signal sequence of the neutral proteinase itself (28).

Integration vectors. As noted above, plasmid instability in lactococci is the basis for rapid loss of several traits important in milk fermentation. Several groups have investigated chromosomal integration as a means to genetically stabilize these fermentation functions (3, 17). In addition to stabilization, such strategies might also facilitate gene amplification or effect random or specific mutations in the lactococcal chromosome. All methods that have been successful thus far use plasmids (from both gram-positive and gram-negative organisms) that do not replicate in lactococci and are endowed with an antibiotic resistance gene that can be selected for in *L. lactis*. In these plasmids, DNA from the chromosome or from a temperate bacteriophage provided homology with the lactococcal chromosome. The Cm^r gene from *S. aureus* used in several of the vectors immediately gave rise to amplification of the integrated structure as a result of poor expres-

sion of the gene in lactococci (17). The amplification was lost upon growth under nonselective conditions, but one copy of the plasmid was stably maintained. In an extensive survey of the effects of different replicons and antibiotic resistance genes on the integration events, Leenhouts et al. were able to produce stable single and multicopy integrations via a Campbell-like mechanism (18). Using one of the integration vectors, the proteinase genes from *L. lactis* Wg2 were also stably integrated in single and multiple gene copies in the chromosome of *L. lactis* (K. Leenhouts, Ph.D. thesis, University of Groningen, Groningen, The Netherlands, 1990). Integration via a double-crossover mechanism was also observed (3), and recently a cloned chromosomally located peptidase gene was used to examine the parameters for this type of integration (Leenhouts, Ph.D. thesis, 1990). It was also shown that it is possible to isolate integrants with a nonselectable phenotype by using a replicating plasmid for the initial selection of lactococcal transformants.

Conclusions

The last 5 years have seen considerable progress in development of the tools for genetic manipulation of lactococci. Easily selectable, small plasmid vectors have been developed for general cloning purposes. These vectors were optimized by inserting MCSs and by introducing genes suitable for cloning with marker inactivation. The nucleotide sequences of most of the vectors are known, thus facilitating the analyses of cloning experiments. The development and use of promoter- and terminator-screening vectors and of vectors that select for signal sequences have resulted in the construction of expression and secretion vectors for lactococci. An increasing number of homologous and heterologous genes are being (over)expressed in *L. lactis* with use of these plasmids. An important observation concerns the fact that these lactococcal plasmids also replicate in various species of the genus *Streptococcus* and in *B. subtilis* and *E. coli*, thereby directly coupling the advanced genetic engineering protocols available in the latter two organisms to the genetic research of lactococci, streptococci, and enterococci. The development of a strategy to stably integrate genetic information in the chromosome of *L. lactis* led to the construction of a strain in which proteinase gene stability was significantly improved, an observation of eminent importance to future strain improvement programs. The vectors constructed so far constitute what can be called the first generation of cloning vehicles. These vectors are extremely useful for fundamental research. The next challenge will be to construct a second generation of vectors that can be used in industrial strains to improve these strains or endow them with new properties: the so-called food-grade vectors. All of the materials needed are at hand, and we can be confident that the first examples of such vectors will become available in the near future.

Maps of some of the plasmids mentioned in this paper are given in Appendix 4 of this volume.

LITERATURE CITED

1. **Achen, M. G., B. E. Davidson, and A. J. Hillier.** 1986. Construction of plasmid vectors for the detection of streptococcal promoters. *Gene* **45:**45–49.

2. **Behnke, D., M. S. Gilmore, and J. Ferretti.** 1981. Plasmid pGB301, a new multiple resistance streptococcal cloning vehicle and its use in cloning of the gentamycin/kanamycin resistance determinant. *Mol. Gen. Genet.* **182:**414–421.

2a. **Bojovic, B., G. Djordjevic, and L. Topisirovic.** 1991. Improved vector for promoter screening in lactococci. *Appl. Environ. Microbiol.* **57:**385–388.

3. **Chopin, M.-C., A. Chopin, A. Rouault, and N. Galleron.** 1989. Insertion and amplification of foreign genes in the *Lactococcus lactis* subsp. *lactis* chromosome. *Appl. Environ. Microbiol.* **55:**1769–1774.

4. **Dao, M. L., and J. J. Ferretti.** 1985. *Streptococcus-Escherichia coli* shuttle vector pSA3 and its use in the cloning of streptococcal genes. *Appl. Environ. Microbiol.* **49:**115–119.

5. **Davies, F. L., and M. J. Gasson.** 1981. Reviews of the progress of dairy science: genetics of lactic acid bacteria. *J. Dairy Res.* **48:**363–376.

6. **de Vos, W. M.** 1986. Gene cloning in lactic streptococci. *Neth. Milk Dairy J.* **40:**141–154.

7. **de Vos, W. M.** 1987. Gene cloning and expression in lactic streptococci. *FEMS Microbiol. Rev.* **46:**281–295.

8. **de Vos, W. M., and G. Simons.** 1988. Molecular cloning of lactose genes in dairy lactic streptococci: the phospho-β-galactosidase and β-galactosidase genes and their expression products. *Biochimie* **70:**461–473.

9. **Gasson, M. J.** 1983. Genetic transfer systems in lactic acid bacteria. *Antonie van Leeuwenhoek J. Microbiol. Serol.* **49:**275–282.

10. **Gasson, M. J., and P. H. Anderson.** 1985. High copy number plasmid vectors for use in lactic streptococci. *FEMS Microbiol. Lett.* **30:**193–196.

11. **Kok, J., K. J. Leenhouts, A. J. Haandrikman, A. M. Ledeboer, and G. Venema.** 1988. Nucleotide sequence of the cell wall proteinase gene of *Streptococcus cremoris* Wg2. *Appl. Environ. Microbiol.* **54:**231–238.

12. **Kok, J., J. M. B. M. van der Vossen, and G. Venema.** 1984. Construction of plasmid cloning vectors for lactic streptococci which also replicate in *Bacillus subtilis* and *Escherichia coli*. *Appl. Environ. Microbiol.* **48:**726–731.

13. **Kondo, J. K., and L. L. McKay.** 1982. Transformation of *Streptococcus lactis* protoplasts by plasmid DNA. *Appl. Environ. Microbiol.* **43:**1213–1215.

14. **Kondo, J. K., and L. L. McKay.** 1984. Plasmid transformation of *Streptococcus lactis* protoplasts: optimization and use in molecular cloning. *Appl. Environ. Microbiol.* **48:**252–259.

15. **Kondo, J. K., and L. L. McKay.** 1985. Gene transfer systems and molecular cloning in group N streptococci: a review. *J. Dairy Sci.* **68:**2143–2159.

16. **Lakshmidevi, G., B. E. Davidson, and A. J. Hillier.** 1990. Molecular characterization of promoters of the *Lactococcus lactis* subsp. *cremoris* temperate bacteriophage BK5-T and identification of a phage gene implicated in the regulation of promoter activity. *Appl. Environ. Microbiol.* **56:**934–942.

17. **Leenhouts, K. J., J. Kok, and G. Venema.** 1989. Campbell-like integration of heterologous plasmid DNA into the chromosome of *Lactococcus lactis* subsp. *lactis*. *Appl. Environ. Microbiol.* **55:**394–400.

18. **Leenhouts, K. J., J. Kok, and G. Venema.** 1990. Stability of integrated plasmids in the chromosome of *Lactococcus lactis*. *Appl. Environ. Microbiol.* **56:**2726–2735.

19. **Macrina, F. L., J. A. Tobian, K. R. Jones, R. P. Evans, and D. B. Clewell.** 1982. A cloning vector able to replicate in *Escherichia coli* and *Streptococcus sanguis*. *Gene* **19:**345–353.

20. **McKay, L. L.** 1983. Functional properties of plasmids in lactic streptococci. *Antonie van Leeuwenhoek J. Microbiol. Serol.* **49:**259–274.

21. **Romero, D. A., P. Slos, C. Robert, I. Castellino, and A. Mercenier.** 1987. Conjugative mobilization as an alternative vector delivery system for lactic streptococci. *Appl. Environ. Microbiol.* **53:**2405–2413.

22. **Schleifer, K. H., and R. Kilpper-Bälz.** 1987. Molecular and chemotaxonomic approaches to the classification of streptococci, enterococci and lactococci: a review. *Syst. Appl. Microbiol.* **10:**1–19.

23. **Simon, D., and A. Chopin.** 1988. Construction of a vector plasmid family and its use for molecular cloning in *Streptococcus lactis*. *Biochimie* **70:**559–566.

24. **Simon, D., A. Rouault, and M.-C. Chopin.** 1986. High-efficiency transformation of *Streptococcus lactis* protoplasts by plasmid DNA. *Appl. Environ. Microbiol.* **52:**394–395.

24a. **Simons, G., H. Buys, R. Hogers, E. Koenhen, and W. M. de Vos.** 1990. Construction of a promoter probe vector for lactic acid bacteria using the *lacG* gene of *Lactococcus lactis*. *Dev. Ind. Microbiol.* **31:**31–39.

25. **Simons, G., G. Rutten, M. Hornes, and W. M. de Vos.** 1988. Production of bovine prochymosin by lactic acid bacteria, p. 183–187. *In* H. Breteler, P. H. van Lelieveld, and K. C. A. M. Luyben (ed.), *Proceedings of the 2nd Netherlands Biotechnology Congress*. Netherlands Biotechnological Society, Zeist, The Netherlands.

26. **Smith, H., S. Bron, J. van Ee, and G. Venema.** 1987. Construction and use of signal sequence selection vectors in *Escherichia coli* and *Bacillus subtilis*. *J. Bacteriol.* **169:**3321–3328.

27. **van Asseldonk, M., G. Rutten, M. Oteman, R. J. Siezen, W. M. de Vos, and G. Simons.** 1990. Cloning of *usp45*, a gene encoding a secreted protein from *L. lactis*. *Gene* **95:**155–160.

28. **van de Guchte, M., J. Kodde, J. M B. M. van der Vossen, J. Kok, and G. Venema.** 1990. Heterologous gene expression in *Lactococcus lactis* subsp. *lactis*: synthesis, secretion, and processing of the *Bacillus subtilis* neutral proteinase. *Appl. Environ. Microbiol.* **56:**2606–2611.

29. **van de Guchte, M., J. M. B. M. van der Vossen, J. Kok, and G. Venema.** 1989. Construction of a lactococcal expression vector: expression of hen egg white lysozyme in *Lactococcus lactis* subsp. *lactis*. *Appl. Environ. Microbiol.* **55:**224–228.

30. **van der Vossen, J. M. B. M., J. Kok, and G. Venema.** 1985. Construction of cloning, promoter-screening, and terminator-screening shuttle vectors for *Bacillus subtilis* and *Streptococcus lactis*. *Appl. Environ. Microbiol.* **50:**540–542.

31. **van der Vossen, J. M. B. M., D. van der Lelie, and G. Venema.** 1987. Isolation and characterization of *Streptococcus cremoris* Wg2-specific promoters. *Appl. Environ. Microbiol.* **53:**2452–2457.

32. **Vos, P., G. Simons, R. J. Siezen, and W. M. de Vos.** 1989. Primary structure and organization of the gene for a procaryotic, cell envelope-located serine proteinase. *J. Biol. Chem.* **264:**13579–13585.

33. **Williams, R. D., D. I. Young, and M. Young.** 1990. Conjugative plasmid transfer from *Escherichia coli* to *Clostridium acetobutylicum*. *J. Gen. Microbiol.* **136:**819–826.

Physical and Genetic Mapping of the *Lactococcus lactis* Chromosome

BARRIE E. DAVIDSON,[1] LLOYD R. FINCH,[1] DEBRA L. TULLOCH,[1] ROXANNA LLANOS,[1]
AND ALAN J. HILLIER[2]

*Russell Grimwade School of Biochemistry, University of Melbourne, Parkville, Victoria 3052,[1] and
Commonwealth Scientific and Industrial Research Organization, Division of Food Processing, Dairy Research
Laboratory, Highett, Victoria 3190,[2] Australia*

Lactic acid bacteria are used extensively for the manufacture of cheese, fermented milks, and other processed dairy products. *Lactococcus lactis* subspp. *cremoris* and *lactis* are the major starter organisms used in cheese making, particularly for Cheddar types of cheese. A broad range of different strains of these bacteria has evolved during the many years of cheese making. These strains can differ in a number of industrially important properties such as rate of acid production, susceptibility to different phage types, ability to coexist with other cheese starter strains, and flavor- and texture-conferring properties. Since there is always a need for strain improvement, considerable effort has been devoted in recent years toward using advanced techniques of molecular genetics to achieve controlled improvement in strain performance (25).

The Lactococcal Genome

The lactococcal genome, like those of many other eubacteria, comprises both plasmids and a chromosome. Our knowledge of the molecular genetics of the plasmids is much further advanced than that of the chromosome for a number of reasons. First, the smaller size of plasmids compared with that of the chromosome has made them more accessible to analysis by the available technology. Second, development of genetic transfer mechanisms that are suitable for the analysis of the lactococcal chromosome has been limited. The absence of suitable generalized transducing phages, Hfr-like strains, or comparable genetic tools has meant that there is no genetic map of *L. lactis*. Third, plasmids have been observed to encode essential genetic determinants for a number of industrially important properties, thereby stimulating interest in them for practical reasons. Thus, a number of different plasmids that confer a Lac$^+$ phenotype have been characterized, and the nucleotide sequence of *lacG*, the gene encoding phospho-β-galactosidase in these plasmids, has been reported (3, 8). Plasmids and genes encoding proteinases have also been analyzed at the molecular level (18, 31), and much progress has been made in elucidating the properties of some plasmids that confer reduced bacteriophage sensitivity on their host (21).

The available information about the lactococcal chromosome can easily be summarized. Jarvis and Jarvis (13) determined the genome sizes of *L. lactis* subsp. *cremoris* AM$_1$, *L. lactis* subsp. *lactis* ML$_3$, and *L. lactis* subsp. *lactis* ML$_8$ to be 3.1, 2.85, and 2.75 Mbp, respectively. Even though plasmids larger than 100 kbp have been observed in the lactococci (11), the majority of this DNA must be chromosomal. The chromosomal DNA, like that of the streptococci and lactobacilli, has a relatively high A+T content of around 64% (16). Reports on the cloning of chromosomal DNA are limited to one on the cloning and sequencing of five promoters from *L. lactis* subsp. *cremoris* Wg2 (29) and a second on the cloning of a gene that complements *lacZ* in *Escherichia coli* (24). Promoters have also been cloned from *L. lactis* subsp. *lactis* C2, and their sequences and start points of transcription in *L. lactis* subsp. *lactis* have been determined (M. G. Achen, Ph.D. thesis, University of Melbourne, Melbourne, Australia, 1986).

At the Second International American Society for Microbiology Conference on Streptococcal Genetics in 1986, Fitzgerald et al. described experiments using transposon mutagenesis as a tool for advancing our knowledge of the lactococcal chromosome (10). The recent development of techniques for the preparation and endonucleolytic digestion of intact chromosomes (26) and for the separation of large DNA molecules by pulsed-field gel electrophoresis (PFGE) (6) provides another powerful method for the study of the lactococcal chromosome. Given that chromosomally borne determinants also contribute to the industrial usefulness of the lactococci (e.g., the genes for the majority of the enzymes involved in the conversion of lactose to lactic acid, genes encoding protein-processing enzymes, and perhaps genes encoding phage receptors), we decided to use PFGE to investigate the molecular and genetic organization of the lactococcal chromosome. At this stage we have constructed a physical map of the chromosome, estimated the number of rRNA operons that it contains, and located the operons on the map (28b). Preliminary data from PFGE of some lactic acid bacteria

103

TABLE 1. Digestion products of high-molecular-weight DNA from *L. lactis* subsp. *lactis* LM0230

Enzyme	Fragment sizes from gel[a] (kbp)
*Not*I	2,000, 225, 180
*Sfi*I	1,900, 510
*Sma*I	540, 360, 320, 220, 2 × 182, 134, 128, 69, 49, 2 × 46, 43, 35, 2 × 26, 20, 18, 12

[a]Determined by PFGE in a Bio-Rad CHEF-DRII apparatus over a range of ramped pulse times.

has also been reported by Le Bourgeois et al. (20). The different PFGE *Sma*I digestion patterns yielded by different *L. lactis* strains form the basis of a new and powerful technique for strain identification (28a).

Physical Map of *L. lactis* subsp. *lactis* LM0230

We started our map construction with a commonly used, plasmid-cured laboratory strain (*L. lactis* subsp. *lactis* LM0230 [9]) in order to avoid any possible complicating effects of plasmid molecules on the interpretation of PFGE separation patterns. High-molecular-weight DNA was prepared (26) and digested with a range of restriction endonucleases, and the digestion products were separated by using the contour-clamped homogeneous electric field (CHEF) mode of PFGE. The sizes of the products of *Not*I, *Sfi*I, and *Sma*I digestion are shown in Table 1. Double digestions with *Not*I and *Sfi*I enabled the map of the LM0230 chromosome for these two enzymes to be readily deduced (Fig. 1). Extension of the map to include the *Sma*I sites was made difficult because of the large difference between the numbers of *Not*I (or *Sfi*I) sites and *Sma*I sites.

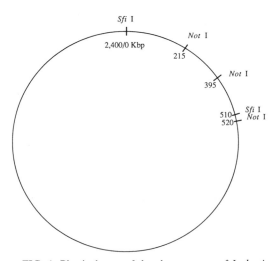

FIG. 1. Physical map of the chromosome of *L. lactis* subsp. *lactis* LM0230.

Physical Map of *L. lactis* subsp. *lactis* DL11

To circumvent the aforementioned problem, digestion patterns of a number of different strains were examined to identify any strain(s) that had a more satisfactory distribution of digestion sites for the three endonucleases. The strain chosen, *L. lactis* subsp. *lactis* DL11 (a Prt⁻ derivative of ATCC 11454 [19]), had 6 sites each for *Not*I and *Sfi*I and 21 for *Sma*I (Table 2). The fact that this strain was not plasmid-free did not cause major problems in the development of a map. We constructed a map of the *Not*I and *Sma*I sites in the DL11 chromosome (Fig. 2A) by using a number of approaches. First, we used two-dimensional separations of reciprocal *Not*I and *Sma*I digests (2). In these experiments, high-molecular-weight DNA was digested with one enzyme, the digestion products were separated by PFGE through ultrapure agarose, and the lane containing the separated fragments was excised from the gel, redigested with the other enzyme, and electrophoresed again at right angles to the direction of the first electrophoresis separation. Second, we used two-dimensional separations of *Not*I-partial *Sma*I digests. Third, to map all of the *Sma*I sites in two of the *Not*I fragments (NtA and NtE), it was necessary to use another enzyme, *Sal*I. We purified NtA and NtE by PFGE, excised the region of the gel containing them, and subjected them separately to two-dimensional separations of reciprocal *Sma*I and *Sal*I digests as described above. These experiments also indicated the location of many of the *Sal*I sites in both NtA and NtE. Fourth, to ascertain the location on the map of the smallest *Sma*I fragment (SmO), we made use of the observation that SmO hybridized with a probe composed of rDNA from the bacterium *Mycoplasma mycoides*. We did not attempt to map the *Sfi*I sites in *L. lactis* subsp. *lactis* DL11, because the digestion patterns that we obtained with this enzyme showed a small but significant number of minor bands that would have complicated the interpretation of two-dimensional separations. These bands may have arisen as a consequence of different rates of digestion by *Sfi*I at sites with different nucleotides in the redundant recognition sequence GGCCNNNNNGGCC.

TABLE 2. Digestion products of high-molecular-weight DNA from *L. lactis* subsp. *lactis* DL11

Enzyme	Fragments from digestion	
	No.	Size from gel[a] (kbp)
*Not*I	6	750, 600, 450, 310, 260, 160
*Sma*I	21	810, 260, 3 × 135, 130, 2 × 125, 105, 100, 2 × 85, 75, 3 × 50, 45, 27, 22, 4.0[b], 3.5[b]

[a]Determined by PFGE in a Bio-Rad CHEF-DRII apparatus over a range of ramped pulse times.
[b]From fixed-field gel electrophoresis.

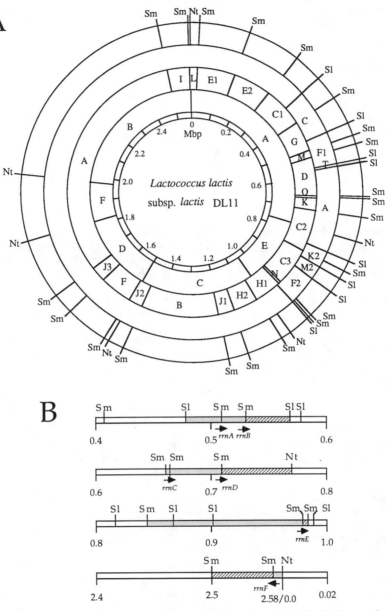

FIG. 2. (A) Physical map of the chromosome of *L. lactis* subsp. *lactis* DL11. Radiating out from the center, the five annuli show the scale (in megabase pairs) and digestion sites for *Not*I (Nt), *Sma*I (Sm), *Sal*I (Sl), and all three enzymes, respectively. Only the *Sal*I sites located between map units 0.345 and 0.990 are shown. The *Not*I site between NtA and NtB has been arbitrarily defined as map unit 0.0. (B) Map of the putative rRNA operons of *L. lactis* subsp. *lactis* DL11. Fragments that hybridized with both the pMC5 probe and a 16S rDNA probe (polymerase chain reaction product) are stippled, and those that hybridized only with the pMC5 probe are shown by diagonal lines. In each case, only the minimum-size fragment that hybridized is marked. The locations of putative rRNA operons (*rrnA* to *rrnF*) are shown by arrows pointing in the direction of transcription of the operon.

The map of the *L. lactis* subsp. *lactis* DL11 chromosome is circular, indicating that the chromosome itself is circular. The sum of the sizes of the DNA fragments that comprise the map, 2.58 Mbp, gives a reliable value for the size of the chromosome. As might be expected, this value is a little smaller than the values for total genome size that were obtained by Jarvis and Jarvis using renaturation kinetics (13). The size of the *L. lactis* subsp. *lactis* DL11 chromosome is around the

TABLE 3. Sizes of chromosomes and numbers of rRNA operons in eubacteria

Organism	Size[a] (Mbp)	No. of rRNA operons (N)	Size/N	Reference
Pseudomonas aeruginosa	5.9	4	1.48	12, 23
Bacillus subtilis	4.7	10	0.47	28, 30
Escherichia coli	4.55	7	0.65	17, 27
Clostridium perfringens	3.58	9	0.44	5
Lactococcus lactis subsp. *lactis*	2.58	6	0.43	This work
Haemophilus parainfluenzae	2.34			14
Haemophilus influenzae	1.98			15
Mycoplasma mycoides	1.2	2	0.60	1, 22
Ureaplasma urealyticum	0.9	2	0.45	7

[a] Only sizes determined by physical mapping are included.

middle of the range of sizes for bacterial chromosomes that have been determined by physical mapping techniques (Table 3). Its smaller size by comparison with the chromosomes of *Pseudomonas aeruginosa*, *Bacillus subtilis*, and *E. coli* is consistent with the notion that its relatively fastidious growth requirements are due to the absence of the appropriate genes.

Localization of rRNA Operons

To commence the construction of a genetic map of *L. lactis* subsp. *lactis*, we decided to determine the location of rRNA operons on the physical map by using Southern blots of PFGE separations. In the absence of molecular clones of *L. lactis* subsp. *lactis* rRNA genes, we made probes of *Mycoplasma capricolum* ribosomal DNA (Fig. 3). In PFGE separations, the larger (operon) probe hybridized with six, five, four, six, three, and five bands in digests by *Apa*I, *Bgl*I, *Bss*HII, *Sac*II, *Sal*I, and *Xba*I, respectively (Fig. 4). More than six bands in *Sma*I digestion patterns hybridized with this probe, and the intensity of hybridization varied considerably (Fig. 4). Double digests with *Apa*I plus *Sal*I, *Bgl*I plus *Sal*I, and *Bgl*I plus *Bss*HII each yielded six bands that hybridized with the probe. These data were taken to indicate that there are six rRNA operons in the *L. lactis* subsp. *lactis* DL11 chromosome and that each operon contains a *Sma*I site. Additional evidence favoring this interpretation was obtained with the smaller (16S) probe, which hybridized equally with only six *Sma*I fragments and

with six *Eco*RI fragments (not shown). On the basis of genome size, the rRNA content of *L. lactis* subsp. *lactis* DL11 is comparable to that of other eubacteria (Table 3).

Direction of Transcription of rRNA Operons

Since the 16S probe was near one end of the rRNA operon (Fig. 3), the differential hybridization patterns obtained with the two probes enabled the orientation of the rRNA operons to be deduced (Fig. 2B). We concluded that the five operons that are clustered in the 20% of the chromosome between map units 0.5 and 1.0 are transcribed codirectionally and that the sixth operon is transcribed in the opposite direction. Brewer (4) has suggested that for *E. coli* there is a selective advantage in the transcription and replication of rRNA operons being codirectional. If a similar form of selection has operated in *L. lactis* subsp. *lactis*, then these data lead to the prediction that the origin of replication of the *L. lactis* subsp. *lactis* DL11 chromosome is located in the 0.5 Mbp of DNA between *rrn*F and *rrn*A (Fig. 2A).

Future Prospects

Two aspects of this work that can be expected to progress rapidly in the immediate future are the location of specific genetic loci on the map and an increase in the resolution of the map by the inclusion of sites for other enzymes. We are currently mapping the sites for the initiation and termination of chromosome replication and for

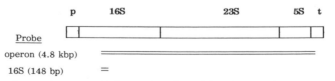

FIG. 3. Details of probes used to locate *L. lactis* subsp. *lactis* DL11 rRNA operons. The map shows the typical organization of a eubacterial rRNA operon (12) showing the promoter (p) and the terminator (t). The regions included in the so-called operon and 16S probes are shown by the double lines.

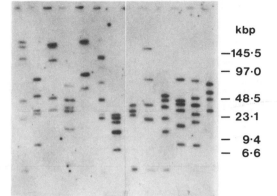

kbp

—145·5

— 97·0

— 48·5

— 23·1

— 9·4
— 6·6

FIG. 4. Southern blot hybridizations of a PFGE separation of digests of *L. lactis* subsp. *lactis* DL11 DNA, using pMC5 as a probe. Restriction endonuclease digests of high-molecular-weight DL11 DNA in agarose were separated by PFGE (pulse time ramped from 1 to 20 s), and a Southern blot was hybridized with ^{32}P-labeled DNA from the insert in pMC5 (1). The lanes contain the following digests: 1, *Apa*I; 2, *Bgl*I; 3, *Bss*HII; 4, *Sac*II; 5, *Sal*I; 6, *Sma*I; 7, *Xba*I; 8, *Bgl*I and *Apa*I; 9, *Bss*HII and *Apa*I; 10, *Apa*I and *Sal*I; 11, *Bss*HII and *Bgl*I; 12, *Bgl*I and *Sal*I; 13, *Sal*I and *Bss*HII. Numbers on the side are the sizes of molecular weight markers.

genes that encode enzymes involved in lactic acid production. By using polymerase chain reaction amplification, we have localized genes encoding lactate dehydrogenase (*ldh*), pyruvate kinase (*pyk*), phosphofructokinase (*pfk*), and glyceraldehyde-3-phosphate dehydrogenase (*gpd*). In addition to these aspects of the study of the lactococcal chromosome, an understanding of the molecular bases for the differences in restriction digestion patterns of different *L. lactis* strains and the evolution of these differences would be of fundamental and practical interest. With the rapid development in techniques for mapping and sequencing DNA, we are clearly at the beginning of a period of rapid expansion in our knowledge and understanding of the structure and organization of the lactococcal chromosome.

This work was supported by grants from the Australian Research Council and the Australian Dairy Research and Development Council.

We are grateful to P. Arnold for the synthesis of the oligonucleotides, to J. Whitley for advice, and to E. Tanskanen for assistance and advice.

LITERATURE CITED

1. **Amikan, D., S. Razin, and G. Glaser.** 1982. Ribosomal RNA genes in Mycoplasma. *Nucleic Acids Res.* **10**:4215–4222.
2. **Bautsch, W.** 1988. Rapid physical mapping of the *Mycoplasma mobile* genome by two-dimensional field inversion gel electrophoresis techniques. *Nucleic Acids Res.* **16**:11461–11467.
3. **Boizet, B., D. Villeval, P. Slos, M. Novel, G. Novel, and A. Mercenier.** 1988. Isolation and structural analysis of the phospho-β-galactosidase gene from *Streptococcus lactis* Z268. *Gene* **62**:246–261.
4. **Brewer, B.** 1988. When polymerases collide: replication and the transcriptional organization of the E. coli chromosome. *Cell* **53**:679–686.
5. **Canard, B., and S. T. Cole.** 1989. Genome organization of the anaerobic pathogen *Clostridium perfringens*. *Proc. Natl. Acad. Sci. USA* **86**:6676–6680.
6. **Chu, G., D. Vollrath, and R. W. Davis.** 1986. Separation of large DNA molecules by contour-clamped homogeneous electric fields. *Science* **234**:1582–1585.
7. **Cocks, B. G., L. E. Pyle, and L. R. Finch.** 1989. A physical map of the genome of *Ureaplasma urealyticum* 960^T with ribosomal loci. *Nucleic Acids Res.* **17**:6713–6719.
8. **de Vos, W. M., and M. J. Gasson.** 1989. Structure and expression of the *Lactococcus lactis* gene for phospho-β-galactosidase (*lacG*) in *Escherichia coli* and *L. lactis*. *J. Gen. Microbiol.* **135**:1833–1846.
9. **Efstathiou, J. D., and L. L. McKay.** 1977. Inorganic salts resistance associated with a lactose-fermenting plasmid in *Streptococcus lactis*. *J. Bacteriol.* **130**:257–265.
10. **Fitzgerald, G. F., C. Hill, E. Vaughan, and C. Daly.** 1987. Tn*919* in lactic streptococci, p. 238–241. *In* J. J. Ferretti and R. Curtiss III (ed.), *Streptococcal Genetics*. American Society for Microbiology, Washington, D.C.
11. **Harmon, K. S., and L. L. McKay.** 1987. Restriction enzyme analysis of lactose and bacteriocin plasmids from *Streptococcus lactis* subsp. *diacetylactis* WM$_4$ and cloning of *Bcl*I fragments coding for bacteriocin production. *Appl. Environ. Microbiol.* **53**:1171–1174.
12. **Hartmann, R. K., H. Y. Toschka, N. Ulbrich, and V. A. Erdmann.** 1986. Genome organization of rDNA in *Pseudomonas aeruginosa*. *FEBS Lett.* **195**:187–193.
13. **Jarvis, A. W., and B. D. W. Jarvis.** 1981. Deoxyribonucleic acid homology among lactic streptococci. *Appl. Environ. Microbiol.* **41**:77–83.
14. **Kauc, L., and S. H. Goodgal.** 1989. The size and a physical map of the chromosome of *Haemophilus parainfluenzae*. *Gene* **83**:377–380.
15. **Kauc, L., M. Mitchell, and S. H. Goodgal.** 1989. Size and physical map of the chromosome of *Haemophilus influenzae*. *J. Bacteriol.* **171**:2474–2479.
16. **Kilpper-Bälz, R., G. Fischer, and K. H. Schleifer.** 1982. Nucleic acid hybridization of group N and group D streptococci. *Curr. Microbiol.* **7**:245–250.
17. **Kiss, A., B. Sain, and P. Venetianer.** 1977. The number of rRNA genes in *Escherichia coli*. *FEBS Lett.* **79**:77–79.
18. **Kok, J., K. J. Leenhouts, A. J. Haandrikman, A. M. Ledeboer, and G. Venema.** 1988. Nucleotide sequence of the cell wall proteinase gene of *Streptococcus cremoris* Wg2. *Appl. Environ. Microbiol.* **54**:231–238.
19. **LeBlanc, D. J., V. L. Crow, L. N. Lee, and C. F. Garon.** 1979. Influence of the lactose plasmid on the metabolism of galactose by *Streptococcus lactis*. *J. Bacteriol.* **137**:878–884.
20. **Le Bourgeois, P., M. Mata, and P. Ritzenthaler.** 1989. Genome comparison of *Lactococcus* strains by pulsed-field gel electrophoresis. *FEMS Microbiol. Lett.* **59**:65–70.
21. **McKay, L. L., M. J. Bohanon, K. M. Polzin, P. L. Rule, and K. A. Baldwin.** 1989. Localization of separate genetic loci for reduced sensitivity towards small isometric-headed bacteriophage sk1 and prolate-headed bacteriophage c2 on pGBK17 from *Lactococcus lactis* subsp. *lactis* KR2. *Appl. Environ. Microbiol.* **55**:2702–2709.
22. **Pyle, L., and L. R. Finch.** 1988. A physical map of the genome of *Mycoplasma mycoides* subspecies *mycoides* Y with some functional loci. *Nucleic Acids Res.* **16**:6027–6039.
23. **Römling U., D. Grothues, W. Bautsch, and B. Tümmler.** 1989. A physical genome map of *Pseudomonas aeruginosa* PAO. *EMBO J.* **8**:4081–4089.
24. **Ross, R., F. O'Gara, and S. Condon.** 1989. Cloning of chromosomal genes of *Lactococcus* by heterologous com-

plementation: partial characterization of a putative lactose transport gene. *FEMS Microbiol. Lett.* **61**:183–188.

25. **Sandine, W. E.** 1987. Looking backward and forward at the practical applications of genetic researches on lactic acid bacteria. *FEMS Microbiol. Rev.* **46**:205–220.

26. **Smith, C. L., and C. R. Cantor.** 1987. Purification, specific fragmentation, and separation of large DNA molecules. *Methods Enzymol.* **155**:449–467.

27. **Smith, C. L., J. G. Econome, A. Schutt, S. Klco, and C. R. Cantor.** 1987. A physical map of the *Escherichia coli* K12 genome. *Science* **236**:1448–1453.

28. **Stewart, G. C., F. E. Wilson, and K. F. Bott.** 1982. Detailed physical mapping of the ribosomal RNA genes of *Bacillus subtilis*. *Gene* **19**:153–162.

28a. **Tanskanen, E. I., D. L. Tulloch, A. J. Hillier, and B. E. Davidson.** 1990. Pulsed-field gel electrophoresis of *Sma*I digests of lactococcal genomic DNA, a novel method of strain identification. *Appl. Environ. Microbiol.* **56**:3105–3111.

28b. **Tulloch, D. L., L. R. Finch, A. J. Hillier, and B. E. Davidson.** 1991. Physical map of the chromosome of *Lactococcus lactis* subsp. *lactis* DL11 and localization of six putative rRNA operons. *J. Bacteriol.* **173**:2768–2775.

29. **van der Vossen, J. M. B. M., D. van der Lelie, and G. Venema.** 1987. Isolation and characterization of *Streptococcus cremoris* Wg2-specific promoters. *Appl. Environ. Microbiol.* **53**:2452–2457.

30. **Ventra, L., and A. S. Weiss.** 1989. Transposon-mediated restriction mapping of the *Bacillus subtilis* chromosome. *Gene* **78**:29–36.

31. **Vos, P., G. Simons, R. J. Siezen, and W. M. de Vos.** 1989. Primary structure and organization of the gene for a prokaryotic, cell envelope-located serine proteinase. *J. Biol. Chem.* **264**:13579–13585.

Structure, Expression, and Evolution of the Nisin Gene Locus in *Lactococcus lactis*

MARK STEEN AND J. NORMAN HANSEN

Department of Chemistry and Biochemistry, University of Maryland, College Park, Maryland 20742

Nisin is a gene-encoded peptide antibiotic that is produced by many strains of *Lactococcus lactis*. The structural gene of the nisin precursor peptide has recently been cloned, which has permitted a partial characterization of the locus of the nisin gene. The nisin gene appears to be encoded as part of a polycistronic operon with a promoter at least 2.5 kb upstream from the nisin gene. There are open reading frames (ORFs) on both sides of the nisin gene; the upstream ORF has a strong homology to the transposase gene of the *Escherichia coli* IS2 insertion element, and the downstream ORF consists of more than 500 amino acids. The presence of a transposase-like gene suggests that the nisin locus is, or at one time was, part of an insertional element such as a transposon. Sequence homologies with other lanthionine-containing antibiotics indicate a common, though evolutionarily distant, origin. It seems likely that a complete characterization of the nisin operon will reveal a variety of genes which encode proteins that carry out the posttranslational processing (dehydrations, cross-linking, cleavage of leader sequence, and export) that must occur to give the mature, secreted antibiotic. In contrast to expectation from earlier reports, the nisin gene is located in the chromosome of *L. lactis* ATCC 11454, as established by orthogonal-field agarose gel electrophoresis. The nisin gene was located on a *Not*I-generated restriction fragment with a size of 625 kb, which is sufficiently large to be unambiguously of chromosomal origin.

Nisin is one of the longest-known antibiotics, having been observed in the 1930s (19) and studied since the 1940s (14). Since nisin is a product of milk-fermenting *L. lactis*, it has long played an important role in dairy fermentations (10), and it is now recognized as an effective but nontoxic natural food preservative. Nisin is an antibiotic peptide that is active against closely related bacterial strains that are nonproducers of nisin, so it is appropriate to refer to nisin as a bacteriocin. An important advance in our understanding of nisin has been the cloning and characterization of the gene that encodes its structure (5). Although the unusual amino acids (lanthionine, β-methyllanthionine, dehydroalanine, and dehydrobutyrine) that constitute 13 of its 34 amino acid residues suggest a nonribosomal mechanism of nisin biosynthesis, the cloned gene sequence demonstrates that the unusual amino acids are formed by posttranslational processing of serines, threonines, and cysteines. This is accomplished by steps involving dehydration, Michael-type addition of cysteine across the double bond of either dehydroalanine or dehydrobutyrine, and cleavage of a leader sequence (5). The cloning of genes for other lanthionine-containing antibiotics such as subtilin (2) and epidermin (1) has established that this mechanism of antibiotic synthesis is not unique, but dispersed among widely disparate species of gram-positive bacteria.

Unless posttranslational processing of the nisin precursor peptide occurs spontaneously and without assistance from maturation proteins, which is unlikely, the producing organism has to have genes for proteins to carry out the various processing steps as well as genes to confer nisin resistance. It is commonly observed that genes associated with biosynthesis of a particular antibiotic are clustered together, often as an operon (8). It would not be surprising if the genes for nisin resistance and processing were all part of an operon. Indeed, numerous reports have demonstrated that the genetic information required for nisin biosynthesis can be transferred among nisin-producing and -nonproducing strains by conjugation (6, 7, 15, 16, 18). Moreover, nisin resistance and nisin production have been associated with particular plasmids (6, 7, 15).

The availability of the cloned structural gene for nisin allows a direct determination of gene location, rather than the indirect inferences that can be made from genetic experiments. It also allows the direct observation of gene expression and polycistronic gene organization by means of characterization of transcripts.

Evolutionary Relationships among Cloned Lanthionine Antibiotic Genes

Figure 1 shows the gene and amino acid sequences for the precursor peptides for nisin, subtilin, and epidermin that were characterized in *L. lactis* ATCC 11454 (5), *Bacillus subtilis* ATCC 6633 (2), and *Staphylococcus epidermidis* Tu3298 (1), respectively. Although there are extensive structural homologies among these antibiotics (5), they are highly divergent at both the amino acid and nucleic acid sequence levels. Indeed, the

Amino Acid Homologies

```
          (Leader Region)          |       (Structural Region)
     MSTK DFNLDLVSVSKKDSGASPR|ITSISLCTPGCKTGALMGCNMKTATCHCSIHVSK Nisin
          --KFD--D--V-K---Q--KIT-Q|WK-E------B-V----QT-FLQ-L--N-K-   -- Subtilin
MEAVKEKNDLFNLDVKVNAKESN----E--|-A-KFI-----AKTGSFNSYCC            Epidermin
```

Nucleic Acid Homologies

```
Nisin:      Leader Region-->  ATGAGT...ACAAAAGATTTTAACTTGGATTTGGTATCTGTT
Subtilin:                     ---TCAAAGTTCG-T-----CG-T------G-T--GAAA--C
Epidermin:  ATGGAAGCAGTAAAAGAA-AA-A-GATCTTTTTA--C--G-TG-TA-AG-TAATG-AAAA


Nisin:      TCGAAGAAAGATTCAGGTGCATCACCACGC  <--Leader Region
Subtilin:   --T--AC----C---AAAATCA-T--G-AA   Structural Region-->
Epidermin:  GAATCT--C--------A--TGA----A-A


Nisin:      ATTACAAGTATTTCGCTATGTACACCCGGTTGTAAAACAGGAGCTCTGATGGGTTGTAACATGAAA
Subtilin:   TGG-A----GAA--A--T--------A--A---GT---T--T--AT--CAAAC---CTT-C-TC--
Epidermin:  ---G-T----AA-TTA-------T--T-A---GC--A-AC--G-AGTT-TAACA--T-TTGTTGT


Nisin:      ACAGCAACTTGTCATTGTAGTATTCACGTAAGC
Subtilin:   ---CT--C----A-C-GC-AA-TC......TCT
```

Per Cent Homology Among Regions

	NL		NS		SL		SS	
	NA	AA	NA	AA	NA	AA	NA	AA
SL	50	54						
SS			51	56				
EL	42	25			34	20		
ES			48	32			41	23

FIG. 1. Homologies of nisin amino acid and nucleic acid sequences to subtilin and epidermin. Top shows amino acid homologies among nisin, subtilin, and epidermin. A single-residue gap in the nisin leader and a two-residue gap in the subtilin structural region are inserted to improve homology. Middle shows homologies of the nisin, subtilin, and epidermin genes, using the same gaps as above. Sequences of the nisin (2), subtilin (2), and epidermin (5) genes were determined previously. The values for percent homology between amino acid sequences (AA) and nucleic acid sequences (NA) are calculated between the nisin leader (NL), nisin structural region (NS), and the corresponding regions for subtilin (SL, SS) and epidermin (EL, ES). (Data from reference 5.)

silent positions in the codons for the conserved amino acids appear to be random, indicating that these genes have been evolving separately for a long time, and the sequence homologies that remain are the consequence of conservation of essential amino acid functions.

Characteristics of mRNA Transcripts Produced from the Nisin Gene

Information about the mRNA transcripts is important for several reasons. The time at which mRNA transcripts appear constitutes the onset of nisin expression. The sizes and termini of these transcripts also provide crucial information about the organization of the nisin gene within a polycistronic operon. Depending on the size of the transcript and the rate at which it is processed (or degraded), one can use the transcripts to define the operon. S1 mapping data that suggest that the nisin gene is transcribed as a polycistronic mRNA (Fig. 2). This mRNA transcript contained RNA

processing signals that resulted in its being cleaved at a site about 25 nucleotides upstream from the ribosome binding site and about 40 nucleotides downstream from the termination codon. This cleavage yielded a processed mRNA transcript of about 267 nucleotides in length; stability measurements showed that this transcript has a half-life of about 7 to 10 min (5). The existence of a larger polycistronic mRNA was further confirmed by primer extension, in which we used an oligomeric sequence from within the nisin gene to prime reverse transcriptase copying of total cellular mRNA. The longest cDNA product had a size of 2.5 kb, from which we conclude that the nisin gene is transcribed from a promoter that is at least 2.5 kb upstream from the nisin gene. Since the mRNA template may be partially degraded, the actual promoter may be even farther upstream than this.

The time at which the nisin gene is expressed has also been determined. Although Hurst (9)

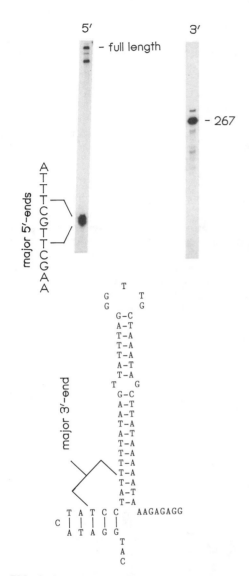

FIG. 2. S1 mapping of 5' and 3' ends of the nisin mRNA. The 5' and 3' ends of transcripts that contain sequences encompassing the nisin gene were mapped by protection against S1 nuclease in the RNA-DNA hybrid between the nisin mRNA and its cDNA. The sequence that contains two major cutting sites that are in close proximity to the nisin gene sequence is shown. The position marked as full length corresponds to a protected fragment that would extend to the end of the cDNA, or a size of about 1,000 nucleotides. The 3' lane shows the S1 protection pattern from which the location of the 3' end was determined and is shown on the sequence of the inverted repeat. (Data from reference 5.)

has shown that nisin is produced mainly during late growth phases, we have found that the nisin gene is expressed (i.e., transcribed) throughout exponential phase and during stationary phase (5). Although it is possible that the early tran-

scripts are not translated, a more likely possibility is that the nisin precursor peptide accumulates until late growth phase, whereupon one or more maturation steps convert it to mature antibiotic. This is supported by observations by Hurst and Peterson (11) that nisin precursor forms are present in the cell envelope. Although there is as yet no direct experimental evidence, an attractive hypothesis is that the leader region of the nisin precursor peptide directs the precursor to a site in the cell envelope, such as the membrane, to await the appearance of one or more critical processing proteins that carry out the final processing steps that lead to release of active antibiotic.

The Nisin Gene Is Located in the Chromosome of *L. lactis* ATCC 11454

About 15 years ago, it was observed that nisin-producing strains of *Streptococcus lactis* that had been cured of plasmids lost their ability to make nisin (13). Since then, many papers have reported a link between nisin production and occurrence of particular plasmids (see above), although conclusive proof of a plasmid origin of nisin production has been elusive (e.g., references 9 and 16). Now that the nisin gene is available, a demonstration of the location of the gene either on the plasmid DNA or on the chromosome should be straightforward. However, nisin-producing strains of *L. lactis* typically contain many plasmids, some of which are quite large. In our initial experiments to determine plasmid location, the nisin probe hybridized with total DNA isolated from strain ATCC 11454 having a size that corresponded to the upper fractionation limit of the agarose gel that we were using, which was about 25 kb. This could not discriminate between the signal being toward chromosomal DNA or a large plasmid. Kaletta and Entian (12) attempted to resolve this question by fractionating chromosomal from plasmid DNA on CsCl gradients and concluded that the nisin gene was of plasmid origin in *L. lactis* 6F3. We attempted to repeat this experiment on strain ATCC 11454 and found that both the plasmid fraction and the linear DNA fraction hybridized to the nisin probe. Considering that one cannot be certain how very large plasmids might band in CsCl and that we could not rule out cross-contamination of bands, we concluded that CsCl fractionation did not lead to a definitive determination of the nisin gene location for this strain.

One unambiguous approach is the use of orthogonal-field gel electrophoresis, which can resolve DNA restriction fragments as large as 10,000 kb (4, 17). Using this technique, one can distinguish between large chromosomal DNA fragments and large plasmids. A nisin gene probe was hybridized against undigested total cellular DNA and the same DNA that had been digested with the infrequently cutting *Not*I restriction enzyme (Fig. 3). Under orthogonal-field conditions, covalently closed plasmid circles migrate anomalously and should remain in the wells (3). We noted that there was no hybridization signal with material in the wells in the undigested DNA lane, but there was a single strong hybridization signal at the position of undigested chromosomal DNA. In the *Not*I lane, there was a single hybridization signal associated with a band corresponding to a size of 625 kb. The position of this band shifted considerably with respect to the undigested lane, demonstrating that cleavage by the enzyme had occurred, that the digestion product was a linear form of DNA, and that the position of the band reflected its actual size. Any plasmids that do not possess a *Not*I site should still remain in the wells, whereas those that do possess a *Not*I site should be cleaved and have migrated far down or off the gel. These results are fully consistent with the nisin gene being of chromosomal origin. The only way that these results can be consistent with a

plasmid origin is to invoke the idea that the hybridization signal in the *Not*I lane is toward a *Not*I fragment of a megabase-size plasmid, which is unprecedented as far as we know. The location of the nisin gene on the *Not*I fragment and the restriction map in the immediate vicinity of the nisin gene were determined (Fig. 4). Summing the sizes of the *Not*I restriction fragments predicts a genome size of about 2,495 kb for strain 11454. This is consistent with restriction patterns from other enzymes (data not shown).

ORFs Flanking the Nisin Gene

Primer extension experiments (see above) established that the nisin gene is part of a polycistronic operon with a promoter at least 2.5 kb upstream from the nisin gene. Inspection of the nucleic acid sequences that flank the nisin gene suggests that the downstream region is also part of the transcriptional unit in that there is a downstream ORF with its own ribosome binding site but no promoter, indicating that it is transcribed

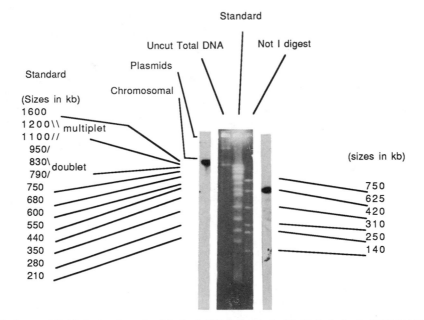

FIG. 3. Orthogonal-field electrophoresis of *L. lactis* DNA digested with *Not*I. *L. lactis* ATCC 11454 DNA was isolated from cells immobilized in agarose plugs, digested with *Not*I restriction enzyme, and electrophoresed on an agarose gel, using a Bio-Rad CHEF-DRII orthogonal-field apparatus. The middle three lanes are a photograph of the stained gel, containing uncut total cellular DNA, a yeast chromosomal DNA standard, and a *Not*I digest of total cellular DNA, respectively. The far left lane is the hybridization pattern of the uncut DNA, hybridized with a labeled synthetic 20-mer oligonucleotide that is complementary to the nisin gene. Hybridization occurs only with the large species marked "Chromosomal." Covalently closed circular plasmid DNA does not migrate significantly under these conditions (3), and the expected locations of uncut plasmids are indicated. The far right lane shows the hybridization pattern of total cellular DNA after treatment with *Not*I restriction enzyme. Total uncut cellular DNA is in the left lane. Fragment sizes on the left are of the yeast standard, and those on the right are the estimated sizes of the *Not*I fragments. The shift in the size of the hybridizable band proves that the DNA contained a restriction site, and the digested hybridizable fragment is a linear form of DNA with a size of 625 kb.

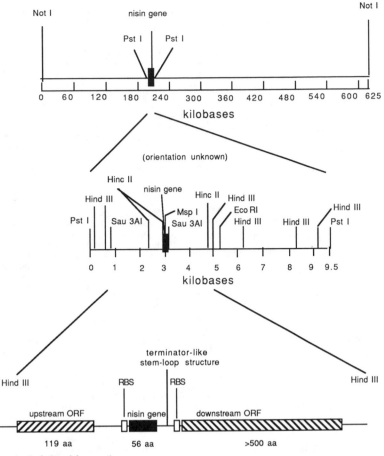

FIG. 4. Organization of the nisin gene locus within the *Not*I restriction fragment. (Top) Location of the nisin gene within the 625-kb *Not*I fragment; (middle) a restriction map of the nisin gene within the 9.5-kb *Pst*I fragment; (bottom) organization of ORFs.

by readthrough from the nisin gene (5). We have carried out sequence analysis within this ORF and found that it encodes a protein of more than 500 amino acids, with a size of 75 to 80 kDa, or more. We have not yet been able to associate this sequence with a function. The complete upstream ORF has not yet been obtained, because we have sequenced to the end of the cloned DNA fragment on which it lies. This partial ORF possesses a strong homology to the transposase of the *E. coli* IS*2* insertion element (5). The presence of this transposase homology leads one to suspect that the nisin gene is, or at one time was, associated with an insertional genetic element such as a transposon. If this is true, it provides a mechanism by which the nisin-producing trait was originally disseminated among various lactococci, and some ancestral form of this transposonlike element could have been the origin of all the known lanthionine antibiotics.

Implications for Future Applications of Nisin and Other Lanthionine Antibiotics

Ribosomally synthesized peptide antibiotics that contain unusual amino acids are not common, and accordingly they possess unique advantages in comparison with other antibiotic systems. An important aspect of drug design involves synthesis of structural analogs of known antibiotics to improve their properties. Since most common antibiotics are small organic molecules that are synthesized by multistep enzyme pathways, the only practical approach to making analogs is by chemical synthesis. An alternative exists for antibiotics whose structures are gene encoded in that they can be modified by mutagenesis, which is much simpler. Whereas nature may provide only a few such antibiotics, the number of structural variations that can be made by using mutagenesis is enormous. There is a caveat in that the precursors of the structural variants will have to be recognized by the posttranslational processing system in order for the appropriate maturation steps to be made. The question then becomes one of understanding the processing signals that are present in the precursor peptide sequence so that they can be appropriately incorporated into the analogs. Another feature of this system is that the posttranslational processing that converts ordinary amino acids into unusual ones has the effect of sidestepping the limitations generally imposed by the genetic code, allowing the introduction of amino acids with novel properties into what would otherwise be ordinary proteins. This has the effect of expanding the repertoire of chemical functionalities that can be engineered into proteins, which could prove to be highly useful. Nisin and its biosynthesis seem to provide an appropriate model system with which to explore these and other possibilities.

This project was supported by Public Health Service grant RO1-AI24454, by a grant from the National Dairy Promotion and Research Board, and by Applied Microbiology, Inc., New York.

We thank Sue Mischke, Biocontrol Laboratory, U.S. Department of Agriculture, Beltsville, Md., for use of the CHEF apparatus.

LITERATURE CITED

1. **Allgaier, H., G. Jung, R. G. Werner, U. Schneider, and H. Zahner.** 1986. Epidermin: sequencing of a heterodetic tetracyclic 21-peptide amide antibiotic. *Eur. J. Biochem.* **160:**9–22.

2. **Banerjee, S., and J. N. Hansen.** 1988. Structure and expression of a gene encoding the precursor of subtilin, a small protein antibiotic. *J. Biol. Chem.* **263:**9508–9514.

3. **Beverly, S. M.** 1988. Characterization of the 'unusual' mobility of large circular DNAs in pulsed field-gradient electrophoresis. *Nucleic Acids Res.* **16:**925–939.

4. **Birren, B. W., E. Lai, S. M. Clark, L. Hood, and M. I. Simon.** 1988. Optimized conditions for pulsed field gel electrophoretic separations of DNA. *Nucleic Acids Res.* **16:**7563–7582.

5. **Buchman, G. W., S. Banerjee, and J. N. Hansen.** 1988. Structure, expression, and evolution of a gene encoding the precursor of nisin, a small protein antibiotic. *J. Biol. Chem.* **263:**16260–16266.

6. **Gasson, M. J.** 1984. Transfer of sucrose fermenting ability, nisin resistance and nisin production in *Streptococcus lactis* 712. *FEMS Microbiol. Lett.* **21:**7–10.

7. **Gonzales, C. F., and B. S. Kunka.** 1985. Transfer of sucrose-fermenting ability and nisin production phenotype among lactic streptococci. *Appl. Environ. Microbiol.* **49:**627–633.

8. **Hopwood, D. A., F. Malpartida, H. M. Kieser, H. Ikeda, J. Duncan, I. Fujii, B. A. M. Rudd, and H. G. Floss.** 1985. Prediction of "hybrid" antibiotics by genetic engineering. *Nature* (London) **314:**642–644.

9. **Hurst, A.** 1966. Biosynthesis of the antibiotic nisin and other basic peptides by *Streptococcus lactis* grown in batch culture. *J. Gen. Microbiol.* **45:**503–513.

10. **Hurst, A.** 1981. Nisin. *Adv. Appl. Microbiol.* **27:**85–123.

11. **Hurst, A., and G. M. Peterson.** 1971. Observations on the conversion of an inactive precursor protein to the antibiotic nisin. *Can. J. Microbiol.* **17:**1379–1384.

12. **Kaletta, C., and K. D. Entian.** 1989. Nisin, a peptide antibiotic: cloning and sequencing of the NisA gene and posttranslational processing of its peptide product. *J. Bacteriol.* **171:**1597–1601.

13. **Kozak, W., M. Rajchert-Trzpil, and W. T. Dobrzanski.** 1974. The effect of proflavin, ethidium bromide and an elevated temperature on the appearance of nisin-negative clones in nisin-producing strains of *Streptococcus lactis*. *J. Gen. Microbiol.* **83:**295–302.

14. **Mattick, A. T. R., and A. Hirsch.** 1944. A powerful inhibitory substance produced by group N streptococci. *Nature* (London) **154:**551.

15. **McKay, L. L., and K. A. Baldwin.** 1984. Conjugative 40-megadalton plasmid in *Streptococcus lactis* subsp. *diacetylactis* DRC3 is associated with resistance to nisin and bacteriophage. *Appl. Environ. Microbiol.* **47:**68–74.

16. **McKay, L. L., K. A. Baldwin, and P. M. Walsh.** 1980. Conjugal transfer of genetic information in group N streptococci. *Appl. Environ. Microbiol.* **40:**84–91.

17. **Schwartz, D. C., and C. R. Cantor.** 1984. Separation of yeast chromosome-sized DNAs by pulsed field gradient gel electrophoresis. *Cell* **37:**67–75.

18. **Steele, J. L., and L. L. McKay.** 1986. Partial characterization of the genetic basis for sucrose metabolism and nisin production in *Streptococcus lactis*. *Appl. Environ. Microbiol.* **51:**57–64.

19. **Whitehead, H. R.** 1933. A substance inhibiting bacterial growth produced by certain strains of lactic streptococci. *Biochem. J.* **27:**1793–1800.

Production, Processing, and Engineering of the *Lactococcus lactis* SK11 Proteinase

WILLEM M. DE VOS, INGRID BOERRIGTER, PIETER VOS, PAUL BRUINENBERG,
AND ROLAND J. SIEZEN

Molecular Genetics Group, Department of Biophysical Chemistry, Netherlands Institute for Dairy Research, NIZO, 6710 BA Ede, The Netherlands

Lactococci that are used in industrial dairy fermentations contain an active proteolytic system for the degradation of caseins, the major milk proteins, into small peptides and free amino acids that support cell growth (23). This cascade of proteolytic reactions is initiated by a cell envelope-located serine proteinase of approximately 135 kDa (10, 11). The active proteinase is essential for growth in milk; moreover, it produces peptides that contribute to the development of flavor in fermented milk products (24). Despite their functional similarities, a remarkable genetic and immunological heterogeneity has been found between the proteinases of different lactococcal strains. In all natural *Lactococcus lactis* strains, proteinase production is encoded by plasmids that may differ largely in size and genetic organization (3). In addition, poly- and monoclonal antibodies have been used to group proteinases of different strains (15, 19, 23). A functional classification into two main groups has been proposed on the basis of the differences in caseinolytic specificity: the type I proteinase that predominantly degrades β-casein and is found in the vast majority of lactococcal strains (12), and the type III proteinase that degrades α_{s1}-, β-, and κ-casein (25). Model studies have shown that the type III proteinase produces fewer bitter-tasting peptides from casein than does the type I proteinase (26).

The structural proteinase (*prtP*) genes of representatives of both groups, the type I proteinase from strains Wg2 (17, 18) and NCDO 763 (16) and the type III proteinase of strain SK11 (8, 27), have been cloned and sequenced. Expression studies in different *L. lactis* strains showed that the unique caseinolytic specificity is determined by the SK11 *prtP* gene and its product but not by the lactococcal host (8). In addition, proteinase production in the industrial strain SK11 (4) appears to be regulated as a function of the medium composition. Therefore, we have used the nonbitter SK11 strain as a model for analyzing the organization, structure, and function of the proteinase genes and their products.

Organization of the SK11 Proteinase Genes

Proteinase production in *L. lactis* subsp. *cremoris* SK11 is encoded by a 78-kb plasmid, pSK111 (4, 7). A region of approximately 10 kb encoding proteinase production has been cloned, sequenced, and expressed in *Escherichia coli* and *L. lactis* (8, 27, 28; Fig. 1). Expression and deletion studies have shown that the region contains two oppositely oriented genes, *prtP* and *prtM*, that are both required for the production of an active proteinase (27, 28). The *prtP* gene is the structural proteinase gene and encodes a 206-kDa serine proteinase precursor that is synthesized as a preproproteinase (27; see below). The *prtM* gene codes for a 33-kDa *trans*-acting lipoprotein that is involved in the maturation of the proteinase precursor into an active enzyme (28). An identical protein is also required for activation of the type I proteinase from strain Wg2 (13). Between the *prtM* and *prtP* genes is located a highly A + T-rich 0.3-kb region that contains the overlapping promoters of the divergently transcribed *prt* genes (9, 27). Transcription initiation occurs at oppositely located nucleotides in the central part of this region, which shows an unusual rotational symmetry (9, 21).

The proteinase regulon is flanked by copies of an *iso*-ISS*1* element. Sequence comparison of those copies, designated ISS*1*-N1 (14) and ISS*1*-N2, showed that they differed only slightly from that of the originally described ISS*1* sequence (20); moreover, both contained intact open reading frames (ORFs) for the putative transposase (Fig. 2). These results suggest that the SK11 proteinase regulon is part of a composite transposon, which has been designated Tn*5277*. Although the functionality of this proteinase transposon has yet to be determined, the presence of the *iso*-ISS*1* copies may explain some of the instability and rearrangements that have been observed with metabolic plasmids in lactococci (14).

Production of the SK11 Proteinase

A variety of plasmids have been constructed that contain both the *prtP* and *prtM* genes and express a functional proteinase when introduced in lactococcal strains. One such plasmid, pNZ521, is based on the heterogramic, 3-kb plasmid vector pNZ122, which has a higher copy number in *L. lactis* than do all natural proteinase plasmids (8, 28). Upon introduction of pNZ521 into the Prt⁻

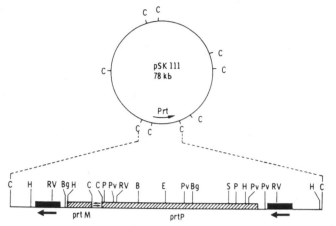

FIG. 1. Transposon-type organization of the SK11 *prt* regulon on plasmid pSK111. Hatched bars indicate coding sequences for the *prtP* and *prtM* genes (28). Arrows indicate the divergent promoters of both genes. The two *iso*-ISS*1* copies, ISS*1*-N1 (left) and ISS*1*-N2 (right), are indicated by the black bars. The orientation of the deduced transposase is indicated by the arrows. B, Bg, C, E, H, P, Pv, and RV indicate recognition sites for the restriction enzymes *Bam*HI, *Bgl*II, *Cla*I, *Eco*RI, *Hin*dIII, *Pst*I, *Pvu*I, and *Eco*RV, respectively.

Lac⁺ *L. lactis* strain MG1820 (5), we observed greater production of proteinase than with the natural lactose-proteinase plasmid pLP712 (8; Table 1). Interestingly, this overproduction resulted in a higher specific growth rate in milk (Table 1). Other plasmids (P. Bruinenberg et al., unpublished data) with different proteinase expression levels have been analyzed, and the results have confirmed that the amount of proteinase produced per cell determines the growth rate and acid formation in milk.

In the course of studying *prtP* gene expression, we observed that proteinase production in strain SK11 is regulated by the medium composition (W. M. de Vos, unpublished observations), in contrast to strain Wg2, in which proteinase synthesis seems to be constitutive. Comparison of the structures of the divergent proteinase gene promoters of both strains showed that the extensive dyad symmetry (see above) found in pSK111

is not present in the Wg2 proteinase promoter region (17) because of a 5-bp deletion in the latter sequence (23). This may indicate that this structure is involved in controlling the coordinate expression of the *prtP* and *prtM* genes as with other divergently transcribed systems (1). Preliminary experiments using pNZ521 and derivative plasmids have shown that induction is observed only in SK11 and not in MG1820 derivatives containing those plasmids. This finding indicates that there is a specific host function involved in the induction of proteinase production in strain SK11.

Structure, Location, and Processing of the SK11 Proteinase

A schematic representation of the primary structure of the 1,962-amino-acid serine proteinase as deduced from the nucleotide sequence of the *prtP* gene is shown in Fig. 3. The size of 135

```
ORF-N2  MNHFKGKQFKKDFIIVAVGYYLRYNLSYREVQKLLYDRGINVCHTTIYRWVQEYSKVLYD  60
ORF-N1      R       V        V          E

ORF-N2  LWKKKNRQSFYSWKMDETYIKIKGRWHYLYRAIDADGLTLDIWLRKKKRETQAAYAFLKRL  120
ORF-N1

ORF-N2  HKQFGEPKAIVTDKAPSLGSAFIKLQSVGLYTKTEHRTVKYLNNLIEQDHRPIKRRIKFY  180
ORF-N1

ORF-N2  QSLRTASSTIKGMETLRGIYKNNRRNGTLFGFSVSTEIKVLMGITA              226
ORF-N1                       K
```

FIG. 2. Alignment of deduced amino acid sequences of the putative transposase ORFs present on the *iso*-ISS*1* elements of the SK11 *prt* regulon (ORF-N1 from ISS*1*-N1 and ORF-N2 from ISS*1*-N2). Only nonidentical residues in ORF-N1 (14) are shown.

TABLE 1. Proteinase production and generation times in milk of *L. lactis* strains containing different or no proteinase plasmids

Strain	Proteinase plasmid	Proteinase production[a]	Generation time (min) in milk[b]
NCDO 4109	pLP712	+	78
MG1820	None	−	ND
MG1820	pNZ521	+ +	66

[a] Estimated by isolating the proteinase and separation by sodium dodecyl sulfate-polyacrylamide gel electrophoresis (8, 27). +, Normal proteinase production in wild-type strain NCDO 4109; + +, twofold overproduction compared with lactococcal cells harboring the wild-type proteinase plasmid pLP712.

[b] In milk plus 0.1% Casitone, all cells grew with a generation time of 60 min. ND, No detectable growth.

kDa observed with the purified SK11 proteinase (8) is not in agreement with its calculated molecular size of 206 kDa, and the proteinase is also much smaller than the expression product observed in an *E. coli* T7 expression system (27, 28). These results suggest that the primary translation product is processed to a smaller, active enzyme. N-terminal amino acid sequencing of the mature proteinase and gene fusion studies using *E. coli lacZ* (6), *Bacillus stearothermophilus* α-amylase (22), and bovine prochymosin (9) genes have shown that the proteinase is synthesized as a preproprotein with a 33-amino-acid signal sequence and a 154-amino-acid pro sequence (29). The N-terminal part of the SK11 proteinase shows significant homology to the serine proteinases of the subtilisin family (28). This homology is highest around the residues of the active-site triad, Asp, His, and Ser, and the oxyanion hole Asn, but it also includes the substrate-binding region of the subtilisinlike proteinases (28; Fig. 3). Site-directed mutagenesis experiments strongly suggest that these regions have a similar function in the lactococcal proteinase (see below). An important difference from most other members of the subtilisin family is the presence in the SK11 proteinase of a large, additional C-terminal domain of more than 1,200 residues. Recently, we observed (R. J. Siezen et al., unpub-

lished data) that this extra domain is partially homologous with a similar domain in the C5a serine proteinase of *Streptococcus pyogenes* (2). However, the function of this extra domain is unknown, with the exception of the 30 most C-terminal residues that are homologous to membrane anchor sequences identified in a great number of cell envelope-located proteins from other gram-positive bacteria (27). Deletion of this membrane anchor or introduction of a charged residue into the hydrophobic membrane-spanning domain of this anchor results in the secretion of the SK11 proteinase (8, 28; W. M. de Vos et al., unpublished observations).

The processing events required for the formation of an active but shortened proteinase have been investigated by using active-site mutant forms of the SK11 proteinase. An example is the mutant SK11 proteinase in which the putative active-site residue Ser is replaced by an Ala residue (S433A) and the membrane anchor has been deleted. As predicted from the homology with the subtilisin family of serine proteinases (Fig. 3), this mutant proteinase did not show any proteolytic activity (Fig. 4). Analysis of the extracellular proteins of a strain containing the mutant S433A proteinase shows the persistence of a 60-kDa secreted Usp protein, the product of the *usp45* gene (24). This Usp protein is a natural substrate for the SK11 proteinase and is degraded in strains containing secreted, active proteinases. The size of the S433A proteinase is in agreement with the expected size of a secreted, unprocessed *prtP* expression product and much larger than that of the active proteinases. This indicates that the processing of the lactococcal proteinase is an autocatalytic event (Fig. 4). Because the size difference between the unprocessed mutant proteinase (220 kDa) and active proteinases (135 kDa) is much larger than may be explained by removal of the 154-amino-acid residue pro sequence alone, C-terminal processing also has to occur. This conclusion is supported by the observation that C-terminal deletion of either 190 or 402 amino acids resulted in a secreted proteinase of similar size, 135 kDa (Fig. 4).

FIG. 3. Comparison of the SK11 proteinase and subtilisin. The dotted N-terminal segments represent the signal sequence. The dotted C-terminal segment of the lactococcal proteinase indicates the cell membrane anchor (29). Homologous active-site regions between the SK11 proteinase and subtilisin are shown in black; D, H, and S indicate the active-site aspartic acid, histidine, and serine, respectively (28, 29). B indicates homologous substrate-binding regions (horizontally striped). Vertical lines above the SK11 proteinase indicate differences between the proteinases of strains SK11 and Wg2 (29). The box indicates the position of the duplication in the SK11 proteinase that is absent in the Wg2 proteinase.

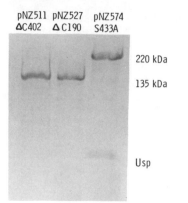

FIG. 4. Sizes of secreted engineered proteinases. Secreted proteinases were isolated from *L. lactis* strains
harboring plasmids pNZ511 (27) and pZN527 (encoding
SK11 proteinase with 402- and 190-amino-acid C-terminal deletions, respectively) and pNZ574, a derivative
of pNZ527 containing the S433A mutation. Proteinases
were subsequently analyzed by sodium dodecyl sulfate-
polyacrylamide gel electrophoresis (27).

In conclusion, these results show that the SK11
proteinase is processed in an autocatalytic process
at both the N and C termini. A unique feature
of this process is the involvement of the *trans*-
acting maturation protein, encoded by *prtM*.

Engineering of the SK11 Proteinase Caseinolytic Specificity

Despite the marked difference in caseinolytic
specificity, the type III SK11 proteinase differs
from type I Wg2 proteinase by only 44 amino acid
residues (27). In addition, the SK11 proteinase
contained a 60-residue duplication near the C terminus (Fig. 3). Since C-terminal deletions of
about 300 amino acids that include the duplication and 7 of the 44 amino acid substitutions do
not affect the caseinolytic specificity (3, 8), the
remaining 37 differences must be responsible for
the differences in casein degradation by both enzymes. To analyze the specific contributions of
these and other residues on the specificity, activity, and stability of the lactococcal proteinase, we
have constructed (i) a series of hybrid proteinases
between the SK11 and Wg2 proteinase (P. Vos,
I. Boerrigter, A. J. Haandrikman, M. Nijhuis,
M. B. deReuver, R. J. Siezen, G. Venema,
W. M. de Vos, and J. Kok, *Protein Eng.*, in press)
and (ii) a number of site-specific mutations in the
SK11 proteinase based on knowledge-based
modeling with serine proteinases of the subtilisin
family (P. Vos et al., submitted for publication).

The results show that the lactococcal proteinase
has an additional region involved in substrate
binding in comparison with the related subtilisins.
In addition, by using both approaches we were
able to create proteinases with novel caseinolytic

specificities. Finally, several residues have been
identified that contribute to the stability and specificity of the SK11 proteinase.

LITERATURE CITED

1. **Beck, C. F., and R. A. J. Warren.** 1988. Divergent promoters, a common form of gene regulation. *Microbiol. Rev.*
52:318–335.
2. **Chen, C., and P. P. Cleary.** 1990. Complete nucleotide
sequence of the streptococcal C5a peptidase gene from
Streptococcus pyogenes. J. Biol. Chem. **265:**3161–3167.
3. **de Vos, W. M.** 1987. Gene cloning and expression in lactic
streptococci. *FEMS Microbiol. Rev.* **46:**281–295.
4. **de Vos, W. M., and F. L. Davies.** 1984. Plasmid DNA in
lactic streptococci: bacteriophage resistance and proteinase
plasmids in *S. cremoris* SK11, p. 201–205. *In Third European Congress on Biotechnology*, vol. III. Verlag Chemie,
Weinheim, Federal Republic of Germany.
5. **de Vos, W. M., and M. J. Gasson.** 1989. Structure and
expression of the *Lactococcus lactis* gene for phospho-β-
galactosidase (*lacG*) in *Escherichia coli* and *L. lactis. J.
Gen. Microbiol.* **135:**1833–1846.
6. **de Vos, W. M., and G. Simons.** 1988. Molecular cloning
of lactose genes in dairy lactic streptococci: the phospho-
β-galactosidase and β-galactosidase genes and their expression products. *Biochimie* **70:**461–473.
7. **de Vos, W. M., H. U. Underwood, and F. L. Davies.** 1984.
Plasmid encoded bacteriophage resistance in *Streptococcus
cremoris* SK11. *FEMS Microbiol. Lett.* **23:**175–178.
8. **de Vos, W. M., P. Vos, H. de Haard, and I. Boerrigter.**
1989. Cloning and expression of the *Lactococcus lactis*
subsp. *cremoris* SK11 gene encoding an extracellular serine
proteinase. *Gene* **85:**169–176.
9. **de Vos, W. M., P. Vos, G. Simons, and S. David.** 1990.
Gene organization and expression in mesophilic lactic acid
bacteria. *J. Dairy Sci.* **72:**3398–3405.
10. **Exterkate, F. A., and G. J. C. M. de Veer.** 1985. Partial
isolation and degradation of caseins by cell wall protein-
ase(s) of *Streptococcus cremoris* HP. *Appl. Environ. Microbiol.* **49:**328–332.
11. **Geis, A., W. Bockelman, and M. Teuber.** 1985. Simultaneous extraction and purification of a cell-wall associated
peptidase and a β-casein specific proteinase from *Streptococcus cremoris* AC1. *Appl. Microbiol. Technol.* **23:**79–84.
12. **Geis, A., B. Kiefer, and M. Teuber.** 1986. Proteolytic activities of lactic streptococci isolated from dairy starter cultures. *Chem. Mikrobiol. Technol. Lebensm.* **10:**93–95.
13. **Haandrikman, A. J., J. Kok, H. Laan, S. Soemitro, A.
Ledeboer, W. N. Konings, and G. Venema.** 1989. Identification of a gene required for maturation of an extracellular
lactococcal serine proteinase. *J. Bacteriol.* **171:**2789–2794.
14. **Haandrikman, A. J., C. van Leeuwen, J. Kok, P. Vos,
W. M. de Vos, and G. Venema.** 1990. Insertion elements
on lactococcal plasmids. *Appl. Environ. Microbiol.*
56:1890–1896.
15. **Hugenholtz, J., F. A. Exterkate, and W. N. Konings.** 1984.
The proteolytic systems of *Streptococcus cremoris*: an immunological analysis. *Appl. Environ. Microbiol.* **53:**853–
859.
16. **Kiwaki, M., H. Ikemura, M. Shimuza-Kadota, and A.
Harashima.** 1989. Molecular characterization of a cell wall-
associated proteinase gene from *Lactococcus lactis* NCDO
763. *Mol. Microbiol.* **54:**239–369.
17. **Kok, J., C. J. Leenhouts, A. J. Haandrikman, A. M.
Ledeboer, and G. Venema.** 1988. Nucleotide sequence of
the gene for the cell wall bound proteinase of *Streptococcus
cremoris* Wg2. *Appl. Environ. Microbiol.* **54:**231–238.
18. **Kok, J., J. M. van Dijl, J. M. B. M. van der Vossen, and
G. Venema.** 1985. Cloning and expression of a *Streptococcus cremoris* proteinase in *Bacillus subtilis* and *Streptococcus lactis. Appl. Environ. Microbiol.* **50:**94–101.
19. **Laan, H., E. J. Smid, L. de Leij, E. Schwander, and
W. N. Konings.** 1988. Monoclonal antibodies to the cell

wall-associated proteinase of *Lactococcus lactis* subsp. *cremoris* Wg2. *Appl. Environ. Microbiol.* **54:**2250–2256.

20. **Polzin, K. M., and M. Shimizu-Kadota.** 1987. Identification of a new insertion element, similar to gram-negative IS*26*, on the lactose plasmid of *Streptococcus lactis* ML3. *J. Bacteriol.* **169:**5481–5488.

21. **Simons, G., H. Buys, R. Hogers, E. Koehnen, and W. M. de Vos.** 1990. Construction of a promoter-probe vector for lactic acid bacteria using the *lacG* gene of *Lactococcus lactis. Dev. Ind. Microbiol.* **31:**31–39.

22. **Simons, G., M. van Asseldonk, G. Rutten, M. Nijhuis, M. Hornes, and W. M. de Vos.** 1990. Analysis of secretion signals of Lactococci, p. 290–294. *In* C. Christianse, L. Munck, and J. Villadsen (ed.), *Proceedings of the Fifth European Congress on Biotechnology.* Munksgaard International Publisher, Copenhagen.

23. **Thomas, T. D., and G. G. Pritchard.** 1987. Proteolytic enzymes of dairy starter cultures. *FEMS Microbiol. Rev.* **46:**245–268.

24. **van Asseldonk, M., G. Rutten, M. Oteman, R. J. Siezen,** W. M. de Vos, and G. Simons. 1990. Cloning of *usp45*, a gene encoding a secreted protein from *Lactococcus lactis* subsp. *lactis* MG1363. *Gene* **95:**155–160.

25. **Visser, S., F. A. Exterkate, C. J. Slangen, and G. J. C. M. de Veer.** 1986. Comparative study of action of cell wall proteinases from various strains of *Streptococcus cremoris* on bovine α_{s1}-, β-, and κ-casein. *Appl. Environ. Microbiol.* **52:**1162–1166.

26. **Visser, S., G. Hup, F. A. Exterkate, and J. Stadhouders.** 1983. Bitter flavour in cheese. *Neth. Milk Diary J.* **29:**319–334.

27. **Vos, P., G. Simons, R. J. Siezen, and W. M. de Vos.** 1989. Primary structure and organization of the gene for a prokaryotic, cell-envelope-located serine proteinase. *J. Biol. Chem.* **264:**13579–13585.

28. **Vos, P., M. van Asseldonk, F. van Jeveren, R. J. Siezen, G. Simons, and W. M. de Vos.** 1989. A maturation protein is essential for production of active forms of *Lactococcus lactis* SK11 serine proteinase located or secreted from the cell envelope. *J. Bacteriol.* **171:**2795–2802.

Cloning and Nucleotide Sequence of the X-Prolyl Dipeptidyl Aminopeptidase Gene from *Lactococcus lactis*

MARIE-CHRISTINE CHOPIN,[1] JEAN-CLAUDE GRIPON,[2] AND ALAIN CHOPIN[1]

Laboratoire de Génétique Microbienne-Institut de Biotechnologie[1] and Station de Recherches Laitières,[2] Institut National de la Recherche Agronomique, 78352 Jouy-en-Josas, France

Lactococci used by the dairy industry have complex nutritional requirements, including free amino acids and small peptides (23, 24). These compounds are available in limited amounts in milk (19, 30), in which amino nitrogen is available mainly in the form of proteins (whey proteins and caseins). Therefore, lactococcal growth depends on proteolytic enzymes, which hydrolyze milk proteins to peptides and free amino acids. In addition, these proteolytic enzymes remain active after growth has ceased and contribute to cheese ripening by producing free amino acids that are cheese aroma precursors (17) and by degrading bitter peptides responsible for cheese defects (35).

The proteolytic system of lactococci is composed of cell wall proteinases that hydrolyze caseins to oligopeptides (7 to 19 amino acid residues) (20) and of peptidases that hydrolyze oligopeptides to small peptides and free amino acids. Only small peptides (up to five amino acid residues) are able to be translocated across the cell membrane (16, 24, 29, 31). Therefore, some peptidases must be extracellular or cell wall located, while others must be intracellular. Several peptidases have now been characterized, but their localization has not been clearly established.

Proteinases have been characterized in detail by biochemical (8, 21) and genetic (14) studies. More limited knowledge is available on the peptidases. Some have been isolated, and their biochemical properties have been determined (Table 1). Five aminopeptidases that release N-terminal amino acids from oligopeptides have been described. Two of them are able to hydrolyze a large number of substrates (22, 28). The three others are more specific. A glutamyl aminopeptidase releases only glutamic or aspartic acid (7). Two X-prolyl dipeptidyl aminopeptidases (X-PDAP) releasing dipeptides with the sequence X-Pro from the amino terminus of peptide chains have been isolated from *Lactococcus lactis* subsp. *lactis* (36) and *L. lactis* subsp. *cremoris* (12). In addition to these aminopeptidases, one tripeptidase (5) and two dipeptidases (9, 32) able to hydrolyze a large number of substrates have been characterized. A specific dipeptidase (prolidase) able to hydrolyze X-Pro dipeptides is produced by an *L. lactis*

subsp. *cremoris* strain (10; V. Monnet et al., personal communication). However, it is not yet known how many peptidases are synthesized, to what extent they participate in casein degradation and cheese ripening, where they are located in the cell, and how they are processed and regulated. Genetic studies, currently in progress in several laboratories, are needed to better characterize the lactococcal peptidolytic system.

The lactococcal proteinases are mainly active on β-casein (34) and liberate proline-rich peptides. These peptides are the main source of proline for *L. lactis*, in which no significant uptake of free proline occurs (27). Since hydrolysis of peptide bonds involving proline requires specific peptidases, such enzymes must be an important component of nitrogen catabolism in lactococci. As X-PDAP is present in several genera of lactic acid bacteria, including *Lactobacillus*, *Streptococcus*, and *Lactococcus* (1, 3, 11, 18), in which it has been found to be the major proline peptidase activity (3), our interest has been focused on this enzyme. We report here on the cloning and sequencing of an X-PDAP gene from the *L. lactis* chromosome.

Biochemical Characterization of X-PDAP

Two enzymes with similar biochemical properties have been purified from a crude cellular extract of *L. lactis* subsp. *lactis* ML3 (36) and *L. lactis* subsp. *cremoris* P8-2-47 (12), respectively. The molecular size of the X-PDAP enzyme from strain ML3 was estimated to be 190 kDa by gel filtration and 85 kDa by sodium dodecyl sulfate-polyacrylamide gel electrophoresis, which indicates that the native enzyme is a dimer with two identical subunits. Enzyme activity was maximal at pH 8.5 and 40 to 45°C. It was inactivated by serine proteinase inhibitors such as diisopropyl-fluorophosphate and phenylmethylsulfonyl fluoride. Inhibitors of metalloenzymes or sulfhydryl group reagents had no effect.

The purified enzyme hydrolyzed substrates with an X-Pro N-terminal sequence. On synthetic substrates such as *p*-nitroanilide (pNA) derivatives, highest activity was recorded with Arg-Pro-pNA, but Ala-Pro-pNA and Gly-Pro-pNA were

TABLE 1. Peptidases purified from lactococci

Enzyme	Substrate specificity	Strain	Inhibitor[a]	Size (kDa)	Reference
Aminopeptidases	X↓-Y-····	AM2	pHMB	300	22
		Wg2	EDTA	95	28
	Glu↓-Y-····	HP	EDTA	130	6
	X-Pro↓-Z-····	ML3	DFP	190	36
		P8-2-47	DFP	180	12
Di- and tripeptidases	X↓-Y	H61	EDTA	100	9
		Wg2	pHMB	49	32
	X↓-Pro	H61	EDTA	43	10
	X↓-Y-Z	CNRZ267	EDTA	75	5

[a]pHMB, p-Hydroxymercuribenzoate; DFP, diisopropylfluorophosphate.

also efficiently hydrolyzed. There was no activity if the N terminus was blocked or if there was no proline in the penultimate position.

Cloning of the X-PDAP Determinant from *L. lactis*

The chromosomal X-PDAP gene from strain ML3 was cloned by complementation of an *L. lactis* X-PDAP-deficient strain (21a). In an independent study, the X-PDAP gene from strain P8-2-47 was cloned in *Escherichia coli* and detected by using an oligonucleotide synthetic probe deduced from the N-terminal sequence of the purified enzyme (17a).

Total DNA prepared from *L. lactis* subsp. *lactis* ML3 was cloned into pIL252. pIL252 is a low-copy-number plasmid conferring erythromycin resistance and is derived from pAMβ1 (25). The ligation mixture was used to transform an X-PDAP-deficient mutant of ML3. The erythromycin-resistant transformants were screened for X-PDAP activity by using an enzymatic plate assay. One colony with the restored parental X-PDAP phenotype was isolated and contained, in addition to its original complement, a plasmid of 18 kb, designated pTIL1 (Fig. 1). A subfragment cloned in both orientations on pIL252 fully restored the X-PDAP phenotype of the ML3 mutant, suggesting that the gene was expressed from its own signals. The recombinant plasmid conferred the X-PDAP phenotype to *Bacillus subtilis*, suggesting that it carried the structural gene for X-PDAP activity.

The X-PDAP gene of *L. lactis* subsp. *cremoris* P8-2-47 was cloned in *E. coli* by using pUC18 (17a). A degenerate oligonucleotide mix, deduced from X-PDAP, was used to screen the transformants. As *E. coli* does not have X-PDAP activity, the positive clones were also tested for this activity in an enzymatic plate assay. Expression of the gene in this heterologous host indicated that the structural gene had been cloned. The DNA fragment was subcloned on pGKV2, a pWV01-derived cloning vector (33). The recombinant plasmid fully restored the X-PDAP activity in a negative *Lactococcus* mutant.

In both cases, the X-PDAP activity was not increased although the cloned gene was present in 5 to 10 copies per chromosome equivalent, as contrasted with a single copy on the chromosome of the parental strain. These results suggested that the X-PDAP gene is regulated.

DNA Sequence Analysis of the X-PDAP Determinants

The independently derived sequences (17a, 21a) showed an almost complete homology. An open reading frame of 763 codons specified a protein with an N-terminal amino acid sequence identical to that determined on the purified X-PDAP enzyme. The molecular sizes of the de-

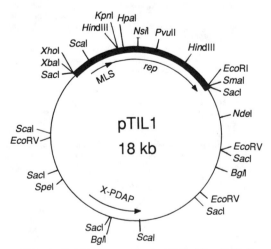

FIG. 1. Restriction map of pTIL1. The light line represents the DNA segment cloned from the *L. lactis* ML3 chromosome; the heavy line represents pIL252 DNA. The X-PDAP-coding region is indicated by an arrow.

duced proteins (88 kDa) were in agreement with those of the purified enzymes (85 and 90 kDa, respectively) (12, 36). These results indicate that the enzyme is not processed at the N terminus and is probably not exported. This view is in agreement with other results indicating that the enzyme from ML3 is located in the cytoplasmic soluble fraction (36) but inconsistent with the description of the X-PDAP from P8-2-47 as an extracellular enzyme (12). Direct visualization of the enzyme in the cell by electron microscopy after immunogold labeling with specific antibodies may help to answer the question of X-PDAP localization.

The X-PDAP genes from ML3 and P8-2-47 exhibited more than 99% homology in both the nucleotide and amino acid sequences. No significant homology was detected between the X-PDAP proteins and those in the protein data bases. These observations raised the question of X-PDAP gene conservation in lactococcal strains. Southern hybridization experiments were performed on chromosomal DNA of several *L. lactis* subspp. *lactis* and *cremoris* strains, using the X-PDAP gene from ML3 or P8-2-47 as a probe. Conservation and expression of the gene were observed in all strains.

Conclusions

Sequencing data obtained thus far suggest that X-PDAP is probably an intracellular protein. X-PDAP could therefore play a role in nitrogen nutrition of the lactococci by hydrolyzing prolyl peptides small enough to be transported across the membrane, i.e., those with a maximum size of four to five residues (16, 24, 29, 31). Such peptides could be produced by the combined action of proteinases and peptidases associated with the cell envelope according to the scheme proposed by Laan et al. (15) and Thomas and Pritchard (30). The X-Pro dipeptides released by X-PDAP could be further hydrolyzed by a prolidase that has been detected in lactococci (10; Monnet et al., personal communication). However, X-PDAP-deficient mutants are still able to grow, albeit slowly, in milk, which suggests that another intracellular peptidase able to hydrolyze prolyl peptides exists. The iminopeptidase activity previously detected in *L. cremoris* (6) could play this role.

The X-PDAP from *L. lactis* ML3 and P8-2-47 are inhibited by diisopropylfluorophosphate, which is regarded as presumptive evidence for the involvement of a serine residue in the catalytic mechanism. Despite different specificities, several prokaryotic serine proteinases have conserved amino acid sequences around their active sites (4, 13, 26). Such conserved sequences were not found in the X-PDAP gene, and a possible reactive serine could not be identified. A similar observation has been made for a serine proteinase from *Staphylococcus aureus* V8 (2) which shows no significant homology to the proteinases of this group. These results suggest that the X-PDAP gene may have a different evolutionary origin.

We thank B. Mayo, J. Kok, K. Venema, W. Bockelmann, and G. Venema for communicating results prior to publication. We are grateful to J. Kok for critical reading of the manuscript.

Part of this research was supported by the Centre National de la Recherche Scientifique (Biotechnology Programme, Lactic Acid Bacteria).

LITERATURE CITED

1. **Atlan, D., P. Laloi, and R. Portalier.** 1989. Isolation and characterization of aminopeptidase-deficient *Lactobacillus bulgaricus* mutants. *Appl. Environ. Microbiol.* **55:**1717–1723.

2. **Carmona, C., and G. L. Gray.** 1987. Nucleotide sequence of the serine protease gene of *Staphylococcus aureus*, strain V8. *Nucleic Acids Res.* **16:**6757.

3. **Casey, M. J., and J. Meyer.** 1985. Presence of an X-prolyl dipeptidyl peptidase in lactic acid bacteria. *J. Dairy Sci.* **68:**3212–3215.

4. **Chen, C. C., and P. Cleary.** 1990. Complete nucleotide sequence of the streptococcal C5a peptidase gene of *Streptococcus pyogenes*. *J. Biol. Chem.* **265:**3161–3167.

5. **Desmazeaud, M. J., and C. Zevaco.** 1979. Isolation and general properties of two intracellular aminopeptidases of *Streptococcus diacetylactis*. *Milchwissenschaft* **34:**606–610.

6. **Exterkate, F. A.** 1984. Location of peptidases outside and inside the membrane of *Streptococcus cremoris*. *Appl. Environ. Microbiol.* **47:**177–183.

7. **Exterkate, F. A., and G. J. C. M. de Veer.** 1987. Purification and some properties of a membrane-bound aminopeptidase A from *Streptococcus cremoris*. *Syst. Appl. Microbiol.* **9:**183–191.

8. **Geis, A., W. Bockelman, and M. Teuber.** 1985. Simultaneous extraction and purification of a cell wall-associated peptidase and B-casein specific protease from *Streptococcus cremoris* AC1. *Appl. Microbiol. Biotechnol.* **23:**79–84.

9. **Hwang, I. K., S. Kaminogawa, and K. Yamauchi.** 1981. Purification and properties of a dipeptidase from *Streptococcus cremoris* H61. *Agric. Biol. Chem.* **45:**159–165.

10. **Kaminogawa, S., N. Azuma, I. K. Wang, Y. Suzuki, and K. Yamauchi.** 1984. Isolation and characterization of a prolidase from *Streptococcus cremoris* H61. *Agric. Biol. Chem.* **48:**3035–3040.

11. **Khalid, N. M., and E. H. Marth.** 1990. Purification and partial characterization of a prolyl-dipeptidyl aminopeptidase from *Lactobacillus helveticus* CNRZ32. *Appl. Environ. Microbiol.* **56:**381–388.

12. **Kiefer-Partsch, B., W. Bockelmann, A. Geis, and M. Teuber.** 1989. Purification of an X-prolyl-dipeptidyl aminopeptidase from the cell wall proteolytic system of *Lactococcus lactis* subsp. *cremoris*. *Appl. Microbiol. Biotechnol.* **31:**75–78.

13. **Kok, J., K. J. Leenhouts, A. J. Haandrikman, A. M. Ledeboer, and G. Venema.** 1988. Nucleotide sequence of the cell wall proteinase gene of *Streptococcus cremoris* Wg2. *Appl. Environ. Microbiol.* **54:**213–238.

14. **Kok, J., and G. Venema.** 1988. Genetics of proteinases of lactic acid bacteria. *Biochimie* **70:**475–488.

15. **Laan, H., E. J. Smid, P. S. T. Tan, and W. N. Konings.** 1989. Enzymes involved in the degradation and utilisation of casein in *Lactococcus lactis*. *Neth. Milk Dairy J.* **43:**327–345.

16. **Law, B. A.** 1978. Peptides utilisation by group N streptococci. *J. Gen. Microbiol.* **105:**113–118.

17. **Law, B. A.** 1984. Flavour development in cheeses, p. 187–208. *In* F. L. Davies and B. A. Law (ed.), *Advances in the Microbiology and Biochemistry of Cheese and Fermented Milk*. Elsevier Applied Science Publishers, London.

17a. **Mayo, B., J. Kok, K. Venema, W. Bockelmann, M. Teuber, H. Reincke, and G. Venema.** 1991. Molecular cloning and sequence analysis of the X-prolyl dipeptidyl aminopeptidase gene from *Lactococcus lactis* subsp. *cremoris. Appl. Environ. Microbiol.* **57:**38–44.

18. **Meyer, R. J., and R. Jordi.** 1987. Purification and characterization of an X-prolyl dipeptidyl aminopeptidase from *Lactococcus lactis* and *Streptococcus thermophilus. J. Dairy Sci.* **70:**738–745.

19. **Mills, O. E., and T. D. Thomas.** 1981. Nitrogen sources for growth of lactic acid streptococci in milk. *N.Z. J. Dairy Sci. Technol.* **15:**43–45.

20. **Monnet, V., W. Bockelmann, J. C. Gripon, and M. Teuber.** 1989. Comparison of cell-wall bound proteinases from *Lactococcus lactis* subsp. *cremoris* AC1 and *Lactococcus lactis* subsp. *lactis* NCDO763. II. Specificity toward bovine β-casein. *Appl. Microbiol. Biotechnol.* **31:**112–118.

21. **Monnet, V., D. Le Bars, and J. C. Gripon.** 1987. Purification and characterization of the cell wall proteinase from *Streptococcus lactis* NCDO763. *J. Dairy Res.* **54:**247–255.

21a. **Nardi, M., M. C. Chopin, A. Chopin, M. M. Cals, and J. C. Gripon.** 1991. Cloning and DNA sequence analysis of an X-prolyl dipeptidyl aminopeptidase gene from *Lactococcus lactis* subsp. *lactis* NCDO 763. *Appl. Environ. Microbiol.* **57:**45–50.

22. **Neviani, E., C. Y. Boquien, V. Monnet, L. Phan Thanh, and J. C. Gripon.** 1989. Purification and characterization of an aminopeptidase from *Lactococcus lactis* subsp. *cremoris* AM2. *Appl. Environ. Microbiol.* **55:**2308–2314.

23. **Reiter, B., and J. D. Oram.** 1962. Nutritional studies on cheese starters. I. Vitamin and amino acid requirements of single strain starters. *J. Dairy Res.* **29:**63–77.

24. **Rice, G. H., F. H. C. Stewart, A. J. Hillier, and G. R. Jago.** 1978. The uptake of amino acids and peptides by *Streptococcus lactis. J. Dairy Res.* **45:**93–107.

25. **Simon, D., and A. Chopin.** 1988. Construction of a vector plasmid family for molecular cloning in *Streptococcus lactis. Biochimie* **70:**559–566.

26. **Sloma, A., G. A. Rufo, C. F. Rudolph, B. J. Sullivan, K. A. Theriaut, and J. Pero.** 1990. Bacillopeptidases of *Bacillus subtilis*: purification of the protein and cloning of the gene. *J. Bacteriol.* **172:**1470–1477.

27. **Smid, E. J., R. Plapp, and W. N. Konings.** 1989. Peptide uptake is essential for growth of *Lactococcus lactis* on the milk protein casein. *J. Bacteriol.* **171:**6135–6140.

28. **Tan, P. S. T., and W. N. Konings.** 1990. Purification and characterization of an aminopeptidase from *Lactococcus lactis* subsp. *cremoris* Wg2. *Appl. Environ. Microbiol.* **56:**526–532.

29. **Thomas, T. D., and O. E. Mills.** 1981. Proteolytic enzymes of starter bacteria. *Neth. Milk Dairy J.* **35:**255–273.

30. **Thomas, T. D., and G. G. Pritchard.** 1987. Proteolytic enzymes of dairy starter cultures. *FEMS Microbiol. Rev.* **46:**245–268.

31. **Van Boven, A., and W. N. Konings.** 1988. Utilization of dipeptides by *Lactococcus lactis* subsp. *cremoris. Biochimie* **70:**535–542.

32. **Van Boven, A., P. S. T. Tan, and W. N. Konings.** 1988. Purification and characterization of a dipeptidase from *Streptococcus cremoris* Wg2. *Appl. Environ. Microbiol.* **54:**43–49.

33. **Van der Vossen, J. M. B. M., J. Kok, and G. Venema.** 1985. Construction of cloning, promoter-screening, and terminator-screening shuttle vectors for *Bacillus subtilis* and *Lactococcus lactis* subsp. *lactis. Appl. Environ. Microbiol.* **50:**540–542.

34. **Visser, S., F. A. Exterkate, C. J. Slangen, and G. J. C. M. de Veer.** 1986. Comparative study of action of cell wall proteinases from various strains of *Streptococcus cremoris* on bovine α_{s1}-, β-, and κ-casein. *Appl. Environ. Microbiol.* **52:**1162–1166.

35. **Visser, S., G. Hup, F.A. Exterkate, and J. Stadhouders.** 1983. Bitter flavour in cheese. 2. Model studies on the formation and degradation of bitter peptides by proteolytic enzymes from calf rennet, starter cells and starter cell fractions. *Neth. Milk Dairy J.* **37:**169–180.

36. **Zevaco, C., V. Monnet, and J. C. Gripon.** 1990. Intracellular X-prolyl dipeptidyl peptidase from *Lactococcus lactis* subsp. *lactis:* purification and properties. *J. Appl. Bacteriol.* **68:**357–366.

Molecular Analysis of pTR2030 Gene Systems That Confer Bacteriophage Resistance to Lactococci[†]

TODD R. KLAENHAMMER,[1,2] DENNIS ROMERO,[2‡] WES SING,[2§] AND COLIN HILL[1‖]

Departments of Food Science[1] and Microbiology,[2] Southeast Dairy Foods Research Center, North Carolina State University, Raleigh, North Carolina 27695-7624

One of the most exciting areas that has emerged through genetic analysis of the dairy-related lactococci is the discovery and characterization of gene systems that provide phage resistance to these industrially important bacteria. Naturally occurring strains that are phage insensitive are known to harbor multiple defense systems that can act at different points of the lytic cycle to prevent the successful adsorption, infection, or replication of virulent phages (18). Although phage attack of lactococcal starter cultures is still a major problem for the cultured dairy product industries, this difficulty can now be successfully addressed by using genetic approaches to construct strains that are resistant to phages most often encountered in the industry (27, 29). The practical use and successful tenure of pTR2030 in transconjugants in the commercial environment has fueled greater interest in understanding lactococcal phage defense systems at the molecular level.

The self-transmissible, phage resistance plasmid pTR2030 is present as a high-molecular-weight multimer in the prototype phage-insensitive strain *Lactococcus lactis* ME2 (C. Hill, L. Miller, and T. Klaenhammer, *Plasmid*, in press), in which it is one component of the multiple phage defense system harbored by this strain (18). Of the numerous phage resistance plasmids that have now been detected and characterized in lactococci (for reviews, see references 4, 17, 27, and 31; for recent reports, see references 2, 5, 6, 15, 22, and 32), pTR2030 has at this juncture received the most thorough investigation of its genetic organization, mechanisms of phage resistance, conjugative ability, and practical utility. Following the introduction of pTR2030 transconjugants into the industry in 1985, it has also become possible to investigate at the molecular level the counter-defenses of virulent phages that occur in one of the most dynamic environments of phage adaptation, the cheese plant. This report briefly describes our most recent work on the molecular analysis of pTR2030 and its phage defense systems.

Phenotypes and Location of the Phage Resistance Region

Plasmid pTR2030 was first discovered in lactose-fermenting (Lac[+]), phage-resistant transconjugants formed by matings with the prototype phage-insensitive strain, *L. lactis* ME2. The plasmid is a 46.2-kb self-transmissible conjugal element that encodes two phage resistance phenotypes: Hsp[+], abortive infection of phage described initially as a heat-sensitive reduction in the burst size and efficiency of plaquing (EOP) of the prolate-headed phage c2 (19); and R/M[+], restriction and modification activity, first recognized via host-dependent replication of the small isometric phage nck202.48 (φ48) on *L. lactis* LMA12, a pTR2030 transconjugant of an industrial cheese starter culture (12, 29). The phage resistance phenotypes exhibited by pTR2030 are shown in Table 1 in two distinct lactococcal host backgrounds that comprise the isogenic strain groupings of *L. lactis* NCK203/L2FA/LMA12/TL2F1 and *L. lactis* LM0230/LM2301/C2/712/ML3. Molecular characterization of pTR2030 has localized the region responsible for the Hsp and R/M activities (12, 13, 18). Figure 1 shows the physical map of pTR2030 and locations of the R/M and Hsp regions. Initially, the region responsible for the phage resistance was localized by mapping an 11.5-kb deletion in pTR2030 that occurred when a phage-resistant transconjugant of the industrial strain LMA12-4 became sensitive to phage (29). The deleted plasmid, pTR2023 (Fig. 1), retained its conjugative ability but failed to confer phage resistance in either the *L. lactis* NCK203 or LM2301 host background. A 13.6-kb fragment that spanned the deletion was cloned, first in lambda EMBL3 and then in the *Escherichia coli-L. lactis* shuttle vector pSA3 (13). The recombinant plasmid, pTK6, exhibited both Hsp[+] and R/M[+] activities, albeit to a lesser degree than did pTR2030 (Table 1). It is interesting to note that conjugal mobilization of pTK6 via conduction by a self-transmissible plasmid in *L. lactis* MG1363 (7) and the recombination-deficient strain MMS362(pRS01) (1) can reestablish

[†]Paper no. FSR91-13 of the journal series of the Department of Food Science, North Carolina State University, Raleigh, NC 27695-7624.

[‡]Present address: Promega, Madison, WI 53711-5399.

[§]Present address: Vivolac, Indianapolis, IN 46201.

[‖]Present address: National Dairy Research Centre, Moorepark, Ireland.

TABLE 1. Phage EOP and plaque sizes on lactococci harboring pTR2030 or its derivatives cloned in vivo or in vitro

Plasmid	Phenotype	Phage EOP		
		Small isometric		Prolate, c2[b]
		φ31[a]	sk1[b]	
pTR2030	Hsp+ R/M+	$<10^{-10}$	$<10^{-10}$	0.3 (SP)[c]
pTR2023	Hsp− R/M−	1.0 (LP)[c]	1.0 (LP)	1.0 (LP)
pTK6	Hsp+ R/M+	10^{-4} (SP)	0.1 (PP)[c]	0.5 (SP)
pTRK18	Hsp+ R/M−	0.1 (SP)	0.25 (PP)	0.5 (SP)

[a] Host background *L. lactis* NCK203.
[b] Host background *L. lactis* LM0230.
[c] SP, Small plaques, 0.2 mm; LP, large plaques, 1.5 to 3.0 mm; PP, pinpoint plaques.

phage resistance to levels equal to or exceeding those conferred by pTR2030 (25). Formation of cointegrates between phage resistance plasmids and conjugal elements has important practical implications in the construction of phage-insensitive strains for use in the industry, particularly when the level of phage resistance is elevated as a consequence of the recombination event. At this time, however, it remains unclear why these dramatic differences in phenotype occur between pTR2030, pTK6, and pTK6::pRS01 cointegrates, especially since subsequent work has defined the essential structural regions for Hsp and R/M

within this 13.5-kb region cloned in pTK6. Detailed investigations on the effects of copy number, formation of high-molecular-weight multimers, and regulation of phage resistance gene systems in lactococci are needed to resolve these questions.

pTR2030 contains two functional copies of IS*946*, an 808-bp element that shares 96% homology to IS*S1* (24, 26). One copy of IS*946* is located near the *Xho*I-*Ava*I site within the 13.5-kb fragment of pTR2030 cloned in pTK6 (Fig. 1). This copy of IS*946* mediates cointegrate formation during mobilization of pTK6 by self-trans-

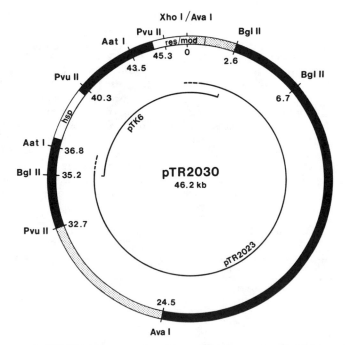

FIG. 1. Circular map of pTR2030. The general positions of the *hsp* and *res/mod* determinants are indicated. The inner lines denote the 13.6-kb *Bgl*II fragment which was cloned into pSA3 (10.2 kb) to form pTK6 (23.8 kb) and the extent of pTR2030 remaining in pTR2023 (34.7 kb) after an 11.5-kb deletion of the region encoding phage resistance determinants. The stippled areas between map positions 0 and 2.6 and positions 24.5 and 32.7 indicate regions that contain one copy of IS*946*. (Reprinted from Romero and Klaenhammer [26].)

missible plasmids in lactococci (25) and promotes recombinational events by pTR2030 during mobilization of nonconjugative plasmids and cloning vectors (26). The position of the insertion sequence, relative to the Hsp and R/M regions defined on pTR2030, explains the deletion events that have occurred in vivo during the formation of pTR2023 (Hsp⁻ R/M⁻) and pTRK18 (Hsp⁺ R/M⁻) (12, 26). DNA sequencing of the IS946 junctions on pTRK18 and pTK6 revealed that 8-bp direct repeats were not present in either plasmid. This finding suggests that transposition of a second copy of IS946 occurred into pTK6 and was followed by intramolecular recombination between the two insertion sequence elements to delete the 3.5-kb intervening R/M region (26). We suspect that a similar recombination event generated the deletion which spanned the Hsp and R/M regions during formation of pTR2023 (Hsp⁻ R/M⁻).

Molecular Analysis and Function of Hsp

The essential region for Hsp has been defined by Tn5 mutagenesis of the Hsp⁺ R/M⁻ plasmid pTRK18 (13). Four Tn5 insertions within a 3-kb region eliminated the Hsp⁺ phenotype. Intermediate phenotypes were not observed and in vitro deletions of flanking regions confirmed that a single locus was responsible for Hsp⁺. The 3-kb region has been completely sequenced, and a structural gene of 1,887 bp was identified which could encode a protein with a predicted molecular size of 73.8 kDa (11). The open reading frame for the hsp structural gene spanned the four Tn5 insertions that inactivated the phenotype. Expression signals upstream of hsp directed the constitutive expression of cat-86 on pGKV210 (33), suggesting that the transcription of hsp is not induced in response to phage infection. The DNA sequence for hsp was recently confirmed on pCI829 (A. Coffey, C. Daly, and G. F. Fitzgerald, Abstr. 3rd Int. Am. Soc. Microbiol. Conf. Streptococcal Genet., A/49, 1990), an abortive plasmid that was recognized some time ago through physical maps and phage resistance

phenotypes to be largely identical to pTR2030 (2, 3).

Phage infection is aborted in cells that express Hsp⁺ (19, 30). For small isometric phages susceptible to Hsp, the EOP is reduced to <10⁻⁹ even though infective centers are formed at efficiencies of 10⁻² per input phage (Table 2). Examination of DNA and protein sequence data bases (GenBank version 63) did not reveal any significant relatedness between hsp and other genes or classes of proteins that have been sequenced previously; therefore, the specific role of the hsp protein in conferring phage resistance to lactococci remains to be elucidated. Using a rapid method to evaluate intracellular phage DNA replication (Fig. 2), we have obtained recent evidence suggesting that the action of Hsp occurs at the level of phage DNA replication (10). pTR2030 does not retard phage adsorption or injection of phage DNA into the cell (19, 30). These observations were confirmed by noting that after injection into NCK204 (harboring pTR2030), phage DNA (modified so as to be insensitive to pTR2030-encoded R/M) remained intact over the course of infection (Fig. 2). The DNA was neither degraded nor replicated. In the absence of pTR2030, phage DNA replication occurred in NCK203 cells within 45 min after infection. These simple experiments suggest that Hsp acts to retard phage DNA replication. Interference at this point would be expected to stop lytic development completely or severely limit the number of progeny phage that burst from Hsp⁺ infected cells. Such responses have been reported for both prolate- and small isometric-headed phages in lactococci harboring Abi and Hsp plasmids (2, 9, 14, 19, 22, 23, 30, 31). The specific mechanism through which the product of the hsp gene interferes with phage DNA replication or redirection of the host's replicative machinery is not yet known. In this light, it is interesting that at least two genotypes representing abortive mechanisms exist in lactococci (11, 23). Perhaps numerous targets exist for the action of Abi-type mechanisms or, as is the case with many classes

TABLE 2. Separate and combined effects of Hsp and R/M on lactococcal bacteriophages

Phage[a]	Plasmid	Defense[b]	EOP	COI[c]/PFU	% Cell death[d]
p2	pTR2030	Hsp	<10⁻⁹	10⁻²	78
p2	pTRK30	R/M	10⁻⁴	<10⁻⁴	<1
p2	pTR2030 pTRK30	Hsp R/M	<10⁻⁹	<10⁻⁴	8
φ31	pTR2030	Hsp + R/M	<10⁻⁹	<10⁻⁵	8
mod31[e]	pTR2030	Hsp	<10⁻⁹	10⁻²	91

[a] Phages p2 and φ31 are small isometric-headed phages.

[b] Phage p2 is not susceptible to restriction by the R/M system encoded by pTR2030.

[c] COI, Centers of infection.

[d] CFU per milliliter after phage infection/CFU per milliliter before phage infection. The multiplicity of infection was 10. Survivors were plated 5 or 10 min postinfection with phage p2 or φ31, respectively (30).

[e] mod31 is modified phage φ31 prepared by propagation through L. lactis NCK203 bearing pTRK70 (R⁻/M⁺) (12).

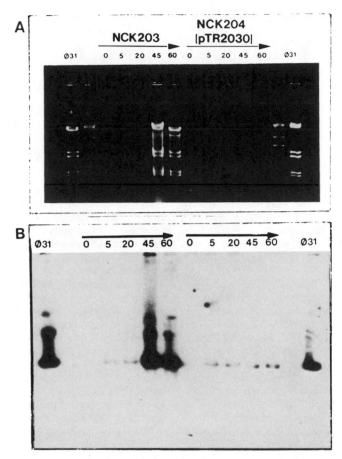

FIG. 2. Effect of pTR2030 (Hsp⁺) on internal phage DNA replication in *L. lactis* NCK203. (A) *Eco*RI digestion of *L. lactis* NCK203 (not carrying pTR2030) and NCK204(pTR2030) total DNA isolated from cells before (0 min) and during the course of an infection (5 to 60 min) with phage mod31. This phage was modified so as not to be restricted by the R/M system encoded by pTR2030 by propagation through NCK216(pTRK70, R⁻/M⁺). (B) Autoradiogram after hybridization with ³²P-labeled pTRK139; pBluescript::3.0-kb *Eco*RI fragment of φ31. Lanes designated φ31 show the restriction patterns for phage φ31 DNA, which was extracted by traditional techniques from high-titer purified phage preparations. See Hill et al. (10) for details. (Reprinted from Hill et al. [10].)

of proteins and enzymes, *abi* gene products are structurally heterogeneous but maintain a common functional domain and act at a common point to interfere with phage DNA replication.

Molecular Analysis of R/M

An R/M system is also present within the 13.5-kb fragment of pTR2030 that was cloned on pTK6 (Fig. 1; 12). R/M⁺ activities were detected via host-dependent replication of phage φ31 in *L. lactis* NCK203 but were not observed when pTK6 or pTR2030 was introduced into the *L. lactis* LM2301 or MG1363 host background (Table 1). It is not clear why this difference in the detection of R/M activity occurs, but the variation could reflect the relative susceptibilities of the different phages to R/M activities due to antirestric-

tion mechanisms (20). Alternatively, if *L. lactis* LM0230 and MG1363 already exhibit a Mod⁺ activity capable of protecting sites targeted by pTR2030-encoded R/M, phages propagated on these hosts would not be restricted subsequently in isogenic derivatives bearing pTR2030 or pTK6. This latter possibility is supported by the fact that in order to detect the R/M⁺ activity encoded by pBF61 in *L. lactis* LM0230, Froseth et al. (8) first had to propagate c2 phage through *L. cremoris* KH.

The location of the R/M region on pTK6 was defined by an in vivo deletion in pTK6 responsible for loss of R/M⁺ (12; D. A. Romero, Ph.D. thesis, North Carolina State University, Raleigh, 1990). Through molecular analysis of this region via subcloning, expression, and DNA sequencing, we have defined a gene encoding a type II

A-methylase, designated *Lla*I (C. Hill, L. Miller, and T. R. Klaenhammer, *Abstr. 3rd. Int. Am. Soc. Microbiol. Conf. Streptococcal Genet.*, A/52, 1990). The *Lla*I gene is 1,866 bp in length and encodes a protein of 622 amino acids with a predicted molecular mass of 72.5 kDa. Lauster (21) previously identified two consensus sequences found in all type II A-methylases. Each of these consensus sequences was found twice within the *Lla*I gene (Fig. 3). The first consensus sequence, found in all type II methylases, is located at the amino terminus of each methylase domain in *Lla*I. Two D-P-P-Y consensus sequences, located in the carboxyl domain of only type II A-methylases (21), were also present in *Lla*I. This organization is striking in that two methylase domains are organized in tandem. In this regard, the protein sequence of *Lla*I shows considerable homology to that of an atypical type II A-methylase, *Fok*I (16). *Fok*I contains two functional methylase domains which recognize nonpalindromic sequences and act as an asymmetric dimer to modify the complementary sequences on opposite strands. We conclude from these observations that pTR2030 encodes a type II A-methylase with potentially two functional methylase domains.

We have obtained phenotypic evidence for one functional methylase activity at the amino terminus of *Lla*I. Molecular characterization of an industrial phage that is not susceptible to restriction by pTR2030 has revealed that the phage genome harbors 1,273 bp of the *Lla*I region from pTR2030, including the upstream expression signals and the first 344 codons of the methylase gene (Fig. 3; C. Hill, L. A. Miller, and T. R. Klaenhammer, submitted for publication). The phage is therefore capable of self-methylation in any propagating host provided that the amino domain of *Lla*I is functional. A 4.5-kb fragment from phage nck202.50 (phage φ50) which encoded the amino domain of the *Lla*I methylase, designated *Lla*PI ("P" indicating the phage locus of the methylase gene), was cloned into pSA3 to form pTRK103. When R/M-susceptible phages were propagated on *L. lactis* NCK203 harboring pTRK103, the progeny were afforded partial protection from pTK6-encoded R/M activities. Phage modified by pTRK103 in *trans* plaques at EOPs approaching 0.1 relative to EOPs of 10^{-3} for phage propagated in the absence of pTRK103. Therefore, the amino terminus of *Lla*I, represented by the phage gene *Lla*PI, has functional methylase activity.

We are continuing to analyze downstream sequences of the R/M region to identify the *res* gene(s) that complements the *Lla*I methylase. It is significant at this juncture, however, that definition of *Lla*I as a type II A-methylase provides the first evidence that type II R/M systems function in vivo to direct host-dependent replication of phage in lactococci. The type II restriction endonuclease *Scr*FI isolated from *Lactococcus cremoris* F failed to exhibit such a role in the in vivo restriction of phage kh (4). The DNA sequences of *Lla*I and *Lla*PI and a detailed analysis of the R/M region of pTR2030 will be published elsewhere (Hill et al., submitted).

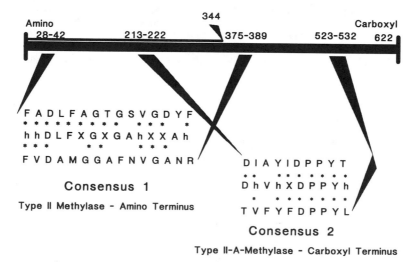

FIG. 3. General structural organization of the type II A-methylase, *Lla*I. Consensus sequences 1 and 2 of Lauster (21) are denoted for comparison as the middle line of the three sequences presented for each consensus region defined on *Lla*I. Single-letter amino acid codes are used throughout; h = any hydrophobic amino acid and X = any amino acid. Amino acid positions corresponding to the consensus regions are numbered above the major line representing the 622 amino acids of *Lla*I. The exact DNA sequence of *Lla*I carried on phage nck202.50 is denoted by the thin uppermost line from 0 to 344 amino acids in the amino domain.

Combined Effects of R/M and Hsp

Combinations of Abi and R/M defense systems have been identified in lactococci. The practical significance of any synergism between Abi and R/M can now be acknowledged, given the longevity at which strains carrying such combinations survive in the industry (18, 27, 29). Within any given strain, these resistance mechanisms are often found together on a single plasmid or detected as combinations of different replicons carrying Abi or R/M defenses (3, 8, 9, 12, 15, 18, 22, 28). It is now clear that the two pTR2030 defense mechanisms complement each other by acting at different points of the phage lytic cycle. The separate and combined effects of Hsp and R/M on EOP, formation of infective centers, and cell death occurring after phage infection are shown in Table 2. For cases in which Hsp is the sole effective defense system operating (see the examples for phages p2 and modified φ31 on pTR2030 hosts), substantial levels of cell death occur as a result of the aborted phage infection. While in this case only 1 cell in 100 will form an infective center and release progeny phage, 80 to 90% of the remaining infected cells will nevertheless die. In contrast, cell death after a phage infection is reduced to less than 10% when R/M activities are combined with Hsp (Table 2). R/M action serves as a first line of phage defense and maximizes the chance for survival in cells with an abortive defense system. When incoming phage DNA is restricted, the course leading to Hsp-induced phage abortion and cell death is avoided. The Abi/Hsp functions also serve to complement R/M functions by severely limiting the potential of infected cells to release modified progeny phage. When a phage escapes restriction, abortive actions leading to cell death prevent the infected cells from releasing modified progeny phage into the environment. Together, these mechanisms create a powerful barrier to phage proliferation and thereby also minimize the potential for phage adaptation through mutation, recombination, or host-dependent modification.

Summary

Molecular characterization of pTR2030 has revealed two defense systems that cooperate effectively to prevent phage infection and proliferation. Combinations of such phage defenses are found naturally in lactococci (9, 12, 15, 22, 28) and most likely reflect the adaptation of these bacteria to a phage-contaminated environment that must be highly selective for sophisticated defense systems. In addition, pTR2030 and many of the other conjugative phage resistance plasmids found in the lactococci carry insertion sequences that mediate recombination and cointegration events (1, 7, 24, 26, 32). Their ability to conduct replicon fusions via insertion sequence-mediated recombination might be anticipated given that their associated phage resistance gene systems are vital to cell survival in milk. As replicons and genes encoding phage defenses have been characterized over the past decade to the molecular level, it has become clear that phage resistance in lactococci is both fundamentally important and industrially significant. The lactococci may well represent the most dynamic species in which to investigate phage-host interactions in prokaryotes and study cycles of bacterial defense and phage counterdefense. Looking ahead, our efforts to understand and control the expression and regulation of genes encoding phage resistance, particularly in terms of cooperation between gene systems, will have a dramatic impact on the effectiveness and industrial longevity of starter cultures engineered for phage insensitivity.

This work has been supported in part by the USDA-Animal Molecular Biology Program grants 85-CRCR-1-2547 and 87-CRCR-1-2457 and Miles, Inc., Marschall Dairy Products Division and Biotechnology Products Divisions.

LITERATURE CITED

1. **Anderson, D. G., and L. L. McKay.** 1984. Genetic and physical characterization of recombinant plasmids associated with cell aggregation and high-frequency conjugal transfer in *Streptococcus lactis* ML3. *J. Bacteriol.* **158**:954–962.

2. **Coffey, A. G., G. F. Fitzgerald, and C. Daly.** 1989. Identification and characterization of a plasmid encoding abortive infection from *Lactococcus lactis* ssp. *lactis* UC811. *Neth. Milk Dairy J.* **43**:229–244.

3. **Coffey, A. G., M. Murphy, V. Costello, S. Lennon, C. Daly, and G. F. Fitzgerald.** 1987. Characterization of three plasmid-coded bacteriophage resistance systems in lactic streptococci. *FEMS Microbiol. Rev.* **46**:P46.

4. **Daly, C., and G. Fitzgerald.** 1987. Mechanisms of bacteriophage insensitivity in the lactic streptococci, p. 259–268. In J. J. Ferretti and R. Curtiss III (ed.), *Streptococcal Genetics.* American Society for Microbiology, Washington, D.C.

5. **de Vos, W. M.** 1989. On the carrier state of bacteriophages in starter lactococci: an elementary explanation involving a bacteriophage resistance plasmid. *Neth. Milk Dairy J.* **43**:221–227.

6. **Dunny, G. M., D. A. Krug, C.-L. Pan, and R. A. Ledford.** 1988. Identification of cell wall antigens associated with a large conjugative plasmid encoding phage resistance and lactose fermentation ability in lactic streptococci. *Biochimie* **70**:443–450.

7. **Fitzgerald, G. F., and M. J. Gasson.** 1988. In vivo gene transfer systems and transposons. *Biochimie* **70**:489–502.

8. **Froseth, B. R., S. K. Harlander, and L. L. McKay.** 1988. Plasmid-mediated reduced phage sensitivity in *Streptococcus lactis* KR5. *J. Dairy Sci.* **71**:275–284.

9. **Gautier, M., and M.-C. Chopin.** 1987. Plasmid-determined systems for restriction and modification activity and abortive infection in *Streptococcus cremoris*. *Appl. Environ. Microbiol.* **53**:923–927.

10. **Hill, C., I. J. Massey, and T. R. Klaenhammer.** 1991. Rapid method to characterize lactococcal bacteriophage genomes. *Appl. Environ. Microbiol.* **57**:283–288.

11. **Hill, C., L. A. Miller, and T. R. Klaenhammer.** 1990. Nucleotide sequence and distribution of the pTR2030 resistance determinant (*hsp*) which aborts bacteriophage infection in lactococci. *Appl. Environ. Microbiol.* **56**:2255–2258.

12. **Hill, C., K. Pierce, and T. R. Klaenhammer.** 1989. The conjugative plasmid pTR2030 encodes two bacteriophage defense mechanisms in lactococci, restriction modification (R⁺/M⁺) and abortive infection (Hsp⁺). *Appl. Environ. Microbiol.* **55:**2416–2419.

13. **Hill, C., D. A. Romero, D. S. McKenney, K. R. Finer, and T. R. Klaenhammer.** 1989. Localization, cloning, and expression of genetic determinants for bacteriophage resistance (Hsp) from the conjugative plasmid pTR2030. *Appl. Environ. Microbiol.* **55:**1684–1689.

14. **Jarvis, A. W.** 1989. Bacteriophages of lactic acid bacteria. *J. Dairy Sci.* **72:**3406–3428.

15. **Josephsen, J., and F. K. Vogensen.** 1989. Identification of three different plasmid-encoded restriction/modification systems in *Streptococcus lactis* subsp. *cremoris* W56. *FEMS Microbiol. Lett.* **59:**161–166.

16. **Kita, K., H. Kotani, H. Sugisake, and M. Takanami.** 1989. The *Fok*I restriction-modification system. I. Organization and nucleotide sequences of the restriction and modification genes. *J. Biol. Chem.* **264:**5751–5756.

17. **Klaenhammer, T. R.** 1987. Plasmid-directed mechanisms for bacteriophage defense in lactic streptococci. *FEMS Microbiol. Rev.* **46:**313–325.

18. **Klaenhammer, T. R.** 1989. Genetic characterization of multiple mechanisms of phage defense from a prototype phage-insensitive strain, *Lactococcus lactis* ME2. *J. Dairy Sci.* **72:**3429–3442.

19. **Klaenhammer, T. R., and R.B. Sanozky.** 1985. Conjugal transfer from *Streptococcus lactis* ME2 of plasmid encoding phage resistance, nisin resistance and lactose-fermenting ability: evidence for a high-frequency conjugative plasmid responsible for abortive infection of virulent bacteriophage. *J. Gen. Microbiol.* **131:**1531–1541.

20. **Kruger, D. H., and T. A. Bickle.** 1983. Bacteriophage survival: multiple mechanisms for avoiding the deoxyribonucleic acid restriction systems of their hosts. *Microbiol. Rev.* **47:**345–360.

21. **Lauster, R.** 1989. Evolution of type II DNA methyltransferases. A gene duplication model. *J. Mol. Biol.* **206:**313–321.

22. **McKay, L. L., M. J. Bohanon, K. M. Polzin, P. L. Rule, and K. A. Baldwin.** 1989. Localization of separate genetic loci for reduced sensitivity towards small isometric-headed bacteriophage sk1 and prolate-headed bacteriophage c2 on pGBK17 from *Lactococcus lactis* subsp. *lactis* KR2. *Appl. Environ. Microbiol.* **55:**2702–2709.

23. **Murphy, M. C., J. L. Steele, C. Daly, and L. L. McKay.** 1988. Concomitant conjugal transfer of reduced-bacteriophage-sensitivity mechanisms with lactose- and sucrose-fermenting ability in lactic streptococci. *Appl. Environ. Microbiol.* **54:**1951–1956.

24. **Polzin, K. M., and M. Shimizu-Kadota.** 1987. Identification of a new insertion element, similar to gram-negative IS26, on the lactose plasmid of *Streptococcus lactis* ML3. *J. Bacteriol.* **169:**5481–5488.

25. **Romero, D. A., and T. R. Klaenhammer.** 1990. Abortive phage infection and restriction/modification activities directed by pTR2030-determinants are enhanced by recombination with conjugal elements in lactococci. *J. Gen. Microbiol.* **36:**1817–1824.

26. **Romero, D. A., and T. R. Klaenhammer.** 1990. Characterization of gram-positive insertion sequence IS946, an iso-IS5J element, isolated from the conjugative lactococcal plasmid pTR2030. *J. Bacteriol.* **172:**4151–4160.

27. **Sanders, M. E.** 1988. Phage resistance in lactic acid bacteria. *Biochimie* **70:**411–421.

28. **Sanders, M. E., and T. R. Klaenhammer.** 1984. Phage resistance in a phage-insensitive strain of *Streptococcus lactis*: temperature-dependent phage development and host-controlled phage replication. *Appl. Environ. Microbiol.* **47:**979–985.

29. **Sanders, M. E., P. J. Leonhard, W. D. Sing, and T. R. Klaenhammer.** 1986. Conjugal strategy for construction of fast acid-producing, bacteriophage-resistant lactic streptococci for use in dairy fermentations. *Appl. Environ. Microbiol.* **52:**1001–1007.

30. **Sing, W. D., and T. R. Klaenhammer.** 1990. Characteristics of phage abortion conferred in lactococci by the conjugal plasmid pTR2030. *J. Gen. Microbiol.* **136:**1807–1815.

31. **Sing, W. D., and T. R. Klaenhammer.** 1990. Plasmid-induced abortive infection in lactococci: a review. *J. Dairy Sci.* **73:**2239–2251.

32. **Steele, J. L., M. C. Murphy, C. Daly, and L. L. McKay.** 1989. DNA-DNA homology among lactose- and sucrose-fermenting transconjugants from *Lactococcus lactis* strains exhibiting reduced bacteriophage sensitivity. *Appl. Environ. Microbiol.* **55:**2410–2413.

33. **van der Vossen, J. M. B. M., D. van der Lelie, and G. Venema.** 1987. Isolation and characterization of *Streptococcus cremoris* Wg2-specific promoters. *Appl. Environ. Microbiol.* **53:**2452–2457.

Plasmid-Encoded Bacteriophage Insensitivity in Members of the Genus *Lactococcus*, with Special Reference to pCI829

AIDAN COFFEY, VERONICA COSTELLO, CHARLES DALY, AND GERALD FITZGERALD

Food Microbiology Department, University College, Cork, Ireland

Members of the genus *Lactococcus* have key roles in the production of high-quality fermented milk products. A major challenge to the industry is the provision of stable strains that function consistently in large-scale industrial fermentations; therefore, during the past 20 years these bacteria have become the target of intense genetic studies. The demonstration that several key traits such as lactose and citrate utilization, proteinase activity, and insensitivity to bacteriophage attack are usually plasmid encoded has provided an explanation for the instability frequently observed with these traits and has also facilitated genetic and molecular analysis. Given the significance of phage as a major inhibitor of culture activity in commercial practice, it is not surprising that considerable research effort has focused on plasmid DNA-encoded phage insensitivity mechanisms observed in a variety of strains worldwide (Table 1).

Although significant research effort is needed to provide precise details of the mechanisms involved in phage insensitivity within lactococci, these mechanisms can be divided into three types. The insensitivity may be similar to classical resistance whereby adsorption of phage to specific receptors on the host cell surface does not occur or occurs at a greatly reduced efficiency (6, 23). Phage restriction and modification systems, whereby phage DNA entering the host cell is damaged by restriction enzymes, are also common in lactococci (15, 21). The third mechanism has been termed abortive infection, and its expression results in the inability or reduced ability of the phage to produce progeny phage particles within the cell (15, 21).

An abortive infection mechanism was identified on pNP40 in *Lactococcus lactis* subsp. *lactis* biovar diacetylactis DRC3 by McKay and Baldwin (19). The phage insensitivity associated with this plasmid, which was also implicated in resistance to nisin, was manifested by the total inhibition of plaque-forming ability of prolate-headed phage c2 at 30°C but not at 37°C in the plasmid-free strain *L. lactis* subsp. *lactis* LM2301. Since then a large number of plasmids encoding this type of phage insensitivity have been identified (Table 1). The elegant work of Klaenhammer's group, including the localization, cloning, and sequencing of the *hsp* gene (encoding abor-

tive infection) from *L. lactis* subsp. *lactis* ME2, is reviewed elsewhere in this volume. It is important that researchers examine all available phage defense systems. Detailed and specific analyses of the genes involved and their products will allow comparison between phage insensitivity determinants from different sources and provide insight into the diversity of available genetic material that may be used to provide new, genetically manipulated strains for industrial fermentations.

This report summarizes data on three plasmids mediating an abortive infection mechanism(s), isolated from three different *Lactococcus* strains in our laboratory: pCI528 from *L. lactis* subsp. *cremoris* UC503, pCI750 from *L. lactis* subsp. *cremoris* UC653, and pCI829 from *L. lactis* subsp. *lactis* UC811. The distinct restriction maps of these different-size plasmids are shown (Fig. 1), and their properties are summarized and compared (Table 2).

pCI528

Plasmid pCI528 appears to encode two mechanisms of phage insensitivity. It was initially implicated in adsorption blocking following comparison of the adsorption efficiency of a number of phage on *L. lactis* subsp. *cremoris* UC503 with that of the same phage on pCI528-cured derivatives of UC503. In addition, cell surface characteristics of pCI528-cured derivatives were different from those of the parent. However, an abortive infection system was also associated with strains harboring pCI528. The abortive infection mechanism associated with pCI528 was identified following analysis of the plaquing efficiencies of phage that adsorbed normally to UC503. For example, the small isometric-headed phage uc2021 exhibited a fivefold lower burst size when plaqued on UC503 than it did on variants of this strain lacking pCI528. The latent period remained unchanged. pCI528 confers a powerful insensitivity to both prolate- and small isometric-headed phage when transferred to lactococcal hosts and is therefore an excellent target for continued genetic analysis (unpublished data). However, the determinants for the two-phage defense mechanisms directed by this plasmid have not yet been separated.

TABLE 1. Phage insensitivity plasmids reported to date

Plasmid	Size (kb)	Conjugal transfer ability[a]	Original host[b]	Determinant cloned[c]	Reference(s)
Adsorption inhibition					
pCI528	46	−	C. UC503	−	Costello[d]
pME0030	48	−	L. ME2	−	23
pSK112	54	−	C. SK11	−	6
pKC50	80	NR	L. 57150	−	Tortorello et al.[e]
Restriction/modification					
pME100	17.5	−	C. KH	−	22
pLR1020	32	−	C. M12R	−	27
pTN20	28	+	L. N1	−	16
pTR2030	48	+	L. ME2	+	16, 17
pTRK68	46	NR	C. NCK202	−	12
p3085-2	15.5	NR	L. 3085	−	28
pIL6	28	+	L. IL594	−	3
pIL7	33	−	L. IL594	+	3; Gautier et al.[f]
pIL103	5.7	NR	C. IL964	+	10; Gautier et al.
pIL107	15.2	NR	C. IL964	−	10
pKR223	38	−	D. KR2	+	18, 20
pBF61	42	−	L. KR5	−	9
pJW563	12	NR	C. W56	−	14
pJW566	25	NR	C. W56	−	14
pJW565	14	NR	C. W56	−	14
Abortive infection					
pTR2030	48	+	L. ME2	+	16, 17
pNP40	64	+	D. DRC3	−	19
pBU1-8	64	+	D. BU1	−	29
pIL105	8.7	NR	C. IL964	+	10; Gautier et al.
pCI528	46	−	C. UC503	−	Costello
pCI750	65	+	C. UC653	+	1, 25
pCI829	44	+	L. UC811	+	4
pAJ1106	106	+	D. 4942	−	13
pKR223	38	−	D. KR2	+	18, 20
pBF61	42	−	L. KR5	−	9
pCLP51R	90	+	L. 33-4	−	8

[a]Presence (+) or absence (−) of evidence in support of conjugal plasmid transfer. NR, Not reported.
[b]Abbreviations: L., L. lactis subsp. lactis; C., L. lactis subsp. cremoris; D., L. lactis subsp. lactis biovar diacetylactis.
[c]Determinant cloned (+) or not yet reported (−).
[d]V. Costello, Ph.D. thesis, National University of Ireland, Cork, 1988.
[e]M. Tortorello, P. Chang, R. Ledford, and G. Dunny, Abstr. 3rd Int. Am. Soc. Microbiol. Conf. Streptococcal Genet., A/50, 1990.
[f]M. Gautier, M. Veaux, and M.-C. Chopin, Abstr. 2nd Symp. Lactic Acid Bacteria, C12, 1987.

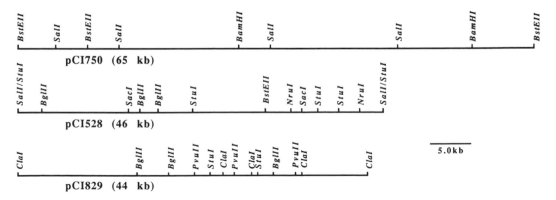

FIG. 1. Restriction maps of phage insensitivity plasmids pCI750, pCI528, and pCI829.

TABLE 2. Comparison between pCI829, pCI750, and pCI528

			Characteristic							
Plasmid	Size (kb)	Self-transmissible	Extent of sensitivity		Mechanism involved	Compatibility with other phage insensitivity plasmids	Temp (37°C) effect on mechanism	Conjugative transfer to:		
			Small isometric	Prolate				*L. lactis* subsp. *lactis*	*L. lactis* subsp. *cremoris*	*L. lactis* subsp. *lactis* biovar diacetylactis
pCI829	44	Yes	Total	Partial	Abortive infection	pI750	None	Yes	Yes	Yes
pCI750	65	Yes	Total	Partial	Abortive infection	pCI829	None	Yes	Yes	Yes
pCI528	46	No	Total	Partial	Abortive infection and absorption inhibition	Unknown	None	Yes	Yes	No

pCI750 and pCI829

The conjugative nature of pCI829 and pCI750 has facilitated their transfer to a number of strains of lactococci. Individually, these plasmids provided total insensitivity to the small isometric-headed phage 712 and partial insensitivity to the prolate-headed phage c2 in an *L. lactis* subsp. *lactis* MG1363Sm background (a plasmid-free derivative of strain 712, sensitive to both of these phages). The partial insensitivity to phage c2 was manifested by significant reductions in burst size and efficiency of plaquing. It was noteworthy that a strain containing both plasmids, designated *L. lactis* subsp. *lactis* AC003, became completely insensitive to phage c2 (4). This finding highlights the additive effect of stacking phage defense mechanisms in selected lactococcal strains.

Molecular cloning of the abortive infection determinants from pCI750 and pCI829 has been achieved. In the case of pCI750, the gene was cloned on a 13.9-kb *Bcl*I fragment by using the streptococcal vector pGB301, giving rise to the recombinant plasmid pMM1 (25). In an *L. lactis* subsp. *lactis* LM0230 background, pMM1 provided total insensitivity to small isometric-headed phage and partial insensitivity to prolate-headed phage in that there was a reduction in plaque size. However, this level of insensitivity mediated by the recombinant plasmid was not as strong as that conferred by the original plasmid pCI750 (unpublished data). A similar observation was made in the case of pCI829. The region encoding abortive infection (Abi) from this plasmid was cloned on a 6.2-kb *Stu*I fragment into *Escherichia coli* by using the *E. coli-L. lactis* shuttle vector pSA3. The recombinant plasmid, designated pCI816, was subsequently transformed into *L. lactis* subsp. *lactis* MG1363Sm, giving rise to strain AC816 (A. Coffey, G. Fitzgerald, and C. Daly, *Abstr. Commission Eur. Communities Meet. Contractors*, p. 37, 1989). The observed reduced level of insensitivity to prolate-headed phage mediated by pMM1 or pCI816, in comparison with pCI750 and pCI829, is most likely due to the low copy number of the cloning vectors used.

Nucleotide Sequence of *abi* from pCI829

The nucleotide sequence analysis of *abi* from pCI829 was facilitated by localizing the gene to a 3.8-kb *Hin*dII fragment situated within the two *Stu*I sites. This 3.8-kb *Hin*dII segment encoded phage insensitivity when cloned in both orientations, indicating that the entire gene was present on this fragment. *Bal*31 analysis showed that approximately 2.0 kb of DNA was essential to confer the phage insensitivity phenotype. Following the identification of the exact location of the *abi* determinant, restriction fragments from within this region were inserted into pUC18 or pUC19. Nucleotide sequences of both strands were determined by using T7 DNA polymerase. An open reading frame of 1,887 bp was observed which was capable of encoding a peptide with a predicted molecular size of 73.8 kDa. Upstream of the open reading frame we identified a putative promoter sequence and ribosome binding site that exhibited similarity to consensus *E. coli* and *Bacillus subtilis* transcription and translation signals. When the nucleotide sequence of the *abi* gene was presented at this meeting, it became apparent that it was identical to the sequence of the *hsp* region from plasmid pTR2030 (11).

Relationship between Plasmid DNA-Encoded Abortive Infection Systems Observed in Lactococci

The rapidly occurring advances in the molecular analysis of genetic determinants encoding abortive infection in lactococci in several laboratories allow comparison between the genes isolated from different sources. The fact that pCI750 and pCI829 can exist unaltered in the same strain suggests that they are unrelated, and indeed, the restriction maps of the cloned DNA from each plasmid did not indicate any similarities. In addition, no homology was detected when 1.0- and 0.3-kb fragments from within the *abi* region of pCI829 were probed against the 13.9-kb *Bcl*I fragment of pMM1, indicating that these two plasmids harbor physically distinct but phenotypically similar mechanisms of phage insensitivity. When Steele et al. (25) used the 13.9-kb fragment of pMM1 as a probe against eight plasmids encoding abortive infection, homology was observed to six of these, including pTR2030. However, caution must be exercised in interpreting these results; the large size of the probe suggests the presence of a large amount of DNA that is not involved in abortive infection. Hill et al. (11) have described similar experiments in which an intergenic *hsp* probe from pTR2030 was used. No homology was detected to pNP40 or to total DNA from 17 other lactococcal strains examined, 11 of which had been designated phage insensitive. These observations suggest that there exist a number of genetically distinct types of abortive infection systems in lactococci.

As noted above, the *hsp* gene from pTR2030 (11) and the *abi* gene from pCI829 (this report) are identical. These plasmids were isolated from different *Lactococcus* strains (based on comparison of plasmid profiles) at geographically distinct locations. It will be interesting to observe the precise relationship between additional abortive infection mechanisms as sequence data from other plasmid-encoded systems become available. It is noteworthy, however, that although *lac* genes identified to date among lactococci have also proved to be identical (2, 5), analysis of proteinase systems has revealed subtle but significant

differences within genes that shared extensive homology and had almost identical restriction maps (7). The data available to date suggest that a number of genetically distinct abortive infection systems are present in different strains of lactococci. This conclusion is likely to be supported by the availability of further sequence data and will augur well for the application of genetic knowledge to the construction of strains with superior phage resistance for commercial use. Already the introduction of pTR2030 into factory cultures and the commercial use of these strains have been described (16, 24, 26). The availability of distinct insensitivity determinants also provides the attractive possibility of stacking defense mechanisms in individual strains to strengthen their phage insensitivity even further.

The data now available suggest that the abortive infection systems identified in lactococci have very similar phenotypes. If this observation is confirmed by more specific experiments, it may be worthwhile to agree on a standard abbreviation for abortive infection, such as Abi, which was originally used by McKay et al. (20). Other terms such as lpr (lytic phage resistance [8]), Rbs (reduced bacteriophage sensitivity [25]) and Hsp (heat-sensitive phage defense [17]) have been used to describe phenotypes that appear to involve abortive infection.

LITERATURE CITED

1. **Baumgartner, A., M. Murphy, C. Daly, and G. F. Fitzgerald.** 1986. Conjugative co-transfer of lactose and bacteriophage resistance plasmids from *Streptococcus cremoris* UC653. *FEMS Microbiol. Lett.* **35**:233–237.
2. **Boizet, B., D. Villeval, P. Slos, M. Novel, G. Novel, and A. Mercenier.** 1988. Isolation and structural analysis of the phospho-β-galactosidase from *Streptococcus lactis* Z268. *Gene* **62**:249–261.
3. **Chopin, A., M.-C. Chopin, A. Moillo-Batt, and P. Langella.** 1984. Two plasmid determined restriction and modification systems in *Streptococcus lactis*. *Plasmid* **11**:260–263.
4. **Coffey, A. G., G. F. Fitzgerald, and C. Daly.** 1989. Identification and characterisation of a plasmid encoding abortive infection from *Lactococcus lactis* ssp. *lactis* UC811. *Neth. Milk Dairy J.* **43**:229–244.
5. **de Vos, W. M., and M. J. Gasson.** 1989. Structure and expression of the *Lactococcus lactis* gene for phospho-β-galactosidase (*lacG*) in *Escherichia coli* and *L. lactis*. *J. Gen. Microbiol.* **135**:1833–1846.
6. **de Vos, W. M., H. M. Underwood, and F. L. Davies.** 1984. Plasmid encoded bacteriophage resistance in *Streptococcus cremoris* SK11. *FEMS Microbiol. Lett.* **23**:175–178.
7. **de Vos, W. M., P. Vos, G. Simons, and S. David.** 1989. Gene organization and expression in mesophilic lactic acid bacteria. *J. Dairy Sci.* **72**:3398.
8. **Dunny, G. M., D. A. Krug, C.-L. Pan, and R. A. Ledford.** 1988. Identification of cell wall antigens associated with a large conjugative plasmid encoding phage resistance and lactose fermentation ability in lactic streptococci. *Biochimie* **70**:443–450.
9. **Froseth, B. R., S. K. Harlander, and L. L. McKay.** 1988. Plasmid-mediated reduced phage sensitivity in *Streptococcus lactis* KR5. *J. Dairy Sci.* **71**:275–284.
10. **Gautier, M., and M.-C. Chopin.** 1987. Plasmid-determined systems for restriction and modification activity and abor-

tive infection in *Streptococcus cremoris*. *Appl. Environ. Microbiol.* **53**:923–927.

11. **Hill, C., L. Miller, and T. R. Klaenhammer.** 1990. Nucleotide sequence of the pTR2030 resistance determinant (*hsp*) which aborts phage infection in lactococci. *Appl. Environ. Microbiol.* **56**:2255–2258.

12. **Hill, C., K. Pierce, and T. R. Klaenhammer.** 1989. The conjugative plasmid pTR2030 encodes two bacteriophage defense mechanisms in lactococci, restriction modification (R$^+$/M$^+$) and abortive infection (Hsp$^+$). *Appl. Environ. Microbiol.* **55**:2416–2419.

13. **Jarvis, A. W.** 1988. Conjugal transfer in lactic streptococci of plasmid-encoded insensitivity to prolate- and small isometric-headed bacteriophages. *Appl. Environ. Microbiol.* **54**:777–783.

14. **Josephsen, J., and F. K. Vogensen.** 1989. Identification of three different plasmid-encoded restriction/modification systems in *Streptococcus lactis* subsp. *cremoris* W56. *FEMS Microbiol. Lett.* **59**:161–166.

15. **Klaenhammer, T. R.** 1987. Plasmid directed mechanisms for bacteriophage defense in lactic streptococci. *FEMS Microbiol. Rev.* **46**:313–325.

16. **Klaenhammer, T. R.** 1989. Genetic characterization of multiple mechanisms of phage defense from a prototype phage-sensitive strain, *Lactococcus lactis* ME2. *J. Dairy Sci.* **72**:3429–3443.

17. **Klaenhammer, T. R., and R. B. Sanozky.** 1985. Conjugal transfer from *Streptococcus lactis* ME2 of plasmids encoding phage resistance, nisin resistance and lactose-fermenting ability: evidence for a high-frequency conjugative plasmid responsible for abortive infection of virulent bacteriophage. *J. Gen. Microbiol.* **131**:1531–1541.

18. **Laible, N. J., P. L. Rule, S. K. Harlander, and L. L. McKay.** 1987. Identification and cloning of plasmid deoxyribonucleic acid coding for abortive phage infection from *Streptococcus lactis* ssp. *diacetylactis* KR2. *J. Dairy Sci.* **70**:2211–2219.

19. **McKay, L. L., and K. A. Baldwin.** 1984. Conjugative 40-megadalton plasmid in *Streptococcus lactis* subsp. *diacetylactis* DRC3 is associated with resistance to nisin and bacteriophage. *Appl. Environ. Microbiol.* **47**:68–74.

20. **McKay, L. L., M. J. Bohanon, K. M. Polzin, P. L. Rule, and K. A. Baldwin.** 1989. Localization of separate genetic loci for reduced sensitivity towards small isometric-headed bacteriophage sk1 and prolate-headed bacteriophage c2 on pGBK17 from *Lactococcus lactis* subsp. *lactis* KR2. *Appl. Environ. Microbiol.* **55**:2702–2709.

21. **Sanders, M. E.** 1988. Phage resistance in lactic acid bacteria. *Biochimie* **70**:411–422.

22. **Sanders, M. E., and T. R. Klaenhammer.** 1981. Evidence for plasmid linkage of restriction and modification in *Streptococcus cremoris* KH. *Appl. Environ. Microbiol.* **42**:944–950.

23. **Sanders, M. E., and T. R. Klaenhammer.** 1983. Characterization of phage-sensitive mutants from a phage-insensitive strain of *Streptococcus lactis*: evidence for a plasmid determinant that prevents phage adsorption. *Appl. Environ. Microbiol.* **46**:1125–1133.

24. **Sanders, M. E., P. J. Leonhard, W. E. Sing, and T. R. Klaenhammer.** 1986. Conjugal strategy for construction of fast acid-producing, bacteriophage-resistant lactic streptococci for use in dairy fermentations. *Appl. Environ. Microbiol.* **52**:1001–1007.

25. **Steele, J. L., M. C. Murphy, C. Daly, and L. L. McKay.** 1989. DNA-DNA homology among lactose- and sucrose-fermenting transconjugants from *Lactococcus lactis* strains exhibiting reduced bacteriophage sensitivity. *Appl. Environ. Microbiol.* **55**:2410–2413.

26. **Steenson, L. R., and T. R. Klaenhammer.** 1985. *Streptococcus cremoris* M12R transconjugants carrying the conjugal plasmid pTR2030 are insensitive to attack by lytic bacteriophages. *Appl. Environ. Microbiol.* **50**:851–858.

27. **Steenson, L. R., and T. R. Klaenhammer.** 1986. Plasmid heterogeneity in *Streptococcus cremoris* M12R: effects on proteolytic activity and host-dependent phage replication. *J. Dairy Sci.* **69**:2227–2236.

28. **Teuber, M.** 1986. Final report of the achievements of the research programme on construction of phage resistant dairy starter cultures, p. 539–547. *In* E. Magnien (ed.), *Biomolecular Engineering in the European Community*. Martinus Nijhoff, Dordrecht, The Netherlands.

29. **Wetzel, A., H. Neve, A. Geis, and M. Teuber.** 1986. Transfer of plasmid-mediated phage resistance in lactic acid streptococci. *Chem. Mikrobiol. Technol. Lebensm.* **10**:86–89.

Lactococcal Plasmid Replicon: Vector Construction and Genetic Organization

FENGFENG XU,[1] PAK-LAM YU,[1] AND LINDSAY E. PEARCE[2]

Department of Biotechnology, Massey University,[1] and New Zealand Dairy Research Institute,[2]
Palmerston North, New Zealand

The application of recombinant DNA technology to the lactococci has greatly increased our understanding of these dairy bacteria. Construction of stable vectors with high transformation efficiency is essential for cloning genes, including those for industrially important traits such as lactose utilization, proteinase activity, and phage resistance.

Vectors based on nonlactococcal replicons are generally unstable in lactococci (14). Recently, constructs derived from the homologous lactococcal plasmids pWV01 and pSH71 have been described (5). These vectors, which are based solely on lactococcal plasmid replicons, replicate in many gram-positive organisms as well as in *Escherichia coli*, and they have proved to be most useful in lactococcal gene cloning studies.

Although there is no difficulty in introducing lactococcal vectors into plasmid-free strains by electroporation, transformation of wild-type lactococci, which carry an average of four to five plasmids, has been much less successful in this laboratory. Plasmid incompatibility will clearly be an important consideration if cloning vectors are to be used with such strains (16). For this reason, it would be useful to be able to select from a variety of vectors based on different lactococcal replicons. Plasmid pDI25 (3, 19) was chosen as the basis to construct a series of cloning vectors and to analyze the lactococcal plasmid replicon at the molecular level. The complete genetic organization of a lactococcal replicon has not yet been reported; such information would be useful for further lactococcal vector construction and plasmid replication studies.

Construction of a New Family of Lactococcal Vectors

The 5.5-kb high-copy-number cryptic plasmid pDI25 from *Lactococcus lactis* subsp. *lactis* 5136 (3, 19) was isolated and used to construct a series of vectors. A recombinant plasmid (pFX1, 5.5 kb) was first made by ligating the 4.5-kb *Hpa*II-*Mbo*I fragment of pDI25 to the 1.0-kb chloramphenicol transacetylase structural gene from the staphylococcal plasmid pC194 (8, 19; Fig. 1). pFX1 was further modified by deleting a nonessential 1.9-kb *Cla*I region to construct pFX2 (3.6

kb). By further deletions, the essential region for plasmid replication was located within the 1.2-kb *Cfo*I-*Tha*I-*Cfo*I region (19a).

pFX3 was constructed by incorporating the α fragment of the *E. coli lacZ* structural gene, a multiple cloning region, and the T7 and T3 promoters from pUBS (constructed by R. L. S. Forster) into pFX2 (Fig. 1). By cloning into one of the multiple cloning sites, recombinant plasmids could be directly selected in the appropriate *E. coli* hosts by inactivation of the *lacZ* function. Recombinant plasmids isolated from white colonies on 5-bromo-4-chloro-3-indolyl-β-D-galactopyranoside (X-Gal) plates could then be electroporated into lactococci. pFX3 can also be directly used for transcription studies or DNA sequencing of cloned inserts.

A set of translational gene fusion vectors was constructed by incorporating the *E. coli lacZ* gene fusion system (pNM480, -481, and -482 [10]) into pFX2 (Fig. 1). These constructions, pFX4, pFX5, and pFX6, permit the fusion of cloned genes to *lacZ* in all three reading frames. The *lacZ* expression requires a promoter on the cloned fragment in the correct translational reading frame. Gene expression can be readily and quantitatively monitored by measuring β-galactosidase activity. The sizes of pFX constructs are shown in Fig. 2. All of the pFX vectors could be efficiently introduced into lactococci and *E. coli* by electroporation (10^4 to 10^6 CFU/μg of DNA in each host) and maintained stably in both organisms (>95% cells carrying the Cm marker after 100 generations of growth without drug selection).

Lactococcal Gene Cloning and Transcription Studies

Plasmid pDI21 encodes a PI-type proteinase in its original host, *L. lactis* subsp. *cremoris* H2 (19). The *prt* genes were located on a 6.5-kb *Hind*III fragment and were first cloned into *E. coli* with λNM1149 (19). Using pFX1, the *prt* gene fragment was recloned and directly electroporated into lactococci, in which it was efficiently expressed. No rearrangements in either the insert or the vector DNA were detected after electro-

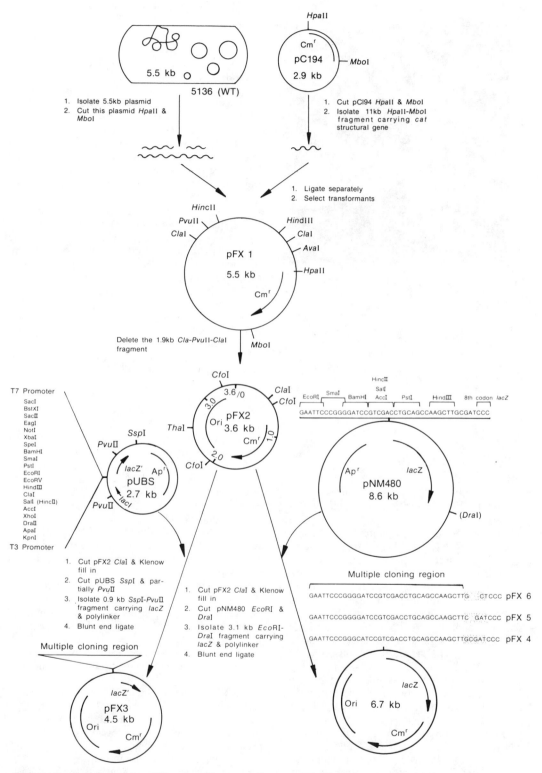

FIG. 1. Construction of the pFX series of lactococcal vectors. The shaded nucleotides mark the only differences between pFX4, pFX5, and pFX6 vectors.

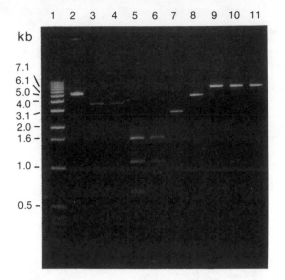

FIG. 2. Sizes of the pFX series constructs after restriction digestions. Lanes: 1, Bethesda Research Laboratories 1-kb DNA fragment ladder as molecular size standard; 2, pFX1 (5.5 kb) digested with *Hin*dIII; 3, pFX1 digested with *Cla*I; 4, pFX2 (3.6 kb) digested with *Cla*I; 5, pFX2 digested with *Cfo*I; 6, pFX7 (2.9 kb, pFX2 with the 0.7-kb *Cfo*I-*Cla*I-*Cfo*I region deleted) digested with *Cfo*I; 7, pFX7 digested with *Hpa*II; 8, pFX3 (4.5 kb) digested with *Eco*RI; 9 to 11, pFX4, pFX5, and pFX6 (all 6.7 kb), respectively, digested with *Sma*I.

poration. Greater than 99% of cells still carried the *prt* recombinant plasmid after growth of approximately 100 generations in broth without drug selection.

The tagatose-1,6-bisphosphate aldolase gene of *L. lactis* subsp. *lactis* 4560 is located on the 4.4-kb *Eco*RI fragment of plasmid pDI1 and is expressed in *E. coli* (9, 21). This 4.4-kb *Eco*RI fragment was cloned into pFX3 and transformed into *E. coli*. The recombinant plasmid pFX301 could then be electroporated into *L. lactis* subsp. *lactis* 4125, in which it was stably maintained. The expression of this gene in lactococci was not examined. Using the translational fusion vectors pFX4, pFX5, and pFX6, the 6.5-kb *Hin*dIII *prt* gene fragment of pDI21 was identified as having two promoters with opposite orientations (19a). This result was consistent with the published sequences of other lactococcal *prt* genes (7, 18). The 2.0-kb *Eco*RI fragment of pDI21 encoding the N-terminal part of galactose-6-phosphate isomerase was also cloned. Gene expression was directed toward the D-tagatose-6-phosphate kinase gene and likely to be controlled by the isomerase promoter (19a, 20).

Genetic Organization of Lactococcal Plasmid pFX2

The complete DNA sequence of the lactococcal portion of pFX2 (*Hpa*II-*Mbo*I, 2,508 bp; Fig. 1) was determined by the dideoxy sequencing method (13), and the genetic organization was analyzed. A 215-bp region that included the recombination site A sequence was found to be 100% homologous to the staphylococcal plasmid pE194 (8). The 215-bp sequence was in a nonessential 0.7-kb *Cfo*I-*Cla*I-*Cfo*I region that could be deleted in pFX2 to give pFX7 (Fig. 2, lane 6). A putative plus *ori* site with strong homology to those of the pE194 group plasmids (6) and two open reading frames (ORFs) were located within the 1.2-kb *Cfo*I-*Tha*I-*Cfo*I minimum replicon. ORF1 (part of *cop* in Fig. 3) was preceded by a consensus lactococcal promoter and ribosome binding site and encoded a 53-amino-acid peptide which had an α helix-turn-α helix motif, a geometry typical of proteins that act as DNA-binding repressors (12). The two ORFs were separated by 66 bp (also part of *cop* in Fig. 3) containing an inverted repeat of 23 bp. ORF2 (*rep* in Fig. 3) encoded a 233-amino-acid peptide, which showed strong homology at that protein level to the whole sequence of the pLS1 replication initiation protein (4) and to the N-terminal regions of replication initiation proteins of pE194 (17), pLB4 (1), and pADB201 (2). There appear to be two possible mechanisms for controlling the transcription of *rep*. First, the DNA-binding repressor (ORF1) could bind at a putative operator site, a region immediately upstream of itself that has strong homology to the pLS1 (4) operator. Second, the RNA transcribed from the inverted repeat between ORF1 and ORF2 could be capable of forming either attenuator- or terminator-type configurations.

The individual components of the replication apparatus of pFX2 have homology with those of plasmids from other bacterial genera, including pMV150 (parent plasmid of pLS1; 15; *Streptococcus* sp.), pE194 (8, 17; *Staphylococcus* sp.), pLB4 (1; *Lactobacillus plantarum*), and pADB201 (2; *Mycoplasma* sp.). These replicons are very similar in genetic organization (Fig. 3). The region between the *ori* site and the replication initiation protein-coding region comprises the negative control system (*cop*) (11) in these plasmids.

The pFX series of vectors provides an efficient means for gene cloning and the study of gene expression in lactococci. Cloning can be in either *E. coli* or lactococci, and the vectors are stably maintained in each host. The sequencing results have provided an insight into the genetic organization of a lactococcal plasmid replicon. This information will assist further plasmid studies with these industrially important bacteria.

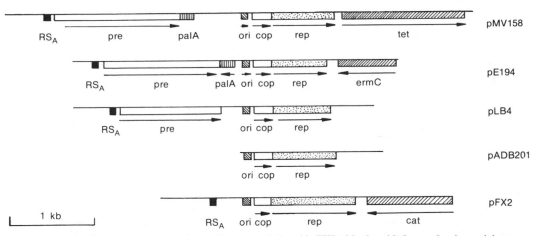

FIG. 3. Genetic organization comparison of lactococcal plasmid pFX2 with plasmids from other bacterial genera. Arrows represent direction of transcription for major ORFs and functional orientation for other sequence elements. Abbreviations: RS$_A$, recombination site A; *pre*, plasmid recombination enzyme; *palA*, lagging-strand conversion signal; *cop*, replication control system; *rep*, replication initiation protein; *tet*, tetracycline resistance; *ermC*, rRNA methylase; *cat*, chloramphenicol transacetylase.

We thank R. L. S. Forster for plasmid pUBS and N. P. Minton for plasmids pNM480, pNM481, and pNM482.

LITERATURE CITED

1. **Bates, E. E. M., and H. J. Gilbert.** 1989. Characterization of a cryptic plasmid from *Lactobacillus plantarum. Gene* **85:**253–258.
2. **Bergemann, A. D., J. C. Whitly, and L. R. Finch.** 1989. Homology of mycoplasma plasmid pADB201 and staphylococcal plasmid pE194. *J. Bacteriol.* **171:**593–595.
3. **Crow, V. L., G. P. Davey, L. E. Pearce, and T. D. Thomas.** 1983. Plasmid linkage of the D-tagatose 6-phosphate pathway in *Streptococcus lactis*: effect on lactose and galactose metabolism. *J. Bacteriol.* **153:**76–83.
4. **de la Campa, A. G., G. H. de Solar, and M. Espinosa.** 1990. Initiation of replication of plasmid pLS1. The initiator protein repB acts on two distant DNA regions. *J. Mol. Biol.* **213:**247–262.
5. **de Vos, W. M.** 1987. Gene cloning and expression in lactic streptococci. *FEMS Microbiol. Rev.* **46:**281–285.
6. **Gruss, A., and S. D. Ehrlich.** 1989. The family of highly interrelated single-stranded deoxyribonucleic acid plasmids. *Microbiol. Rev.* **53:**231–241.
7. **Haandrikman, A. J., J. Kok, H. Laan, S. Soemitro, A. M. Ledeboer, W. N. Konings, and G. Venema.** 1989. Identification of a gene required for maturation of an extracellular lactococcal serine proteinase. *J. Bacteriol.* **171:**2789–2794.
8. **Horinouchi, S., and B. Weisblum.** 1982. Nucleotide sequence and functional map of pE194, a plasmid that specifies inducible resistance to macrolide, lincosamide, and streptogramin type B antibiotics. *J. Bacteriol.* **150:**804–814.
9. **Limsowtin, G. K. Y., V. L. Crow, and L. E. Pearce.** 1986. Molecular cloning and expression of the *Streptococcus lactis*, 1,6-bisphosphate aldolase gene in *Escherichia coli. FEMS Microbiol. Lett.* **33:**79–83.
10. **Minton, N. P.** 1984. Improved plasmid vectors for the isolation of translational *lac* gene fusions. *Gene* **31:**269–273.
11. **Novick, N. P.** 1989. Staphylococcal plasmids and their replication. *Annu. Rev. Microbiol.* **43:**537–565.

12. **Pabo, C. D., and R. T. Sauer.** 1984. Protein-DNA recognition. *Annu. Rev. Biochem.* **53:**293–321.
13. **Sanger, F., F. Nicklen, and A. R. Coulson.** 1977. DNA sequencing with chain-terminating inhibitors. *Proc. Natl. Acad. Sci. USA* **74:**5463–5467.
14. **Simon, D., and A. Chopin.** 1988. Construction of a vector plasmid family and its use for molecular cloning in *Streptococcus lactis. Biochimie* **70:**559–566.
15. **van der Lelie, D., S. Bron, G. Venema, and L. Oskam.** 1989. Similarity of minus origin of replication and flanking open reading frames of plasmids pUB110, pTB913 and pMV158. *Nucleic Acids Res.* **18:**7283–7294.
16. **van der Lelie, D., J. M. B. M. van der Vossen, and G. Venema.** 1988. Effect of plasmid incompatibility on DNA transfer to *Streptococcus cremoris. Appl. Environ. Microbiol.* **54:**865–871.
17. **Villafane, R., C. H. Bechofer, C. S. Narayanan, and D. Dubnau.** 1987. Replication control genes of plasmid pE194. *J. Bacteriol.* **169:**4822–4829.
18. **Vos, P., M. van Asseldonk, F. van Jeveren, R. Siezen, G. Simons, and W. M. de Vos.** 1989. A maturation protein is essential for production of active forms of *Lactococcus lactis* SK11 serine proteinase located in or secreted from the cell envelope. *J. Bacteriol.* **171:**2795–2802.
19. **Xu, F., L. E. Pearce, and P.-L. Yu.** 1990. Molecular cloning and expression of a proteinase gene from *Lactococcus lactis* subsp. *cremoris* H2 and construction of a new lactococcal vector pFX1. *Arch. Microbiol.* **154:**99–104.
19a. **Xu, F., L. E. Pearce, and P.-L. Yu.** 1991. Construction of a family of lactococcal vectors for gene cloning and translational fusions. *FEMS Microbiol. Lett.* **77:**55–60.
20. **Yu, P.-L., R. D. Appleby, G. G. Pritchard, and G. K. Y. Limsowtin.** 1989. Restriction mapping and localization of the lactose-metabolizing genes of *Streptococcus cremoris* pDI21. *Appl. Microbiol. Biotechnol.* **30:**71–74.
21. **Yu, P.-L., G. K. Y. Limsowtin, V. L. Crow, and L. E. Pearce.** 1988. In vivo and in vitro expression of tagatose 1,6-bisphosphate aldolase gene of *Streptococcus lactis. Appl. Microbiol. Biotechnol.* **28:**471–473.

Pulsed-Field Gel Electrophoresis as a Tool for Studying the Phylogeny and Genetic History of Lactococcal Strains

P. LE BOURGEOIS, M. MATA, AND P. RITZENTHALER

Centre de Recherche de Biochimie et Génétique Cellulaires, CTBM-INSA, Toulouse, France

Very little information concerning the genome organization of the lactic acid bacteria is available. The aim of this study was to estimate the sizes of some lactococcal chromosomes and to compare the genomic restriction patterns of various *Lactococcus lactis* strains with each other or with those of other species more or less related to *Lactococcus* species.

Genome Size of Some Streptococcal Strains

A previously described method (2, 8, 11) was used to prepare and to digest genomic DNAs and to separate the resulting fragments by pulsed-field gel electrophoresis (PFGE). The two restriction enzymes *Sma*I and *Apa*I were found to produce distributions of fragments useful for genome analysis of streptococcal strains (Table 1). In contrast, *Not*I, a restriction enzyme with an eight-base recognition sequence, produced only a small number of fragments (Fig. 1). The lactococcal chromosomes were cut into three or four *Not*I fragments except for strains A15, Z254, Z146, and 188, which gave five or six *Not*I restriction fragments (Fig. 1, lanes 1, 2, 3, and 16); the *Enterococcus faecalis* genome contained more than 10 fragments (lane 7).

The molecular sizes of the chromosomal DNAs were determined by adding the sizes of all of the restriction fragments generated by *Sma*I and *Apa*I endonuclease digestions (Table 1). The lactococcal chromosomes had sizes comparable to those of the *E. faecalis* and *Streptococcus sanguis* chromosomes (2.3 to 2.6 megabases [Mb]). The genome size of *Streptococcus thermophilus* ST1 was significantly lower (1.7 Mb). In comparison, the genome size of *Escherichia coli* is 4.55 Mb (16). Plasmids of different sizes are present in some lactococcal strains (C2, Z268, 187, BK5, H2, and F7/2). These plasmids are potentially able to produce bands indistinguishable from chromosomal bands. Almost all of the small plasmid open circular and covalently closed circular forms as well as large plasmid covalently closed circular forms were eliminated during preparation of the agarose inserts by a preelectrophoresis (8). Large plasmid open circular forms did not migrate during PFGE, but they could interfere in the genome size determination only if they had been cut into fragments by digestion with *Apa*I or *Sma*I. However, comparison between the plasmid-containing strain C2 and its plasmid-cured derivative LM2301 showed that the genome size determination was not significantly affected by the presence of plasmids. It is noteworthy that the unique streptococcal species used in this study and known to be the least nutritionally fastidious species among the streptococci (17), i.e., *S. equinus* (formerly *S. bovis*) ATCC 33317, had a genome size similar to that of *S. thermophilus*. Usually, streptococcal strains have numerous growth requirements; it is unlikely that the small size of the streptococcal genomes is due to large deletions in gene clusters involved in amino acid, vitamin, or nucleotide biosynthesis pathways. This hypothesis is supported by preliminary work on amino acid biosynthesis in *L. lactis* IL1403 showing that in this strain the *his* and *trp* genes were present but silent (P. Renault and C. Delorme, *Abstr. 3rd Int. Am. Soc. Microbiol. Conf. Streptococcal Genet.*, abstr. A/67, p. 27, 1990).

Comparison of Lactococcal Lysogenic Strains and Phage-Propagating (Indicator) Host Strains

In a previous study (8), we have shown that the genomic restriction pattern analysis has no taxonomic value but enables the rapid detection of genetically related strains; for example, *L. lactis* BK5 and its propagating host H2 were shown to be closely related. To determine whether this observation extends to other lactococcal strains or whether BK5 and H2 constitute an isolated nonrepresentative case, *Sma*I restriction patterns from 18 strains (lysogenic and indicator strains) were examined (Table 2; Fig. 2 and 3). Groups of genetically related strains were determined as follows: when the restriction patterns of various strains were compared in a previous study (8), about 80% comigrating fragments were observed for isogenic or closely related strains. This percentage fell to 20 to 40% for strains of the same species that were not closely related as well as for strains of different species. Lactococcal strains were classified as being in the same genetic group when about 60% of their *Sma*I fragments comigrated.

By comparing the *Sma*I restriction profiles (Fig. 2 and 3), three genetic groups could be constituted (Table 3). The lysogenic strains 187, 188, 189, BK5, and C2 have been shown to contain genetically closely related prophages (15). All of

TABLE 1. Sizes of restriction fragments of genomic DNA from *L. lactis*, *Enterococcus*, and *Streptococcus* strains[a]

Fragment size (kbp)

L. lactis																		*E. faecalis*		*S. sanguis*		*S. equinus* ATCC 33317			*S. thermophilus* ST1	
C2		LM2301		MG1363		IL1403		187		BK5		H2		F166		F7/2		JH2		BM4154					ST1	
ApaI	SmaI	ApaI	SmaI	ApaI	SmaI	ApaI	SmaI	ApaI	SmaI	ApaI	SmaI	ApaI	SmaI	ApaI	SmaI	ApaI	SmaI	ApaI	SmaI	ApaI	SmaI	ApaI	SmaI	SacII	ApaI	SmaI
205	550	205	550	205	610	225	765	300	560	225	540	270	500	215	620	220	720	290	365	320	280	500	1200	195	175	310
200	325	200	325	200	325	225	235	230	335	215	435	225	435	150	580	215	235	150	300	240	270	420	180	175	165	215
175	290	175	290	175	290	190	145	220	290	215	220	195	220	150	245	195	210	140	290	200	225	360	140	145	150	205
165	220	160	220	160	220	180	135	180	230	195	210	185	210	135	190	175	115	125	240	185	210	170	135	130	130	175
115	220	120	180	145	180	150	120	175	185	185	170	170	170	115	185	150	110	125	210	165	155	160	125	125	130	155
102	180	115	180	115	180	140	110	170	165	170	150	135	150	115	130	140	105	110	165	145	145	125	60	90	105	145
97	130	102	120	97	130	130	105	125	150	130	150	120	150	105	115	130	95	110	140	115	140	70	45	86	97	110
85	120	99	120	85	120	115	95	98	105	120	87	90	87	97	70	115	80	100	130	105	130	55		85	90	95
85	120	85	105	83	105	102	80	96	67	90	67	79	82	80	63	100	80	95	125	105	110	26		80	72	95
80	105	83	63	80	49	98	75	94	67	79	67	79	75	80	63	88	75	85	115	97	85			72	67	80
76	63	80	49	76	44	92	73	80	63	79	63	77	75	74	55	81	66	80	110	85	75			69	61	70
74	60	75	44	74	41	88	68	78	58	77	60	77	67	69	44	73	66	73	110	85	70			59	51	55
69	49	73	44	73	33	80	64	76	53	77	56	70	63	66	42	58	58	71	78	71	52			55	50	47
67	45	71	33	71	24	75	47	70	44	70	44	60	63	61	42	58	58	70	70	54	47			55	49	45
59	43	67	23	67	23	56	47	70	40	60	40	60	60	59	33	55	49	68	70	43	44			51	42	40
56	33	59	19	67	18	51	41	60	35	60	35	50	56	53	32	48	49	66	54	43	37			47	40	12
50	23	50	18	59	16	48	38	47	35	59	25	49	44	50	24	46	44	60	43	30	33			40	40	6
49	19	49	16	50	12	45	24	46	25	45	21	44	40	43	23	44	40	56	40	26	25			37	32	
45	16	45	12	50	3	41	20	44	23	44	16	43	26	42	19	41	23	54	38	22	25			33	26	
41	12	41	3	49		33	5	43	21	44	14	41	25	38	18	38	22	50	36	20	20			29	26	
40	3	40		45		25	4	41	16	43	14	40	21	37	14	34	20	46	36	18	16			29	24	
37		37		41		20		39	14	41	12	39	16	37	12	26	5	43	15	18	15			26	19	
37		33		40		20		38	13	41	9	38	15	35	4	24	4	40	11		5			21	19	
35		32		37		17		35	9	39	5	36	12	34		22		37	9					20	17	
34		29		37		17		34	7	38	4	35	9	33		20		35						18	15	
33		26		33		14		29	4	35		34	5	32		19		32						15	13	
32		24		32		12		28		34		31	4.5	30		16		30						14	9	
29		24		29		9		26		28		28	4	29		15		28						9		
26		20		26				25		24		24		27				25						6		
26		19		24				24		23		23		26				23								
25		17		24				23		23		17		26				22								
24		17		20				23		17		14		25				20								
24		12		19				18		14		10		24				20								
20				17				15		10		7		24				19								
19				16				7		7		4		23				18								
17				12						4				22				17								
16				7										20				15								
15				4										19				14								
12				2										17				13								
														17				11								
														12				10								
														7				7								
														6												
														4												
														2												
Total[b] 2,438	2,588	2,303	2,358	2,446	2,530	2,298	2,296	2,684	2,614	2,660	2,500	2,499	2,684.5	2,359	2,623	2,531	2,469	2,503	2,730	2,334	2,214	1,886	1,885	1,816	1,744	1,765

[a] Each restriction digest was analyzed with six different pulse times (1, 2.5, 6, 15, 25, and 40 s) to obtain optimal separation in the range of 2 to 1,200 kbp.
[b] The error of the genome size was estimated to be ±100 kbp.

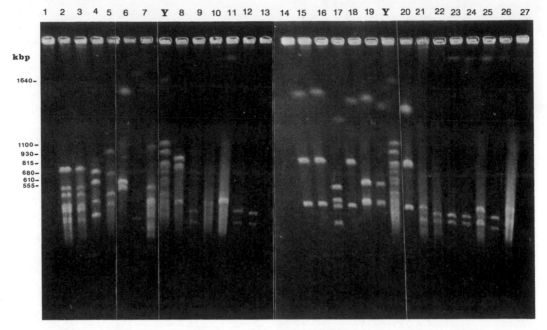

FIG. 1. *Not*I restriction patterns of *Lactococcus*, *Enterococcus*, and *Streptococcus* strains. Lanes: 1, *L. cremoris* A15; 2, *L. lactis* Z254; 3, *L. lactis* Z146; 4, *L. lactis* A45; 5, *L. lactis* F7/2; 6, *S. thermophilus* ST1; 7, *E. faecalis* JH2; 8, *L. lactis* A100; 9, *L. lactis* Z251; 10, *L. lactis* A61; 11, *L. lactis* Z304; 12, *L. lactis* MG1363; 13, *L. lactis* Z268; 14, *L. cremoris* 187; 15, *L. cremoris* 189; 16, *L. cremoris* 188; 17, *L. lactis* 205; 18, *L. cremoris* BK5; 19, *L. cremoris* H2; 20, *L. cremoris* 1-S; 21, *L. lactis* NCDO 505 (C2); 22, *L. lactis* NCDO 2031 (C2); 23, *L. lactis* C2; 24, *L. lactis* LM2301; 25, *L. lactis* NCDO 712; 26, *L. lactis* A311; 27, *L. lactis* NCDO 763. Lanes Y (yeast chromosomal DNA of strain 334) yielded the size standards shown on the left. PFGE conditions were 150 V for 33 h with a 150-s pulse time in 1% agarose and 0.05 M TBE (1 M TBE is 1 M Tris, 1 M boric acid, and 0.02 M EDTA), using an LKB-Pharmacia CHEF Pulsaphor apparatus.

these strains except C2, as well as the corresponding phage-propagating hosts 205, H2, and 1-S, could be classified into a unique genetic group. Strains obtained from the CNRZ-INA collection (France) were also analyzed (Table 2). These lysogenic strains produced temperate phages be-

longing to a single DNA homology group (7) and could be differentiated into two distinct genetic groups (B and C; Table 3). Z151 and A100 had atypical *Sma*I restriction patterns (Fig. 2). The indicator strain A45 was in the same group as its corresponding prophage-containing strains, con-

TABLE 2. Strains analyzed[a]

Lysogenic strains		Propagating host	
Designation	Origin (reference)	Designation	Origin (reference)
187	Starter, U.S.A. (5)	205	Starter, U.S.A. (5)
188	Starter, U.S.A. (5)	205	
189	Starter, U.S.A. (5)	205	
BK5	Starter, U.S.A. (5)	H2	Starter, U.S.A. (5)
C2	From NCDO 712 (5)	1-S	Starter, U.S.A. (5)
Z151	Starter, France (7)	Z268	Natural starter, France (7)
Z304	NCDO 700, U.K., 1950 (7)	Z268	
A61	Raw milk, France, 1960 (7)	Z268	
A311	Raw milk, France, 1970 (7)	Z268	
A100	Starter, France, 1965 (7)	A45	Starter, Ressons, France, 1958 (7)
A15	Raw milk, France, 1958 (7)	A45	
Z146	NCDO 509, Australia, 1955 (7)	A45	
Z254	Roquefort cheese, France (7)	A45	

[a] The following strains were also used: *L. lactis* F166, plasmid-free derivative of Z268 (14); F7/2 (9); IL1403 (1); NCDO 712 (3) and derivatives; *L. lactis* subsp. *cremoris* S77; *E. faecalis* JH2 (6); *S. sanguis* BM4154 (10); *S. equinus* ATCC 33317; and *S. salivarius* subsp. *thermophilus* ST1.

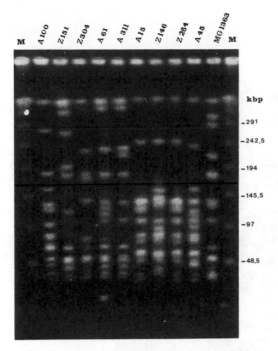

FIG. 2. PFGE separation of *Sma*I digests of lactococcal chromosomes. Lanes M (phage λ DNA concatemers) yielded the size standards indicated on the right. Running conditions were 275 V for 14 h with a 20-s pulse time.

trary to the other host strain Z268 (or its plasmid-free derivative F166), which had a distinct *Sma*I restriction profile.

These results show that (i) a lysogenic strain and the corresponding indicator strain do not necessarily have similar restriction profiles and (ii) most often, lysogenic strains containing related prophages belong to the same genetic group.

Comparison of Strain *L. lactis* 712 and Its Derivatives

Davies et al. suggested that strains 712, ML3 and C2 were closely related on the basis of their similar plasmid profiles (3). Examination of National Collection of Dairy Organisms (NCDO) records (3) showed that strain 712 was first deposited in the NCDO in 1954 and distributed to different laboratories, becoming relabeled as ML3 and C2. We have followed the genetic history of strain 712 by analyzing the restriction patterns of strains NCDO 712, NCDO 763 (ML3), NCDO 505 (C2), NCDO 2031 (C2), and C2. This last strain without other mentioned indication came from the University of Minnesota (L. L. McKay) via J. L. Parada (Argentina). All of these strains were expected to be isogenic, but they

have been handled in different laboratories for several years. Strains MG1363 and LM2301 were also included in this study. MG1363 originated from NCDO 712 (4), and LM2301 originated from NCDO 2031 (12, 13). Both strains were cured of their prophages and plasmids by different treatments.

The close relationship of these strains was confirmed, since the *Not*I and *Sma*I profiles of all of the strains were very similar (Fig. 1 and 3). Only minor variations were observed. The *Sma*I restriction pattern of NCDO 763 showed the largest difference compared with that of the other strains (Fig. 3). In addition to the variations observed in the small-size fragments, the 325-kb *Sma*I fragment common in all of the profiles was absent only from NCDO 763, whereas a 160-kb band was observed only in the profile of this strain. These variations suggest that NCDO 763 has diverged more from strain NCDO 712 than have the other strains.

The 175-kb *Not*I band present in strains C2, NCDO 2031, and LM2301 was replaced by a 230-kb *Not*I band in the profiles of NCDO 712, NCDO 505, and MG1363 or by a 210-kb *Not*I band in NCDO 763 (Fig. 1). The same strains diverged in their *Sma*I restriction patterns: the largest fragment (610 kb) present in the profile of NCDO 712, NCDO 505, and MG1363 (or the 590-kb fragment in the NCDO 763 profile) was absent in those of C2, LM2301, and NCDO 2031, and an additional 550-kb fragment appeared. Preliminary restriction mapping data indicated that all three *Not*I restriction sites of these strains are located in the same 610-kb *Sma*I fragment. The difference observed in the *Sma*I and *Not*I profiles between NCDO 712 and its derivatives suggested that in some derivatives (C2, NCDO 2031, and LM2301), a deletion of about 60 kb occurred. This deletion is located on the 230-kb *Not*I fragment included in the 610-kb *Sma*I fragment. M. J. Gasson (*Abstr. 3rd Int. Am. Soc. Microbiol. Conf. Streptococcal Genet.*, abstr. 3, p. 10, 1990) showed by DNA-DNA hybridizations that a sex factor was located on the largest *Sma*I fragment of strain NCDO 712. The deletion of 60 kb occurring in the 610-kb *Sma*I fragment in strains C2, LM2301, and NCDO 2031 could be corre-

TABLE 3. Classification of strains according to *Sma*I restriction profiles

Genetic group	Strains
A	187, 188, 189, BK5, 205, H2, 1-S
B	A15, Z146, Z254, A45
C	Z304, A61, A311

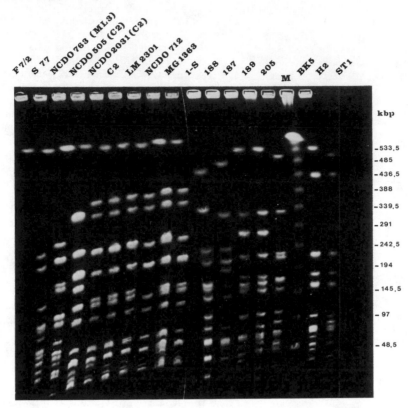

FIG. 3. PFGE of lactococcal chromosomes digested with SmaI. Lanes M (phage λ DNA concatemers) yielded the size standards indicated on the right. Running conditions were 275 V for 12 h with a 15-s pulse time.

lated with the lost of a chromosomally located sex factor. The DNA rearrangements occurring in the derivatives of strain NCDO 712 remain to be more precisely determined.

This work was supported by grants from the Centre National de la Recherche Scientifique (LP8201, Action "Bactéries Lactiques" du programme Biotechnologies). We thank M.-C. Chopin, M. Desmazeaud, G. Novel, and W. Sandine for providing strains and information concerning the origin of some lactococcal strains used in this work and M. Coddeville for technical assistance.

LITERATURE CITED

1. **Chopin, A., M.-C. Chopin, A. Moillo-Batt, and P. Langella.** 1984. Two plasmid determined restriction and modification systems in *Streptococcus lactis. Plasmid* **11**:260–263.
2. **Chu, G., D. Vollrath, and R. W. Davis.** 1986. Separation of large DNA molecules by contour-clamped homogeneous electric fields. *Science* **234**:1582–1585.
3. **Davies, F. L., H. M. Underwood, and M. J. Gasson.** 1981. The value of plasmid profiles for strain identification in lactic streptococci and the relationship between *Streptococcus lactis* 712, ML3 and C2. *J. Appl. Bacteriol.* **51**:325–337.
4. **Gasson, M. J.** 1983. Plasmid complements of *Streptococcus lactis* NCDO 712 and other lactic streptococci after protoplast-induced curing. *J. Bacteriol.* **154**:1–9.
5. **Huggins, A. R., and W. E. Sandine.** 1977. Incidence and

6. **Jacob, A. E., and S. J. Hobbs.** 1974. Conjugal transfer to plasmid-borne multiple antibiotic resistance in *Streptococcus faecalis* var. *zymogenes. J. Bacteriol.* **117**:360–372.
7. **Lautier, M., and G. Novel.** 1987. DNA-DNA hybridizations among lactic streptococcal temperate and virulent phages belonging to distinct lytic groups. *J. Ind. Microbiol.* **2**:151–158.
8. **Le Bourgeois, P., M. Mata, and P. Ritzenthaler.** 1989. Genome comparison of *Lactococcus* strains by pulsed-field gel electrophoresis. *FEMS Microbiol. Lett.* **59**:65–70.
9. **Loof, M., J. Lembke, and M. Teuber.** 1983. Characterization of the genome of the *Streptococcus lactis* subsp. *diacetylactis* bacteriophage P008 widespread in German cheese factories. *Syst. Appl. Microbiol.* **4**:413–423.
10. **Macrina, F. L., K. R. Jones, and P. H. Wood.** 1980. Chimeric streptococcal plasmids and their use as molecular cloning vehicles in *Streptococcus sanguis* (Challis). *J. Bacteriol.* **143**: 1425–1435.
11. **McClelland, M., R. Jones, Y. Patel, and M. Nelson.** 1987. Restriction endonucleases for pulsed-field mapping of bacterial genomes. *Nucleic Acids Res.* **15**: 5985–6005.
12. **McKay, L. L., K. A. Baldwin, and J. D. Efstathiou.** 1976. Transductional evidence for plasmid linkage metabolism in *Streptococcus lactis* C2. *Appl. Environ. Microbiol.* **32**:45–52.
13. **McKay, L. L., K. A. Baldwin, and P. M. Walsh.** 1980. Conjugal transfer of genetic information in group N streptococci. *Appl. Environ. Microbiol.* **40**:84–91.
14. **Ramos, P., M. Novel, M. Lemosquet, and G. Novel.** 1983. Fragmentation du plasmide lactose-protéase chez les dérivés lactose-négatifs de *Streptococcus lactis* et de *S. lactis*

ssp. *diacetylactis. Ann. Microbiol. Inst. Pasteur* (Paris) **134B:**387–399.

15. **Relano, P., M. Mata, B. Bonneau, and P. Ritzenthaler.** 1987. Molecular characterization and comparison of 38 virulent and temperate bacteriophages of *Streptococcus lactis. J. Gen. Microbiol.* **133:** 3053–3063.

16. **Smith, C. L., J. G. Econome, A. Schutt, S. Klco, and C. R. Cantor.** 1987. A physical map of the *Escherichia coli* K12 genome. *Science* **236:**1448–1453.

17. **Sneath, P. H. A., N. S. Mair, M. E. Sharpe, and J. G. Holt (ed.).** 1986. *Bergey's Manual of Systematic Bacteriology*, vol. 2. Williams & Wilkins, Baltimore, Md.

IV. Structure and Evolution of the M-Protein Gene Family

A Virulence Regulon in *Streptococcus pyogenes*

P. PATRICK CLEARY, DIQUI LaPENTA, DAVID HEATH,[†] ELIZABETH J. HAANES,[‡] AND
CECIL CHEN

Department of Microbiology, University of Minnesota, Minneapolis, Minnesota 55455

The capacity of group A streptococci to invade and persist in human tissue depends on a variety of cell surface and excreted macromolecules. The surface of virulent strains is covered with a fibrillar layer composed of M protein, T antigen, streptococcal C5a peptidase (SCP) (17, 19), and immunoglobulin G (IgG) and/or IgA Fc receptors (FcRA). The M protein, of which there are more than 80 antigenically distinct forms, is the most studied and is now the focus of vaccine development. This antiphagocytic protein blocks the interaction of bound C3b with polymorphonuclear leukocyte (PMN) receptors (13) and specifically binds factor H to speed the decay of deposited C3b opsonin (14). The biological and biochemical properties of these proteins have been reviewed in depth (6).

SCP inactivates the chemotactic response that normally attracts PMNs to the site of infection (25, 26). Endoproteolytic destruction of C5a, one of the primary mediators of chemotaxis in human tissues, retards the recruitment of phagocytes (20). C5a is specifically cleaved in the PMN binding site to leave a 67-residue N-terminal fragment that is unable to interact with PMNs. The protein, as deduced from the DNA sequence of the SCP gene (*scpA*), is an enzyme of 128 kDa. This endopeptidase has significant sequence homology with the catalytic domains of *Bacillus* subtilisins (3) and protease III from lactococci (de Vos et al., this volume). Abrogation of chemotaxis has been postulated to augment infection by providing streptococci a period within which to change their microenvironment in order to gain a foothold before the onslaught of the inflammatory response (20, 25). Studies from our laboratory and others suggest that this novel virulence mechanism may be an invasive strategy common to a variety of human streptococcal pathogens. Human isolates of group B (11; Survorov et al., this volume) and group G (P. Cleary et al., unpublished data) were also discovered to produce SCP-like enzymes and to have nucleotide sequences highly homologous to that of *scpA*.

Receptors for Fc domains of IgG and IgA are commonly expressed by virulent strains of group A streptococci (18). We cloned and sequenced a type II receptor from an M76 group A culture (9, 10), and other receptor genes from this species have examined by Lindahl et al. (this volume). The protein deduced from the DNA sequence exhibits remarkable similarities to M proteins. It is extensively helical and has an internal repeat sequence that contains the IgG binding site (8a). Even more impressive is the high degree of similarity in the primary sequence of their signal sequences of C termini (10). The importance of Fc receptors to virulence is unknown. We and others have postulated that they may prevent effective opsonization by covering the cell surface with up-ended immunoglobulin (5). The nearly universal association of immunoglobulin receptors with virulent bacteria suggests that they must play an important role in the biology of the genus.

Organization of the Virulence Regulon

Little is known about the genetic control of surface-bound proteins in gram-positive bacteria. The *agr* locus of *Staphylococcus aureus* regulates the expression of more than 12 extracellular proteins, including the IgG receptor protein A (21). Although this locus is known to be composed of at least four genes, details of their interaction with environmental signals or with the genes that they control are unavailable (15). The M[+] phenotype of group A streptococci has long been known to be genetically unstable. A systematic analysis of M[−] variants of the M12 strain, CS24, recognized that cultures undergo high-frequency on-off variation (23). By using colony morphology to select M[−] variants, three phenotypes were identified: transparent M[−] colonies which retained the ca-

[†]Present address: Department of Oral Biology, The Dental Research Institute, University of Michigan, Ann Arbor, MI 48209.

[‡]Present address: Department of Medicine, University of Minnesota, Minneapolis, MN 55455.

pacity to revert to the M⁺ state; transparent M⁻ colonies which were phase locked yet, like the former, had no detectable DNA rearrangements; and M⁻ phase-locked colonies which harbored deletions upstream of the M12 gene (*emm12*). The latter have been key to identifying components of the regulatory circuit. These deletions defined a locus that maps upstream of *emm12* which is required for its transcription (22, 23). Caparon and Scott later showed that insertion of TN*916* upstream of *emm6* eliminated transcription of this gene (2). It is not known whether this locus, named *mry*, corresponds to the *virR* locus, even though their positions relative to the M genes are similar.

Wexler et al. were the first to recognize that M⁻ cultures express smaller amounts of SCP protein than do their M⁺ parent strain (26). This finding was confirmed by Simpson et al. (24). Both phase-locked and phase-switching strains produced diminished amounts of SCP antigen and specific mRNA. SCP expression phase varied concurrently with M-protein biosynthesis. Moreover, the failure of variants with deletions in the *virR* locus to produce SCP-specific mRNA indicated that SCP expression is also dependent on this locus. The SCP structural gene, *scpA*, is 3.5 kb and is located downstream of *emm12* (4) and *emm49* (8), oriented in the same direction as these genes.

Group A streptococcal serotypes consist of two lineages that were originally distinguished by the production of serum opacity factor (27) (Fig. 1). The M proteins of serum opacity-negative (OF⁻) strains, such as the M12, M1, M5, M6, and M24 strains, have been sequenced and found to have a high degree of homology in their signal sequences, C-terminal cell wall contact regions, and membrane anchor domains (6). The only M protein from an OF⁺, serotype to be sequenced thus far is from an M49 strain (8). Progressive sequence alignment compared the *emm49* gene with those from OF⁻ strains and constructed a phylogenetic tree (8). This analysis of conserved regions placed *emm49* on a separate branch from the node of origin, distant from the three other M-protein genes of OF⁻ serotypes. Furthermore, the OF⁺ M49 strain contains three tandemly arranged M-protein-like genes. An M-protein-like gene, named *ennX*, is immediately downstream of *emm49*, and an IgG receptor gene, *fcrA*, is upstream of *emm49*. The *ennX* gene was found to be transcriptionally silent; therefore, its function has not been defined (D. LaPenta and P. Cleary, unpublished data). On the other hand, *fcrA* encodes an IgG receptor that is antigenically similar to the FcRA76 protein previously described by our laboratory (unpublished data). The location of *fcrA* upstream of *emm49* and *emm76* was established by Southern hybridization of chromosomal DNA (E. J. Haanes et al., submitted for publication).

The linkage of *fcrA* to *emm49* suggested that it too may be subject to phase variation and coordinately expressed with M protein. Because the M12 strain and its variants lack an *fcrA* gene, we turned our analysis to the M49 strain, CS101, and an isogenic M49⁻ variant, CS103. RNAs extracted from both strains were examined by dot-blot hybridization using a DNA probe that corresponded to an N-terminal *fcrA*-specific sequence (Fig. 2). RNA extracted from M49⁺ cells hybridized strongly to the probe, whereas that from strain CS103 M⁻ cells gave a very weak signal. As expected, RNA from the M12 strain CS24, which does not produce a type II Fc receptor, did not hybridize to the probe. Other pairs of M⁺ and M⁻ cultures, representing OF⁺

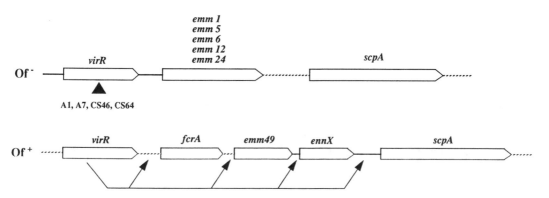

FIG. 1. *Streptococcus pyogenes* virulence regulon. The gene organization of OF⁻ and OF⁺ lineages of group A streptococci is depicted. Boxes indicate the positions of structural genes; heavy solid lines indicate regions of known sequence; dashed lines show regions where the sequence is unknown; the solid triangle marks the position of deletion mutations in the M12 strain CS24. The *virR* gene product is postulated to act at the promoters of each structural gene (arrows).

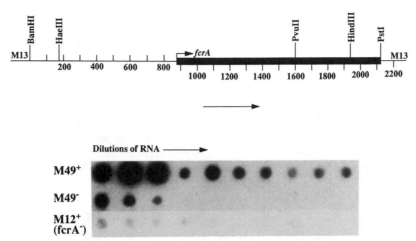

FIG. 2. Relative concentrations of *fcrA* mRNA from M⁺ and M⁻ cultures. Total RNA hybridized to an internal *fcrA* probe at 85% stringency. The initial concentration was 12 μg, and twofold dilutions were made for each sample. The arrow below the map represents the DNA insert used as probe.

serotypes, gave similar results. These results suggest that transcription of *fcrA* is controlled by the same phasing switch that controls M-protein and SCP expression. Results of hybridization experiments suggest that strains CS101 and CS103 have a genetic locus that is partially homologous to *virR*, but until the appropriate genetic experiments have been performed, we cannot be certain that *fcrA*, *emm49*, and *scpA* are under its control.

Comparison of RNA Transcripts and Putative Promoters

To determine whether M, FcRA, and SCP genes are transcribed from independent promoters and thus compose a regulon, the sizes of their respective mRNAs were determined. Both *emm12* (22) and *emm6* (12) mRNAs were previously shown be 2.0 kb, a size consistent with that of the native proteins. SCP is translated from a transcript of approximately 4.0 kb, which is also consistent with the size of the SCP protein. Northern (RNA) analysis of RNA from strain CS101 using the *fcrA* probe depicted in Fig. 2 showed that the *fcrA* transcript is 1.2 kb, a size adequate to encode the Fc receptor protein (10). Detection of mRNAs which correspond in size to each protein strongly suggests that transcription of each gene is initiated from its own promoter. Posttranscriptional processing of a single larger mRNA is a remote but untested explanation for these findings.

Comparison of the DNA sequence upstream of *fcrA*, *emm12*, and *scpA* is also consistent with the conclusion that these genes are transcribed from separate promoters (Fig. 3; 1). Two transcription start sites were experimentally observed for the *emm12* gene (21). In general, the nucleotide sequences within 200 bp and 5′ of each gene have remarkable similarities. The putative −10 and −35 consensus sequences 5′ to each start site

```
                                         -35                        -35 -10
scpA.upstream    -196   AACTCTTAAAAAGCTGACCTTTACTAATAATCGTCTTTTTTTTATAATAAAGATG
                        ||  || ||||||||| ||||||  |  |  |  || |||| |||| |  | |
                                                                             *
emm12.upstream   -124   AAAGCTAAAAAAGCTGGTCTTTACCTTTTGGCTTATATTATTTACAATAGAATTA
                          |    |||||| |||| ||||||||||||||||| |||||||| |||| | |||
fcrA.upstream    -115   CAGCTCAAAAAAACTGGCCTTTACCTTTTGGCTTTTTTTTATTTAGAATAATTTA
                                               -10
scpA.upstream           TTAGTAATATAATTGATAAATGAGATACATTTAATCATTATGGCAAAAGCAAGA  ┐
                        |||| || |  |||||||  *  ||| || ||| |||  | ||                │ 58%
emm12.upstream          TTAGAGTTAAACCCTGAAAATGAGGGTTTTTTCCTAAAAATGATAACATAAGGA  ┘
                        || ||| |  |   || | ||    ||| || |||| | | ||||||||          ┐ 66%
fcrA.upstream           TTGGAGAGATGCTTAATAATTTAAGCACAATTCTTAGAAATTAAGAAATAAGGC  ┘
```

FIG. 3. Comparison of *fcrA*, *scpA*, and *emm12* promoter sequences. Sequences that correspond to DNA 5′ to each structural gene are shown. Coordinates on the left indicate the number of bases from the first codon of each structural gene, *scpA* (3), *emm12* (22), and *fcrA* (10). Starred bases show experimentally determined transcription start sites (22). Sequences corresponding to two −35 and −10 consensus promoter regions are shown. The degrees of similarity of *emm12* to *fcrA* and *scpA* are indicated on the right.

of *emm12* are also highly conserved in sequences upstream of and adjacent to *scpA* and *fcrA*. Sequences corresponding to the most 5' promoter show the greatest degree of homology.

The *virR* Locus

We previously demonstrated that transcription of the *emm12* and *scpA* genes required an intact *virR* locus which was defined by deletion mutations (22, 23). Deletions in a region 400 bp upstream of *emm12* eliminated expression of both M12 and SCP proteins (22, 23). A region of DNA 2.9 kb immediately 5' of *emm12* has now been cloned and sequenced (C. Chen and P. Cleary, unpublished data). A single open reading frame of 1,497 bp was identified that would encode a protein of 58.8 kDa. This protein possesses some interesting features: it is located 279 bp upstream of *emm12*; it contains overlap deletion mutations A1, A7, CS46, and CS64; its molecular size is 58.8 kDa. Secondary structure deduced by the algorithm of Garnier et al. (7) revealed a region with the helix-turn-helix motif typical of many DNA-binding proteins. Although this region lacked significant sequence homology with other DNA-binding proteins, the relative positions of the helices and key amino acids aligned perfectly with those of a variety of *Escherichia coli* and phage repressor proteins. Analysis of hydropathicity plots by using the algorithm and parameters suggested by Kyte and Doolittle (16) did not show a membrane stop region, suggesting that the *virR* product is not a membrane protein. A search of the gene bank did not identify significant sequence similarity to other genes.

Conclusion

Our studies of the M-protein gene cluster suggest that the M protein, SCP, and immunoglobulin receptor proteins compose a virulence regulon. Transcription of each gene from individual promoters is dependent on the VirR protein, which appears to be a DNA-binding protein. The discovery that M and SCP genes undergo simultaneous phase variation between on and off stages suggests that the genetic switch which controls variation may regulate the expression of *virR*. Experiments are now in progress to identify both *cis*- and *trans*-acting components of that switch.

The ultimate goal of our studies is to understand the importance of phase variation to the infectious process. It is possible that the on state is favored at one stage of infection, whereas the off state provides a selective advantage to streptococci at some other stage of infection. The recent literature is replete with reports of virulence operons and regulons in gram-negative pathogens (17). Most are controlled by a two-component regulator: one component receives environmental signals and then transduces those signals to a DNA-binding protein that directly controls expression of the genes in question. Other systems achieve a similar end with a single protein component. Neither in vivo nor in vitro conditions that alter the level of M protein or SCP expression nor the frequency of phase variation have been identified. One presumes that these bacteria have the capacity to sense changes in the external environment. The variety of macromolecules required for streptococci to persist on the human skin, in blood, or on mucosal surfaces must surely vary. The VirR protein or some other global regulatory protein that interacts with VirR could direct the physiological changes required for these bacteria to persist in these dramatically different compartments of the human body.

This work was supported by Public Health Service grants AI 16722 and AI 20016 and by The Minnesota Medical Foundation grant SMF-623-89.

LITERATURE CITED

1. **Altschul, S. F., and B. W. Erickson.** 1986. Optimal sequence alignment using affine gap costs. *Bull. Math. Biol.* **48:**603–616.
2. **Caparon, M., and J. Scott.** 1987. Identification of a gene that regulates expression of M protein, the major virulence determinant of group A streptococci. *Proc. Natl. Acad. Sci. USA* **84:**8677–8681.
3. **Chen, C., and P. Cleary.** 1990. Complete nucleotide sequence of the streptococcal C5a peptidase gene of *Streptococcus pyogenes*. *J. Biol. Chem.* **265:**3161–3167.
4. **Chen, C., and P. P. Cleary.** 1989. Cloning and expression of the streptococcal C5a peptidase gene in *Escherichia coli*; linkage to the type 12 M protein gene. *Infect. Immun.* **57:**1740–1745.
5. **Cleary, P., and D. Heath.** 1990. Type II immunoglobulin receptor and its gene, p. 83–99. *In* M. Boyle (ed.), *Bacterial Immunoglobulin-Binding Protein*, vol. I. Academic Press, New York.
6. **Fischetti, V. A.** 1989. Streptococcal M protein: molecular design and biological behavior. *Clin. Microbiol. Rev.* **2:**285–315.
7. **Garnier, J., D. J. Osguthorpe, and B. Robson.** 1978. Analysis of the accuracy and implications of simple methods for predicting the secondary structure of globular proteins. *J. Mol. Biol.* **120:**97–120.
8. **Haanes, E., and P. Cleary,** 1989. Identification of a divergent M protein gene and an M protein-related gene family in *Streptococcus pyogenes* serotype M49. *J. Bacteriol.* **171:**6397–6408.
8a. **Heath, D. G., M. D. Boyle, and P. P. Cleary.** 1990. Isolated DNA repeat region from *fcrA*76, the Fc-binding protein gene from an M-type 76 strain of group A streptococci, encodes a protein with Fc-binding activity. *Mol. Microbiol.* **4:**2071–2079.
9. **Heath, D. G., and P. P. Cleary.** 1987. Cloning and expression of the gene for an immunoglobulin G Fc receptor protein from group A streptococci. *Infect. Immun.* **55:**1233–1238.
10. **Heath, D. G., and P. P. Cleary.** 1989. Fc-receptor and M-protein genes of group A streptococci are products of gene duplication. *Proc. Natl. Acad. Sci. USA* **86:**4741–4745.
11. **Hill, H. R., J. F. Bohnsack, E. Z. Morris, N. H. Augustine, C. J. Parker, P. Cleary, and J. T. Wu.** 1988. Group B streptococci inhibit chemotactic activity of the fifth component of complement. *J. Immunol.* **141:**3551–3556.
12. **Hollingshead, S. K.** 1987. Nucleotide sequences that signal the initiation of transcription for the gene encoding type 6

M protein in *Streptococcus pyogenes*, p. 98–100. *In* J. J. Ferretti and R. Curtiss III (ed.), *Streptococcal Genetics*. American Society for Microbiology, Washington, D.C.

13. **Hortsmann, R. D., H. J. Sievertsen, J. Knobloch, and V. A. Fischetti.** 1988. Antiphagocytic activity of streptococcal M protein: selective binding of complement control protein factor H. *Proc. Natl. Acad. Sci. USA* **85:**1657–1661.

14. **Jacks-Weis, J., Y. Kim, and P. P. Cleary,** 1982. Restricted deposition of C3 on M⁺ group A streptococci: correlation with resistance to phagocytosis. *J. Immunol.* **128:**1897–1902.

15. **Krieswirth, B. N.** 1989. Genetics and expression of toxic shock syndrome toxin 1: overview. *Rev. Infect. Dis.* **11:**S97–100.

16. **Kyte, J., and R. F. Doolittle.** 1982. Analysis of the accuracy and implications of simple methods for predicting the secondary structure of globular proteins. *J. Mol. Biol.* **157:**105–132.

17. **Miller, J. F., J. J. Mekalanos, and S. Falkow.** 1989. Coordinated regulation and sensor transduction in the control of bacterial virulence. *Science* **243:**916–921.

18. **Myhre, E. B., and G. Kronvall.** 1982. Immunoglobulin specificities of defined types of streptococcal Ig receptors, p. 209–210. *In* S. E. Holm and P. Christensen (ed.), *Basic Concepts of Streptococci and Streptococcal Diseases*. Reedbooks Ltd., Chertsey, England.

19. **O'Connor, S. P., and P. P. Cleary.** 1986. Localization of the streptococcal C5a peptidase to the surface of group A streptococci. *Infect. Immun.* **53:**432–434.

20. **O'Connor, S. P., and P. P. Cleary.** 1987. *In vivo Strepto-*

coccus pyogenes C5a peptidase activity: analysis using transposon and nitrosoguanidine-induced mutants. *J. Infect. Dis.* **156:**495–504.

21. **Recsei, P., B. Kreiswirth, M. O'Reilly, P. Schlievert, A. Gruss, and R. P. Novick.** 1986. Regulation of exoprotein gene expression in *Staphylococcus aureus* by *agr*. *Mol. Gen. Genet.* **202:**58–61.

22. **Robbins, J. C., J. G. Spanier, S. J. Jones, W. J. Simpson, and P. P. Cleary.** 1987. *Streptococcus pyogenes* type 12 M protein gene regulation by upstream sequences. *J. Bacteriol.* **169:**5633–5640.

23. **Simpson, W. J., and P. P. Cleary.** 1987. Expression of M type 12 protein by a group A streptococcus exhibits phase-like variation: evidence of coregulation of colony opacity determinants and M protein. *Infect. Immun.* **55:**2448–2455.

24. **Simpson, W. J., D. LaPenta, C. Chen, and P. Cleary.** 1990. Coregulation of type 12 M protein and streptococcal C5a peptidase genes in group A streptococci: evidence for a virulence regulon controlled by the *virR* locus. *J. Bacteriol.* **172:**696–700.

25. **Wexler, D. E., D. E. Chenoweth, and P. P. Cleary.** 1985. Mechanism of action of the group A streptococcal C5a inactivator. *Proc. Natl. Acad. Sci. USA* **82:**8144–8148.

26. **Wexler, D. E., R. D. Nelson, and P. P. Cleary.** 1983. Human neutrophil chemotactic response to group A streptococci: bacteria-mediated interference with complement-derived chemotactic factors. *Infect. Immun.* **39:**239–246.

27. **Widdowson, J. P., W. R. Maxted, and D. L. Grant.** 1970. The production of opacity in serum by group A streptococci and its relationship with the presence of M antigen. *J. Gen. Microbiol.* **61:**343–353.

Surface Proteins from Gram-Positive Cocci Have a Common Motif for Membrane Anchoring

OLAF SCHNEEWIND, VIJAYKUMAR PANCHOLI, AND VINCENT A. FISCHETTI

The Rockefeller University, New York, New York 10021

Information regarding the presentation of antigens on the bacterial cell surface may be an important factor in our understanding of microbial pathogenesis. However, the mechanism by which gram-positive bacteria anchor their surface proteins to the cell is not completely known. Proteins generally can be anchored to membranes by a variety of chemical structures. These include stretches of hydrophobic amino acids, with either mono- or bipolar membrane spanning (1), covalent or noncovalent binding to integral membrane molecules (15), acylation at the N or C terminus (8, 9), and finally anchoring through a glycosyl-phosphatidylinositol anchor complex (5). With the exception of the latter, which has not been described to anchor proteins in prokaryotes, all are found in both eukaryotes and prokaryotes. These different types of anchoring motifs not only participate in regulating the expression of individual proteins at specific subcellular locations (14) but may also offer cells the opportunity to control protein turnover (2) through hydrolysis of either a peptide, ester, or phosphodiester bond.

First clues that the mode of attachment for gram-positive surface proteins is through the cytoplasmic membrane were recently reported by Pancholi and Fischetti (13). They presented evidence that M6 protein, an α-helical coiled-coil molecule found on the surface of group A streptococci (6), may be found either as an integral membrane protein or as a released form, depending on the pH condition used to enzymatically remove the bacterial wall. The release of M6 protein from the protoplast membrane was found to have characteristics of an enzymatic reaction (13). Using the M protein as a model system, we attempted to dissect the process involved in anchoring the M molecule to the cytoplasmic membrane.

Surface Proteins from Gram-Positive Cocci Have a Common C-Terminal Arrangement of Amino Acids

Sequence analyses of M proteins and all known surface molecules from gram-positive cocci show a similar C-terminal arrangement of amino acids comprising 15 to 20 hydrophobic residues followed by 4 to 7 charged amino acids at the C-terminal end (7). Despite this similarity, the most striking identity among these proteins is a hexapeptide with the consensus sequence LPSTGE located about nine residues preceding the C-terminal hydrophobic domain (7). While serine (S) and glutamic acid (E) residues at positions 3 and 6, respectively, may be somewhat varied in some proteins, the remainder of the hexapeptide sequence does not change. This highly conserved arrangement of amino acids suggests a common mechanism of membrane anchoring for these different surface molecules.

Examination of the DNA sequence coding for the LPSTGE motif revealed significant homology among the different surface proteins (7). By using the evolutionary tree program of Feng and Doolittle (4), the proteins may be grouped according to base substitutions in this region. Most differences occurred within the third position of codon triplets in the LPSTGE motif, which did not cause amino acid changes. On this basis, we could identify evolutionary relationships among the different surface proteins. We believe that the conservation of the LPSTGE hexapeptide sequence on both the DNA and protein levels reveals the enormous pressure for its preservation during the evolutionary development of these molecules and indicates the importance of this sequence in the process of anchoring surface proteins in gram-positive cocci.

Role of the C-Terminal Hydrophobic Domain in Membrane Anchoring

The influence of the C-terminal hydrophobic domain on protein anchoring is yet to be completely understood. We analyzed the sequences of the hydrophobic domains of several surface proteins from gram-positive cocci for their hydrophobic moments according to the Eisenberg algorithm (3) (Table 1). In only two cases (protein A and WapA) did proteins exhibit values higher than the speculated limit necessary for membrane anchoring (≥ 0.68 for monomers; ≥ 1.10 for dimers) (3), which questions the capability of this region for direct membrane attachment. Considering that the M6 protein is a coiled-coil dimer, we tested the ability of the C-terminal hydrophobic domain, with a hydrophobic moment of 1.18, to anchor the M molecule to the bacterial membrane.

TABLE 1. Hydrophobic moment plot (3) analysis of the C-terminal hydrophobic domain of surface proteins from gram-positive cocci[a]

Protein	Monomer
WapA	0.76
Protein A	0.70
Protein G	0.65
Wg2	0.65
T6	0.65
PAc	0.63
M6	0.59 (1.18)
Protein H	0.59
M49	0.56
Immunoglobulin A binding	0.56
ScpA	0.54
Fibrinogen binding	0.46

[a]Values are given as mean hydrophobicity ($\langle H \rangle$) for monomers; the dimer value for M6 protein is in parentheses. Cutoff for membrane anchoring: $\langle H \rangle$ 0.68 (monomer)/1.10 (dimer).

The LPSTGE Motif Plays a Determining Role in Membrane Anchoring

When cloned and overexpressed in *Escherichia coli*, M6 protein is found both bound to the cytoplasmic membrane and secreted into the periplasm (O. Schneewind and V. A. Fischetti, unpublished data). This observation suggests that the M molecule could also be an integral membrane protein in *E. coli* as it is in *Streptococcus pyogenes*. Thus, we chose the cytoplasmic membrane of *E. coli* as a model system with which to examine the properties of the M protein LPSTGE motif and C-terminal hydrophobic domain with specifically constructed mutants.

The *emm6.1* gene and its derivatives were selectively expressed with T7 polymerase in *E. coli* HMS174(DE3) (16). Plasmid pM6.1 contains the complete *emm6.1* reading frame. The LPSTGE motif and C-terminal hydrophobic domain from position 407 onward are deleted in pM6.1$_{1-406}$. In pM6.1$_{\Delta LPSTGE}$, only the LPSTGE sequence is deleted and replaced by a two-amino-acid substi-

tution (GS). All three constructs were expressed in *E. coli* HMS174(DE3) and K38, labeled with [^{35}S]methionine, subjected to cell fractionation, and immunoprecipitated with M6-specific antibodies. The wild-type M6 protein was found only in the membrane fraction and exhibited its typical multiple banding pattern on sodium dodecyl sulfate-polyacrylamide gel electrophoresis (SDS-PAGE) (Fig. 1). This banding pattern was found associated with the C-terminal half of the molecule upon its isolation from the streptococcal wall (12). It is also present when the C-terminal half of the M6 protein is expressed in *E. coli*, suggesting a posttranslational modification of the peptide chain (Schneewind and Fischetti, unpublished data). The truncated M6 protein (pM6.1$_{1-406}$) and the deletion mutant (pM6.1$_{\Delta LPSTGE}$) were both secreted into the periplasmic space and appeared only as single bands (Fig. 1). When we tested *E. coli* K38 for the expression and subcellular location of our M6 clones, the M6 protein was found to be secreted along with its mutant derivatives (data for M6 derivatives not shown). The reason for this difference in cellular location between these two *E. coli* strains is unknown.

To investigate whether the bound M6 molecule behaved as an integral membrane protein in strain HMS174, the membranes were washed with either 100 mM Na$_2$CO$_3$ or 1.2 M NaCl. In both cases M6 protein was not released from the membranes, establishing its integral membrane behavior. A proteinase accessibility assay was used to determine the side of the cytoplasmic membrane on which the M6 protein was localized. *E. coli* spheroplasts were exposed to trypsin in the absence or presence of either 1% Triton X-100 or trypsin inhibitor. The results indicate that the M6 protein was digested by trypsin after exposure of either intact or Triton X-100-lysed spheroplasts, an effect that was inhibited by trypsin inhibitor. We concluded from this finding that the native M6 protein in *E. coli* HMS174(DE3)

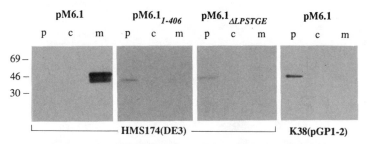

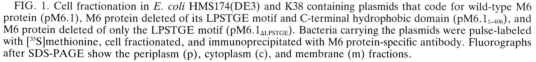

FIG. 1. Cell fractionation in *E. coli* HMS174(DE3) and K38 containing plasmids that code for wild-type M6 protein (pM6.1), M6 protein deleted of its LPSTGE motif and C-terminal hydrophobic domain (pM6.1$_{1-406}$), and M6 protein deleted of only the LPSTGE motif (pM6.1$_{\Delta LPSTGE}$). Bacteria carrying the plasmids were pulse-labeled with [^{35}S]methionine, cell fractionated, and immunoprecipitated with M6 protein-specific antibody. Fluorographs after SDS-PAGE show the periplasm (p), cytoplasm (c), and membrane (m) fractions.

is located on the outer side of the cytoplasmic membrane.

Our data suggest that the LPSTGE motif is required for the proper anchoring of M protein and by analogy other surface proteins from gram-positive cocci in the cytoplasmic membrane. It remains to be investigated whether this peptide sequence functions either as a signal for a post-translational modification or as a retention motif. A known paradigm for retention is the signals for soluble (10) or transmembrane (11) proteins resident on the endoplasmic reticulum. Alternatively, the LPSTGE motif could be required for the correct folding and assembly within the lipid bilayer. Perhaps the presence of both the C-terminal hydrophobic domain and the LPSTGE motif in surface proteins from gram-positive cocci is necessary as a complex signal for both membrane integration and a posttranslational modification of the polypeptide chain, as suggested by the multiple banding pattern.

This work was supported by Public Health Service grant AI11822 from the National Institutes of Health and by a grant from the Mallinckrodt Foundation. O.S. was supported by grant Schn 327/1-1 from the Deutsche Forschungsgemeinschaft.

LITERATURE CITED

1. **Blobel, G.** 1980. Intracellular protein topogenesis. *Proc. Natl. Acad. Sci. USA* **77:**1496–1500.
2. **Cross, G. A. M.** 1990. Glycolipid anchoring of plasma membrane proteins. *Annu. Rev. Cell Biol.* **6:**1–31.
3. **Eisenberg, D., E. Schwarz, M. Komaromy, and R. Wall.** 1984. Analysis of membrane and surface protein sequences with the hydrophobic moment plot. *J. Mol. Biol.* **179:**125–142.
4. **Feng, D. F., and R. F. Doolittle.** 1987. Progressive sequence alignment as a prerequisite to correct phylogenetic trees. *J. Mol. Evol.* **25:**351–360.
5. **Ferguson, M. A. J., and A. F. Williams.** 1988. Cell-surface anchoring of proteins via glycosyl-phosphatidylinositol structures. *Annu. Rev. Biochem.* **57:**285–320.
6. **Fischetti, V. A.** 1989. Streptococcal M protein: molecular design and biological behavior. *Clin. Microbiol. Rev.* **2:** 285–314.
7. **Fischetti, V. A., V. Pancholi, and O. Schneewind.** 1990. Conservation of a hexapeptide sequence in the anchor region of surface proteins of gram-positive cocci. *Mol. Microbiol.* **4:**1603–1605.
8. **Grand, R. J. A.** 1989. Acylation of viral and eukaryotic proteins. *Biochem. J.* **258:**625–638.
9. **Kaufman, J. F., M. S. Krangel, and J. L. Strominger.** 1984. Cysteines in the transmembrane region of major histocompatibility complex antigens are fatty acylated via thioester bonds. *J. Biol. Chem.* **259:**7230–7238.
10. **Munro, S., and H. R. B. Pelham.** 1987. A C-terminal signal prevents secretion of luminal ER proteins. *Cell* **48:**899–907.
11. **Nilsson, T., M. Jackson, and P. A. Peterson.** 1989. Short cytoplasmic sequences serve as retention signals for transmembrane proteins in endoplasmic reticulum. *Cell* **58:**707–718.
12. **Pancholi, V., and V. A. Fischetti.** 1988. Isolation and characterization of the cell-associated region of group A streptococcal M6 protein. *J. Bacteriol.* **170:**2618–2624.
13. **Pancholi, V., and V. A. Fischetti.** 1989. Identification of an endogeneous membrane anchor-cleaving enzyme for group A streptococcal M protein. *J. Exp. Med.* **170:**2119–2133.
14. **Rogers, J., P. Early, C. Carter, K. Calame, M. Bond, L. Hood, and R. Wall.** 1980. Two mRNAs with different 3′ ends encode membrane-bound and secreted forms of immunoglobulin M chain. *Cell* **20:**303–312.
15. **Singer, S. J., and G. L. Nicolson.** 1972. The fluid mosaic model of the structure of cell membranes. *Science* **175:**720–731.
16. **Studier, F. W., and B. A. Moffatt.** 1986. Use of bacteriophage T7 RNA polymerase to direct selective high-level expression of cloned genes. *J. Mol. Biol.* **189:**113–130.

Genetics and Biochemistry of Protein Arp, an Immunoglobulin A Receptor from Group A Streptococci

GUNNAR LINDAHL,[1] BO ÅKERSTRÖM,[2] LARS STENBERG,[1] ELISABET FRITHZ,[3] AND
LARS-OLOF HEDÉN[3]

*Department of Medical Microbiology,[1] Department of Physiological Chemistry,[2] and Department of
Microbiology,[3] University of Lund, S-22362 Lund, Sweden*

Certain strains of group A streptococci are known to express a cell surface receptor that binds immunoglobulin (Ig) A or IgG (3). The biological function of these Ig receptors is still unclear, and they have therefore been studied less extensively than the M proteins, the well-known cell surface proteins that are important virulence factors of group A streptococci (5). However, much information has recently become available concerning the biochemistry and genetics of the Ig receptors, and it now seems possible that these proteins contribute to bacterial virulence (3). Studies of the Ig receptors are therefore important for an understanding of the mechanisms by which group A streptococci establish an infection. Apart from their biological function, these Ig receptors are of interest as potential reagents in immunochemical work and as model systems for studies of the binding of Ig to specific receptors. This chapter summarizes work done with two receptors that bind IgA. These IgA-binding proteins are of particular interest, since IgA is the major Ig class on mucous membranes, the most common site for infections with group A streptococci.

Most Clinical Isolates of Group A Streptococci Bind IgA and/or IgG

To study the distribution of Ig receptors among routine clinical isolates of group A streptococci, we tested a large number of strains for the ability to bind radiolabeled IgA and IgG (16). Results obtained with 194 throat strains are shown in Fig. 1, which demonstrates that binding of Igs is a common property among these strains. For 82% of the strains, there was significant binding of IgA or IgG, and 53% of the strains had the capacity to bind both. Binding of IgA alone was not observed for any strain, but 29% of the strains showed binding of IgG only. Similar results were obtained with a series of septicemia strains (16), and similar results have also been reported by Bessen and Fischetti (1).

Cloning of IgA Receptor Genes from Two Different Strains of Group A Streptococci

For the work described below, we used two strains that bind IgA well but bind IgG only weakly (Fig. 1). The IgA-binding protein expressed by these strains is referred to as protein Arp (for IgA receptor protein). By using bacteriophage lambda as the vector, the gene for protein Arp was cloned from the two strains, one of type M4 and the other of type M60 (13, 14).

Protein Arp corresponding to the two cloned genes has been purified after expression in *Escherichia coli*, and these two proteins are referred to as Arp4 and Arp60, respectively. Analysis of both proteins by Western immunoblotting demonstrated a limited size heterogeneity, with an apparent molecular weight of about 42,000 (13, 14).

Sequence Analysis of Protein Arp4 Demonstrates Extensive Homology to Streptococcal M Proteins

The complete nucleotide sequence of the gene for protein Arp4 has been determined (6). The deduced amino acid sequence of 386 residues includes a signal sequence of 41 amino acids and a putative membrane anchor region, both of which are homologous to similar regions in other streptococcal surface proteins. The processed form of the IgA receptor is 345 amino acids long and has a calculated molecular weight of 39,544.

Analysis of the sequence of protein Arp4 shows that several different regions can be distinguished. Furthermore, there are striking sequence homologies to the five streptococcal M proteins for which the sequence is known, the M5, M6, M12, M24, and M49 proteins (8, 11, 18–20). The structure of protein Arp4 is schematically illustrated in Fig. 2a.

The processed form of protein Arp4 starts with a unique sequence of 111 amino acids, which represents the IgA-binding region (unpublished results). This N-terminal sequence is followed by a region that contains three units with extensive internal homology. These three units also show strong homology to the C repeats found in all M proteins sequenced. In protein Arp4, there are 35 amino acid residues in the first unit and 42 residues in the second and third units. The C repeats of protein Arp4 show more than 50% amino acid sequence homology to the C repeats of all M proteins studied (6).

Following the three C repeats, there is a 54-residue region which also shows extensive ho-

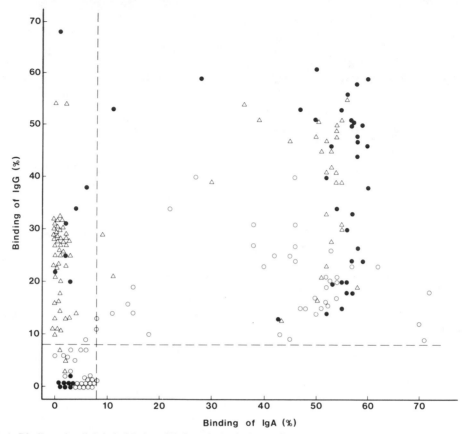

FIG. 1. Binding of radiolabeled IgA and IgG to throat isolates of group A streptococci. Each symbol represents one strain. The different symbols correspond to three different series of strains, isolated during different time periods and in different geographical areas of Sweden (16). The dashed line indicates the position of the cutoff value, 8% binding, used to subdivide the strains into binders and nonbinders.

mology (more than 90%) to the corresponding region in the five M proteins. The next region, 24 amino acids, shows only limited homology to the M proteins (except for the M49 protein; see below), but the C-terminal sequence (37 residues) shows more than 50% homology to the M proteins. This region may be required for membrane anchorage (11).

The extensive sequence homology to M proteins is not unique to protein Arp. Recent work has demonstrated that protein H, an IgG receptor from a streptococcal strain of type M1, is structurally very similar both to M proteins and to protein Arp (7). A more limited sequence homology to M proteins has also been found for the *fcrA76* gene product, an IgG receptor from a type M76 strain (10).

Comparison of the Sequences of Proteins Arp4 and Arp60: Variable and Conserved Regions

When the sequence of protein Arp4 became available, it was found that the N-terminal se-

quence of this protein was different from that of protein Arp60, which had been previously analyzed by amino acid sequencing of the purified protein (14). On the other hand, a rabbit antiserum against protein Arp4 reacted equally well with protein Arp60, suggesting extensive structural similarity (15). This result indicated that there are variable and conserved regions in protein Arp, as previously described for M proteins (5). The complete sequence of protein Arp60 has now been determined (L.-O. Hedén, E. Frithz, and G. Lindahl, submitted for publication), and a comparison with protein Arp4 indeed shows that the sequences can be subdivided into variable and conserved regions. The structure of protein Arp60 is schematically illustrated in Fig. 2b.

The overall structure of protein Arp60 is very similar to that of protein Arp4, described above. The two proteins also show extensive sequence homology. The sequences of the two proteins are virtually identical in a long region corresponding to about three-quarters of the molecule, starting

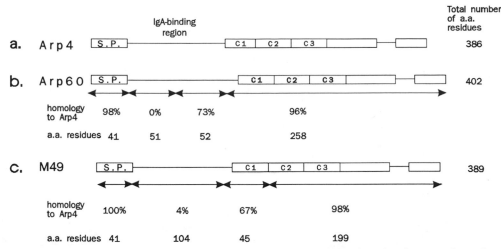

FIG. 2. Schematic representations of proteins Arp4, Arp60, and M49. Regions showing extensive amino acid sequence homology (>50%) to the M5, M6, M12, and M24 proteins are boxed. S.P., Signal peptide. The figure is based on sequence data for Arp4 (6), for Arp60 (Hedén et al., submitted), and for M49 (8). The subdivision of the Arp60 and M49 sequences into regions with different degrees of amino acid (a.a.) sequence homology to the Arp4 sequence is indicated by horizontal arrows; the limits of these regions were chosen to optimize the difference in homology between the different regions.

from the C-terminal end. In contrast, there is no sequence homology in the N-terminal region of the processed form of the protein. Between these two extremes, there is a region with about 73% homology. This comparison between Arp4 and Arp60 shows that the distinction between variable and conserved regions is very striking in these two proteins and more clear-cut than for the M proteins. It has been suggested that the structural variation seen in the N-terminal region of M proteins may be the result of a selective pressure that causes this part of the molecule to evolve faster in response to immunological selection by the host (12). Such a mechanism may also contribute to the structural variation in protein Arp. However, the fact that the C-terminal parts of Arp4 and Arp60 are virtually identical implies that the two molecules are closely related from an evolutionary point of view. This view is difficult to reconcile with the complete lack of homology in the N-terminal region. This apparent paradox can be explained by assuming that recombination contributes to the generation of the N-terminal sequence variation (9). However, the nature of such a recombination event is not understood.

The M49 protein (8) is also very similar in sequence to the two proteins Arp (Fig. 2c). The processed forms of the three proteins have essentially no homology in the N-terminal regions, but the C-terminal parts are almost identical. This striking similarity between M49 and two proteins Arp further supports the hypothesis that the N-terminal sequence variation is generated by recombination events.

Is Protein Arp an M Protein with IgA-Binding Capacity?

There are at least three types of structural similarity between protein Arp and streptococcal M proteins: (i) protein Arp shows strong sequence similarity to M proteins, (ii) both protein Arp and M proteins show N-terminal sequence variation, and (iii) for both protein Arp and M proteins, there is a seven-residue periodicity in the distribution of nonpolar and charged amino acids in the N-terminal region (5, 6); this periodicity is most striking for protein Arp60, which has a direct repeat of a seven-residue sequence at the N terminus (14). These structural similarities between the two types of protein indicate that protein Arp might also be a virulence factor, like the M proteins. If this is the case, one could regard protein Arp as a type of M protein that has also acquired the capacity to bind Ig. However, attempts to demonstrate that protein Arp is a virulence factor have been unsuccessful (13). On the other hand, genetic analysis has shown that both the Arp4 and Arp60 genes are closely linked to another gene, which may code for an M protein (13). This linked gene codes for a protein that binds fibrinogen, a characteristic property of M proteins (22). Bacterial mutants that have simultaneously lost the capacity to express both proteins have also been isolated (13). These mutants are avirulent, as measured in a phagocytosis assay. The lack of virulence of these mutants could be due to the lack of expression of either protein or both proteins or to the absence of some other cell surface protein that is required for virulence.

The nature of the mutants is not yet understood, but their properties can be simply explained by assuming that they are affected in a regulatory element required for the expression of several cell surface components. Such a regulatory mechanism has been postulated to exist in other strains of group A streptococci (4, 20).

Close linkage between a putative M-protein gene and the gene for an Ig-binding protein has recently been proposed to exist also in streptococcal strains of types M1, M49, and M76 (7, 8, 10; L. Björck, personal communication). Taken together, the available data suggest that the presence of linked M-protein and Fc receptor genes is very common in group A streptococci. Such genetic linkage may explain why Ig binding is common in streptococcal strains of certain M types but not in strains of other M types (21).

Binding Properties of Purified Protein Arp

As mentioned above, the streptococcal strains that express protein Arp bind IgG weakly in addition to binding IgA. Purified protein Arp also binds IgG weakly (13, 14), so the binding properties of the bacterial strains can be fully accounted for by the expression of a single Ig receptor, protein Arp. There is some evidence that IgA and IgG bind to separate sites in protein Arp, but conclusive evidence for this is still missing (14; B. Åkerström, A. Lindqvist, and G. Lindahl, *Mol. Immunol.*, in press).

Purified protein Arp binds IgA of both subclasses and binds both serum IgA and secretory IgA (13, 14). The binding of IgA, as well as the weak binding of IgG, is to the Fc region of the Ig molecules (14, 21).

The affinity constant for the binding reaction between protein Arp and human serum IgA was determined to be 9.7×10^8 M^{-1} for protein Arp4 and 5.6×10^8 M^{-1} for protein Arp60 (14; Åkerström et al., in press). Unexpectedly, the affinity of protein Arp was found to be more than 10-fold lower for secretory IgA than for serum IgA (Fig. 3). The significance of this difference in affinities is not yet understood, and it was a surprising finding, since group A streptococci usually establish infections on mucous membranes where secretory IgA predominates.

Evolutionary Aspects

There is evidence that strains of group A streptococci can usually be assigned to one of two major classes. Thus, streptococcal strains can be classified as OF positive or OF negative, according to their capacity to produce serum opacity factor (17). These two classes of strains can also be differentiated by immunological means, since OF-negative strains are usually recognized by certain monoclonal antibodies directed against epitopes in the C repeats of the M6 protein, whereas

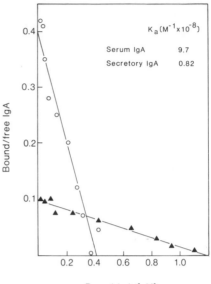

FIG. 3. Scatchard plots of the binding between protein Arp4 and two different molecular forms of IgA. The data were obtained in a solid-phase radioimmunoassay as described elsewhere (14; Åkerström et al., in press).

OF-positive strains are not recognized by these monoclonal antibodies (1, 2). Sequence comparisons indicate that these two classes of streptococcal strains express two distinct classes of M proteins with separate evolutionary lineages (8). The relationship of Ig receptor genes to these two hypothetical lineages is not clear. Strains of the OF-positive class usually bind Ig, but strains of the OF-negative class may or may not express an Ig receptor (1). The two IgA-binding strains that we have studied are OF positive, as is the M49 strain, which expresses an M protein that is structurally very similar to protein Arp. It can therefore be concluded that these three OF-positive strains, which express surface proteins with very similar structures (Fig. 2), belong to the same evolutionary lineage. The available evidence also suggests that such OF-positive strains express at least two linked genes of the same family, coding for an Ig receptor and an M protein, respectively (8, 13). These two genes may have arisen through gene duplication (8).

Concluding Remarks

It is now clear that most strains of group A streptococci express a cell surface receptor that binds to the Fc region of IgA or IgG. Characterization of several such Ig-binding proteins has shown that they have a number of remarkable properties. In particular, studies on both IgA and IgG receptors have demonstrated extensive sequence homology to streptococcal M proteins. For protein Arp, the IgA receptor studied by us, comparison of two different Arp genes has also shown that the sequence can be divided into a variable N-terminal region and a conserved C-terminal region, as previously described for M proteins. The available evidence suggests that this variability in the N-terminal region of protein Arp has resulted from a recombinational event.

All Ig-binding strains that have been studied in detail express at least two cell surface proteins, an Ig receptor and a protein that probably is the M protein of the strain (7, 8, 10, 13). Moreover, these proteins are encoded by closely linked genes. An important task now is to obtain conclusive evidence concerning the biological functions of these two proteins. Considering the structural similarity between Ig receptors and M proteins, it seems possible that most strains of group A streptococci express two M proteins, one of which has the capacity to bind Ig. In this context, it is of interest to note that almost 30 years ago, certain strains of group A streptococci that apparently express two different M proteins were described (23). At the time, such strains were regarded as unusual exceptions, but the available evidence suggests that they may actually be very common.

This work was supported by grants from the Swedish Medical Research Council (B90-16X-9096-IA and projects 7144 and 7480); by the foundations of Emil and Vera Cornell, Crafoord, Johansson, Kock, Lars Hiertas Minne, and Österlund; and by Hightech Inc.

LITERATURE CITED

1. **Bessen, D., and V. A. Fischetti.** 1990. A human IgG receptor of group A streptococci is associated with tissue site of infection and streptococcal class. *J. Infect. Dis.* **161:**747–754.
2. **Bessen, D., K. F. Jones, and V. A. Fischetti.** 1989. Evidence for two distinct classes of streptococcal M protein and their relationship to rheumatic fever. *J. Exp. Med.* **169:**269–283.
3. **Boyle, M. D. P. (ed.).** 1990. *Bacterial Immunoglobulin-Binding Proteins*, vol. I. Academic Press, Inc., New York.
4. **Caparon, M. G., and J. R. Scott.** 1987. Identification of a gene that regulates expression of M protein, the major virulence determinant of group A streptococci. *Proc. Natl. Acad. Sci. USA* **84:**8677–8681.
5. **Fischetti, V. A.** 1989. Streptococcal M protein: molecular design and biological behavior. *Clin. Microbiol Rev.* **2:**285–314.
6. **Frithz, E., L. O. Hedén, and G. Lindahl.** 1989. Extensive sequence homology between IgA receptor and M proteins in *Streptococcus pyogenes*. *Mol. Microbiol.* **3:**1111–1119.
7. **Gomi, H., T. Hozumi, S. Hattori, C. Tagawa, F. Kishimoto, and L. Björck.** 1990. The gene sequence and some properties of protein H, a novel IgG-binding protein. *J. Immunol.* **144:**4046–4052.
8. **Haanes, E. J., and P. P. Cleary.** 1989. Identification of a divergent M protein gene and an M protein-related gene family in *Streptococcus pyogenes* serotype 49. *J. Bacteriol.* **171:**6397–6408.
9. **Haanes-Fritz, E., W. Kraus, V. Burdett, J. B. Dale, E. H. Beachey, and P. Cleary.** 1988. Comparison of the leader sequences of four group A streptococcal M protein genes. *Nucleic Acids Res.* **16:**4667–4677.
10. **Heath, D. G., and P. P. Cleary.** 1989. Fc-receptor and M-protein genes of group A streptococci are products of gene duplication. *Proc. Natl. Acad. Sci. USA* **86:**4741–4745.
11. **Hollingshead, S. K., V. A. Fischetti, and J. R. Scott.** 1986. Complete nucleotide sequence of type 6 M protein of the group A *Streptococcus*. Repetitive structure and membrane anchor. *J. Biol. Chem.* **261:**1677–1686.
12. **Jones, K. F., B. N. Manjula, K. H. Johnston, S. K. Hollingshead, J. R. Scott, and V. A. Fischetti.** 1985. Location of variable and conserved epitopes among the multiple serotypes of streptococcal M protein. *J. Exp. Med.* **161:**623–628.
13. **Lindahl, G.** 1989. Cell surface proteins of a group A streptococcus type M4: the IgA receptor and a receptor related to M proteins are coded for by closely linked genes. *Mol. Gen. Genet.* **216:**372–379.
14. **Lindahl, G., and B. Åkerström.** 1989. Receptor for IgA in group A streptococci: cloning of the gene and characterization of the protein expressed in *Escherichia coli*. *Mol. Microbiol.* **3:**239–247.
15. **Lindahl, G., B. Åkerström, J.-P. Vaerman, and L. Stenberg.** 1990. Characterization of an IgA receptor from group B streptococci: specificity for serum IgA. *Eur. J. Immunol.* **20:**2241–2247.
16. **Lindahl, G., and L. Stenberg.** 1990. Binding of IgA and/or IgG is a common property among clinical isolates of group A streptococci. *Epidemiol. Infect.* **105:**87–93.
17. **Maxted, W. R., and J. P. Widdowson.** 1972. The protein antigens of group A streptococci, p. 251–266. *In* L. W. Wannamaker and J. M. Matsen (ed.), *Streptococci and Streptococcal Diseases*. Academic Press, Inc., New York.
18. **Miller, L., L. Gray, E. Beachey, and M. Kehoe.** 1988. Antigenic variation among group A streptococcal M proteins. Nucleotide sequence of the serotype 5 M protein gene and its relationship with genes encoding types 6 and 24 M proteins. *J. Biol. Chem.* **263:**5668–5673.
19. **Mouw, A. R., E. H. Beachey, and V. Burdett.** 1988. Molecular evolution of streptococcal M protein: cloning and nucleotide sequence of the type 24 M protein gene and relation to other genes of *Streptococcus pyogenes*. *J. Bacteriol.* **170:**676–684.
20. **Robbins, J. C., J. G. Spanier, S. J. Jones, W. J. Simpson, and P. P. Cleary.** 1987. *Streptococcus pyogenes* type 12 M protein gene regulation by upstream sequences. *J. Bacteriol.* **169:**5633–5640.
21. **Schalén, C.** 1980. The group A streptococcal receptor for human IgA binds IgA via the Fc-fragment. *Acta Pathol. Microbiol. Scand. Sect. C* **88:**271–274.
22. **Whitnack, E., and E. H. Beachey.** 1985. Biochemical and biological properties of the binding of human fibrinogen to M protein in group A streptococci. *J. Bacteriol.* **164:**350–358.
23. **Wiley, G. G., and A. T. Wilson.** 1961. The occurrence of two M antigens in certain group A streptococci related to type 14. *J. Exp. Med.* **113:**451–465.

M Proteins of the Equine Group C Streptococci

JOHN F. TIMONEY, MAOWIA MUKHTAR, JORGE GALÁN, AND JIABING DING

*Department of Veterinary Microbiology, Immunology and Parasitology, Cornell University,
Ithaca, New York 14853*

The Lancefield group C streptococci *Streptococcus equi* subsp. *equi* and *S. equi* subsp. *zooepidemicus* are important pathogens of members of the *Equidae* that, although closely related, exhibit major differences in biology. *S. equi* subsp. *equi* is highly host adapted and causes strangles, a contagious upper respiratory tract disease characterized by pharyngitis and regional lymphadenitis. It is quickly cleared from the nasopharynx of the convalescent horse and rarely is isolated from the normal nasopharynx. In contrast, *S. equi* subsp. *zooepidemicus* is less host adapted and less virulent and occurs as a mucosal commensal of all normal horses. An opportunist invader, it complicates viral respiratory tract infections of foals and yearlings and is also frequently isolated from uterine infections of mares. *Streptococcus equisimilis*, another group C species, is occasionally isolated from abscesses in lymph nodes and other tissues of horses. The three streptococci are distinguished by fermentation behavior in lactose, sorbitol, trehalose, and ribose and by the serologic reactivity of their M proteins.

Nucleic acid hybridization studies have proved the close relatedness of the subspecies of *S. equi* (8). *S. equisimilis*, however, is less closely related to the subspecies of *S. equi* than the subspecies are to each other.

M Protein of *S. equi* subsp. *equi*

Antiphagocytic M proteins are widely recognized as important virulence factors and protective antigens of the group A streptococci. The genetics and structures of these surface proteins have been the subject of prolonged intensive study, and they are therefore among the most completely characterized of the medically important bacterial antigens (2). The genetic basis and control of their antigenic and size variation is of particular interest because of its potential value in understanding the mechanism by which group A streptococci elude protective immune responses.

Unlike the group A streptococci, the group C streptococcus *S. equi* subsp. *equi* exists as a single M-protein type and shows no antigenic variation (3, 11). Its M protein, although found as a dimer or trimer with a molecular weight (MW) of between 56,000 and 60,000, shows remarkable size homogeneity (3). In acid extracts, the protein occurs in a series of defined fragments, the most

conspicuous of which are of MW about 41,000 and 46,000 (Fig. 1). DNA fingerprinting of *S. equi* subsp. *equi* isolates from the United States and Europe and Southern hybridization analysis with an *S. equi* subsp. *equi* gene probe support the conclusion that *S. equi* subsp. *equi* is highly conserved and shows no variation (3).

In contrast to *S. equi* subsp. *equi*, *S. equi* subsp. *zooepidemicus* strains from horses have been shown to be serologically heterogeneous (11). At least 15 serologic variants of surface protein antigens extracted by hot acid were described on a series of equine isolates. This observation together with the variable DNA fingerprints of a series of isolates from human and animal hosts suggest that *S. equi* subsp. *zooepidemicus* is genetically variable (12). To prove this for a set of equine strains, we compared the DNA restriction profiles of a series of isolates from the upper respiratory tract (Fig. 2). The results confirm the DNA fingerprint diversity of *S. equi* subsp. *zooepidemicus* from the same host location. Furthermore, we have noted a similar diversity among strains isolated from the same outbreak of rhinitis and pharyngitis in yearlings (M. Mukhtar and J. F. Timoney, unpublished data).

Is this diversity reflected in differences in percentage DNA homology? Thermal melting profiles of labeled duplexes of DNA from *S. equi* subsp. *equi* and *S. equi* subsp. *zooepidemicus* isolates of equine origin (Table 1) revealed DNA homology of greater than 97%, a level of homology even greater than previously reported (8) and possibly reflecting the common source (equine respiratory tract) of the strains in our study.

M Proteins of *S. equi* subsp. *zooepidemicus*

Characterization of an M-like protein on an equine strain of *S. equi* subsp. *zooepidemicus* (strain 631) was accomplished by immunoblotting the proteins separated by sodium dodecyl sulfate-polyacrylamide gel electrophoresis (SDS-PAGE) with an opsonic equine serum, purifying each of the reactive proteins by preparative electrophoresis, and immunizing guinea pigs. The resulting antisera were screened for the ability to opsonize *S. equi* subsp. *zooepidemicus* 631 and other strains (Table 2). Antisera against 58,000- and 56,000-MW bands were each opsonic for 7 of 16 equine strains, suggesting that the M protein oc-

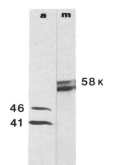

FIG. 1. M protein of *S. equi* subsp. *equi* following separation by SDS-PAGE and immunoblotting with M-protein-specific antiserum. Lanes: a, the most prominent fragments of the M protein following extraction with hot acid; m, the dimeric form of the protein as it occurs in a mutanolysin extract. K, Kilodaltons.

curs both as a dimer and in more than one M type. Antisera reacted with similar-size bands in the same strain, but the reactive bands varied greatly in MW from strain to strain (Fig. 3). Interestingly, strains opsonized by antiserum to the 58,000-MW M-protein band also showed a variety of sizes of reactive bands, suggesting that the same opsonophagocytic epitope could be carried on proteins of different sizes.

The variety of M proteins of *S. equi* subsp. *zooepidemicus* strains shows a remarkable parallel to that of *Streptococcus pyogenes*, the M proteins of which vary in size and differ in opsonophagocytic epitopes (2). The genetic basis of this variation appears to be homologous recombination between intragenic repeats, with mutations triggered by deletions of these repeats (6, 7). The relationship of these recombinations to

reported variations in DNA fingerprints of different M types (1) is not fully understood, but obviously any rearrangement of genomic DNA may cause changes in restriction fragment profile.

M-Protein Variation in the Group C Streptococci: a Role for Repeat Sequences?

What is the genetic basis of M-protein variation in *S. equi* subsp. *zooepidemicus* and why is the closely related *S. equi* subsp. *equi* so conserved? The explanation will probably be found in the sequences of the M-protein structural genes of each of these subspecies. Repeated attempts in our laboratory and elsewhere to subclone and sequence these genes have been frustrated by the toxicity of the M proteins for *Escherichia coli*, in which transformants always show DNA rearrangements (J. Ding and J. F. Timoney, J. Galán and J. F. Timoney, and M. Kehoe, unpublished data). However, DNA sequence analysis of *S. equi* subsp. *equi* DNA inserts in two clones, pJEG34 (5) and pSE21 (Ding and Timoney, unpublished data), has revealed an association of tandem repeat sequences with frameshift mutations that result invariations in the protein produced. The clones were independently selected from different lambda gt11 libraries of genomic DNA from the same isolate of *S. equi* subsp. *equi*. Recombinant clones were selected by using opsonic antiserum to the M protein of *S. equi* subsp. *equi* and then subcloned into the vectors pACYC184 and pWM401, yielding plasmids pSE21 and pJEG34, respectively. pSE21 encoded a 78-kDa protein, whereas pJEG34 encoded a protein of about 58 kDa. DNA sequence analysis of pSE21 revealed the presence of 23 repetitions of a nine-base repeat located within the open reading frame between bases 1434 and 1639 (Fig.

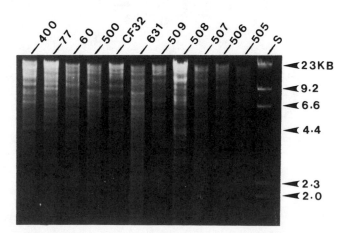

FIG. 2. DNA fingerprints of *S. equi* subsp. *equi* (CF32) and strains of *S. equi* subsp. *zooepidemicus*. Genomic DNAs were digested with *Eco*RI. Lane S, Lambda phage DNA digested with *Hin*dIII. Strain numbers are shown above the other lanes.

TABLE 1. DNA-DNA homology of *S. equi* subsp. *equi* and *S. equi* subsp. *zooepidemicus*

Hybrid pair	% Homology[a]
subsp. *zooepidemicus* (77) × subsp. *zooepidemicus* (77)	100
subsp. *zooepidemicus* (77) × subsp. *equi* (E23)	97.8
subsp. *zooepidemicus* (77) × subsp. *equi* (H113)	97.6
subsp. *equi* (E23) × subsp. *equi* (E23)	100
subsp. *zooepidemicus* (631) × subsp. *zooepidemicus* (631)	100
subsp. *zooepidemicus* (631) × subsp. *equi* (E23)	98.5

[a]Percent base pair mismatching was calculated from thermal melting profiles of labeled duplexes (10).

TABLE 2. Bactericidal activities of guinea pig antisera to the 58,000- and 56,000-MW dimeric forms of the M protein of *S. equi* subsp. *zooepidemicus* 631 for equine strains of *S. equi* subsp. *zooepidemicus*

Strain	CFU after 90-min incubation in horse blood		
	Normal guinea pig serum	Guinea pig antiserum to:	
		58,000-MW protein	56,000-MW protein
631	91, 69, 76*	2, 1, 5[a]	12, 11, 9[a]
66	40, 51, 61	45, 39, 49	42, 68, 52
78	41, 56, 31	39, 52, 36	55, 67, 23
44	41, 38, 29	36, 29, 31	33, 20, 28
77	48, 58, 55	18, 11, 7	13, 16, 19[a]
321	89, 81, 77	78, 80, 75	82, 89, 90
30	145, 178, 198	37, 46, 56[a]	33, 31, 40[a]
600	>300, >300, >300	59, 66, 72[a]	67, 66, 74[a]
60	72, 84, 91	35, 22, 34[a]	38, 31, 40[a]
324	97, 102, 113	35, 62, 47[a]	41, 53, 52[a]
08	185, 198, 205	189, 178, 184	190, 176, 86
505	52, 60, 78	51, 49, 72	49, 41, 61
500	90, 176, 92	1, 25, 9[a]	13, 31, 17[a]
506	64, 67, 82	58, 61, 77	63, 68, 72
400	161, 190, 220	143, 168, 190	158, 170, 182
13	60, 88, 78	45, 54, 62	49, 71, 53

[a]$P < 0.05$ versus normal guinea pig serum (modified Friedman's rank sum test).

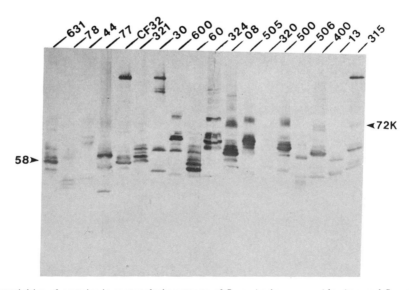

FIG. 3. Reactivities of proteins in mutanolysin extracts of *S. equi* subsp. *zooepidemicus* and *S. equi* subsp. *equi* (CF32) following separation by SDS-PAGE and immunoblotting with antiserum to the M protein of strain 631. Strain numbers are shown above the lanes. K, Kilodaltons.

4a). Analysis of pJEG34 revealed a region of sequence identical to that of pSE21 but with the stop codon located at base 1383 and upstream from the tandem repeat region (Fig. 4b). pJEG34 apparently arose as a result of three frameshift mutations caused by deletions and insertions of bases adjacent to the repeat sequences. The molecular sizes of the proteins predicted from the DNA sequences in pSE21 and pJEG34 were 62.7 and 50.1 kDa, respectively. The 62.7-kDa but not the 50.1-kDa protein reacted strongly with convalescent horse serum, an indication that the additional peptide sequence carries an epitope to which the horse makes a strong humoral response following infection. The 50.1-kDa recombinant

protein did not protectively immunize mice against experimental challenge with *S. equi* subsp. *equi* and so lacks the opsonic epitope characteristic of M-like proteins. Antiserum to the recombinant 62.7-kDa protein reacted with a protein of 90 kDa in a mutanolysin extract of *S. equi* subsp. *equi*. This finding suggests that only a part of the structural gene for the 90-kDa protein was present in pSE21 and that the DNA sequences of the inserts in both pJEG34 and pSE21 have been rearranged.

The surprising finding of a tandem repeat nucleotide sequence which undergoes rearrangement in the structural gene of an immunologically reactive protein of *S. equi* subsp. *equi* shows that

a

```
                    ▼
CAG CTT ATG CTC AGG AGA CAG GAA CCG ACA GCT ATT ACT AAC AGC TTG GTT AAG GTG TTA *1386
Gln Leu Met Leu Arg Arg Gln Glu Pro Thr Ala Ile Thr Asn Ser Leu Val Lys Val Leu

                      ▼                                              ▼
AAT GCT AAG AAA TCC CTC TCA GAT GCC AAG CAG CCT TGG TTG CTA AAC GGT CGA TCC AGT *1446
Asn Ala Lys Lys Ser Leu Ser Asn Ala Lys Gly Pro Trp Leu Leu Asn Gly Arg Ser Ser

AGA TCC AGT AGA CCC AGT GAT CCG GTA GAC CCA GTA GAT CCG GTA GAC CCA TGT GAT CCG *1506
Arg Ser Ser Arg Pro Ser Asp Pro Val Asp Pro Val Asp Pro Val Asp Pro Val Asp Pro

GTA GAC CCA GTG GAT CCA GTA GAC CCA GTA GAC CCA GTA GAC CCA GTG GAT CCG GTA GAC *1566
Val Asp Pro Val Asp Pro Val Asp Pro Val Asp Pro Val Asp Pro Val Asp Pro Val Asp

CCA GTG GAT CCG GTA GAC CCG GTC GAT CCA ATC GAC CCA GCG GAT CCA GTA AAA CCA TCA *1626
Pro Val Asp Pro Val Asp Pro Val Asp Pro Ile Asp Pro Ala Asp Pro Val Lys Pro Ser

GAT CCT GAG GTT AAG CCA GAG CCT AAA CCA GAA TCT AAG CCT GAA GCT AAG AAG GAG GAC *1686
Asp Pro Glu Val Lys Pro Glu Pro Lys Pro Glu Ser Lys Pro Glu Ala Lys Lys Glu Asp

AAG AAA GCA TAC ACT AAT CAC CTC ACA TCA TTA TTA TAC CAT ACT AGC TTA AAT TAG GCT *1746
Lys Lys Ala Tyr Thr Asn His Leu Thr Ser Leu Leu Tyr His Thr Ser Leu Asn - - -

GAAAATGAGGATTCCTGCTGCAACTGCAACATTTAAGCTTTCAGCTTGTCCAGGCATGCTAATATG *1812

AATCAGCTGATCAGCCAAAGCTGCCATCTCAGGACTAATGCCCTGACCTTCATTTCCTAAAACAA *1877
```

b

```
    1330         1340         1350         1360         1370         1380
     •            •            •            •            •            •
CAG CTT ATG CTC AGG AGA CAG AAC CGA CAG CTA TTA CTA ACA GCT TGG TTA AGG TGT
Gln Leu Met Leu Arg Arg Gln Asn Arg Gln Leu Leu Leu Thr Ala Trp Leu Arg Cys

    1390         1400         1410         1420         1430         1440
     •            •            •            •            •            •
TAA ATG CTA AGA AAT CCC TCT CAG ATG CCA AGG CAG CCT TGG TTG CTA AAC CGG TCG
- - -

        1450         1460         1470         1480         1490
         •            •            •            •            •
ATC CAG TAG ATC CAG TAG ACC CAG TGA TCC GGT AGA CCC AGT AGA TCC GGT AGA CCC

  1500         1510         1520         1530         1540         1550
   •            •            •            •            •            •
AGT GGA TCC GGT AGA CCC AGT GGA TCC AGT AGA CCC AGT AGA CCC AGT AGA CCC AGT

    1560         1570         1580         1590         1600         1610
     •            •            •            •            •            •
GGA TCC GGT AGA CCC AGT GGA TCC GGT AGA CCC GGT CGA TCC AAT CGA CCC AGC GGA

      1620         1630         1640         1650         1660
       •            •            •            •            •
TCC AGT AAA ACC ATC AGA TCC TGA GGT TAA GCC AGA GCC TAA ACC AGA ATC TAA GCC

  1670         1680         1690         1700         1710         1720
   •            •            •            •            •            •
TGA AGC TAA GAA GGA GGA CAA GAA AGC ATA CAC TAA TCA CCT CAC ATC ATT ATT ATA
```

FIG. 4. (a) DNA sequence showing frameshift mutations (arrows), nine-base repeat region, and stop codon in pSE21. This plasmid contains the same 5.1-kb insert as in pJEG34 but encodes a 62.7-kDa protein that reacts with antiserum to the M protein of *S. equi* subsp. *equi*. (b) DNA sequence showing stop codon and downstream nine-base repeat region in pJEG34. This plasmid contains a 5.1-kb insert of *S. equi* subsp. *equi* DNA that encodes a 50.1-kDa protein reactive with antiserum to the M protein of *S. equi* subsp. *equi*.

genomic DNA in this subspecies could generate size and antigenic variation in surface proteins. Failure to observe antigenic variation in *S. equi* subsp. *equi* suggests that rearrangement in the M-protein gene is in some way repressed or defective in the native host. It is tempting to speculate that this function is still operational in *S. equi* subsp. *zooepidemicus*, which has been shown by Southern blotting in our laboratory to hybridize with a synthetic oligonucleotide replica of the nine-base repeat in pSE21. Diversity of DNA fingerprints and of antigenicity and size of M proteins among variants of *S. equi* subsp. *zooepidemicus* could result from recombination between repeats such as these.

S. equi subsp. *equi* may represent a recombination-deficient mutant of *S. equi* subsp. *zooepidemicus* similar to those known to occur in *S. pyogenes* K56 (9). However, definitive proof of this must await direct sequencing of genomic DNA of *S. equi* subsp. *equi*.

LITERATURE CITED

1. **Cleary, P. P., E. L. Kaplan, C. Livdahl, and S. Skjold.** DNA fingerprints of *Streptococcus pyogenes* are M type specific. *J. Infect. Dis.* **158:**1317–1323.
2. **Fischetti, V. A.** 1989. Streptococcal M protein: molecular design and biological behavior. *Clin. Microbiol. Rev.* **2:**285–314.
3. **Galán, J. E., and J. F. Timoney.** 1985. Immunologic and genetic comparison of *Streptococcus equi* isolates from the United States and Europe. *J. Clin. Microbiol.* **26:**1142–1146.
4. **Galán, J. E., and J. F. Timoney.** 1987. Molecular analysis of the M protein of *Streptococcus equi* and cloning and expression of the M protein gene in *Escherichia coli*. *Infect. Immun.* **55:**3181–3187.
5. **Galán, J. E., J. F. Timoney, and R. Curtiss III.** 1988. Expression and localization of the *Streptococcus equi* M protein in *Escherichia coli* and *Salmonella typhimurium*, p. 34–40. *In* D. G. Powell (ed.), *Equine Infectious Diseases V*. Proceedings of the Fifth International Conference. The University Press of Kentucky, Lexington.
6. **Hollingshead, S. K., V. A. Fischetti, and J. R. Scott.** 1987. Size variation in group A streptococcal M protein is generated by homologous recombination between intragenic repeats. *Mol. Gen. Genet.* **207:**196–203.
7. **Jones, K. F., S. K. Hollingshead, J. R. Scott, and V. A. Fischetti.** 1988. Spontaneous M6 protein size mutants of group A streptococci display variation in antigenic and opsonogenic epitopes. *Proc. Natl. Acad. Sci. USA* **85:**8271–8275.
8. **Klipper-Balz, R., and K. H. Schleifer.** 1984. Nucleic acid hybridization and cell wall composition studies of pyogenic streptococci. *FEMS Microbiol. Lett.* **24:**355–364.
9. **Malke, H.** 1975. Recombination-deficient mutants of *Streptococcus pyogenes* K56. *Z. Allg. Mikrobiol.* **15:**31–37.
10. **Marmur, J., and P. Doty.** 1962. Determination of the base composition of deoxyribonucleic acid from its thermal denaturation temperature. *J. Mol. Biol.* **5:**109–118.
11. **Moore, B. O., and J. T. Bryans.** 1970. Type specific antigenicity of group C streptococci from diseases of the horse, p. 231–238. *In* J. T. Bryans and H. Gerber (ed.), *Proceedings of the Second International Conference on Equine Infectious Disease*. S. Karger, Basel.
12. **Skjold, S. A., P. G. Quie, C. A. Fries, M. Barnham, and P. P. Cleary.** 1987. DNA fingerprinting of *Streptococcus zooepidemicus* (Lancefield group C) as an aid to epidemiologic study. *J. Infect. Dis.* **155:**1145–1150.

Expression of Group A Streptococcal M Protein in Live Oral Vaccines

THOMAS P. POIRIER,[1,2,3] RONALD K. TAYLOR,[4] MICHAEL A. KEHOE,[5] MILOS RYC,[6†]
BRUCE A. D. STOCKER,[7] VICKERS BURDETT,[8] ELLEN WHITNACK,[1,2] JAMES B. DALE,[1,2]
AND EDWIN H. BEACHEY[1,2†]

Veterans Affairs Medical Center,[1] Departments of Medicine[2] and Microbiology/Immunology,[4] University of Tennessee, and Baptist Memorial Hospital,[3] Memphis, Tennessee 38104; Department of Microbiology, University of Newcastle upon Tyne, Newcastle upon Tyne, England[5]; Institute of Hygiene and Epidemiology, Prague, Czechoslovakia;[6] Department of Medical Microbiology, Stanford University, Stanford, California 94305[7]; and Department of Microbiology/Immunology, Duke University, Durham, North Carolina 27710[8]

Group A streptococci can be subclassified into 80 distinct serotypes based on the presence of the dimeric (coiled-coil), alpha-helical, cell surface polypeptide known as M protein (13, 18). M protein confers on the streptococcus the ability to resist opsonization by complement and therefore phagocytosis by neutrophils. This resistance can be overcome only by type-specific antibodies directed toward protective epitopes of M protein (13, 26). Therefore, as the major virulence factor for *Streptococcus pyogenes*, M protein is a prime candidate for a vaccine. However, efforts to develop an M-protein vaccine have been hampered by the finding that in many instances the molecule possesses tissue cross-reactive epitopes that could conceivably be involved in the pathogenesis of rheumatic fever and glomerulonephritis (8–10).

To date, numerous approaches, including the use of enzyme-extracted M protein, synthetic M peptides, or recombinant M protein, have been employed to gain a better understanding of the immunochemical properties of M protein and the relationship of the various epitopes to protective immunity and to the pathogenesis of acute rheumatic fever and glomerulonephritis (1). The ability to clone and express several M proteins in *Escherichia coli* (12, 17, 19, 23, 24) allows for comparative genetic analysis of M-protein nucleotide sequences and arrangements, both serotype specific and non-type specific, as they pertain to immunoprotective and tissue cross-reactive epitopes (6, 11, 19). Detailed genetic characterization makes possible further subcloning into other nonrelated bacteria to be used as vaccine delivery vehicles. *Salmonella* spp. (14, 20, 25) and *Streptococcus sanguis* (21, 22, 27) are potentially suitable vaccine delivery vehicles, since both naturally colonize mucosal surfaces, which is an accepted prerequisite for initiating mucosal immunity (2, 5, 7, 20, 25), and both can be readily transformed with foreign DNA (15, 25).

Expression of M Protein in Gram-Negative Bacteria

It has been previously demonstrated that *Salmonella* spp. can be readily attenuated and used effectively as live oral vaccines in both animal and human models (14, 20, 25). In previous studies, we introduced the cloned M5 gene into an attenuated *aroA* strain of *Salmonella typhimurium*, which was then used to immunize mice both intraperitoneally and orally (20). The immunized mice were completely protected from intranasal challenge with the homologous M5 streptococci, as well as the virulent wild-type *S. typhimurium*, but not from challenge with heterologous M24 streptococci. The immunized mice developed opsonic anti-M5 antibodies in their serum as well as secretory anti-M5 immunoglobulin A antibodies in their saliva. More recently, Stocker et al. have obtained similar results in mouse protection studies using an oligonucleotide representing 15 amino acids of the amino-terminal portion of the M5 molecule substituted into the flagellin gene *H1-d* of *Salmonella dublin*. The mutant organisms have functional flagella with M5 immunoprotective properties on the surface of the bacterium (unpublished data).

Although M protein has been cloned and expressed in other genera of bacteria (12, 17, 19, 23, 24), little is known about the details of M-protein expression, particularly the location of M-protein gene products within the transformed cell and whether they are optimally placed for interaction with the host's immunocytes. To obtain preliminary information concerning the subcellular localization of recombinant M protein, we performed electron microscopy using postembedding indirect immunogold labeling by the method of M. Ryc et al. (unpublished data). Two sera were used as probes: one raised to pepM5, the major pepsin cleavage product of M5 protein, consisting of the amino-terminal half of the molecule, and the other raised to a synthetic peptide represented in the carboxy terminus of the molecule. In *Escherichia coli* LE392 containing

Deceased.

pMK207, expressing the entire M5 protein, the anti-pepM5-reactive material was primarily within the cell envelope and, to a lesser extent, in the cytoplasm (Fig. 1b). In cell fractionation experiments, the anti-pepM5-reactive material copurified with a periplasmic enzyme, alkaline phosphatase (data not shown; 23). However, when the same preparation of cells was reacted with antiserum to a synthetic peptide from the carboxy-terminal region of the molecule, the label was found mainly in the cytoplasm (Fig. 1c). This observation suggests that the M protein expressed by *E. coli* may be subject to some post-translational processing and degradation. *E. coli* containing the parent plasmid lacking the *spm5* gene did not react with either of the antisera (Fig.

1a). In contrast to *E. coli*, in *S. typhimurium* SL3261(pMK207) expression of the M5 protein is apparently cytoplasmic (Fig. 1e). Further, crude cell fractionation of *S. typhimurium* SL3261(pMK207) resulted in the copurification of the M protein with the cytoplasmic enzyme acid phosphatase (data not shown).

Since attenuated strains of *Salmonella typhi* have proven to be both effective and safe for use in oral vaccination against typhoid fever (14), we tested the ability of this organism to express M24 protein cloned in pBR41-L3. This serotype has been previously shown to be both immunogenic and nontoxic in humans (4). Immunoblot analysis of *S. typhi* transformed with pBR41-L3 showed a band pattern (Fig. 2, lane b) similar to

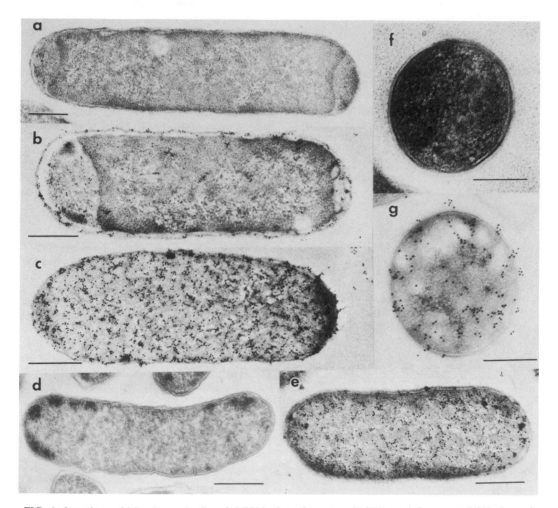

FIG. 1. Locations of M epitopes in *E. coli* LE392, *S. typhimurium* SL3261, and *S. sanguis* V288 shown by postembedding indirect immunogold labeling. (a) *E. coli*(pLG339) (M⁻ control) treated with anti-pepM5; (b) *E. coli*(pMK207) (M5⁺) treated with anti-pepM5; (c) *E. coli*(pMK207) treated with antiserum to a synthetic peptide from the carboxy-terminal region; (d) *S. typhimurium*(pMK207) (M5⁺) treated with preimmune serum; (e) *S. typhimurium*(pMK207) treated with anti-pepM5; (f and g) *S. sanguis*(pVA838) (M⁻ control) (f) and *S. sanguis*(pBK100) (M5⁺) (g) treated with anti-pepM5. Bar = 0.5 μm.

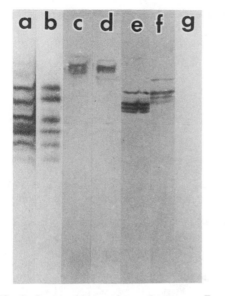

FIG. 2. Immunoblot analyses. Lanes: a, *E. coli* LE392(pBR41-L3) treated with anti-pepM24; b, *S. typhi*(pBR41-L3) treated with anti-pepM24; c, d, e, and g, extracts of *E. coli* expressing pTP200 containing a truncated *spm5* fragment grown under amber suppression (lanes c and d) or nonsuppressed (lanes e and g) conditions (lanes c and e were treated with anti-pepM5; lanes d and g were treated with anti-β-galactosidase); f, *E. coli*(pMK207) treated with anti-pepM5.

that demonstrated in *E. coli* (Fig. 2, lane a; 17), with an M_r of 58,000 for the intact molecule. Oral immunization of rabbits with this preparation resulted in serum enzyme-linked immunosorbent assay titers of 25,600; the serum opsonized type 24 but not type 5 streptococci.

Expression of M Protein in Gram-Positive Bacteria

In 1983, Westergren and Svanberg (27) showed that a nonpathogenic, streptomycin-resistant variant of *Streptococcus sanguis* (strain G26-S) could be maintained in the human oral cavity for more than 3 months after oral inoculation. Since this normally avirulent organism has the ability to colonize oral mucosal surfaces, it is a potential vaccine delivery vehicle. Therefore, we cloned M protein into *S. sanguis* for three reasons: (i) to determine whether cloned M protein is more likely to be expressed on the surface of grampositive than of gram-negative bacteria; (ii) to test a shuttle vector system that had been developed for transformation between *E. coli* and *S. sanguis*; and (iii) to compare the efficacy of an organism that colonizes the oral mucosa with that of one that invades the intestinal mucosa in delivering cloned immunoprotective M-protein epitopes.

Previous studies had shown that M5 protein is expressed on the surface of *S. sanguis* harboring pBK100 containing the M5 gene (21, 22). Further investigation by postembedding indirect immunogold labeling using anti-pepM5 antibody demonstrated the M5 antigen to be present both on the cell surface and in the cytoplasm of *S. sanguis* (pBK100) (Fig. 1g). This procedure failed to label control cells harboring the M-negative parent plasmid, pVA838 (Fig. 1f). The transformed cells had a three- or fourfold prolongation of generation time, possibly because the M protein, which is not normally found in *S. sanguis*, may adversely affect vital functions such as cell wall formation or transport mechanisms. Further, the M5 gene was unstably maintained, with eventual spontaneous loss from pBK100 (21, 22).

Before the suitability of *S. sanguis* as an M-protein vaccine delivery vehicle could be further assessed, it was necessary to gain a better understanding of the effects of M protein on the viability of the bacterium. We hypothesized that if the toxicity of M protein were in fact due to interference with cell wall synthesis or metabolite transport, the problem might be overcome if the M molecule were secreted outside the cell. To this end we designed a novel plasmid, pTP100, by modifying the *Bam*HI-*Pst*I polylinker region of pBluescript SK+/− (Stratagene, La Jolla, Calif.) so that gene fragments containing their own promoter, inserted into the polylinker, would restore the β-galactosidase reading frame, permitting the expression of fusion proteins (unpublished data). In addition to the repositioning of the β-galactosidase reading frame, the polylinker was also constructed to contain a TAG sequence in line with the reading frame of β-galactosidase, using the unique sequence 5′-GACATAGTC-3′, which is recognized by the restriction endonuclease *Tth*111I (New England BioLabs, Beverly, Mass.); thus, inserted genes would incorporate a stop codon. M-protein genes

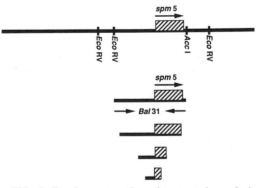

FIG. 3. Random generation of truncated *spm5* via bidirectional *Bal*31 digestion of the *Eco*RV-*Acc*I fragment of pMK207. Arrow represents the direction of transcription of the M5 gene.

were truncated by *Bal*31 digestion (Fig. 3) for insertion into the *Tth*111I site. Theoretically, under conditions of amber suppression (e.g., expression in *E. coli* DH5α), only those M-protein genes truncated within their reading frames would read through the TAG sequence and restore the β-galactosidase reading frame (β-gal⁺). Plasmid DNA isolated from β-gal⁺ colonies and transformed into a suppressor-negative strain of *E. coli*, SG932, would then express the *Bal*31 truncated insert through to the stop codon, yielding truncated peptides with a defined endpoint, i.e., without β-galactosidase fusion peptides.

Preliminary studies using the 4.5-kb *Eco*RV-*Acc*I fragment of pMK207 (12), digested with *Bal*31 and inserted into pTP100, resulted in numerous β-gal⁺ colonies (approximately 1 in 20 transformants) grown under conditions of amber suppression on 5-bromo-4-choro-3-indolyl-β-D-galactopyranoside (X-Gal) medium (26). Immunoblots of extracts of one of the β-gal⁺ colonies, designated pTP200, revealed the characteristic banding pattern for recombinant M5 protein (Fig. 2, lane c), although the molecular size was approximately 8 kDa greater than that of intact M5 protein derived from pMK207 (Fig. 2, lane f). The anti-M5-reactive material in pTP200 also reacted with anti-β-galactosidase (Fig. 2, lane d). Under nonsuppressed conditions, pTP200 expressed M5 peptides with molecular sizes approximately 5 kDa less than that of intact M5 protein (Fig. 2, lane e versus lane f) but no longer expressed anti-β-galactosidase-reactive material (Fig. 2, lane g). These findings suggest that pTP200 produces an M5 peptide that is truncated by approximately 50 amino acids (M_r, 5,000), but its molecular weight is increased by approximately M_r 13,000 via the addition of β-galactosidase when the TAG sequence is suppressed. The DNA sequence encoding the inserted truncated M5 peptide is currently being determined.

Conclusions

We have shown that the M protein of *S. pyogenes* can be cloned and expressed in other species of both gram-positive and gram-negative bacteria. There appears to be much variability in M-polypeptide presentation, i.e., periplasmic in *E. coli*, cytoplasmic in *S. typhimurium*, and on the surface of *S. sanguis*. Although there is a difference in M-antigen distribution in these different bacteria, there is a conservation of the immunologic properties characteristic of native M protein in the ability to evoke protective antibodies. Previous studies have shown that efforts to develop effective M-protein vaccines must include serotype-specific, N-terminal sequences that are known to elicit effective protection (1, 3, 4). To avoid the potential toxicity of host cross-reactive anti-M antibodies, it will be necessary

that M-protein vaccines be well characterized and composed of non-host cross-reactive (type-specific) epitopes. Such epitopes have already been identified in the nonhelical, extreme amino-terminal regions of a number of M proteins. A synthetic trivalent peptide composed of the protective type-specific sequences from the amino termini of the M5, M6, and M24 proteins has been shown to elicit protective antibody to all three M types, demonstrating that epitopes from various M types could be linked to create multivalent immunogens (3). Therefore, as a prerequisite to using M protein expressed by heterologous bacteria as general vaccine delivery systems, further investigation is needed concerning the molecular processes involved in M-protein expression, such as (i) the location of M-protein gene product(s) within the transformed cell, whether it be cell surface, secreted, or cytoplasmic; (ii) the quantity, size, and molecular configuration of the cloned immunogens required for effective elicitation of protective antibody; and (iii) determination of the number of M epitopes that can be effectively incorporated into a multivalent M-protein vaccine as it relates to maximizing immunogen delivery for optimum interaction with the host immune system.

We thank Edna Chiang, Loretta Hatmaker, and Mary Melvin for expert technical assistance. The efforts of Roberta A. Sumrada and the Molecular Resource Center, University of Tennessee, Memphis, in the preparation of oligonucleotides are greatly appreciated. We thank John A. Walker for stimulating discussions.

These studies were supported by Veterans Administration medical research funds and by Public Health Service research grants AI-10085 and AI-13550 from the National Institute of Allergy and Infectious Diseases.

LITERATURE CITED

1. **Beachey, E. H., M. Bronze, J. B. Dale, W. Kraus, T. P. Poirier, and S. Sargent.** 1988. Protective and autoimmune epitopes of streptococcal M proteins. *Vaccine* **6:**192–196.

2. **Beachey, E. H., and H. S. Courtney.** 1987. Bacterial adherence: the attachment of group A streptococci to mucosal surfaces. *Rev. Infect. Dis.* **9:**S475–S481.

3. **Beachey, E. H., J. M. Seyer, and J. B. Dale.** 1987. Protective immunogenicity and T lymphocyte specificity of a tri-valent hybrid peptide containing the NH2-terminal sequences of types 5, 6, and 24 M proteins synthesized in tandem. *J. Exp. Med.* **166:**647–656.

4. **Beachey, E. H., G. H. Stollerman, R. H. Johnson, I. Ofek, and A. L. Bisno.** 1979. Human immune response to immunization with a structurally defined polypeptide fragment of streptococcal M protein. *J. Exp. Med.* **150:**862–877.

5. **Bessen, D., and V. A. Fischetti.** 1988. Influence of intranasal immunization with synthetic peptides corresponding to conserved epitopes of M protein on mucosal colonization by group A streptococci. *Infect. Immun.* **56:**2666–2672.

6. **Bessen, D., K. F. Jones, and V. A. Fischetti.** 1989. Evidence for two distinct classes of streptococcal M protein and their relationship to rheumatic fever. *J. Exp. Med.* **169:**269–283.

7. **Curtiss, R., III, R. Goldschmidt, R. Pastian, M. Lyons, S. M. Michalek, and J. Mestecky.** 1986. Cloning virulence determinants from *Streptococcus mutans* and the use of recombinant clones to construct bivalent oral vaccine strains to confer protective immunity against *S. mutans-*

induced dental caries, p. 173–180. *In* S. Hamada et al. (ed.), *Molecular Microbiology and Immunology of Streptococcus mutans*. Elsevier Science Publishers, New York.

8. **Dale, J. B., and E. H. Beachey.** 1982. Protective antigenic determinant of streptococcal M protein shared with sarcolemmal membrane protein of human heart. *J. Exp. Med.* **156:**1165–1176.

9. **Dale, J. B., and E. H. Beachey.** 1986. Sequence of myosin-cross-reactive epitopes of streptococcal M protein. *J. Exp. Med.* **164:**1785–1790.

10. **Kaplan, M. H.** 1963. Immunologic relation of streptococcal and tissue antigens. I. Properties of an antigen in certain strains of group A streptococci exhibiting an immunologic cross-reaction with human heart tissue. *J. Immunol.* **90:** 595–606.

11. **Kehoe, M. A., L. Miller, T. P. Poirier, E. H. Beachey, M. Lee, and D. Harrington.** 1987. Genetics of type 5 M protein of *Streptococcus pyogenes*, p. 112–116. *In* J. J. Ferretti and R. Curtiss III (ed.), *Streptococcal Genetics*. American Society for Microbiology, Washington, D.C.

12. **Kehoe, M. A., T. P. Poirier, E. H. Beachey, and K. N. Timmis.** 1985. Cloning and genetic analysis of serotype 5 M protein determinant of group A streptococci: evidence for multiple copies of the M5 determinant in the *Streptococcus pyogenes* genome. *Infect. Immun.* **48:**190–197.

13. **Lancefield, R. C.** 1962. Current knowledge of type-specific M antigens of group A streptococci. *J. Immunol.* **89:**307–313.

14. **Levine, M. M., C. Ferreccio, R. E. Black, and R. Germanier.** 1987. Large-scale field trial of TY21A live oral typhoid vaccine in enteric-coated capsule formulation. *Lancet* 9 May 1987:1049–1052.

15. **Macrina, F. L., J. A. Tobian, K. R. Jones, R. P. Evans, and D. B. Clewell.** 1982. A cloning vector able to replicate in *Escherichia coli* and *Streptococcus sanguis*. *Gene* **19:**345–353.

16. **Maniatis, T., E. F. Fritsch, and J. Sambrook.** 1982. Molecular cloning: a laboratory manual. Cold Spring Harbor Laboratory, Cold Spring Harbor, N.Y.

17. **Mouw, A. R., E. H. Beachey, and V. Burdett.** 1988. Molecular evolution of streptococcal M protein: cloning and nucleotide sequence of the type 24 M protein gene and relation to other genes of *Streptococcus pyogenes*. *J. Bacteriol.* **170:**676–684.

18. **Philips, G. N., Jr., P. F. Flicker, C. Cohen, B. N. Manjula, and V. A. Fischetti.** 1981. Streptococcal M-protein: alpha-helical coiled-coil structure and arrangement on the cell surface. *Proc. Natl. Acad. Sci. USA* **78:**4689–4693.

19. **Poirier, T. P., M. A. Kehoe, J. B. Dale, K. N. Timmis, and E. H. Beachey.** 1985. Expression of protective and cardiac tissue cross-reactive epitopes of type 5 streptococcal M protein in *Escherichia coli*. *Infect. Immun.* **48:**198–203.

20. **Poirier, T. P., M. A. Kehoe, and E. H. Beachey.** 1988. Protective immunity evoked by oral administration of attenuated *aroA Salmonella typhimurium* expressing cloned streptococcal M protein. *J. Exp. Med.* **168:**25–32.

21. **Poirier, T. P., M. A. Kehoe, E. Whitnack, and E. H. Beachey.** 1987. Surface expression of type 5 M protein of *Streptococcus pyogenes* in *Streptococcus sanguis*, p. 117–120. *In* J. J. Ferretti and R. Curtiss III (ed.), *Streptococcal Genetics*. American Society for Microbiology, Washington, D.C.

22. **Poirier, T. P., M. A. Kehoe, E. Whitnack, M. E. Dockter, and E. H. Beachey.** 1989. Fibrinogen binding and resistance to phagocytosis of *Streptococcus sanguis* expressing cloned M protein of *Streptococcus pyogenes*. *Infect. Immun.* **57:**29–35.

23. **Scott, J. R., and V. A. Fischetti.** 1983. Expression of streptococcal M protein in *Escherichia coli*. *Science* **221:**758–760.

24. **Spanier, J. G., S. J. C. Jones, and P. Cleary.** 1984. Small DNA deletions creating avirulence in *Streptococcus pyogenes*. *Science* **225:**935–938.

25. **Stocker, B. A. D., and P. H. Makela.** 1986. Genetic determination of bacterial virulence, with special reference to *Salmonella*. *Curr. Top. Microbiol. Immunol.* **124:**149–172.

26. **Stollerma, G. H.** 1975. *Rheumatic Fever and Streptococcal Infection*. Grune and Stratton, New York.

27. **Westergren, G., and M. Svanberg.** 1983. Implantation of transformant strains of the bacterium *Streptococcus sanguis* into adult human mouths. *Arch. Oral Biol.* **28:**729–733.

Molecular Analysis of M-Protein Epitopes

M. A. KEHOE,[1] M. C. ATHERTON,[2] J. A. GOODACRE,[3] M. PINKNEY,[1] M. KOTB,[4]
AND J. H. ROBINSON[2]

Departments of Microbiology,[1] Immunology,[2] and Medicine,[3] Medical School, University of Newcastle upon Tyne, Newcastle upon Tyne NE2 4HH, England, and Veterans Affairs Medical Center, Memphis, Tennessee 38104[4]

Immunity to group A streptococcal infections in humans is due predominantly to serotype-specific antibodies against the antigenically heterogeneous cell surface M proteins (for reviews, see references 5 and 8). In addition to protective antibodies, certain serotypes of M protein can elicit human tissue (e.g., heart) cross-reactive (HCR) antibodies (3), and it has been suggested that these might contribute to the pathogenesis of acute rheumatic fever. To avoid this potential danger, any M-protein vaccine will need to be a defined-epitope vaccine, based on well-characterized, non-HCR, protective M epitopes. The design of such vaccines is further complicated by the fact that over 80 distinct serotypes of M protein exist; with rare exceptions, only one serotype is expressed by each strain, and protective immunity is predominantly M serotype specific (5, 8). Below we review briefly the structural and immunochemical studies undertaken by a number of laboratories in an attempt to overcome these problems and our recent studies on M-protein T-cell epitopes.

Antibody Epitopes in M Proteins

Structural studies on M proteins and cloned M-protein genes have shown that although relationships between individual serotypes differ in detail, in general the C-terminal halves of M proteins, which are closely associated with the cell surface, are very highly conserved. The N-terminal halves, however, which protrude outward from the cell, are highly variable (5). Despite considerable variation in the N-terminal primary sequences, a shared seven-residue periodicity produces a common α-helical, coiled-coil dimeric structure throughout most of the length of these fibrillar molecules, with the extreme N-terminal 10 or so residues being random-coil and cell wall-associated domains at the extreme C terminus (5).

The variable N-terminal halves of M proteins can be released from the streptococcal cell surface by pepsin and are termed pepM antigens. Because they were easier to isolate and purify, before M genes were cloned most immunochemical studies were performed on pepM antigens rather than the intact molecule. Studies with monoclonal antibodies and synthetic peptides defined individual antibody epitopes in several serotypes of pepM antigens and distinguished HCR epitopes from non-HCR epitopes (1, 3–5). A number of non-HCR epitopes capable of eliciting opsonic anti-M antibodies were identified near the N-terminal end of the molecule, including epitopes in the random-coil extreme N-terminal sequences. Although polyclonal anti-pepM5 sera can cross-opsonize some heterologous M types, these cross-reactions are limited (10). Some of these cross-reacting epitopes are also HCR, and to date no widely conserved, non-HCR, opsonogenic epitope has been identified in the variable N-terminal halves of M proteins (10). Further studies on intact M proteins have shown that the conserved sequences in the C-terminal halves of the molecules do not elicit opsonic antibody (7). Nevertheless, a number of studies have shown that intranasal immunization of mice with heat-killed whole streptococci or recombinant vaccinia virus expressing the conserved C-terminal half of M protein elicits a significant reduction in pharyngeal colonization and death upon challenge (2, 6). In these studies no systemic protection was obtained, suggesting that the reduction in deaths was due to localized mucosal immunity. However, although significant, this mucosal immunity reported was limited, and many of the immunized mice died. Type-specific, systemic immunity appears to be much more effective and, in the mouse model, protects fully against challenge doses approximately 10-fold higher than doses that killed intranasally immunized mice. For example, oral immunization of mice with *aroA*-attenuated *Salmonella typhimurium* expressing the entire M5 protein elicits very effective protection (100% survival) against challenge with 4×10^7 M5 streptococci but no protection against challenge with the same dose of a heterologous M24 strain (12; Poirier et al., this volume). These studies suggest that although the conserved C-terminal half of M protein could be a useful component to include in a vaccine designed for mucosal delivery, if used alone it would not elicit a sufficient level of mucosal immunity to provide effective protection. Over 50 years of clinical observations suggest that repeated natural streptococcal infections in humans may induce a certain level of non-type-specific mucosal immunity, but that this is limited and effective protective immunity is due predominantly to systemic, serotype-specific, anti-M antibodies. Thus, an effective M-protein vaccine will need to include defined, type-specific, non-HCR epitopes that are known to elicit effective systemic protection. These defined epitopes will need to be linked to a carrier molecule capable of inducing effective T-cell help for antibody-producing B cells.

M-Protein T-Cell Epitopes

In principle, the most effective defined-epitope vaccine would include T-cell epitopes (in addition to protective B-cell epitopes) from the parent antigen in order to elicit helper T-cell memory to the parent pathogen. Several distinct T-cell epitopes may be required to reduce host restriction (i.e., the failure of certain major histocompatibility complex [MHC] haplotypes to recognize a specific T-cell epitope). Although M-protein antibody (B-cell) epitopes have been studied in detail for many years, studies on T-cell responses have been much more limited, and neither the numbers, types, nor locations of T-cell epitopes in an M protein have been defined. Indeed, there have been few detailed studies of the full range of T-cell epitopes in any bacterial protective antigen. Our understanding of T-cell epitopes is based primarily on model antigens (e.g., ovalbumin), cytotoxic T-cell epitopes in a limited number of viral antigens (e.g., influenza virus hemagglutinin), or recent studies on a limited number of epitopes from an even more limited number of bacterial antigens, such as the 65-kDa heat-shock protein of *Mycobacterium* sp. Other than the general requirement that a B-cell and helper T-cell epitope be located in the same molecule (or molecular complex), we know little about how (or if) specific structural relationships between B- and T-cell epitopes might influence an antibody response.

We have recently initiated a detailed systematic study of T-cell epitopes in serotype 5 M protein. One objective is to determine whether conserved sequences in the C-terminal half of M protein could provide effective, non-type-specific help for type-specific, N-terminal B-cell epitopes. If this were the case, one could attempt to design vaccines based on linking multiple, defined, non-HCR, type-specific, protective B-cell epitopes from heterologous serotypes to a non-type-specific, C-terminal, helper T-cell epitopes carrier. In addition, the link between group A streptococcal infection and rheumatic fever is the only example for which the causative agent of an autoimmune disease has been clearly established, and it would be interesting to explore what role (if any) M-protein T-cell epitopes might play in the pathogenesis of this autoimmune disease.

Antibody responses to M-protein B-cell epitopes have already been well characterized, and it would be interesting to determine whether different T-cell epitopes control antibody responses to different B-cell epitopes (e.g., HCR versus non-HCR; delayed production of opsonic antibody to N-terminal epitopes versus earlier production of nonopsonic antibodies to other epitopes).

To characterize M-protein T-cell epitopes, we have studied the T-cell response by different strains (different MHC haplotypes) of mice to the intact recombinant M5 protein (rM5) purified from *Escherichia coli*. Initial studies on polyclonal T-cell responses demonstrated that these responses are antigen specific and can be totally inhibited by monoclonal antibodies to the relevant MHC class II haplotype, indicating that both pepM5 and the intact rM5 elicit typical class II-restricted helper T-cell responses and that they have no mitogenic activities for mouse T cells. In addition, the N-terminal pepM5 antigen stimulated only part of the intact rM5-specific response by polyclonal T cells from mice that had been immunized with the intact M protein, suggesting that there are T-cell epitopes located both in the variable N-terminal half and in the conserved C-terminal half of the molecule. To examine these responses in detail, over 70 rM5-specific T-cell clones have been isolated and characterized. All of the isolated clones are CD4$^+$, class II-restricted (helper) T cells. The specific epitopes to which these clones are directed have been mapped by examining their responses to a panel of synthetic peptides, covering most (but not all) of the M5 molecule. A small number of clones did not respond to any of the peptides used, and it is likely that these recognize epitopes in a small number of gaps in the sequence to which peptides are currently being synthesized. However, most of the M-protein T-cell epitopes have already been mapped. In some model antigens, most of the T-cell clones isolated appear to recognize a limited number of strongly immunodominant regions (e.g., >90% of mouse T-cell response to ovalbumin can be stimulated with a single 17-residue peptide). In M5, almost every peptide tested to date stimulates two or three specific clones, demonstrating that there are many helper T-cell epitopes located throughout the entire length of the molecule rather than a limited number of strongly immunodominant epitopes. Peptides corresponding to residues 43 to 77 and 70 to 96 at the extreme N-terminal end of the molecule stimulate responses by the largest number of clones (13 of 70 and 11 of 70, respectively), indicating that this region is relatively, but not strongly, immunodominant (note that residues 1 to 42 correspond to the signal peptide that is removed during secretion and that residue 43 corresponds to the N terminus of cell surface M5 [11]). It is interesting to note that these sequences also contain a number of opsonic B-cell epitopes. Further studies will be required to determine whether T-cell help for these opsonic B-cell epitopes is dependent on proximal T-cell epitopes or whether effective help can be provided by epitopes that have been mapped in the conserved C-terminal half of the molecule.

M-Protein Superantigenicity

The recently coined term "superantigen" refers to a special group of mitogens that includes the staphylococcal enterotoxins, toxic shock syndrome toxin, exfoliating toxin, and the streptococcal pyrogenic exotoxins (for a review, see reference 9). These mitogens stimulate proliferation of T cells bearing certain Vβ families of T-cell receptors (TCRs), probably by cross-linking the variable region at the side of the TCR β chain with an MHC class II molecule on an antigen-presenting cell (APC). TCRs are divided into Vβ families on the basis of sequence homologies in the variable regions of the β chain, and each Vβ family contains T cells with specificities for many different antigens and epitopes, determined by the hypervariable sequences at the top of the TCR. Thus, although a superantigen is more selective than many mitogens for certan Vβ families, essentially it is a special type of nonspecific mitogen, which stimulates proliferation of T cells with specificities for a very wide range of different antigens and epitopes. Superantigen interaction with T cells should not be confused with normal antigen presentation. A normal antigen is processed by an APC, and a short peptide, presented in the groove at the top of the APC's class II molecule (or class I in the case of most cytotoxic T cells), is recognized by the hypervariable sequences at the top of the TCR. A specific T cell will respond only to a specific epitope and only if this epitope is presented by a self-MHC class II molecule. In contrast, superantigens act at very low concentrations, to induce proliferation of T cells with many different epitope specificities, are not MHC haplotype restricted, and do not require to be processed by the APC.

Almost 10 years ago, Dale et al. (4a) demonstrated that highly purified pepM5, pepM6, and pepM24 could elicit a nonspecific blastogenic response by human lymphocytes but not by lymphocytes from a number of other animal species. Recent studies in our laboratories have shown that this response is MHC class II dependent and does not require antigen processing. Further, pepM5 selectively stimulates proliferation of certain Vβ families of human T cells (13). Thus, pepM5 has properties that are characteristic of other superantigens, but its mitogenic activity appears to be specific for human T cells. In contrast

to pepM5, which was purified from streptococcal cells, neither intact rM5 purified from *E. coli* nor pepsin-cleaved rM5 elicited a detectable proliferative response by nonimmune human lymphocytes from six different subjects tested (J. A. Goodacre and M. A. Kehoe, unpublished data). At present, we do not understand why pepM5 purified from streptococci can act as a potent human lymphocyte mitogen, whereas neither the intact nor pepsin-cleaved rM5 can. There are several examples in the literature in which mitogenic properties have been assigned incorrectly to highly purified proteins as a result of contamination with trace amounts of other, very potent mitogens. The human species-specific pepM5 mitogenicity suggests that it is unlikely that this finding is due to one of the known pyrogenic toxins, but contamination with trace amounts of an uncharacterized streptococcal mitogen cannot at present be ruled out. An alternative explanation is that the superantigenic properties of M protein may be strongly dependent on conformation and that the rM5 expressed in *E. coli* does not adopt the precise conformation required for superantigenicity, despite its ability to be recognized by a panel of anti-pepM5 monoclonal antibodies and to be cleaved with pepsin. Further studies will be required to determine whether the differences in human lymphocyte responses to pepM5 and rM5 are due to contamination or conformational constraints. If these studies confirm that M protein is a genuine superantigen, this may have significant implications for the pathogenesis of post-streptococcal infection autoimmune diseases such as rheumatic fever and will require that any M-protein-based vaccine be designed to be free of this activity.

Potential Vaccine Design

We are now exploring the design of recombinant, defined-epitope M vaccines that can be delivered to the host by a live oral delivery system (see Poirier et al., this volume). The objective is to link multiple, defined type-specific epitopes to conserved C-terminal sequences containing helper T-cell epitopes, in the absence of sequences encoding HCR epitopes or potential superantigen-MHC and -TCR binding sites. The most difficult problems are the limitations on the number of distinct type-specific epitopes that might be effectively included in such constructs and the danger of novel serotypes arising. Although highly speculative, it would be interesting to determine whether novel sequences might mimic several distinct type-specific epitopes and therefore reduce the number of such epitopes that would need to be included in a vaccine. To examine this question, we have initiated a detailed study of the structural basis for type-specific antibody-epitope recognition.

Work in M.A.K.'s laboratory is funded by grants from The Wellcome Trust (17702/1.5/JGH) and the U.K. Medical Research Council (G814199CA).

We dedicate this contribution to the memory of Ed Beachey, whose recent death prematurely ended a career in which he made an enormous contribution to our understanding of streptococcal M proteins. We will always be grateful for his collaboration, friendship, and encouragement, which have been of invaluable assistance in our own careers.

LITERATURE CITED

1. **Beachey, E. H., J. M. Seyer, and J. B. Dale.** 1987. Protective immunogenicity and T lymphocyte specificity of a trivalent hybrid peptide containing NH$_2$-terminal sequences of type 5, 6 and 24 M proteins synthesized in tandem. *J. Exp. Med.* **166:**647–656.

2. **Bronze, M. S., D. S. McKinsey, E. H. Beachey, and J. B. Dale.** 1988. Protective immunity evoked by locally administered group A streptococcal vaccines in mice. *J. Immunol.* **141:**2767–2770.

3. **Dale, J. B., and E. H. Beachey.** 1985. Multiple heart-crossreactive epitopes of streptococcal M proteins. *J. Exp. Med.* **161:**113–122.

4. **Dale, J. B., and E. H. Beachey.** 1986. Localization of protective epitopes of the amino terminus of type 5 streptococcal M protein. *J. Exp. Med.* **163:**583–591.

4a. **Dale, J. B., W. A. Simpson, I. Ofek, and E. H. Beachey.** 1981. Blastogenic responses of human lymphocytes to structurally defined polypeptide fragments of streptococcal M protein. *J. Immunol.* **126:**1499–1505.

5. **Fischetti, V. A.** 1989. Streptococcal M protein: molecular design and biological behavior. *Clin. Microbiol. Rev.* **2:**285–314.

6. **Fischetti, V. A., W. M. Hodges, and D. E. Hurby.** 1989. Protection against streptococcal pharyngeal colonization with a vaccinia: M protein recombinant. *Science* **244:**1487–1490.

7. **Jones, K. F., and V. A. Fischetti.** 1988. The importance of the location of antibody binding on the M6 protein for opsonization and phagocytosis of group A M6 streptococci. *J. Exp. Med.* **167:**1114–1123.

8. **Lancefield, R. C.** 1962. Current knowledge of the type-specific M antigens of group A streptococci. *J. Immunol.* **89:**307–313.

9. **Marrack, P., and J. Kappler.** 1990. The staphylococcal enterotoxins and their relatives. *Science* **248:**705–711.

10. **Miller, L., V. Burdett, T. P. Poirier, L. D. Gray, E. H. Beachey, and M. A. Kehoe.** 1988. Conservation of protective and nonprotective epitopes in M proteins of group A streptococci. *Infect. Immun.* **56:**2198–2204.

11. **Miller, L., L. Gray, E. Beachey, and M. Kehoe.** 1988. Antigenic variation among Group A streptococcal M proteins. *J. Biol. Chem.* **263:**5668–5673.

12. **Poirier, T. P., M. A. Kehoe, and E. H. Beachey.** 1988. Protective immunity evoked by oral administration of attenuated *aroA Salmonella typhimurium* expressing cloned streptococcal M protein. *J. Exp. Med.* **168:**25–32.

13. **Tomai, M., M. Kotb, G. Majumdar, and E. H. Beachey.** 1990. Superantigenicity of streptococcal M protein. *J. Exp. Med.* **172:**359–362.

Isolation of Spontaneous Antigenic Variants of Group A Streptococci That Exhibit Antigenic Drift in the M6.1 Protein

S. K. HOLLINGSHEAD,[1] K. F. JONES,[2] AND V. A. FISCHETTI[2]

Department of Microbiology, University of Alabama at Birmingham, Birmingham, Alabama 35294,[1] and Laboratory of Bacteriology and Immunology, The Rockefeller University, New York, New York 10021[2]

Antigenic variation is one component of the arsenal used by bacterial pathogens in their encounter with the immune system of a susceptible host. Molecular mechanisms of antigenic variation are understood in depth for relatively few surface proteins of pathogens. In *Neisseria gonorrhoeae* or *Borrelia* spp., the mechanism involves homologous recombination between an expressed copy of the gene encoding the variable surface antigen and a silent extragenic unexpressed gene or partial gene (1, 15). This process leads to the appearance of antigenic variants within a single bacterial culture, often at frequencies approaching 10^{-2}. It seems likely that additional mechanisms for increasing the antigenic diversity of surface proteins are utilized by bacterial pathogens. For many pathogens, antigenic variation is observed to occur only infrequently during the course of a natural infection but is found to be prevalent throughout host populations. This type of variation would be expected to occur by a different mechanism.

A different system of generating antigenic diversity within a cell surface molecule is used by the group A streptococci. In *Streptococcus pyogenes*, several cell surface proteins have been found to have internal repeat domains (3, 5, 9). In the case of Fc receptors, some repeat domains are associated with binding capacities (6, 9), but in many instances the function of the repeats is unknown. Many antigens with repetitive structures are also found in antigenically varied forms in different group A streptococcal strains, suggesting that a mechanism exists that operates to introduce antigenic variation into repetitive molecules. The most extensively studied of these repetitive surface antigens is the M protein, an antiphagocytic coiled-coil dimer protein present on most group A streptococcal isolates.

As a family, M proteins are highly varied surface molecules whose antigenic differences have formed the basis for a classification scheme for group A streptococci. This classification scheme has depended on the use of monospecific sera that are generated against a single M type. There are currently over 80 different serotypes of M protein recognized by such sera, and protection against group A streptococcal infection is dependent on the expression of type-specific antibodies (16).

Sequence Heterogeneity among Heterologous *emm* Genes

Both amino acid and DNA sequences of several antigenically distinct M proteins are now available (7, 19, 20, 22; Fig 1). Each of the M proteins is highly (near 98%) conserved in the carboxy-terminal third of the molecule (10). This homology clearly establishes an evolutionary relationship among these heterologous *emm* genes even though the direct phylogenetic relationship between any two serotypes is unknown. There is somewhat less homology between the carboxy-terminal regions of class I and class II *emm* genes (2, 7). Outside of the coding region, the 5' and 3' DNA sequences of the *emm* locus are also well conserved for class I *emm* genes (8, 20).

The amino-terminal half (approximately) of each M protein is characterized by heterogeneous repeat regions responsible for most of the antigenic diversity observed among serotypes. The sequence in this region of the M molecules is highly varied. Different M proteins have repeat segments that differ in length and primary sequence (Fig. 1). In addition, the repeat segments are surrounded by short lengths of nonrepeated coding DNA which is also distinct for a particular M serotype. When DNA probes from this variable region of *emm* genes are used in Southern hybridization analysis, homologous DNA is only found in group A streptococcal strains of the same M serotype (19, 23). Because each strain lacks DNA encoding the type-specific region of most heterologous serotypes, it seems unlikely that recombination with extragenic sequences will be the major means for antigenic variation in this gene family.

Antigenic shifts in serotype have rarely been reported (18). Although most secondary isolates from patients in carriage studies are of the same M serotype, a small percentage are either found to be of alternative M serotype or considered nontypeable (21, 24; Barry Gray, personal communication). In addition, fewer of the strains isolated recently are typeable (World Health Organization Conference, Tokyo, Japan, 1984). This finding suggests that the M antigen may be evolving at a rapid rate and that this evolution leads to nontypeable strains.

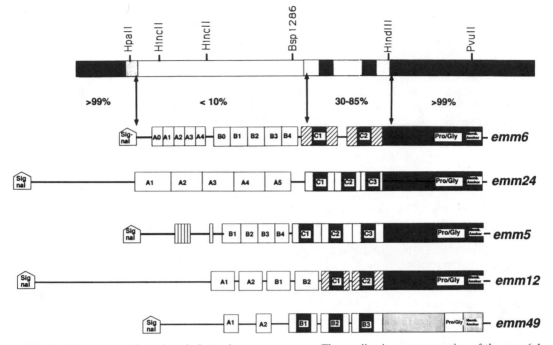

FIG. 1. Alignment and homology in heterologous *emm* genes. The top line is a representation of the *emm6.1* gene showing some of the internal restriction endonuclease cleavage sites. Shading in the boxed areas is indicative of different percent homologies; the range is from white (less than 10%) to black (near 100%). Boxes lettered A, B, and C in the *emm* genes represent intragenic repeats; boxes unlabeled or labeled Pro/Gly or Memb. Anchor encode the carboxy terminus of the M molecules and are not repeated. A or B repeats of different genes are nonhomologous (except for the B repeats in M6 and M5, which are homologous). Thin black lines indicate nonrepeated parts of each gene, the sequences of which are distinct for each *emm* gene. M49, depicted at the bottom, is a class II type M protein and exhibits less homology at the carboxy terminus of the coding region (7, 11, 19–21).

Intragenic Recombination and Antigenic Variation

When one asks what type of genetic mechanisms may have generated the differences now present in heterogeneous *emm* genes, no immediate clear picture emerges. The repeat structure itself suggested the possibility for intragenic recombination, occurring either intramolecularly or between two molecules in the same cell. This is postulated to happen either by slipped-strand mispairing during replication (17) or by unequal crossover during homologous recombination (25). The immediate effect of intragenic recombination within any of the repeated segments of the *emm* genes that have been sequenced is a size change in the encoded M protein because each repeat is a multiple of three base pairs. Size variation has been studied in several strains of the M6 serotype, in which size differences due to intragenic recombination have been found in natural outbreaks and in serial isolates from a single individual (11).

Intragenic recombination creates antigenic differences at the site of the recombination. Because of the amino acid changes, variant M proteins

bind defined antibodies differently, as can be recognized with competition assays (13). It is important that variant M proteins also show differences in opsonization by antibody (K. F. Jones, unpublished data). Thus, the ability of group A streptococci to undergo size variation can be demonstrated to be of survival advantage within an immune environment containing a defined antibody. Intragenic recombination is clearly one factor in generating antigenic variation of M molecules.

To explain the evolution of the M-protein family, additional mechanisms of introducing genetic variation may be required. Some requirements can be estimated by looking at all sequence differences in the variable half of the M molecule in two heterogeneous serotypes (Fig. 1). One must explain, for example, how the primary sequence of a repeat unit could diverge extensively when the carboxy terminus and both upstream and downstream DNA sequences remain highly conserved. A second unanswered question is how a single repeat unit could lengthen during evolution. A third unanswered question is how the nonrepetitive DNA flanking the repeats within the variable half of *emm* genes has also diverged. These nonrepetitive segments of the amino terminus also differ in length of sequence among the heterogeneous *emm* genes. To study the mechanism of variation in the M molecule and to perhaps identify additional genetic mutations contributing to its evolution, we have searched for isogenic antigenic variants within a single culture.

Isolation of Related Antigenic Variants from a Single Culture

Mutants were identified directly on the basis of their differences in antigenicity. Immunodepletion by a procedure called panning (26) was used to enrich for spontaneous antigenic variants of the M6 protein within a laboratory culture of *S. pyogenes* D471. Two purified monoclonal antibodies (MAbs) were used in this process: MAb 3B8 and MAb 10F5. Amino acids forming part of the epitopes recognized by these MAbs have been identified and are shown in Fig. 2 (12; Jones, unpublished data). In both cases, the epitopes identified are repeated sequences. MAb 3B8 is a serotype-specific antibody, recognizing only M molecules of serotype 6. It is the only MAb elicited against an intact native M protein which has been shown to initiate phagocytosis for type 6 streptococci (12). MAb 10F5 was elicited against M6.1 protein purified from *Escherichia coli*; it is a non-type-specific antibody recognizing 30 of 56 different strains tested (14). Antigenic variants were identified by dot blot immunoassay as being single colony isolates that no longer reacted with MAb 3B8 but continued to react with MAb 10F5.

For the panning enrichment, MAb 3B8 was first bound to petri plates in a 2-µg/ml solution in 0.05 M Tris (pH 9.5) buffer. Additional binding sites were then blocked with 1% bovine serum albumin in phosphate-buffered saline. A culture of D471 (10^8 CFU/ml) was washed twice with phosphate-buffered saline, briefly sonicated, pipetted onto the antibody-coated plate, and incubated at 4°C overnight to allow adsorption to the antibody on the plates. The binding capacity of each petri dish prepared in this way was determined to be approximately 5×10^7 CFU. The supernatant was then adsorbed to a second antibody-coated petri dish; three successive enrichment panning passages took place before screening for spontaneous M-protein antigenic variants. Viable cells remaining after each adsorption were quantitated by plating dilutions on Todd-Hewitt plates with 2% added yeast extract. Single colonies on these plates were picked and screened by duplicate dot blot immunoassay with MAbs 3B8 and 10F5.

A positive signal with MAb 10F5 indicated that the variant strain was still producing M protein. Because MAb 10F5 is not type specific, any variants that may have changed the serotype of M protein would still be likely to be recognized by this MAb. Wells showing a negative signal with MAb 3B8 identified the variants in which a change in the antigenic determinant recognized by MAb 3B8 had occurred. Wells showing no reaction with either MAb were later determined to be phase variants that had lost expression of M protein. Phase shifts (Emm$^+$ to Emm$^-$) were found to occur in strain D471 at an estimated frequency of about 10^{-5} isolates per CFU before enrichment.

Antigenic Variants Detected by Panning

Several of the isogenic mutants detected made M proteins of a smaller size than that of the parent strain, D471, when compared with M6.1 on a Western immunoblot (Fig. 3). UAB075, whose M protein is about 3 kDa smaller than that of its

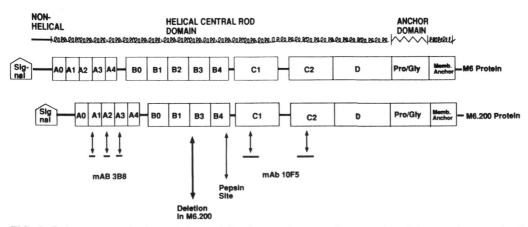

FIG. 2. Epitopes recognized by MAbs used for the panning procedure and site of deletion in an antigenic variant of strain UAB075, M6.200. Domains of M protein are indicated at the top. Boxes A, B, and C are repeat regions. Amino acids contributing to the epitopes for MAbs 3B8 and 10F5 are lined and indicated by an arrow (12, 14). The pepsin cleavage site and deletion site in M6.200 are also shown by arrows.

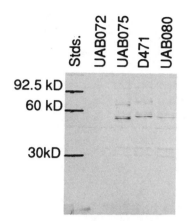

FIG. 3. Western blot analysis of lysin extracts of M6.1 streptococcal strain D471 and three spontaneous antigenic variants showing size variation of the M molecule in strain UAB075. The Western blot was developed with MAb 10F5 of D471. Strain names are above the lanes. UAB072 is an M-negative phase variant; UAB075 and UAB080 are antigenic variants of D471. Sizes of protein standards in kilodaltons (kD) are shown on the left.

parent, is an example of such a mutant. The same-size M variant, and only this size variant, was observed in three separate experiments. The smaller antigenic variant was found at an estimated frequency of about 10^{-6} isolates per CFU before enrichment.

When the *emm* gene in one of the smaller spontaneous variants was sequenced to determine the changes present, a deletion of residues 120 to 144 (25 amino acids) was found. This corresponds to the deletion of a repeat from the B region by intragenic recombination. This mutant was isolated on the basis of its variable antigenicity with MAb 3B8 and thus supports the hypothesis that recombination between intragenic repeats is an important mechanism of antigenic variation in the group A streptococci.

The result was surprising because the B region was not previously thought to be involved in the epitope recognized by the MAb that was used for immunodepletion (MAb 3B8; Fig. 2). This mutation was detected by the panning procedure in three separate experiments. One expected variant that was not observed was one in which all of the A-repeat region had been deleted by intragenic recombination between the two outermost A repeats. Alternatively, this enrichment may have allowed the detection of strains with changes in the type specificity of their M protein by unknown means. Because these types of variants were not detected, we take this as evidence that the frequency for occurrence of these events is less than 10^{-8} in in vitro cultures or that strain D471 lacks the capacity to undergo these events in vitro.

Conformational Aspects of the MAb 3B8 Epitope

The epitope for MAb 3B8 is primarily in the A-repeat region, but the spontaneous antigenic variants were found to have a changed B-repeat region. This finding suggests that changes in the conformation of the M molecule in these mutants may be influencing antibody recognition. Although few differences in primary conformation are projected by using predictive computer algorithms such as that of Garnier et al. (4), the higher-order interaction of the coiled-coil dimer structure may be affected. MAb 3B8 is sensitive to M-protein conformation, as was seen by competition studies between the native antigen present on streptococcal cells and a partial M protein generated by cleavage with pepsin; the pepsin cleavage site is just below the B-repeat region (13). These results also suggest that the epitopes affected by recombination do not have to be exclusively located within the region of the recombination.

Antigenic variation occurring simultaneously with size variation is considered to be antigenic drift because most epitopes of the M molecule are likely to be unaffected. This antigenic drift may have important consequences for antibody recognition and subsequent opsonization of group A streptococci and may be of critical importance to the strain in persisting and avoiding the host immune recognition.

LITERATURE CITED

1. **Barbour, A.** 1989. Antigenic variation in relapsing fever *Borrelia* species: genetic aspects, p. 783–789. *In* D. E. Berg and M. M. Howe (ed.), *Mobile DNA*. American Society for Microbiology, Washington, D.C.
2. **Bessen, D., K. F. Jones, and V. A. Fischetti.** 1989. Evidence for two distinct classes of streptococcal M protein and their relationship to rheumatic fever. *J. Exp. Med.* **169:**269–283.
3. **Frithz, E., L.-O. Hedén, and G. Lindahl.** 1989. Extensive sequence homology between IgA receptor and M proteins in *Streptococcus pyogenes*. *Mol. Microbiol.* **3:**1111–1119.
4. **Garnier, J., D. J. Osguthorpe, and B. Robson.** 1978. Analysis of the accuracy and implications of simple methods for predicting the secondary structure of globular proteins. *J. Mol. Biol.* **120:**97–120.
5. **Gomi, H., T. Hozumi, S. Hattori, C. Tagawa, F. Kishimoto, and L. Bjorck.** 1990. The gene sequence and some properties of protein H: a novel IgG-binding protein. *J. Immunol.* **144:**4046–4052.
6. **Guss, B., M. Eliasson, A. Olsson, M. Uhlen, A.-K. Frej, H. Jornvall, I. Flock, and M. Lindberg.** 1986. Structure of the IgG-binding regions of streptococcal protein G. *EMBO J.* **5:**1567–1575.
7. **Haanes, E. J., and P. P. Cleary.** 1989. Identification of a divergent M protein gene and an M protein-related gene family in *Streptococcus pyogenes* serotype 49. *J. Bacteriol.* **171:**6397–6408.
8. **Haanes-Fritz, E., W. Kraus, V. Burdett, J. Dale, E. Beachey, and P. Cleary.** 1988. Comparison of the leader sequences of four group A streptococcal M protein genes. *Nucleic Acids Res.* **16:**4667–4677.

9. **Heath, D. G., and P. P. Cleary.** 1989. Fc-receptor and M-protein genes of group A streptococci and products of gene duplication. *Proc. Natl. Acad. Sci. USA* **86:**4741–4745.

10. **Hollingshead, S. K., V. A. Fischetti, and J. R. Scott.** 1987. A highly conserved region present in transcripts encoding heterologous M proteins of group A streptococci. *Infect. Immun.* **55:**2237–2239.

11. **Hollingshead, S. K., V. A. Fischetti, and J. R. Scott.** 1987. Size variation in group A streptococcal M protein is generated by homologous recombination between intragenic repeats. *Mol. Gen. Genet.* **207:**196–203.

12. **Jones, K. F., and V. A. Fischetti.** 1988. The importance of the location of antibody binding on the M6 protein for opsonization and phagocytosis of group A M6 streptococci. *J. Exp. Med.* **167:**1114–1123.

13. **Jones, K. F., S. K. Hollingshead, J. R. Scott, and V. A. Fischetti.** 1988. Spontaneous M6 protein size mutants of group A streptococci display variation in antigenic and opsogenic epitopes. *Proc. Natl. Acad. Sci. USA* **85:**8271–8275.

14. **Jones, K. F., S. A. Khan, B. W. Erickson, S. K. Hollingshead, J. R. Scott, and V. A. Fischetti.** 1986. Immunochemical localization and amino acid sequences of cross-reactive epitopes within the group A streptococcal M6 protein. *J. Exp. Med.* **164:**1226–1238.

15. **Koomey, M., E. C. Gotschlich, K. Robbins, S. Bergström, and J. Swanson.** 1987. Effects of *recA* mutations on pilus variation and phase transitions in *N. gonorrhoeae*. *Genetics* **117:**391–398.

16. **Lancefield, R. C.** 1962. Current knowledge of type-specific M antigens of group A streptococci. *J. Immunol.* **89:**307–313.

17. **Levinson, G., and G. A. Gutman.** 1987. Slipped-strand mispairing: a major mechanism for DNA sequence evolution. *Mol. Biol. Evol.* **4:**203–221.

18. **Maxted, W. R., and H. A. Valkenberg.** 1969. Variation in the M-antigen of group A streptococci. *J. Med. Microbiol.* **2:**199–210.

19. **Miller, L., L. Gary, E. Beachey, and M. Kehoe.** 1988. Antigenic variation among group A proteins. *J. Biol. Chem.* **263:**5668–5673.

20. **Mouw, A. R., E. H. Beachey, and V. Burdett.** 1988. Molecular evolution of streptococcal M protein: cloning and nucleotide sequence of the type 24 M protein gene and relation to other genes of *Streptococcus pyogenes*. *J. Bacteriol.* **170:**676–684.

21. **Quinn, R. W., R. V. Zwaag, and P. N. Lowry.** 1985. Acquisition of group A streptococcal M protein antibodies. *Pediatr. Infect. Dis.* **4:**374–378.

22. **Robbins, J. C., J. G. Spannier, S. J. Jones, S. J. Simpson, and P. P. Cleary.** 1987. *Streptococcus pyogenes* type 12 M protein gene regulation by upstream sequences. *J. Bacteriol.* **169:**5633–5640.

23. **Scott, J. R., S. K. Hollingshead, and V. A. Fischetti.** 1986. Homologous regions within M protein genes in group A streptococci of different serotypes. *Infect. Immun.* **52:**609–612.

24. **Siegel, A. C., E. E. Johnson, and G. H. Stollerman.** 1961. Controlled studies of streptococcal pharyngitis in a pediatric population. *N. Engl. J. Med.* **265:**559–571.

25. **Smith, G. P.** 1976. Evolution of repeated DNA sequences by unequal crossover. *Science* **196:**528–535.

26. **Wysocki, L. J., and V. L. Sato.** 1978. Panning for lymphocytes: a method for cell selection. *Proc. Natl. Acad. Sci. USA* **75:**2844–2848.

V. Extracellular Products of Pathogenic Streptococci: Genetics and Regulation

Molecular Analysis of the Group B Streptococcal Capsule Genes

CRAIG E. RUBENS,[1] JANE M. KUYPERS,[1] LAURA M. HEGGEN,[1] DENNIS L. KASPER,[2] AND MICHAEL R. WESSELS[2]

Department of Pediatrics, Division of Infectious Disease, Children's Hospital and Medical Center, Seattle, Washington 98105,[1] and Channing Laboratories, Boston, Massachusetts 02115[2]

Group B streptococci (GBS) are the most common cause of bacteremia, pneumonia, and meningitis in newborn infants (1). GBS serotypes are based on serological identification of the capsular polysaccharides (2, 6, 16). We chose to study type III GBS because this serotype accounts for over two-thirds of the reported infections of newborns (1). The type III capsule is composed of a high-molecular-weight polymer composed of four monosaccharides (glucose, galactose, N-acetyl-glucosamine, and sialic acid) arranged in a pentasaccharide repeating unit (14). Sialic acid is located on the terminal side chain residue of a core polysaccharide structure. The core polysaccharide is identical to that of the pneumococcal type 14 capsular polysaccharide (14). The sialic acid moiety has been shown to have a critical role in the conformation of the type III capsular epitope recognized by protective antibodies (3, 4, 14). This report summarizes our results on experiments deriving mutants in type III capsule expression that were used to identify and characterize the genes involved in capsule biosynthesis. In addition, we have used these mutants to demonstrate the role of the capsular polysaccharide in virulence (see Wessels et al., this volume).

Capsule Mutagenesis

Mutagenesis was performed on two type III clinical isolates (Table 1) that differed in source of isolation, amount of capsule produced, virulence in animal models of GBS disease, and sensitivity to tetracycline. Initially COH31r/s (r/s for rifampin and streptomycin resistant) was used because we needed a tetracycline-sensitive recipient to utilize the tetracycline resistance gene on Tn916 for selection (>90% of GBS are tetracycline resistant) (9). However, COH31r/s was not

a typical strain because it was less virulent and produced less capsule than did strains isolated from infants with invasive disease. To derive capsule mutants in COH1, a more typical virulent clinical isolate, we subsequently replaced the tetracycline resistance gene for the erythromycin resistance gene in Tn916, now designated Tn916ΔE (9).

The general scheme of isolating mutants by transposon mutagenesis in type III capsule expression is shown in Fig. 1. Transposon Tn916 or Tn916ΔE was transferred by conjugation from high-frequency *Enterococcus faecalis* donors CG110 (11) and RH110 (9) into COH31r/s and COH1, respectively. Capsular mutants were identified by their inability to react with anti-type III GBS or anti-type 14 pneumococcal (recognizes the core structure of type III capsule) serum by immunoblot analysis (11). Transconjugants that failed to react with one or both of these antisera were isolated and examined by enzyme-linked immunosorbent assay, immunoelectron microscopy, and structural carbohydrate chemistry to confirm their phenotypes (11, 15). Two mutant phenotypes have been identified and characterized by using representative mutant clones from both strains (Table 2). The first phenotype, represented by COH31-15 and COH1-13, completely lacked any evidence of capsular material and did not react with either antiserum. The second phenotype, represented by COH31-21 and COH1-11, reacted only with the anti-type 14 pneumococcal serum; it was subsequently shown to produce only the core (asialo) polysaccharide (see Wessels et al., this volume). These representative mutants were then subjected to molecular genetic analysis.

Each of these mutants was confirmed to contain the transposon by Southern hybridization analysis

179

TABLE 1. Type III clinical isolates

Strain	Source	Tetracycline	Capsule	50% Lethal dose
COH31r/s	Ulcer	Sensitive	Small	~10^5
COH1	Septic infant	Resistant	Large	~10^2

using Tn*916* as a probe (11, 15). The unencapsulated mutants COH31-15 and COH1-13 contained a single insert in the same location. Nine of 11 of the unencapsulated mutants derived in COH31r/s contained a Tn*916* insert within the same *Eco*RI fragment (see below). Two of the 11 did not have Tn*916* inserts within this chromosomal fragment or within the mapped region discussed below, suggesting that another region of the chromosome may contain genes important for capsule synthesis. All nine unencapsulated mutants in COH1 contained at least one copy of Tn*916*Δ*E* in the same *Eco*RI fragment as observed for the COH31r/s unencapsulated mutants. A single asialo mutant was isolated from the mutagenesis experiments in both strains, and each contained two copies of their respective transposons. However, one of the transposon sites was in the same location in COH31-21 and COH1-11, suggesting that the gene(s) in that location of the chromosome was involved in sialylation of the capsule.

Further mapping studies determined the locations of the transposon sites in the different mutant chromosomes relative to each other (described below). These data demonstrated that a single transposon insertion in the GBS chromosome could completely ablate capsule expression or sialylation of the polysaccharide. What specific enzymes each of these mutations disrupts in the biosynthesis of the capsule is now under study.

Capsule Gene Mapping

The DNA flanking the single Tn*916* site in COH31-15 was cloned into an *Escherichia coli* vector and transformed into DH5 (11). This DNA served as a probe of a GBS genomic cosmid library from COH1 in *E. coli* to identify a contiguous region of the GBS chromosome involved in capsular biosynthesis (5). By chromosome walking, we were able to map a 30-kb region of the type III chromosome surrounding the transposon insertion sites in the mutants derived in COH31r/s and COH1. Figure 2 shows the restriction map of the central 20 kb of this region and the transposon insertion sites for the two different mutant phenotypes observed in both strains. As shown, Tn*916* or Tn*916*Δ*E* inserted within two locations of a 3.0-kb fragment, resulting in the unencapsulated phenotype. Transposon insertion sites within a 7.5-kb *Eco*RI fragment located ~10 kb to the left of the 3.0-kb fragment resulted in the asialo capsular phenotype. These data suggested that some of the capsule genes are regionally clustered.

We have begun to sequence the regions identified by transposon mutagenesis to characterize the genes and their products. A 3.8-kb region of DNA flanking the transposon insert sites for the unencapsulated mutants (Fig. 2, inset) was sequenced, and four open reading frames were identified. The predicted sizes of the gene products are similar to those observed in the gene expression assays (5) described below and are noted in Fig. 2. The predicted direction of transcription of these four open reading frames is from right to left on the map shown, and we are currently in the process of confirming this by Northern (RNA) transcript analysis. Unfortunately, homology searches with each open reading frame in comparison with DNA or protein data base libraries were unsuccessful in identi-

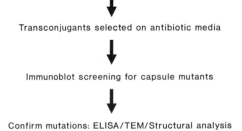

FIG. 1. General scheme for transposon mutagenesis of COH31r/s with Tn*916* and COH1 with Tn*916*Δ*E* and subsequent characterization of capsule mutants. ELISA, Enzyme-linked immunosorbent assay.

TABLE 2. Transposon mutants in the type III GBS capsule

Strain	Phenotype	No. of inserts
COH31r/s	Small capsule	0
COH31-15	Unencapsulated	1
COH31-21	Desialylated	2
COH1	Large capsule	0
COH1-13	Unencapsulated	1
COH1-11	Desialylated	2

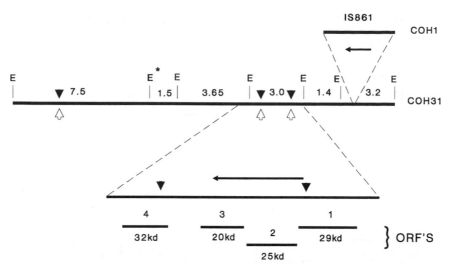

FIG. 2. Restriction enzyme map of GBS type III capsule genes showing the central 20 kb of 30 kb cloned into *E. coli* (5). Notation: E, *Eco*RI sites (numbers between restriction sites indicate sizes of fragments in kilobases). *, Single RFLP seen in COH31r/s and serotype Ia strains but not seen in other type III or type II strains (5); ▼, Tn*916* insert sites in COH31r/s mutants; △, Tn*916*Δ*E* sites in COH1 mutants. See text for further description of mutants. The lower inset shows the nucleotide sequence of the 3.8-kb region described in the text. Four open reading frames (ORF's) and their respective molecular masses (in kilodaltons [kd]) are indicated. The upper inset is an expanded depiction of the 1.4-kb insertion sequence, IS*861*, observed in COH1 only. Arrows in both insets indicate direction of transcription as determined by nucleotide sequencing.

fying similar genes, and their identities are unknown.

Using in vitro gene expression assays and in vivo *E. coli* maxicell gene expression analysis, we have identified the gene products expressed by each of the *Eco*RI fragments cloned into *E. coli* originating from this 30-kb region (5). Several gene products were identified in *E. coli*, some of which appear to contain signal sequences, as indicated by their increased apparent molecular mass in the in vitro expression assays (5). Whether these supposed secreted proteins are involved in some aspect of capsule synthesis beyond the cytoplasmic membrane remains to be determined.

The *Eco*RI fragments from this 30-kb region were subcloned and used as probes in hybridization analysis against *Eco*RI digests of GBS clinical isolates representing serotypes Ia, Ib, II, and III (5). A single *Eco*RI restriction fragment length polymorphism (RFLP) was observed (Fig. 2, asterisk), the significance of which remains unknown. Otherwise, the results from the hybridization studies demonstrated that this region was highly conserved among all serotypes tested and may represent genes common to all serotypes for capsular biosynthesis. The presence of capsule genes conserved among several serotypes has been observed for *E. coli* capsular polysaccharides as well (8, 13). How much of this GBS chromosomal region is involved in capsular biosynthesis is under investigation.

Identification of IS*861*

In addition to the RFLP described above, another difference between COH1 and COH31r/s was observed in the right-hand portion of the map (Fig. 2, inset; 5). A 3.2-kb *Eco*RI fragment in COH31r/s corresponded and hybridized to a 4.6-kb *Eco*RI fragment in COH1. The 4.6-kb fragment from COH1 was also used as a probe against *Eco*RI-digested COH1 and COH31r/s chromosomal DNA. Interestingly, we observed that nine fragments hybridized to this probe from COH1 (Fig. 3, lane A), suggesting the presence of a repetitive sequence, one copy of which was within the 4.6-kb fragment. The same probe hybridized only to a 3.2-kb fragment in COH31r/s. Further mapping analysis determined that an additional 1.4 kb of DNA was present in the 4.6-kb *Eco*RI fragment from COH1 that was not present in the 3.2-kb *Eco*RI fragment from COH31r/s (10). In fact, this additional DNA in COH1 was not present at all in COH31r/s, and hybridization analysis with probes from DNA immediately flanking the 1.4-kb segment revealed significant homology with the 3.2-kb fragment from COH31r/s.

Nucleotide sequencing of the 1.4-kb segment clones from the 4.6-kb *Eco*RI fragment and its flanking DNA revealed an insertion sequence (IS), designated IS*861* (Fig. 2, inset), which was 1,442 bp in length, had two open reading frames, and terminated in 26-bp imperfect inverted repeats (10). The DNA flanking IS*861* was identical

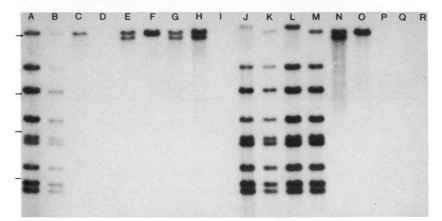

FIG. 3. Southern hybridization of *Eco*RI-digested chromosomal DNA from type III (lanes A to H and J to N), type II (lane O), type Ia/c (lane P), type Ib (lane Q), and type Ia (lanes I and R) clinical isolates with a ^{32}P-labeled internal probe from IS*861*. The autoradiograph demonstrates three hybridization patterns in type III strains. Nine fragments hybridized to the probe in COH1 (lane A) and type III strains in lanes B and J to M in a highly conserved pattern. One or two fragments hybridized in the type III strains in lanes C, E to H, and N and in the single type II strain in lane O. No hybridization was observed with COH31r/s (lane D) or with the type Ia strains in lanes I and P to R. The type III strains were isolated from three different geographic locations over a 10-year period, and all are represented within each hybridization pattern. Bars at the left indicate molecular weight markers (*Hin*dIII-digested phage λ; 23.1, 9.4, 6.6, and 4.4 kb from top to bottom).

to the same region in COH31r/s without the IS element, confirming the hybridization data results discussed above. A 3-bp sequence, ACA, was identified immediately adjacent to the inverted repeats of IS*861* and represents a duplicated target sequence as observed with other IS elements (12).

Several characteristics of IS*861* were observed: 1,442 bp flanked by inverted repeats; 26-bp imperfect inverted repeats and duplicated ACA target sequence; open reading frame 1 encoding a 17-kDa protein; open reading frame 2 encoding a 36-kDa protein (50% identity with IS*150* transposase and 27% identity with IS*3* transposase); putative transcription terminators; active RNA expression in GBS; and multiple copies in a conserved pattern in type III GBS. IS*861* shared significant homology with two *E. coli* IS elements, IS*150* and IS*3* (12), both of which are known to have regulatory effects on adjacent and downstream genes. Significant homology with IS*150* and IS*3* was observed primarily in the open reading frames of the putative transposases and the inverted repeats of IS*3*, suggesting that the mechanism for transposition in IS*861* may be the same as that observed in *E. coli*. The evolutionary significance of similar IS elements in gram-positive and gram-negative organisms remains to be determined.

Significant dyad symmetry was observed close to the inverted repeats on both ends of IS*861*, suggestive of rho-dependent and -independent terminators. With use of an internal probe from IS*861*, Northern hybridization analysis was

performed against total RNA from COH1 and COH31r/s. Only mRNA transcripts from COH1 hybridized to this probe (10). Multiple sizes of mRNA had hybridized, suggesting that mRNA was being synthesized from IS*861* and possibly included transcriptional products from genes adjacent to the nine different copies of IS*861* in COH1. When cloned into *E. coli*, IS*861* expressed two proteins, by maxicell analysis, of 17 and 36 kDa, sizes similar to those predicted by the two open reading frames from nucleotide sequence analysis (10). Whether the mRNA in GBS, which hybridizes to the IS*861* probe, is actively translated remains to be determined. In addition, we are investigating the possibility that IS*861* controls expression of adjacent or downstream genes, particularly the capsule genes described here.

Southern hybridization analysis of several type III clinical isolates has demonstrated three groups of clinical isolates that harbor IS*861*. The majority of strains tested to date contain nine copies of the IS element in a highly conserved pattern throughout the chromosome (Fig. 3). Of interest is that this pattern of hybridization is conserved among strains isolated from widely disparate geographic locations and over a 10-year period. The second group contained one to two copies always in the same *Eco*RI fragment(s), and the third group lacked IS*861*. Among representatives of the other serotypes, only the chromosome of a type II strain hybridized with this sequence and to a similar-size large *Eco*RI fragment observed in the type III strains with multiple copies. We have preliminary data (not shown) that suggest that clinical isolates which contain multiple copies of IS*861* produce more capsule, are more resistant to opsonophagocytosis, and are more virulent in neonatal rats. The pattern of hybridization observed among the type III clinical isolates may therefore be suggestive of specific clones that differ in their virulence traits, e.g., capsule expression, as suggested by multienzyme electrophoresis (7). We are currently pursuing the possibility that type III strains which contain multiple copies of IS*861* represent a clonal population that is more virulent.

Summary

We have derived transposon mutations in type III GBS capsule expression in two different clinical isolates. These mutants have been useful for investigating the role of the GBS type III capsular polysaccharide as a virulence factor and for identifying the genes responsible for capsular biosynthesis. Mutations in two different locations within a 30-kb region in both strains resulted in complete ablation of capsule synthesis or a defect in sialylation of the capsule. The capsule genes identified by transposon mutagenesis have been cloned into *E. coli*, and their protein products have been identified by gene expression analysis techniques. We are now investigating the roles of their gene products in capsule synthesis, transport, and secretion. DNA probes from this region demonstrated by hybridization analysis that this region is highly conserved among all GBS serotypes.

The insertion sequence IS*861* was present in multiple copies in the highly encapsulated, more virulent clinical isolate COH1, and one of the copies in this strain was located within the 30-kb capsule gene region. IS*861* was homologous to the IS*150* superfamily of IS elements that have been shown to regulate downstream and adjacent genes in *E. coli*. Further experiments are in progress to determine whether IS*861* is influencing GBS capsule gene expression. Only type III strains were found to have multiple copies of IS*861*, and their location was highly conserved. Whether strains with multiple copies of this IS element represent a specific clonal population of highly virulent type III GBS remains to be determined.

This work was supported by National Institutes of Health grants AI28040, AI23339, and AI22498. Support was also received from the United Cerebral Palsy and Hartford Foundations to C. Rubens.

We thank Barbara Hensley for help with manuscript preparation.

LITERATURE CITED

1. **Baker, C. J., and M. S. Edwards.** 1983. Group B streptococcal infections, p. 820–881. *In* J. S. Remington and J. O. Klein (ed.), *Infectious Diseases of the Fetus and Newborn Infant.* The W. B. Saunders Co., Philadelphia.
2. **Henrichsen, J., P. Ferrieri, J. Jelinkova, W. Köhler, and W. R. Maxted.** 1984. Nomenclature of antigens of group B streptococci. *Int. J. Syst. Bacteriol.* **34:**500.
3. **Jennings, H. J., C. Lugowski, and D. L. Kasper.** 1981. Conformational aspects critical to the immunospecificity of the type III group B streptococcal polysaccharide. *Biochem. USA* **20:**4511–4518.
4. **Kasper, D. L., C. J. Baker, R. S. Baltimore, J. H. Crabb, G. Schiffman, and H. J. Jennings.** 1979. Immunodeterminant specificity of human immunity to type III group B Streptococcus. *J. Exp. Med.* **149:**327–339.
5. **Kuypers, J. M., L. M. Heggen, and C. E. Rubens.** 1989. Molecular analysis of a region of the group B streptococcus chromosome involved in type III capsule expression. *Infect. Immun.* **57:**3058–3065.
6. **Lancefield, R. C.** 1934. A serological differentiation of specific types of bovine hemolytic streptococci (group B). *J. Exp. Med.* **58:**441–458.
7. **Musser, J. M., S. J. Mattingly, R. Quentin, A. Goudeau, and R. K. Selander.** 1989. Identification of a high-virulence clone of type III Streptococcus agalactiae (group B Streptococcus) causing invasive neonatal disease. *Proc. Natl. Acad. Sci. USA* **86:**4731–4735.
8. **Roberts, I., R. Mountford, N. High, D. Bitter-Suermann, K. Jann, K. Timmis, and G. Boulnois.** 1986. Molecular cloning and analysis of genes for production of K5, K7, K12, and K92 capsular polysaccharide in *Escherichia coli.* *J. Bacteriol.* **168:**1228–1233.
9. **Rubens, C. E., and L. M. Heggen.** 1988. Tn916ΔE: a Tn916 derivative expressing erythromycin resistance. *Plasmid* **20:**137–142.
10. **Rubens, C. E., L. M. Heggen, and J. M. Kuypers.** 1989. IS*861*, a group B streptococcal insertion sequence related to IS*150* and IS*3* of *Escherichia coli.* *J. Bacteriol.* **171:**5331–5335.
11. **Rubens, C. E., M. R. Wessels, L. M. Heggen, and D. L. Kasper.** 1987. Transposon mutagenesis of group B streptococcal type III capsular polysaccharide: correlation of capsule expression with virulence. *Proc. Natl. Acad. Sci. USA* **84:**7208–7212.
12. **Schwartz, E., M. Kroger, and B. Rak.** 1988. IS150: distribution, nucleotide sequence, and phylogenetic relationships of a new E. coli insertion element. *Nucleic Acids Res.* **16:**6789–6802.
13. **Silver, R. P., W. F. Vann, and W. Aaronson.** 1984. Genetic and molecular analyses of *Escherichia coli* K1 antigen genes. *J. Bacteriol.* **157:**568–575.
14. **Wessels, M. R., V. Pozsgay, D. L. Kasper, and H. J. Jennings.** 1987. Structure and immunochemistry of an oligosaccharide repeating unit of the capsular polysaccharide of type III group B Streptococcus. *J. Biol. Chem.* **262:**8262–8267.
15. **Wessels, M. R., C. E. Rubens, V. J. Benedi, and D. L. Kasper.** 1989. Definition of a bacterial virulence factor: sialylation of the group B streptococcal capsule. *Proc. Natl. Acad. Sci. USA* **86:**8983–8987.
16. **Wilkinson, H. W., and R. G. Eagon.** 1971. Type-specific antigens of group B type Ic streptococci. *Infect. Immun.* **4:**596–604.

Expression and Properties of Hybrid Streptokinases Extended by N-Terminal Plasminogen Kringle Domains

HORST MALKE AND JOSEPH J. FERRETTI

Department of Microbiology and Immunology, University of Oklahoma Health Sciences Center, Oklahoma City, Oklahoma 73190

Rationale

Streptokinase (SK), one of the secreted proteins of group A, C, and G streptococci, is the most widely used therapeutic agent for lysing the fibrin matrix of pathologically formed blood clots. In a highly specific and efficient reaction, it interacts with the C-terminal serine protease domain of the plasma proenzyme plasminogen to form a stoichiometric complex which, by binding activation, serves to catalyze the conversion of free plasminogen to plasmin (plasminogen activation). Although fibrin and some of its degradation products can act as cofactors to increase the catalytic efficiency of SK-mediated fibrinolysis, SK itself has no fibrin specificity and thus activates circulating plasminogen as well, leading to undesired systemic fibrinogenolysis resulting from an elevated plasma proteolytic state (14). Imparting fibrin specificity to SK might increase its effective concentration at the clot and thus improve its pharmacological attributes by promoting the activation of plasminogen bound to fibrin. The noncovalent, anisoylated plasminogen-SK activator complex developed by the Beecham Group (20) is targeted to fibrin by way of its plasminogen moiety, and its affinity for fibrin is similar to that of the physiological activator, tissue plasminogen activator (2). The structures in plasminogen responsible for its fibrin affinity are located in the N-terminal five homologous triple-loop, three-disulfide-bridge segments (kringles) which constitute autonomous folding and functional domains having molecular masses of around 10 kDa each (16). Their ligand (lysine) binding sites are not equivalent; there is one strong site on kringle 1 (K1; $K_d = 9$ μM) through which plasminogen is targeted to fibrin. Upon occupation of five additional weaker sites ($K_d = 5$ mM), the prolate ellipsoid three-dimensional plasminogen structure is converted into an open extended form that may be more amenable to activation than the closed form (12). Using the available cDNA for human plasminogen (3; E. Davie, personal communication) and the cloned SK gene (*skc*) from *Streptococcus equisimilis* (10, 11), we took the gene fusion approach to construct and express hybrid *skc* genes extended at their 5′ ends by cDNA segments coding for fibrin binding domains of plasminogen.

Design of Kringle Streptokinases

The open reading frame vector pKM2 (7) was used to produce in-frame fusions between *skc* and kringle cDNA. pKM2 is a pUC18 derivative carrying in the *Hinc*II site the truncated SK gene (*′skc*) lacking, in addition to the expression signals, codons 1 through 39 but retaining the remainder of the *skc* coding sequence together with the transcription terminator. In pKM2, *′skc* is inserted out of frame with respect to *lacZ′*, thus conferring an SK-negative phenotype. Kringle cDNA segments representing open reading frames with a length of 3N + 1 nucleotides inserted in the *Sma*I site of the polylinker between *lacZ′* and *′skc* restore the *′skc* reading frame and result, after transformation of *Escherichia coli* followed by the plasminogen-casein overlay assay for SK (10), in scoreable SK-positive transformant colonies. The source of kringle cDNA was plasmid pFD1 (our designation), provided by E. Davie, which carried the complete human plasminogen cDNA as sequenced by Forsgren et al. (3). Using the restriction sites indicated in Fig. 1, appropriate cDNA segments were isolated from pFD1 and, when necessary, subjected to *Bal*31 digestion and end-filling reactions before ligation into pKM2 cut with *Sma*I. Following transformation of *E. coli* JM109, SK-positive clones were isolated, and appropriate junction fragments were cloned into M13mp18 and -mp19 to determine the termini of the kringle cDNA fragments by nucleotide sequencing. The procedure outlined above resulted in the pKSK series of plasmids specifying (K1)-, (K4)-, (K5)-, (K4 + K5)-, and (K1 + K2 + K3 + K4)SK expressed under *lacPO* and extended at their N termini by the N-terminal hexapeptide of LacZ′ followed by four polylinker-coded amino acids (Fig. 1). In these constructs as well as those described below, the coding sequences for the kringles and *′skc* were linked by vector codons specifying the peptide sequence GGSSR.

In a second system designed for cytoplasmic expression, pKSK56 and pKSK67 were chosen to insert the coding sequences for (K1)SK and (K1 + K2 + K3 + K4)SK into the fusion vector pRX1 (18) cut with *Eco*RI plus *Hin*dIII. The resulting plasmids, pRX56 and pRX67, specified kringle SKs driven by the *trpE* promoter and ex-

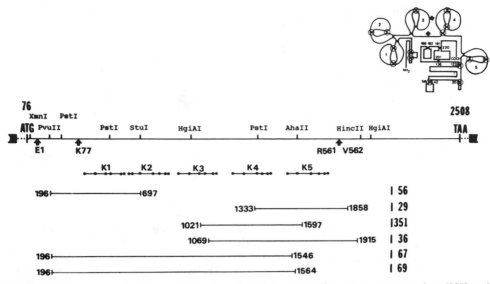

FIG. 1. Physical map of the kringle cDNA inserts (I56 through I69) used to construct the pKSK series of expression plasmids. The upper line represents the complete human plasminogen cDNA, with relevant restriction sites and numbering corresponding to that of the nucleotide sequence (3). E1, Glu-plasminogen; K77, Lys-plasminogen; R561 and V562, activation bond; K1 through K5, kringle domains, with dots representing the Cys residues involved in disulfide bridge formation. The inset at the upper right is a schematic representation of the structure of Glu-plasminogen taken from reference 21.

tended at their N termini by the first 18 amino acid residues of the *trpE* gene product. In both plasmids, the cDNA sequence flanking (K1)DNA at its 5′ end was shorter by the length of the *Pvu*II-*Eco*RI cDNA fragment (17 codons not extending into K1) compared with pKSK56 or pKSK67.

In a third system designed for periplasmic expression, the coding sequences for (K1)SK and (K1 + K2 + K3 + K4)SK were recloned into the secretion vector pIN-III*ompA*1 (4) cut with *Eco*RI plus *Hind*III. The hybrid genes were under the control of the *lpp* and *lac* promoters, and the gene products, after processing of the vector-encoded OmpA signal sequence, were extended by a single Ala residue corresponding to the N-terminal amino acid of mature OmpA protein.

Analysis of pKSK Plasmid-Encoded Kringle SK by Zymography and Western Immunoblotting

To confirm the identity of the translation products and establish some of their basic properties, *E. coli* JM109 strains carrying individual pKSK plasmids were grown in SOC medium (19) containing 2 mM isopropyl-β-D-thiogalactopyranoside (IPTG) as an inducer of the *lac* operon. Sonic extracts of late-exponential-phase cells were subjected to sodium dodecyl sulfate-polyacrylamide gel electrophoresis (SDS-PAGE), and the gels were analyzed by zymography and immunoblotting (7). The zymograph (Fig. 2A) shows that all kringle SKs had retained the plasminogen-acti-

vating potential of the kringleless, truncated SK (Fig. 2A, lane 1) as specified by pKM1 (pKM1 is pUC9 carrying ′*skc* in the *Sma*I site in frame with *lacZ′*). Thus, SK with N-terminal kringle peptide extensions ranging in length from about 180 to as many as 460 amino acid residues continued to be capable of forming active complexes with plasminogen, showing that higher-order structural changes, if any, due to the extension peptides did not abolish the plasminogen-activating function. The zymograph also showed that the molecular masses of the hybrid SKs were consistent with those expected from their respective nucleotide sequences.

The identity of the kringle domains was established by immunodetection with a polyclonal rabbit antiserum to human plasminogen made monospecific by chromatography on plasminogen linked to Affi-Gel 10. All kringle SKs (including those specified by the pRX and pINO plasmids) reacted with this primary antibody in a manner consistent with the zymograph, whereas kringleless SK showed no reactive band in the Western blot (Fig. 2B, lane 1). Both the zymograph and the Western blot also revealed that in addition to the major heaviest band showing specific reactivity, each sonic extract gave rise to lighter SK-reactive or immunoreactive bands. Presumably, these activities reflected proteolytic degradation of the full-length molecules, which proved to be a major problem in the course of this work.

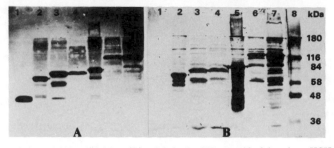

FIG. 2. Zymograph (A) and Western blot (B) of kringle SKs specified by the pKSK series of expression plasmids. Lanes: 1, kringleless SK specified by pKM1; 2, (K5)SK (pKSK29); 3, (K1)SK (pKSK56); 4, (K4)SK (pKSK351); 5, (K4 + K5)SK (pKSK36); 6, (K1 + K2 + K3 + K4)SK (pKSK67); 7, (K1 + K2 + K3 + K4)SK (pKSK69); 8, marker proteins.

Expression as Influenced by Vectors and Host Backgrounds

Attempts at improving expression levels and reducing protein degradation included the use of the pRX and pIN-IIIompA vectors previously shown to support high-level synthesis of other recombinant proteins (4, 18), as well as host strains with htpR, lon (5), lpp, and degP (9, 22) mutations known to promote the stability of foreign protein produced by E. coli (6). The yield and stability of pRX56- and pRX67-specified proteins in sonic extracts of the isogenic strain pair LC133 and LC137 (htpR lon) were dramatically improved by mutations inactivating the heat shock operon both in uninduced cultures and in cultures induced with indole acrylic acid (10 µg/ml) when uniformly grown at 30°C (Fig. 3). In fact, while a proportion of (K1)SK was still detectable in undegraded form in wild-type cultures, (K1 + K2 + K3 + K4)SK was completely degraded to the level of kringleless SK or no longer

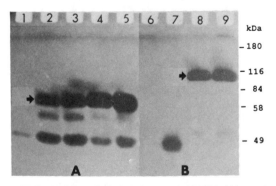

FIG. 3. Degradation patterns of pRX56 (A)- and pRX67 (B)-specified kringle SKs as influenced by heat shock protein mutations in uninduced (lanes 2, 4, 6, and 8) and indole acrylic acid-induced (lanes 3, 5, 7, and 9) cells. Lanes: 1, commercial SK (Sigma) at 0.1 U per lane; 2, 3, 6, and 7, strain LC133; 4, 5, 8, and 9, strain LC137 htpR165 lonR9. Arrows indicate undegraded protein.

detectable at all in such cells. It should be noted, however, that the htpR lon double mutation did not completely prevent the degradation of either (K1)SK or (K1 + K2 + K3 + K4)SK (sonic extracts prepared in the absence of extraneous protease inhibitors). As judged by quantitative values for SK activity determined by the chromogenic method (8) or the well plate assay (10), assuming no differences in specific activities, indole acrylic acid-induced cultures of LC137 (pRX56) and LC137(pRX67) grown to late exponential phase in M9CA medium (19) produced, respectively, 11.6 µg of (K1)SK and 1.7 µg of (K1 + K2 + K3 + K4)SK per ml of culture. These values translated into average yields of 1 and 0.15%, respectively, of the total soluble sonicate protein, indicating that the four-kringle protein was significantly less well expressed than the one-kringle form.

The effect of the degP41 mutation, in conjunction with an lpp mutation, was studied with pINO56 and pINO67. degP41 inactivates a cell envelope endopeptidase indispensable for growth at elevated temperatures, and lpp-5508 relieves the temperature sensitivity due to membrane lipoprotein deficiency leading to the release of periplasmic proteins into the medium (9, 22). In comparison with the degradation pattern of SK present in osmotic shock fluids of the isogenic wild type, KS272, both (K1)SK and (K1 + K2 + K3 + K4)SK showed a considerably lower degree of degradation in shock fluids obtained from the double-mutant strain (KS476 depP41 lpp-5508) (Fig. 4). To assess the expression levels in the latter strain, cultures grown in SOC medium were fractionated (13) and SK activities in the culture supernatant, periplasm, cytoplasm, and membrane fraction were determined (Table 1). The total activities observed fell into the general range of yields of secreted foreign proteins and were similar to the intracellular yields obtained with the pRX plasmids. About half of the total activity was located in the periplasm, but substantial amounts were also detected in the cytoplasm and

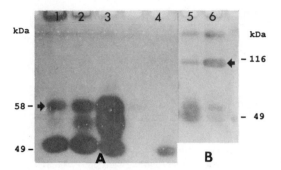

FIG. 4. Degradation patterns of pINO56 (A)- and pINO67 (B)-specified kringle SKs in osmotic shock fluids as influenced by DegP protein. Lanes 1 and 5, KS272; 2, KS303 *lpp-5508*; 3 and 6, KS476 *lpp5508 degP41*; 4, commercial SK at 0.5 U per lane. Arrows indicate undegraded protein.

in the culture supernatant. The release of activity into the medium was clearly supported by the *lpp* mutation, as judged by smaller extracellular activities seen in the isogenic *lpp*⁺ strain. Secretion, however, was incomplete, particularly so with the four-kringle protein.

Binding to Lysine-Sepharose

The kringle domains in plasminogen interact with fibrin by binding to specific lysine side chains exposed at the surface of the clot. In K1 and K4, biophysical studies have identified three aromatic amino acid residues (Trp-62, Phe-64, and Tyr-72 or Trp-72) as key constituents of the lysine-binding pocket, providing a lipophilic surface at its bottom with which the nonpolar part of the ligand interacts (15, 17). Affinity chromatography on L-lysine-substituted Sepharose was therefore used to obtain information about the structural integrity of the lysine-binding site of the kringles attached to SK. Since the periplasmic space provides a more favorable redox potential for correct disulfide bond formation, osmotic shock fluids obtained from cells (KS474 *degP41* and KS476 *degP41 lpp-5508*) carrying pINO56 or pINO67 were used in these experiments. The flowthrough from the reaction mixture contained mainly the low-molecular-weight degradation products present in the starting materials, while significant

amounts of the full-length proteins specified by either plasmid were retained on lysine-Sepharose and could be eluted only with the lysine analog ε-amino caproic acid (EACA) (Fig. 5). However, the eluates also contained traces of the higher-molecular-weight degradation products, which might have retained some lysine-binding activity.

Lysine-Sepharose chromatography carried out in a similar manner with fresh sonic extracts from cells carrying the pKSK or pRX plasmids indicated that kringle SKs produced in the cytoplasm did not bind to immobilized lysine. Since isolated plasminogen kringles are known to be capable of regaining their lysine-binding function from a reduced state after oxidative refolding (23), attempts were made to recover this function by using a denaturation-refolding protocol. In brief, the sonic extracts were denatured in 6 M guanidine-HCl–0.1 mM mercaptoethanol (pH 8.0), cleared by centrifugation, dialyzed against 100 mM Tris-HCl (pH 8.0), and then stirred for 6 h at room temperature in the presence of 1.25 mM each reduced and oxidized glutathione before being reacted with lysine-Sepharose. Column fractions analyzed by SDS-PAGE and zymography revealed that less than 1% of the input activity of undegraded protein was specifically retained on the columns, indicating the need of optimizing the renaturing process for the recovery of the new function of cytoplasmic kringle SKs.

Fibrin Binding

For fibrin-binding experiments, (K1)SK and (K1 + K2 + K3 + K4)SK obtained from osmotic shock fluids of KS476(pINO56) and KS476(pINO67) were partially purified by chromatography on lysine-Sepharose, followed by immunoaffinity chromatography on polyclonal antibodies to plasminogen. The bound proteins were eluted from the antibody column, and neutralized fractions containing mainly undegraded forms as analyzed by SDS-PAGE were concentrated and solvent exchanged by centrifugation using Centricon-30 concentrators (Amicon). Reaction mixtures contained variable concentrations of plasminogen-free fibrinogen, 50 U of the activators per ml, and 100 μg of bovine serum albumin per ml in 38 mM NaCl–50 mM Tris-HCl buffer (pH 7.4). To clot the fibrin-

TABLE 1. Distribution of the plasminogen-activating activity of kringle SKs in cell fractions of KS476 *degP41 lpp-5508* carrying pINO56 or pINO67

Plasmid	Distribution of SK activity (%)[a]			
	Culture supernatant	Periplasm	Cytoplasm	Membrane
pINO56	23	52	18	7
pINO67	12	42	41	5

[a]Total activities determined by pINO56 and pINO67 were 450 and 108 U/ml of culture, corresponding to 0.54 and 0.19%, respectively, of total cell protein.

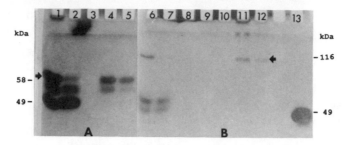

FIG. 5. Zymographic analysis of lysine binding of pINO56 (A)- and pINO67 (B)-specified kringle SKs. Lanes: 1 and 6, before binding to lysine-Sepharose; 2 and 7, column flowthrough after binding to lysine-Sepharose; 3 and 8 to 10, 0.1 M phosphate buffer wash fractions; 4, 5, 11, and 12, EACA eluates; 13, commercial SK at 1 U per lane. Arrows indicate undegraded protein.

ogen, thrombin was added at a final concentration of 1 NIH unit per ml. After a reaction time of 1 h at ambient temperature in the absence or presence of 10 mM EACA, the clots were compacted by centrifugation and unbound activators in the supernatant were quantitated. In the absence of EACA, both (K1)SK and (K1 + K2 + K3 + K4)SK bound to fibrin to similar extents (Fig. 6). In accordance with its known properties, conventional SK showed no fibrin binding. The presence of EACA completely blocked fibrin binding of the kringle SKs, demonstrating the specificity of the reaction. When bound to immobilized fibrin, prepared by thrombin treatment of fibrinogen coupled to the wells of polyvinyl chloride microtiter plates (1), (K1)SK and (K1 + K2 + K3 + K4)SK continued to be capable of plasminogen activation, as determined spectro-photometrically by the plasmin-specific chromogenic method.

Conclusions

The gene fusion technology provides an interesting alternative to the chemical approach (20) of combining the plasminogen-activating potential of SK with the fibrin affinity of plasminogen. When expressed in appropriate hosts, the covalently linked hybrid SKs appear to be bifunctional. Although the genetic design of the kringle SKs described may not be optimal and synthetic or polymerase chain reaction strategies can be envisaged to arrive at tailor-made forms, our primary constructs served to support the principal possibility of engineering SK to obtain covalent hybrids with mixed functions. The combination of such properties in the same molecule may facilitate large-scale production and ultimately improve the therapeutic potential of SK.

This research was supported by grants from the National Institutes of Health (AI19304) and the BOC Group, Murray Hill, N.J.

LITERATURE CITED

1. **Angles-Cano, E.** 1986. A spectrophotometric solid-phase fibrin-tissue plasminogen activator activity assay (SOFIA-tPA) for high-fibrin-affinity tissue plasminogen activators. *Anal. Biochem.* **153:**201–210.
2. **Fears, R.** 1989. Binding of plasminogen activators to fibrin: characterization and pharmacological consequences. *Biochem. J.* **261:**313–324.
3. **Forsgren, M., B. Raden, M. Israelsson, K. Larsson, and L.-O. Heden.** 1987. Molecular cloning and characterization of a full-length cDNA clone for human plasminogen. *FEBS Lett.* **213:**254–260.
4. **Ghrayeb, J., H. Kimura, M. Takahara, H. Hsiung, Y. Masui, and M. Inouye.** 1984. Secretion cloning vectors in *Escherichia coli. EMBO J.* **3:**2437–2442.
5. **Goff, S. A., L. P. Casson, and A. L. Goldberg.** 1984. Heat shock regulatory gene *htpR* influences rates of protein degradation and expression of the *lon* gene in *Escherichia coli. Proc. Natl. Acad. Sci. USA* **81:**6647–6651.
6. **Gottesman, S.** 1989. Genetics of proteolysis in *Escherichia coli. Annu. Rev. Genet.* **23:**163–198.
7. **Klessen, C., K. H. Schmidt, J. J. Ferretti, and H. Malke.** 1988. Tripartite streptokinase gene fusion vectors for gram-

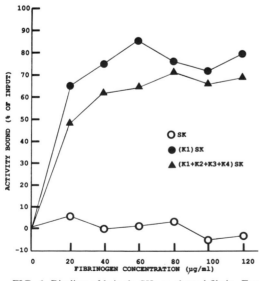

FIG. 6. Binding of kringle SKs to clotted fibrin. For details, see text.

positive and gram-negative procaryotes. *Mol. Gen. Genet.* **212:**295–300.

8. **Kulisek, E. S., S. E. Holm, and K. H. Johnston.** 1989. A chromogenic assay for the detection of plasmin generated by plasminogen activator immobilized on nitrocellulose using a para-nitroanilide synthetic substrate. *Anal. Biochem.* **177:**78–84.

9. **Lipinska, B., S. Sharma, and C. Georgopoulos.** 1988. Sequence analysis and regulation of the *htrA* gene of *Escherichia coli*: a sigma 32-independent mechanism of heat-inducible transcription. *Nucleic Acids Res.* **16:** 10053–10067.

10. **Malke, H., and J. J. Ferretti.** 1984. Streptokinase: cloning, expression, and excretion by *Escherichia coli. Proc. Natl. Acad. Sci. USA* **81:**3557–3561.

11. **Malke, H., B. Roe, and J. J. Ferretti.** 1985. Nucleotide sequence of the streptokinase gene from *Streptococcus equisimilis* H46A. *Gene* **34:**357–362.

12. **Mangel, W. F., B. Lin, and V. Ramakrishnan.** 1990. Characterization of an extremely large, ligand-induced conformational change in plasminogen. *Science* **248:**69–73.

13. **Manoil, C., and J. Beckwith.** 1986. A genetic approach to analyzing membrane protein topology. *Science* **233:**1403–1408.

14. **Marder, V. J., and S. Sherry.** 1988. Thrombolytic therapy: current status. *N. Engl. J. Med.* **318:**1512–1520; 1585–1595.

15. **Motta, A., R. A. Laursen, M. Llinas, A. Tulinsky, and C. H. Park.** 1987. Complete assignment of the aromatic proton magnetic resonance spectrum of the kringle 1 domain from human plasminogen: structure of the ligand-binding site. *Biochemistry* **26:**3827–3836.

16. **Patthy, L.** 1985. Evolution of the proteases of blood coagulation and fibrinolysis by assembly from modules. *Cell* **41:**657–663.

17. **Ramesh, V., A. M. Petros, M. Llinas, A. Tulinsky, and C. H. Park.** 1987. Proton magnetic resonance study of lysine-binding to the kringle 4 domain of human plasminogen: the structure of the binding site. *J. Mol. Biol.* **198:**481–498.

18. **Rimm, D. L., and T. D. Pollard.** 1989. New plasmid vectors for high level synthesis of eukaryotic fusion proteins in *Escherichia coli. Gene* **75:**323–327.

19. **Sambrook, J., E. F. Fritsch, and T. Maniatis.** 1989. *Molecular Cloning: a Laboratory Manual*, vol. 3. Cold Spring Harbor Laboratory, Cold Spring Harbor, N.Y.

20. **Smith, R. A. G., R. J. Dupe, P. D. English, and J. Green.** 1981. Fibrinolysis with acyl-enzymes: a new approach to thrombolytic therapy. *Nature* (London) **290:**505–508.

21. **Sottrup-Jensen, L., H. Claeys, M. Zajdel, T. E. Peterson, and S. Magnusson.** 1978. The primary structure of human plasminogen: isolation of two lysine-binding fragments and one "mini"-plasminogen (MW 38,000) by elastase-catalyzed specific limited proteolysis, p. 191–209. *In* J. F. Davidson, R. M. Rowan, M. M. Samana, and P. C. Desnoyers (ed.), *Progress in Chemical Fibrinolysis and Thrombolysis*, vol. 3. Raven Press, New York.

22. **Strauch, K. L., and J. Beckwith.** 1988. An *Escherichia coli* mutation preventing degradation of abnormal periplasmic proteins. *Proc. Natl. Acad. Sci. USA* **85:**1576–1580.

23. **Trexler, M., and L. Patthy.** 1983. Folding autonomy of the kringle 4 fragment of human plasminogen. *Proc. Natl. Acad. Sci. USA* **80:**2457–2461.

Analysis of the Variable Domain of the Streptokinase Gene from Group A Streptococci by the Polymerase Chain Reaction

K. H. JOHNSTON, J. E. CHAIBAN, AND R. C. WHEELER

Department of Microbiology, Immunology and Parasitology, Louisiana State University Medical Center, New Orleans, Louisiana 70112

Streptokinase, which is produced by the majority of group A, C, and G streptococci, is a 47-kDa extracellular protein that functions in the species-specific conversion of plasminogen to plasmin (12, 18). Genetic (8, 9, 11, 15, 16), chemical (4, 5), and immunological (3, 10, 17) analyses of the different groups and strains of streptococci have suggested that streptokinases as a group of molecules are heterogeneous.

Traditionally, streptokinase as part of a streptokinase-plasmin complex (2) is thought to contribute to the invasiveness of group A streptococci by acting as a spreading factor by virtue of its ability to enzymatically degrade fibrin that had been deposited around the site of the focal infection. Recently, it has been suggested that certain streptokinases from group A streptococci may be involved in the pathogenesis of poststreptococcal glomerulonephritis (PSGN) (10) by virtue of their ability to specifically bind to glomerular structures as a streptokinase-plasmin complex and activate complement in situ. This process triggers the inflammatory response characteristic of PSGN (6, 7). Administration of purified streptokinase isolated from a PSGN-associated group A streptococcal strain to rabbits and mice has been shown to induce both clinical and histopathological signs indicative of human PSGN, whereas streptokinase isolated from non-PSGN-associated group A and group C streptococcal strains does not induce such pathology (A.-M. Bergholm, S. E. Holm, and K. H. Johnston, *Abstr. Xth Lancefield Int. Symp. Streptococci Streptococcal Dis.*, L54, L55, 1987). Further evidence that streptokinase plays a major role in the pathogenesis of PSGN comes from recent experiments in which the streptokinase gene from a nephritogenic group A streptococcus had been excised from the bacterial genome. When this streptokinase-deleted strain was tested in the animal tissue cage model (1), it failed to induce PSGN in experimental animals, whereas the animals infected with the nondeleted parent strain developed PSGN (S. E. Holm, J. J. Ferretti, and K. H. Johnston, *Abstr. XIth Lancefield Int. Symp. Streptococci Streptococcal Dis.*, L40, 1990).

Streptokinase genes from group A streptococci (8, 15) have been cloned and sequenced. Analyses indicated that the streptokinase gene from a group A, M type 1 (15) isolate shared >98% homology with the streptokinase genes from group C (11) and group G (16) isolates, whereas the streptokinase gene from a group A, type M49 (8) isolate shared about 90% nucleotide and 85% protein homology with the other streptokinases. The nonidentical residues were sequestered within two major domains spanning amino acid residues 174 to 244 and 270 to 290; these will be referred to as variable domain one (V1) and variable domain two (V2), respectively. Since the group A, type M49 streptokinase gene was obtained from a strain isolated from a documented case of PSGN and subsequently demonstrated to induce experimental PSGN in rabbits and mice, it was of interest to determine whether the variable domains described above were common to streptokinases secreted by group A streptococci associated with PSGN or simply reflected antigenic variance within group A streptococci. As the V1 domain was the most dissimilar, it was chosen as the region of study.

To accomplish this task, the V1 domain of the streptokinase gene from a battery of PSGN- and non-PSGN-associated streptococcal isolates was amplified by using the polymerase chain reaction (PCR) technique (13). In addition to sequencing of the amplified DNA, restriction enzyme analysis was performed on the amplified DNA, which permitted a more immediate determination of which streptokinase variant was present.

Amplification and Sequencing of the V1 Domain

PCR is a new and extremely powerful technique that has allowed us to bypass the cloning step usually involved in DNA sequencing and to rapidly sequence and compare specific regions of the streptokinase gene from many strains of streptococci. More specifically, PCR is an in vitro method for the enzymatic synthesis of specific DNA sequences, using two oligonucleotide primers that hybridize to opposite strands and which flank the region of interest in the target DNA. A repetitive series of cycles involving template denaturation, primer annealing, and extension of the annealed primers by DNA polymerase results in an exponential accumulation of the specific fragment, whose termini are defined by the 5' ends of the primers.

Streptococcal DNA was incubated with equimolar amounts of the two 21-mers 5'-AACCTTGCCGACCCAACCTGT-3' and 3'-GGCATCGTAAAATGCTTACCT-5'. These primers were selected because they were complementary to the shared domains in both group A and C streptokinase genes (8, 11) that flank the V1 domain. PCR was performed automatically by using a thermal controlling unit programmed to denature the double-stranded DNA at 94°C for 30 s and anneal the oligonucleotide to the DNA template at 55°C for 1 min, with primer extension at 72°C for 1.5 min. This was repeated for 30 cycles. Prior to sequencing, the PCR mixture was resolved by polyacrylamide gel electrophoresis and the amplified double-stranded DNA (335 bp) was isolated by electroelution. Sequencing was performed by the dideoxy method of Sanger et al. (14), using Sequenase. During analysis of a battery of group A streptococcal isolates representing a range of M serotypes, five different sequences became apparent. These were classified into five classes, I to V (Fig. 1). It was also observed that all PSGN

```
                         453
                          ↓
              5'- T TGG AAC GGC TGG GTT GGA GAT - 3' PRIMER 1
                  : ::: ::: ::: ::: ::: ::: :::
  CLASS I       TCG GTA ACC TTG CCG ACC CAA CCT GTC CAA GAA TTT TTG CTA AGT
  CLASS II                                              : ::: ::C
  CLASS III                                            :: G:T :AA
  CLASS IV                                             : T:: ::G
  CLASS V        TCG GTA ACC TTG CCG ACC CAA CCT GTC CAA GAA TTT TT: ::: ::C
  ------------------------------------------------------------------------
  CLASS I        GGG CAT GTG CGC GTT AAA CCG TAT CAA CCT AAA GCC GTT CAC AAC
  CLASS II       ::: ::: ::: ::: ::: :G: ::: ::: ::: ::: ::: ::: ::: ::: :::
  CLASS III      ::A ::: ::: ::: ::: :G: ::A ::: A:: GAA ::: C:A ::A ::A ::T
  CLASS IV       ::A ::: ::: ::: ::: :G: ::A ::A A:: GGA ::: C:A A:A ::A :CT
  CLASS V        ::A ::: ::: ::: ::: :G: ::A ::: A:: GAA ::: C:A A:A ::A :::
  ------------------------------------------------------------------------
  CLASS I        TCT GCT GAA CGC GTT AAC GTC AAC TAT GAA GTG AGC TTT GTC TCC
  CLASS II       ::A ::: ::: ::: ::: ::: ::: ::: ::: ::: ::: ::: ::: ::: :::
  CLASS III      CAA ::A A:: TCT ::: G:T ::A G:A ::: ACT ::A CAG ::: ACT C:T
  CLASS IV       C:A ::A A:: TCT ::: G:T A:A :GA ::: ACT ::A CAG ::: ACT C:T
  CLASS V        CAA ::G A:: TCT ::: G:T ::G G:A ::: ACT ::A CAG ::: ACT C::
  ------------------------------------------------------------------------
  CLASS I        GAA ACA GGA GAT TTA GAC TTT ACA CCG TTG TTA AGA AAC CAA TAC
  CLASS II       ::: ::: ::: A:: ::: ::: ::: ::G ::A :C: ::A: G:A :G: :::
  CLASS III      TT: :AC CCT ::: GAC ::T ::C :G: ::A GG: C:C ::A: G:T ACT A:G
  CLASS IV       TT: :AC CCT ::: GAT ::T ::C :A: ::A G:T C:C ::A: G:T ACT A:A
  CLASS V        TT: :AC CCT ::: GAC ::T ::C :G: ::A GGT C:C ::A: G:T ACT A:G
  ------------------------------------------------------------------------
  CLASS I        CAT TTG ACC ACA CTG GCA GTT GGT GAC TCT CTT TCA TCA CAA GAG
  CLASS II       ::: ::: ::: ::: ::: ::: ::: ::: ::: ::: ::: ::: ::: ::: :::
  CLASS III      :TA ::: :AA ::: ::A ::T A:C ::: ::: A:C A:C A:: ::T ::: ::A
  CLASS IV       :TA ::: :AA ::: ::A ::T A:C ::C A:C A:C A:C A:: ::C ::: ::A
  CLASS V        :TA ::: :AA ::: ::A ::T A:C ::: ::: A:C A:C A:: ::T ::: ::A
  ------------------------------------------------------------------------
  CLASS I        TTA GCA GCA ATT GCC CAA TTT ATC CTA TCA AAA AAA CAT CCA GAT
  CLASS II       ::: ::: ::: ::: ::: ::: ::: ::: ::: ::: ::: G:G ::: ::: :::
  CLASS III      ::: CT: ::T CAA ::A ::: AGC ::T T:: AAC ::: :CC ::C ::: :GC
  CLASS IV       ::G CTG ::T CAA ::T ::: AGC ::T T:: AAC G:: :CC ::: ::: :::
  CLASS V        ::: CT: ::T CAA ::A ::: AGC ::T T:: AAC ::: ::C ::C ::: :GC
  ------------------------------------------------------------------------
  CLASS I        TAT ATC ATT ACA AAA CGT GAC TCC TCA ATC GTC ACT CAT GAC AAT
  CLASS II       ::: ::: ::: ::: ::: ::: ::: ::: ::: ::: ::: ::: ::: ::: :::
  CLASS III      ::: :CG ::: TAT G:: ::: ::: ::: ::: ::: ::: --- --- --- ---
  CLASS IV       ::: :CG ::: TAT G:: ::: ::: ::: ::: ::: ::: --- --- ---
  CLASS V        ::: :CG ::: TAT G:: ::: ::: ::: ::: ::: ::: ::: ::: ::: :::
  ------------------------------------------------------------------------
  CLASS I        GAC ATT TTC CGT ACG ATT TTA CCA ATG GAT CAA GAG TTT ACT TAC
  CLASS V        ::: ::: ::: ::: ::: ::: ::: ::: ::: ::: ::: ::: ::: ::: :::
                 :: ::: ::: ::: ::: ::: :
              3'- AA AAG GCA AGC TAA AAT GGT T -5'   PRIMER 2
                                              ↑
                                             788
```

FIG. 1. Nucleotide sequences of the amplified V1 domains of the streptokinase genes obtained by PCR. :, Homologous nucleotides.

TABLE 1. Nucleotide and amino acid homology between classes of the amplified V1 of the streptokinase genes

Homology	Class	% Homology with:				
		Class I	Class II	Class III	Class IV	Class V
Nucleotide	I		96.1	67.6	66.7	68.8
	II	96.1		68.2	67.0	69.7
	III	67.6	68.2		92.2	96.7
	IV	66.7	67.0	92.2		91.9
	V	68.8	69.7	96.7	91.9	
Protein	I		93.7	63.1	60.4	64.0
	II	93.7		64.0	61.3	64.0
	III	63.1	64.0		89.2	96.4
	IV	60.4	61.3	89.2		90.1
	V	64.0	64.0	96.4	90.1	

isolates possessed either a class I or class II sequence, whereas isolates obtained from non-PSGN sources had either a class III, class IV, or class V sequence. Class I sequence was identical to the group A V1 sequence of strain NZ131 (serotype M49) described by Huang et al. (8), the class III sequence was identical to the group A, serotype M1 V1 sequence described by Walter et al. (15), and the class V sequence was identical to the group C V1 sequence described by Malke et al. (11). Table 1 illustrates the degree of homology between the five classes. Figure 2 depicts the predicted amino acid sequences of the five classes.

Restriction Enzyme Analysis

Analysis of the five sequences with a battery of restriction enzymes indicated that they could be rapidly distinguished from each other. By using the restriction enzymes MluI, PvuII, and DraI, four patterns of enzymic cleavage were observed. A class I sequence was cleaved by MluI but was refractory to PvuII and DraI. A class II sequence was cleaved by MluI and PvuII but not by DraI. Both classes III and IV were not cleaved by MluI and PvuII but were cleaved by DraI to give a distinct restriction pattern (Fig. 3). Class V was not cleaved by either MluI or PvuII but gave a characteristic cleavage pattern with DraI distinct from the patterns of classes III and IV (Fig. 3). The restriction enzyme DdeI was also useful to distinguish the five classes of V1 sequences (Fig. 4). Table 2 summarizes the cleavage products obtained when the amplified V1 domains are analyzed with MluI, PvuII, DraI, and DdeI. This approach of amplification of the V1

```
          152
           ↓
CLASS I    TLPTQPV---LLSGHVRVKPYQPKAVHNSAERVNVNYEVSFVSETGDLDF
CLASS II   TLPTQPV----::::::::R::::::::::::::::::::::::::::N:::
CLASS III  TLPTQPV----VK:::::R::KE:P:Q:Q:KS:D:E:T:Q:TPLNP:D::
CLASS IV   TLPTQPV----:R:::::R::KE:PIQTP:KS:DIR:T:Q:TPLNP:D::
CLASS V    TLPTQPV---:::::::::R::KE:PIQ:Q:KS:D:E:T:Q:TPLNP:D::


CLASS I    TPLLRNQYHLTTLAVGDSLSSQELAAIAQFILSKKHPDYIITKRDSSIVT
CLASS II   ::S:KER:::::::::::::::::::::::::::::::::::::::::::-
CLASS III  R:G:KDTKL:K:::I::TIT::::L:Q::S::N:T::G:T:YE::::::::
CLASS IV   K:V:KDTKL:K:::I:NTIT::::L:Q::S::NET::::T:YE:::::::
CLASS V    R:G:KDTKL:K:::I::TIT::::L:Q::S::N:N::G:T:YE::::::-


CLASS I    ----IFRTILP
CLASS II   ----IFRTILP
CLASS III  ----IFRTILP
CLASS IV   ----IFRTILP
CLASS V    ----IFRTILP
              ↑
             262
```

FIG. 2. Predicted amino acid sequences derived from the nucleotide sequences of the amplified V1 domains of the streptokinase genes. :, Homologous amino acid residues.

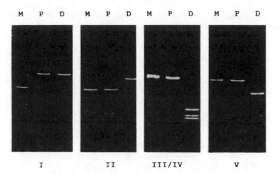

FIG. 3. Restriction enzyme analysis of the amplified V1 domains of the streptokinase genes from group A streptococci. Lanes: M, *Mlu*I; P, *Pvu*II; D, *Dra*I. The labeling at the bottom indicates classes.

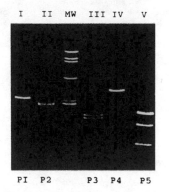

FIG. 4. Restriction enzyme analysis of the amplified V1 domains of the streptokinase genes from group A streptococci with *Dde*I. MW, Molecular weight standards. The labeling at the top and bottom indicates classes and patterns, respectively.

domain and restriction enzyme analysis permitted us to very rapidly analyze a population of streptococcal strains for the presence of a PSGN-associated sequence (class I or II) and a non-PSGN associated sequence (class III, IV, or V).

The streptococcal strains that had been tested in the tissue cage model for their ability to induce experimental PSGN (7) were analyzed. The strains that have been shown to induce experi-

mental PSGN were EF514 (serotype M56), NZ131 (serotype M49), A374 (serotype M12), and D897 (serotype M12). Both EF514 and NZ131 were impetigo isolates, whereas A374 and D897 were pharyngeal isolates. EF514, A374, and D897 possessed a class II streptokinase V1 sequence, whereas NZ131 has a class I sequence. Strains D480 (type M1) and EF445 (type M3), which did not induce experimental PSGN, had a class III (EF445) or a class V (D480) V1 sequence. When a battery of streptococcal isolates representing a range of M types and associated with PSGN were analyzed by PCR and restriction enzyme mapping, it appeared that certain M types had either a class I or class II streptokinase V1 sequence. The M serotypes that had a class I V1 sequence were 4, 5, 14, 32, 33, 34, 39, 49, and 55; those with a class II V1 sequence were 9, 12, 18, 22, 25, 31, 36, 38, 41, 51, 56, 59, and 60. This analysis was based on two isolates of each M serotype. When the analysis is extended to include a larger number of strains of each M serotype, it may then be possible to suggest that there are two major streptokinase populations which may be involved in the induction of PSGN. At present, the V2 domain is under investigation to determine whether it also demonstrates an unique sequence related to the ability of certain streptokinases to induce experimental PSGN. If these domains are involved in the nephrotropic behavior of these streptokinases, particularly those of class I and class II, both domains may be required to impart the structural requirements for binding to glomerular structures.

Summary

The V1 domain of the streptokinase gene from group A streptococci can be classified into at least five separate nucleotide sequences, which were categorized as classes I to V. Classes I and II can be distinguished from classes III, IV, and V by restriction enzyme analysis of the amplified V1 domain with *Mlu*I and *Dra*I (classes I and II cleaved with *Mlu*I and not with *Dra*I; classes III, IV, and V not cleaved with *Mlu*I and cleaved with *Dra*I).

Class I and class II sequences exhibit >96% nucleotide and >94% protein homology and can be distinguished by restriction enzyme analysis

TABLE 2. Fragments generated after enzymatic digestion of the V1 amplified domains of the streptokinase genes from group A streptococci with *Mlu*I, *Pvu*II, *Dra*I, and *Dde*I.

Class	Fragment length (bp)			
	*Mlu*I (5'-A ↓ CGCG-3')	*Pvu*II (5'-CAG ↓ CTG)	*Dra*I (5'-TTT ↓ AAA)	*Dde*I (5'-C ↓ TNA)
I	95, 241	Not cleaved	Not cleaved	35, 300
II	95, 241	89, 246	Not cleaved	35, 50, 250
III	Not cleaved	Not cleaved	132, 114, 89	171, 164
IV	Not cleaved	Not cleaved	132, 114, 89	Not cleaved
V	Not cleaved	Not cleaved	246, 89	35, 136, 164

of the amplified V1 domain by *Pvu*II (class I cleaved; class II not cleaved).

Class I and class II sequences are present in streptokinases that have been demonstrated to induce experimental PSGN.

Class III, class IV, and class V sequences exhibit >90% nucleotide and >89% protein homology with each other but have <70% nucleotide and <64% protein homology with class I and class II V1 sequences. Classes III, IV, and V may be distinguished by restriction enzyme analysis of the amplified V1 domain with *Dra*I and *Dde*I. Class III, class IV, and class V sequences are associated with streptokinases that have been shown not to induce experimental PSGN.

LITERATURE CITED

1. **Bergholm, A.-M., and S. E. Holm.** 1983. Experimental poststreptococcal glomerulonephritis in rabbits. *Acta Pathol. Microbiol. Immunol. Scand. Sect. C* **91**:263–270.
2. **Castellino, F. J., and S. P. Baja.** 1977. Activation of human plasminogen by equimolar levels of streptokinase. *J. Biol. Chem.* **252**:492–498.
3. **Dillon, H. C., and L. W. Wannamaker.** 1965. Physical and immunological difference among streptokinases. *J. Exp. Med.* **121**:351–371.
4. **Gerlach, D., and W. Köhler.** 1977. Studies on the heterogeneity of streptokinases of different origin. *Zentralbl. Bakteriol. Parasitenkd. Infektionskr. Hyg. Abt. 1 Orig. Reihe A* **238**:336–349.
5. **Gerlach, D., and W. Köhler.** 1980. Studies on the heterogeneity of streptokinases. IV. Communication: evidence for isostreptokinases in *Streptococcus pyogenes* type I. *Zentralbl. Bakteriol. Parasitenkd. Infektionskr. Hyg. Abt. 1 Orig. Reihe A* **248**:446–454.
6. **Holm, S. E.** 1989. The pathogenesis of acute post-streptococcal glomerulonephritis in new lights. *Acta Pathol. Microbiol. Immunol. Scand.* **96**:189–193.
7. **Holm, S. E., A.-M. Bergholm, and K. H. Johnston.** 1988. A streptococcal plasminogen activator in the focus of infection and in the kidneys during the initial phase of experimental streptococcal glomerulonephritis. *Acta Pathol. Microbiol. Immunol. Scand.* **96**:1097–1108.
8. **Huang, T.-T., H. Malke, and J. J. Ferretti.** 1989. The streptokinase gene of group A streptococci: cloning, expression in *Escherichia coli*, sequence and analysis. *Mol. Microbiol.* **3**:197–205.
9. **Huang, T.-T., H. Malke, and J. J. Ferretti.** 1989. Heterogeneity of the streptokinase gene in group A streptococci. *Infect. Immun.* **57**:502–506.
10. **Johnston, K. H., and J. B. Zabriskie.** 1986. Purification and partial characterization of the nephritis strain-associated protein from *Streptococcus pyogenes*, group A. *J. Exp. Med.* **163**:697–711.
11. **Malke, H., B. Roe, and J. J. Ferretti.** 1985. Nucleotide sequence of the streptokinase gene from *Streptococcus equisimilis* H46A. *Gene* **34**:357–362.
12. **Marcum, J. A., and D. L. Kline.** 1983. Species specificity of streptokinase. *Comp. Biochem. Physiol.* **75B**:389–394.
13. **Mullis, K., F. Faloona, S. Scharf, R. Saika, G. Horn, and H. Erlich.** 1986. Specific enzymatic amplification of DNA *in vitro*: the polymerase chain reaction. *Cold Spring Harbor Symp. Quant. Biol.* **51**:263–273.
14. **Sanger, F., S. Nicklen, and A. R. Coulson.** 1977. DNA sequencing with chain-terminating inhibitors. *Proc. Natl. Acad. Sci. USA* **74**:5463–5467.
15. **Walter, R., M. Siegel, and H. Malke.** 1989. Nucleotide sequence of the streptokinase gene from a *Streptococcus pyogenes* type 1 strain. *Nucleic Acids Res* **17**:1261.
16. **Walter, R., M. Siegel, and H. Malke.** 1989. Nucleotide sequence of the streptokinase gene from a group G *Streptococcus. Nucleic Acids Res.* **17**:1262.
17. **Weinstein, L.** 1953. Antigenic dissimilarity of streptokinases. *Proc. Soc. Exp. Biol. Med.* **83**:689–691.
18. **Wulf, R. J., and E. T. Mertz.** 1969. Studies on plasminogen. VIII. Species specificity of streptokinase. *Can. J. Biochem.* **47**:927–931.

Molecular Analysis of the Streptococcal Pyrogenic Exotoxins

A. R. HAUSER,[1] S. C. GOSHORN,[1] E. L. KAPLAN,[2] D. L. STEVENS,[3] AND P. M. SCHLIEVERT[1]

Department of Microbiology[1] and Department of Pediatrics, World Health Organization Collaborating Center for Reference and Research on Streptococci,[2] University of Minnesota, Minneapolis, Minnesota 55455, and Division of Infectious Disease, Veterans Affairs Medical Center, Boise, Idaho 83702, and Department of Medicine, University of Washington, Seattle, Washington 98195[3]

A large number of extracellular proteins are elaborated by group A streptococci and may contribute to their ability to cause diseases, which range from relatively mild pharyngitis and impetigo to serious, life-threatening toxic shock-like syndrome (TSLS) and autoimmune diseases such as rheumatic fever and acute glomerulonephritis. Among these extracellular products are the streptococcal pyrogenic exotoxins (synonyms: SPEs, scarlet fever toxins, erythrogenic toxins, lymphocyte mitogens, and blastogen A [type A]). Three serologically distinct SPEs have been purified and characterized, designated types A, B, and C (31). These toxins belong to a larger pyrogenic toxin family thus far limited to group A streptococci and *Staphylococcus aureus* strains. The pyrogenic toxin family includes SPEs, toxic shock syndrome toxin-1, and staphylococcal enterotoxins A to E (4, 31). All members of the family have the ability to induce high fever, enhance host susceptibility to the lethal effects of other agents, and stimulate nonspecifically T-lymphocyte proliferation. In addition, all of the toxins are relatively low-molecular-weight single-peptide chains and are typically heat and acid stable. SPEs also have the ability, not shared with other pyrogenic toxins, to greatly amplify the cardiotoxic properties of other agents. The staphylococcal enterotoxins have the extra property of being able to induce vomiting and diarrhea after oral administration.

Disease Associations of SPEs

SPEs historically are considered the scarlet fever toxins and in that capacity have the ability to induce high fever and erythematous macular rash, a property that appears to depend on SPE amplification of preexisting hypersensitivity (25). Although not conclusively shown, SPE A appears to have been the major toxin type associated with severe scarlet fever present in the early part of this century. More recently, with the decline in severity of scarlet fever, there was a concomitant decline in isolation of SPE A-producing strains (7, 24). SPE B and C have been associated with the milder cases of scarlet fever.

In 1987, Cone and colleagues (7) reported on two cases of TSLS induced by group A streptococci; one of the strains made SPE A. Subsequently, a large number of similar cases were reported to us from across the United States, and again isolated streptococci mainly made SPE A either alone or in combination with SPE B and SPE C (20, 29). Most recently, we have shown that 85% of well-characterized TSLS isolates contained *speA* (A. R. Hauser, D. L. Stevens, E. L. Kaplan, and P. M. Schlievert, submitted for publication). In a small percentage of the cases reported to us, SPE B alone or in combination with SPE C was expressed by the causative bacteria. The association of TSLS with SPE B has also been reported in the United Kingdom (10). Of particular importance is the observation that since 1986 there appears to be a highly significant increase in the occurrence of this severe illness; up to 50% of patients succumb, and many survivors have limbs amputated.

In addition to scarlet fever and TSLS, there are other possible disease associations with SPE production. In 1979, Schlievert et al. (24) noted that 11 of 11 streptococcal strains tested that were isolated from patients with rheumatic fever made SPE C either alone or in combination with other toxin types. In addition, Gray and colleagues (15) have reported that rheumatic heart disease patients have lower T-cell responses to blastogen A (SPE A) than do controls. Schlievert et al. (24) also noted that streptococcal strains isolated from patients with acute glomerulonephritis consistently made SPE B, either alone or in combination with other toxin types.

Biological Properties of SPEs

The ability of SPEs to induce fever appears to depend on their capacity to induce the release of endogenous pyrogens (interleukin-1 and tumor necrosis factor) from macrophages and to directly stimulate the hypothalamic fever response control center (9, 28). This activity is separable from the ability to enhance host susceptibility to lethal endotoxin shock. The latter activity may depend on SPE alteration of liver cell clearance function (26) and results in up to a 100,000-fold reduction in the 50% lethal dose of endotoxin.

Currently the most studied activity of these pyrogenic toxins is their ability to nonspecifically stimulate T-lymphocyte proliferation as Vβ-restricted mitogens (21, 22; B. Leonard, M. K. Jenkins, and P. M. Schlievert, *Abstr. Annu. Meet.*

Am. Soc. Microbiol. 1990, B251, p. 68). T-cell proliferation occurs after interaction of the toxins with class II major histocompatibility products on antigen-presenting cells and interaction of the intact toxin-class II major histocompatibility product complex with T cells. The interaction with T cells is through subsets of the variable regions of the beta chain of the T-cell receptor complex and is without regard for the antigen specificity of the T-cell receptor. The role of this mitogenic activity in disease is not clear, but it has been proposed to mediate rash production and, by way of gamma interferon release, to suppress the development of neutralizing antibodies (4, 23, 25, 31).

Genetics of SPEs

SPE A. SPE A production is a variable trait; this toxin is expressed by some group A streptococcal strains but not by others. Extensive studies of the genetic control as well as cloning and sequencing of the gene, *speA*, encoding SPE A have been performed. The results of those studies were presented in the past conference (27) and thus are only briefly summarized in this report.

In a small percentage of strains, *speA* is contained on the inducible bacteriophage T12 (17, 18, 34). In the remainder of strains, *speA* is associated with bacteriohage T12 DNA, but the phages are not inducible. *speA* is adjacent to the phage attachment site for integration into the chromosome (19). *speA* encodes a protein with a molecular weight of 29,244, and cleavage of a 30-residue signal peptide results in a mature pro-

tein with a molecular weight of 25,787 (32). The amino acid sequence of SPE A is shown in Fig. 1. SPE A shares highly significant sequence similarity with staphylococcal enterotoxins B and C (approximately 50%), and thus it has been proposed that these three toxins form a subfamily within the larger pyrogenic toxin family. Further, it has been suggested that these three toxins have a common ancestor toxin, possibly of staphylococcal origin.

SPE B. The structural gene encoding SPE B, designated *speB*, was cloned from the chromosome of streptococcal isolate 86-858 into *Escherichia coli* and sequenced (6, 16). A 1,194-bp open reading frame encoded a 398-amino-acid protein containing a 27-residue signal peptide (Fig. 2). The 371-residue mature protein (molecular weight, 40,314) was readily proteolytically cleaved to yield a 253-residue peptide (molecular weight, 27,580). This processing was confirmed by amino-terminal sequencing of SPE B purified in the presence and absence of a protease inhibitor.

A comparison of the amino acid sequence of SPE B with the published sequence of streptococcal proteinase precursor (SPP) is shown in Fig. 3. SPP is a cysteine proteinase elaborated by group A streptococci which is activated under reducing conditions (8). Structurally, SPE B and SPP are closely related. We cannot, however, account for the minor sequence differences.

SPE B protein purified from streptococcal isolate 86-858 had no proteinase activity when tested

```
SPE A 1  MENNKKVLKKMVF--F----VLVTFLGLTISQEVFAQQDPDPSQLHRSSLVKNLQ-NIYFLYEGDPVTHENVKSV  68
            :  ::        :    :     ::  ::  :         ::  :  :    :  ::      :   ::::
SEC1  1  M--NKSRFISCVILIFALILVLFTPNVLAESQP-----DPTPDELHKASKFTGLMENMKVLYDDHYVSATKVKSV  68
         :  :  :::  :::::::::  ::::::::::     ::      :::::::: ::::::::::   :::  :::
SEB   1  M--YKRLFISHVILIFALILVISTPNVLAESQP-----DPKPDELHKSSKFTGLMENMKVLYDDNHVSAINVKSI  68


69  DQLLSHDLIYNVSGP---NYDKLKTELKNQEMATLFKDKNVDIYGVEYYHLCYLCENAERSA-------------126
    :  :  :::::  :        ::::: :   :: :: :: ::  ::   :    :
69  DKFLAHDLIYNISDKKLKNYDKVKTELLNEGLAKKYKDEVVDVYGSNYYVNCYFS-----SKDNVG------KVT 131
    : ::  ::::  :  :  ::: :  :   :: ::::  ::: : ::: ::::       : :          :
69  DQFLYFDLIYSIKDTKLGNYDNVRVEFKNKDLADKYKDKYVDVFGANYYYQCYFS-----KKTNDINSHQTDKR- 136


127 ----CIYGGVTNHEGNHLEILKKIVVK--VSIDGIQSLSFDIETNKKMVTAQELDYKVRKYLTDNKQLYTNGPSK 194
        : :::  :  ::::          :   :           :: : :::::::: :  :   :   :  ::
132 GGKTCMYGGITKHEGNHFDNGNLQNVLIRVYENKRNTISFEVQTDKKSVTAQELDIKARNFLINKKNLYEFNSSP 205
    :::::::: : :: :    ::  : :  : :::  ::: :  ::::  :::::::       :::   :::::::
137 --KTCMYGGVTEHNGNQLDKYRSITV--RVFEDGKNLLSFDVQTNKKKVTAQELDYLTRHYLVKNKKLYEFNNSP 206


195 YETGYIKFIPKNKESFWFDFFPEPE--FTQSKYLMIYKDNETLDSNTSQIEVYLTTK    251 SPE A
    ::::::::::  :  :: :  : :  : ::::::: :    :       ::: ::::
206 YETGYIKFIENNGNTFWYDMMPAPGDKFDQSKYLMMYNDNKTVDSKSVKIEVHLTTKNG  266 SEC1
    :::::::::: :  :   : :::::::::::::: : ::::   :::  ::: :::::
207 YETGYIKFIENE-NSFWYDMMPAPGDKFDQSKYLMMYNDNKMVDSKDVKIEVYLTTKKK  266 SEB
```

FIG. 1. Sequence comparison of streptococcal pyrogenic exotoxin type A (SPE A) and staphylococcal enterotoxins B (SEB) and C1 (SEC1). Sequences are the single-amino-acid designations and include the signal peptides of each toxin.

```
-35
GTTGTCAGTG TCAACTAACC GTGTTATTGT CTATTACCAT TCATGGTATC AGCGA
                                         -10

CATCG TATGATAAACC ATACGATTCA GCTAAGTAAG GAGGTGTGTC CAATGTACCG
                                              S.D.

TTAAAAGCAA ATGCAGTAGA TTAACTTATT TTGAAAGAGG TATAAAAAAA ATG
                                                         met

  4: AAT AAA AAG AAA TTA GGT ATC AGA TTA TTA AGT CTT TTA GCA TTA
     asn lys lys lys leu gly ile arg leu leu ser leu leu ala leu

 49: GGT GGA TTT GTT CTT GCT AAC CCA GTA TTT GCC GAT CAA AAC TT
     gly gly phe val leu ala asn pro val phe ala asp gln asn phe

 94: GCT CGT GCA AAA GAA GCA AAA GAT AGC GCT GCT ATC ACA TTT ATC
     ala arg ala lys glu ala lys asp ser ala ala ile thr phe ile

139: CAA AAA TCA GCA GCT ATC AAA GCA GGT GCA CGA GCA GCA GAA GAT
     gln lys ser ala ala ile lys ala gly ala arg ser ala glu asp

184: ATT AAG CTT GAC AAA GTT AAC TTA GGT GGA GAA CTT TCT GGC TCT
     ile lys leu asp lys val asn leu gly gly glu leu ser gly ser

229: AAT ATG TAT GTT TAC AAT ATT TCT GGA GGA TTT GTT ATC GTT
     asn met tyr val tyr asn ile ser gly gly phe val ile val

274: TCA GGA GAT AAA CGT TCT CCA GAA ATT CTA GGA TAC TCT ACC AGC
     ser gly asp lys arg ser pro glu ile leu gly tyr ser thr ser

319: GGA TCA TTT GAC GCT AAC GGT AAA GAA AAC ATT GCT TCC TTC ATG
     gly ser phe asp ala asn gly lys glu asn ile ala ser phe met

364: GAA AGT TAT GTC GAA CAA ATC AAA AAA GAA AAC AAA TTA GAC ACT
     glu ser tyr val glu gln ile lys lys glu asn lys leu asp thr

409: ACT TAT GCT GGT ACC CCT GAG ATT AAA CCA GTT GTT AAA TCT
     thr tyr ala gly thr pro glu ile lys pro val val lys ser

454: CTC CTT GAT TCA AAA GGC ATT CAT TAC AAA GGT AAC CCT TAC
     leu leu asp ser lys gly ile his tyr asn gln gly asn pro tyr

499: AAC CTA TTG ACA CCT GTT ATT GAA AAA GTA AAA CCA GGT GAA CAA
     asn leu thr pro val ile glu lys val lys pro gly glu gln

544: TCT TTT GTA GGT CAA CAT GCA GCT ACA GGA TGT GCT GCT ACT GCA
     ser phe val gly gln his ala ala thr gly cys val ala thr ala

589: ACT GCT CAA ATT ATG AAA TAT CAT AAT TAC CCT AAC AAA GGG TTG
     thr ala gln ile met lys tyr his asn tyr pro asn lys gly leu


634: AAA GAC TAC ACT TAC ACA CTA AGC TCA AAT AAC CCA TAT TTC AAC
     lys asp tyr thr tyr thr leu ser ser asn asn pro tyr phe asn

679: CAT CCT AAG AAC TTG TTT GCA GCT ATC TCT ACT AGA CAA TAC AAC
     his pro lys asn leu phe ala ala ile ser thr arg gln tyr asn

724: TGG AAC AAC ATC CTA CCT ACT TAT AGC GGA GAA AGA GAA TCT AAC GTT
     trp asn asn ile leu pro thr tyr ser gly arg glu arg glu ser asn val

769: CAA AAA ATG GCG ATT TCA GAA TTG ATG GCT GAT GTT GGT ATT TCA
     gln lys met ala ile ser glu leu met ala asp val gly ile ser

814: GTA GAC ATG GAT TAT GGT CCA TCT AGT GGT TCT GCA GGT AGC TCT
     val asp met asp tyr gly pro ser ser gly ser ala gly ser ser

859: CGT GTT CAA AGA GCC TTG AAA GAA AAC TTT GGC TAC AAC CAA TCT
     arg val gln arg ala leu lys glu asn phe gly tyr asn gln ser

904: GTT CAC CAA ATT AAC CGT AGC GAC TTT AGC AAA CAA GAT TGG GAA
     val his gln ile asn arg ser asp phe ser lys gln asp trp glu

949: GCA CAA GAA TTA TCT GAC AAA GAA TTA TCT CAA AAC CAA GTA TAC TAC
     ala gln glu leu ser asp lys glu leu ser gln asn gln pro val tyr tyr

994: CAA GGT GTC GGT AAA GTA GGC GGA CAT GCC TTT GTT ATC GAT GGT
     gln gly val gly lys val gly gly his ala phe val ile asp gly

1039: GCT GAC GGA CGT AAC TTC TAC CAT GTT AAC TGG GGT TGG GGT GGA
      ala asp gly arg asn phe tyr his val asn trp gly trp gly gly

1084: GTC TCT GAC GGC TTC TTC CGT CTT GAC GCA CTA AAC CCT TCA GCT
      val ser asp gly phe phe arg leu asp ala leu asn pro ser ala

1129: CTT GGT ACT GGT GGC GGC GCA GGC GGC TTC AAC GGT TAC CAA AGT
      leu gly thr gly gly gly ala gly gly phe asn gly tyr gln ser

1174: GCT GTT GTA GGC ATC AAA CCT TAG TAT GGAAATGCAT TTCGTTAGAA CA
      ala val val gly ile lys pro ***

GAACTGA GGCACGCGAT AGCTGAAACC TTTTTGTGCC GAAAACACAC AAAGCAA

AAG CCCCTATCGT GTGCAGCAGT TAAGCTATCC ATCAGACACA GATCACTCAG G

TGGTGAGCT ATACATCTAT GCTTTGTCTC CTGCTGGATT TATCATCGTA TCAGGA

GACA CCAGAGCGCA CACCATTTTA GGCTATTCTT TTGATAATAA CCTGGACCTC

AACCATGATA ATGTCAGAAG TATGGTAG
```

FIG. 2. Nucleotide and deduced amino acid sequences of *speB* and SPE B, respectively. Numbering is in reference to the ATG start codon. Possible promoter (−10 and −35) and Shine-Dalgarno (S.D.) sequences are indicated. The probable cleavage site between the signal peptide and the mature protein is marked by the asterisk following residue 27. The amino-terminal sequence of unproteolyzed SPE B is identical to the sequence indicated by the dashed underline, and the amino-terminal sequence of the 253-residue cleavage product is identical to the sequence indicated by the solid underline. Overlined nucleotides indicate palindromic sequences located 3′ of the translation stop codon.

```
SPEB : MNKKKLGIRLLSLLALGGFVLANPVFADQNFARNEKEAKDSAITFIQKSAAIKAGARSAE
SPP  :                   ---------Q----------------------


DIKLDKVNLGGELSGSNMYVYNISTGGFVIVSGDKRSPEILGYSTSGSFDANGKENIASF
----------------------.................................--G


MESYVEQIKENKKLDTTYAGTAEIKQPVVKSLLDSKGIHYNQGNPYNLLTPVIEKVKPGE
----------------------------------------------I-----------


QSFVGQHAATGCVATATAQIMKYHNYPNKGLKDYTYTLSSNNPYFNHPKNLFAAISTRQY
------A-TGH---------------D----N--------PD--D--------------


NWNNILPTYSGRESNVQKMAISELMADVGISVDMDYGPSSGSAGSSRVQRALKENFGYNQ
D-----------Q-QNV-----------------------------------------


SVHQINRSDFSKQDWEAQIDKELSQNQPVYYQGVGKVGGHAFVIDGADGRNFYHVNWGWG
-----D-G----------------------E------------DGA-------D----


GVSDGFFRLDALNPSALGTGGGAGGFNGYQSAVVGIKP
--------------------------E--------
```

FIG. 3. Amino acid sequence similarity between SPE B and SPP. The sequence of SPE B is that inferred from the nucleotide sequence of *speB*. The sequence of SPP is that reported by Tai et al. (30) and Yonaha et al. (33), using protein sequencing techniques. Matched amino acids (−) and gaps (.) are indicated.

in a casein assay similar to that used by Elliott and Liu to characterize SPP (8). Some of the structural differences between the two molecules may have resulted in SPP being proteolytically active and SPE B being inactive. In support of this hypothesis, the active site of SPP differs from the corresponding region of SPE B. Despite the difference in proteolytic activity between SPE B and SPP, we feel that the two molecules are variants of the same protein.

In each of 65 group A streptococcal strains tested, an internal probe, specific for DNA encoding SPE B, hybridized to a single restriction fragment, indicating that a single copy of the gene is present in the chromosome. We suspect that there are at least two subtypes of SPE B, one with and one without proteolytic activity. Both molecules exhibit properties characteristic of pyrogenic toxins: pyrogenicity, skin rash production, and lymphocyte mitogenicity (2, 11).

```
              10        20        30        40        50
SPE A: MENNKKVLKKMVFFVLVTFLGLTISQEVFAQQDPDPSQLHRSSLVKNLQNIYFLYEG---
       :   :   : ::  :           :       :         :: :
SPE C: MK--KINIIKIVFIITV------I-------------------LISTYFTYHQSDS
          10                                            20


          60        70                 80        90
       --DPVTHENVKSVDQLLSHDLIY-------------NVSGP---NYDKLKTELKNQ---
         :   ::::    :        : :       : :    :
       KKDIS---NVKS-------DLLYAYTITPYDYKDCRVNFSTTHTLNIDTQKYRGKDYYIS
       30           40        50        60        70


          100       110       120       130           140
       -----EMATLFK-DKNVDIYGVEYYHLCYLCENAERSACIYGGVT-------NHE--GNH
           :     :: :  :: :  :  :          :      ::::  :   ::  ::
       SEMSYEASQKFKRDDHVDVFGLFYILNSHTGEY------IYGGITPAQNNKVNHKLLGN-
       80        90        100       110           120       130


           150       160       170       180       190
       LEILKKIVVKVSIDGIQSLSFDIETNKKMVTAQELDYKVRKYLTDNKQLY--TNGPSKYE
       : :       :  :: :: : :::: : :  :  :  : :    : :
       LFI-------SGESQQNLNNKIILEKDIVTFQEIDFKIRKYLMDNYKIYDAT---SPYV
                 140       150       160       170        180


        200       210       220       230       240       250
       TGYIKFIPKNKESFWFDFFPEP-EFTQSKYLMIYKDNETLDS-NTSQIEVYLTTK
       : :       : : : :   ::::            ::
       SGRIEIGTKDGKHEQIDLFDSPNEGTRSDIFAKYKDNRIINMKNFSHFDIYLEK
              190       200       210       220       230
```

FIG. 4. Sequence comparison of SPE A and SPE C. Sequences are the single-amino-acid designations and include the signal peptides of each toxin.

TABLE 1. Percentage similarity between aligned pyrogenic toxin amino acid sequences[a]

	SPE B	SPE C	SEA	SEB	SEC1	SEC2	SEC3	SED	SEE	TSST-1
SPE A	22	28	33	48	46	45	46	35	34	25
SPE B		22	22	18	20	21	21	22	21	26
SPE C			27	26	30	31	30	30	27	20
SEA				34	31	32	33	50	70	23
SEB					69	68	70	36	35	23
SEC1						97	94	33	31	22
SEC2							96	33	31	22
SEC3								33	33	22
SED									52	24
SEE										26

[a]Sequences (including signal peptides) were aligned by using the SS2 alignment algorithm of Altschul and Erickson (1). Percentages were calculated by dividing the number of matched residues by the total number of residues in the shorter of the sequences being compared. Alignments and matches reflect identical amino acids but not conservative substitutions.

SPE C. The gene encoding SPE C was cloned from the chromosome of streptococcal strain T18P and sequenced (12, 13). The structural gene *speC* consists of 705 nucleotides and encodes a protein containing 235 amino acids. Removal of the predicted 27-residue signal peptide would result in a mature toxin with a molecular weight of 24,354. The sequence of SPE C is shown in Fig. 4, which also demonstrates the sequence similarity between SPE A and C. The similarity seen between SPE A and C is significantly less than that between SPE A and enterotoxins B and C, even though both are streptococcal toxins.

Goshorn and Schlievert (14) also showed that *specC*, like *speA*, is phage encoded. These investigators cloned *speC* from phage DNA obtained from streptococcal strain CS112. Subsequently, it was shown that *speC* was adjacent to phage DNA in the majority of streptococcal strains that make SPE C, although in most cases phages cannot be induced.

The sequence relatedness of all of the sequenced pyrogenic toxins, as determined by percentage similarity between aligned amino acid sequences, is shown in Table 1. It is noteworthy that among the toxins that have significant similarity, the major sequence relatedness is in the carboxyl halves. Studies have shown that for three of the toxins the carboxyl halves contain the biological activities of the molecules (3, 5), which further suggests that the regions of major similarity may contain the sites responsible for the shared activities. Studies to localize the regions containing the biological as well as antibody reactivity constitute a major future objective.

Molecular Epidemiology

Ouchterlony immunodiffusion assays indicate that levels of SPE A and SPE C production are variable. These assays in addition to hybridization studies indicate a close correlation between SPE A or SPE C production and the presence of *speA* or *speC*, respectively. However, M protein type 1 and 18 isolates are exceptions; many have a SPE A-negative phenotype but contain a DNA sequence which hybridizes to a *speA*-specific probe.

Unlike *speA* and *speC*, DNA sequences that hybridize to a *speB*-specific probe were found in all group A streptococci tested. This, along with SPE B's suggested proteolytic activity and the dissimilar nucleic acid sequence of its gene, indicates that SPE B may have evolved separately from the other pyrogenic toxins. Interestingly, although all tested group A streptococci contain a *speB*-like sequence in their genomes, only approximately half express detectable amounts of SPE B. Several explanations for this are possible. The SPE B-negative strains may express a labile form of the toxin, or they may express a factor that degrades SPE B. Alternatively, the genes encoding SPE B in these strains may contain mutations that eliminate toxin production or may be regulated such that toxin is not produced in vitro.

DNA hybridization experiments indicate that a correlation exists between the restriction fragment length polymorphisms (RFLPs) of *spe* genes and some M protein types of group A streptococci. Of 17 M protein type 18 isolates tested, 16 had identical *spe* RFLPs. All of five M protein type 3 strains tested had identical *spe* RFLPs, but this pattern differed significantly from that seen with the M protein type 18 strains. Clearly, more studies are necessary to understand the relationship between *spe* RFLP and M protein type.

This work was supported by Public Health Service grant HL36611 from the National Heart, Lung, and Blood Institute.

We are indebted to Dwight Johnson for performing the M and T typing and to Tim Leonard for art and photographic work.

LITERATURE CITED

1. **Altschul, S. F., and B. W. Erickson.** 1986. Optimal sequence alignment using affine gap costs. *Bull. Math. Biol.* **48**:603–616.
2. **Barsumian, E. L., C. M. Cunningham, P. M. Schlievert, and D. W. Watson.** 1978. Heterogeneity of group A strep-

tococcal pyrogenic exotoxin type B. *Infect. Immun.* **20:**512–518.

3. **Blomster-Hautamaa, D. A., B. N. Kreiswirth, J. S. Kornblum, R. P. Novick, and P. M. Schlievert,** 1986. The nucleotide and partial amino acid sequence of toxic shock syndrome toxin-1. *J. Biol. Chem.* **261:**15783–15786.

4. **Blomster-Hautamaa, D. A., and P. M. Schlievert.** 1988. Nonenterotoxic staphylococcal toxins, p. 297–330. *In* M. C. Hardegree and A. T. Tu (ed.), *Bacterial Toxins.* Marcel Dekker, Inc., New York.

5. **Bohach, G. A., J. P. Handley, and P. M. Schlievert.** 1989. Biological and immunological properties of the carboxyl terminus of staphylococcal enterotoxin C1. *Infect. Immun.* **57:**23–28.

6. **Bohach, G. A., A. R. Hauser, and P. M. Schlievert.** 1988. Cloning of the gene, *speB*, for streptococcal pyrogenic exotoxin type B in *Escherichia coli. Infect. Immun.* **56:**1665–1667.

7. **Cone, L. A., D. R. Woodard, P. M. Schlievert, and G. S. Tomory.** 1987. Clinical and bacteriologic observations of a toxic shock-like syndrome due to *Streptococcus pyogenes. N. Engl. J. Med.* **317:**146–149.

8. **Elliott, S. D., and T. Y. Liu.** 1970. Streptococcal proteinase. *Methods Enzymol.* **19:**252–261.

9. **Fast, D. J., P. M. Schlievert, and R. D. Nelson.** 1989. Toxic shock syndrome-associated staphylococcal and streptococcal pyrogenic toxins are potent inducers of tumor necrosis factor production. *Infect. Immun.* **57:**291–294.

10. **Gaworzewska, E. T., and G. Hallas.** 1989. Group A streptococcal infections and a toxic shock-like syndrome. *N. Engl. J. Med.* **321:**1546.

11. **Gerlach, D., H. Knoll, W. Köhler, J. H. Ozegowski, and V. Hribalova.** 1983. Isolation and characterization of erythrogenic toxins V. Communication: identity of erythrogenic toxin type B and streptococcal proteinase precursor. *Zentralbl. Bakteriol. Parasitenkd. Infektionskr. Hyg. Abt. 1 Orig. Reihe A* **255:**221–233.

12. **Goshorn, S. C., G. A. Bohach, and P. M. Schlievert.** 1988. Cloning and characterization of the gene, *specC*, for pyrogenic exotoxin type C from *Streptococcus pyogenes. Mol. Gen. Genet.* **212:**66–70.

13. **Goshorn, S. C., and P. M. Schlievert.** 1988. Nucleotide sequence of streptococcal pyrogenic exotoxin type C. *Infect. Immun.* **56:**2518–2520.

14. **Goshorn, S. C., and P. M. Schlievert.** 1989. Bacteriophage association of streptococcal pyrogenic exotoxin type C. *J. Bacteriol.* **171:**3068–3073.

15. **Gray, E. D., L. W. Wannamaker, E. M. Ayoub, E. Kholy, and A. H. Abdin.** 1981. Cellular immune responses to extracellular streptococcal products in rheumatic heart disease. *J. Clin. Invest.* **68:**665–671.

16. **Hauser, A. R., and P. M. Schlievert.** 1990. Nucleotide sequence of the streptococcal pyrogenic exotoxin type B gene and relationship between the toxin and the streptococcal proteinase precursor. *J. Bacteriol.* **172:**4536–4542.

17. **Johnson, L. P., and P. M. Schlievert.** 1983. A physical map of the group A streptococcal pyrogenic exotoxin bacteriophage T12 genome. *Mol. Gen. Genet.* **189:**251–255.

18. **Johnson, L. P., P. M. Schlievert, and D. W. Watson.** 1980. Transfer of group A streptococcal pyrogenic exotoxin production to nontoxigenic strains by lysogenic conversion. *Infect. Immun.* **28:**254–257.

19. **Johnson, L. P., M. A. Tomai, and P. M. Schlievert.** 1986. Bacteriophage involvement in group A pyrogenic exotoxin A production. *J. Bacteriol.* **166:**623–627.

20. **Lee, P. K., and P. M. Schlievert.** 1989. Quantification and toxicity of group A streptococcal pyrogenic exotoxins in an animal model of toxic shock syndrome-like illness. *J. Clin. Microbiol.* **27:**1890–1892.

21. **Marrack, P., and J. Kappler.** 1990. The staphylococcal enterotoxins and their relatives. *Science* **248:**705–711.

22. **Norton, S. D., P. M. Schlievert, R. P. Novick, and M. K. Jenkins.** 1990. Molecular requirements for T cell activation by the staphylococcal toxic shock syndrome toxin-1. *J. Immunol.* **144:**2089–2095.

23. **Poindexter, N. J., and P. M. Schlievert.** 1986. Suppression of immunoglobulin-secreting cells from human peripheral blood by toxic-shock-syndrome toxin-1. *J. Infect. Dis.* **153:**772–779.

24. **Schlievert, P. M., K. M. Bettin, and D. W. Watson.** 1979. Production of pyrogenic exotoxin by groups of streptococci: association with group A. *J. Infect. Dis.* **140:**676–681.

25. **Schlievert, P. M., K. M. Bettin, and D. W. Watson.** 1979. Reinterpretation of the Dick test: role of group A streptococcal pyrogenic exotoxin. *Infect. Immun.* **26:**467–472.

26. **Schlievert, P. M., K. M. Bettin, and D. W. Watson.** 1980. Inhibition of ribonucleic acid synthesis by group A streptococcal pyrogenic exotoxin. *Infect. Immun.* **27:**542–548.

27. **Schlievert, P. M., L. P. Johnson, M. A. Tomai, and J. P. Handley.** 1987. Characterization and genetics of group A streptococcal pyrogenic exotoxins, p. 136–142. *In* J. J. Ferretti and R. Curtiss III (ed.), *Streptococcal Genetics.* American Society for Microbiology, Washington, D.C.

28. **Schlievert, P. M., and D. W. Watson.** 1978. Group A streptococcal pyrogenic exotoxin: pyrogenicity, alteration of blood-brain barrier, and separation of sites for pyrogenicity and enhancement of lethal endotoxin shock. *Infect. Immun.* **21:**753–763.

29. **Stevens, D. L., M. H. Tanner, J. Winship, R. Swarts, K. M. Ries, P. M. Schlievert, and E. Kaplan.** 1989. Severe group A streptococcal infections associated with a toxic shock-like syndrome and scarlet fever toxin A. *N. Engl. J. Med.* **321:**1–7.

30. **Tai, J. Y., A. A. Kortt, T. Y. Liu, and S. D. Elliott.** 1976. Primary structure of streptococcal proteinase. III. Isolation of cyanogen bromide peptides: complete covalent structure of the polypeptide chain. *J. Biol. Chem.* **251:**1955–1959.

31. **Wannamaker, L. W., and P. M. Schlievert.** 1988. Exotoxins of group A streptococci, p. 267–296. *In* M. C. Hardegree and A. T. Tu (ed.), *Bacterial Toxins.* Marcel Dekker, Inc., New York.

32. **Weeks, C. R., and J. J. Ferretti.** 1986. Nucleotide sequence of the type A streptococcal exotoxin (erythrogenic toxin) gene from *Streptococcus pyogenes* bacteriophage T12. *Infect. Immun.* **52:**144–150.

33. **Yonaha, K., S. D. Elliott, and T. Y. Liu,** 1982. Primary structure of zymogen of streptococcal proteinase. *J. Protein Chem.* **1:**317–334.

34. **Zabriskie, J. B.** 1964. The role of temperate bacteriophage in the production of erythrogenic toxin by group A streptococci. *J. Exp. Med.* **119:**761–780.

Extracellular Product Genes of Group A Streptococci

JOSEPH J. FERRETTI, TING-TING HUANG, WAYNE L. HYNES, HORST MALKE,
DANIEL SIMON, ALEXANDER SUVOROV, AND CHANG-EN YU

*Department of Microbiology and Immunology, University of Oklahoma Health Sciences Center,
Oklahoma City, Oklahoma 73190*

Streptococcus pyogenes, a member of the Lancefield group A streptococci, produces a number of extracellular proteins or peptides (Fig. 1), all of which can be identified either by biological activity or by immunological detection with human antiserum. Comprehensive reviews on the streptococcal extracellular products and their properties have been published by Alouf (1) and Ginsburg (7), detailing their pharmacological actions and involvement in the initiation of tissue injury. Although many of these extracellular products have been the object of intense study, their specific roles in the pathogenesis of disease remain to be elucidated.

Molecular genetic techniques have been applied to the study of group A streptococci, resulting in the cloning and sequencing of a number of genes specifying extracellular products. Specific information about some of these extracellular products has now been obtained regarding molecular weight, amino acid sequence, and homology to other proteins. Moreover, studies are now proceeding to understand structure-function relationships of these proteins. The availability of genes specifying these extracellular products has also allowed the construction of isogenic mutants for virulence and pathogenicity studies and specific probes for identification and molecular epidemiology analyses. Additionally, tools are now available to study the regulation of these genes. We present here an overview of molecular studies of several extracellular product genes, including those encoding streptokinase (SK), streptolysin O (SLO), hyaluronidase, proteinase, and erythrogenic toxins.

Streptokinase

SK has long been identified as one of the spreading factors associated with streptococcal infections because of its ability to lyse fibrin-containing clots, thereby disseminating infection (27). Although the exact role of SK in human disease is not yet clear, its presence in patients with streptococcal infections is readily confirmed by high antibody titers. SK binds stoichiometrically with plasminogen to form an SK-plasminogen complex, which then catalyzes the conversion of plasminogen to plasmin. It is plasmin that causes the lysis of clots containing human fibrin. The SK obtained from group C streptococci is one of the major pharmacological agents used in clinical thrombolytic therapy.

The streptokinase gene (*skc*) from a group C strain (*Streptococcus equisimilis*) was the first to be cloned and sequenced (21, 22). *skc* specifies a mature protein of 414 amino acids (47,294 Da) and is preceded by a 26-amino-acid signal peptide. Hybridization studies with *skc* as a probe have shown that it is found only in the pathogenic group A, C, and G streptococci (12). Additionally, *skc* is found in all group A streptococcal M type strains but may exhibit sequence heterogeneity (12). For example, the sequences of SKs specified by *ska1* (29), *skc* (22), and *skg* (30) are more similar to each other than to the SK of *ska49* (13), the latter of which is characterized by two variable regions rich in hydrophilic residues (Fig. 2). Johnston and Zabriskie (16) have suggested that the nephritis strain-associated protein is SK, and Holm (10) has hypothesized that SK has unique domains that allow it to bind to kidney tissue. The possibility that all nephritogenic strains of streptococci possess similar SK sequences that are different from those of SKs of nonnephritogenic strains is now being examined.

The recent development of techniques for introducing DNA into streptococci by electrotransformation provides an important method for the construction of isogenic strains. For example, cloned genes can be specifically engineered to incorporate an antibiotic resistance gene within the putative virulence factor structural gene, thus inactivating it. This engineered DNA fragment

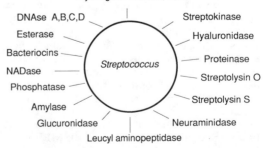

FIG. 1. Extracellular products of group A streptococci.

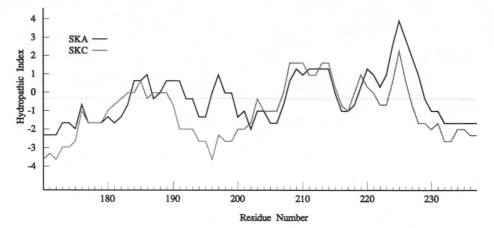

FIG. 2. Hydropathy plot of amino acid residues 173 to 239 of group A streptokinase (SKA) and group C streptokinase (SKC).

can be introduced into a recipient cell by electrotransformation and, following homologous recombination, result in an isogenic strain different from the parental strain by only a single characteristic (Fig. 3). The availability of isogenic *ska* strains is of particular interest in ascertaining the role of SK in the tissue cage model of experimental streptococcal glomerulonephritis described by Holm et al. (11).

Streptolysin O

SLO is produced by the group A, C, and G streptococci and is the toxin primarily associated with the lysis of erythrocytes. It is a potent membrane-damaging agent and acts by the formation of hydrophilic channels in cholesterol-containing cell membranes (2). SLO belongs to a family of related toxins known as thiol-activated toxins; however, recent evidence has shown that the single cysteine found in the protein is not required for cytolytic activity (25).

The streptolysin O gene (*slo*) has been cloned and sequenced by Kehoe and associates (17, 18). *slo* specifies a protein of 538 amino acids preceded by a 33-amino-acid signal peptide. Additional processing of the native SLO appears to occur with removal of an additional 67 amino acids to yield a 53-kDa protein predominantly found in culture supernatants.

The observation that M-protein genes and SK genes both may have considerable sequence heterogeneity prompted us to investigate whether the *slo* gene also showed sequence heterogeneity. A simplified approach to this question was to use the polymerase chain reaction (PCR) and specific oligomeric primers to amplify *slo* from a number of chromosomal DNAs of different M types, including M1, M3, M5, M10, M12, M18, M25, M48, and M49. In all cases a 1.7-kb fragment was obtained, as expected based on the se-

lection of the *slo* oligonucleotide primers. The amplified DNAs were subjected to restriction endonuclease analysis with a number of different enzymes (*Rsa*I, *Taq*I, *Hin*fI, and *Pst*I). The physical maps from all of the amplified *slo* genes except that of M48 were identical to the map previously described by Kehoe et al. (18). Digestion of M48 amplified DNA with either *Hin*fI or *Pst*I gave somewhat different cleavage patterns (Fig. 4). These preliminary findings suggest that *slo* is more highly conserved than some of the other genes represented in strains of different M types.

Hyaluronidase

Hyaluronidase is produced by group A, B, C, G, H, and L streptococci (19) and cleaves hyalu-

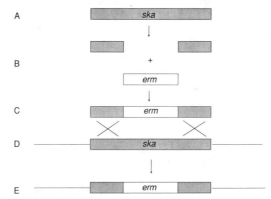

FIG. 3. Construction of a group A *ska* deletion strain. (A) A segment of the *ska* gene is removed. (B) An *erm* gene is ligated into the *ska* deletion site. (C) After being isolated, this engineered DNA fragment is introduced into the recipient strain by electrotransformation. (D) The engineered DNA fragment recombines with the recipient chromosome to form (E) an erythromycin-resistant, streptokinase deletion strain.

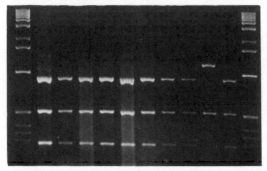

FIG. 4. Agarose gel electrophoresis of PCR-ampli-fied *slo* DNA from different M types after digestion with *Pst*I.

ronic acid to yield the dimer *N*-acetylglucosamine and D-glucuronic acid. It has been identified as a spreading factor since it cleaves hyaluronic acid, a component of the basement membrane of many tissues, allowing its dissemination to mucous membranes and other tissues (5). Two types of hyaluronidase are produced by the group A strep-tococci: a chromosomally specified extracellular protein found in culture supernatants (24) and a phage-encoded enzyme thought to function in the penetration of encapsulated group A organisms (23).

The hyaluronidase gene (*hylP*) of group A bac-teriophage H4489A has been cloned and se-quenced (14). *hylP* specifies a 39.5-kDa protein and, being an integral component of a temperate phage, lacks a signal peptide. Hybridization stud-ies with an internal sequence from *hylP* showed that 54% (196 of 366) of group A strains tested contained a *hylP*. There was no particular asso-ciation with a particular M- or T-type strain among the more than 56 serotype strains tested. Sequence analysis of another *hylP* indicated the presence of substantial heterogeneity at both the nucleotide and amino acid levels. Whether this heterogeneity is due to a *hylP* from a completely different bacteriophage has not yet been deter-mined. Experiments are in progress to obtain the chromosomally determined extracellular hyalu-ronidase.

Proteinase

Proteinase is an extracellular enzyme produced by group A streptococci and is a sulfhydryl pro-tease similar to papain (20). It is secreted as a zymogen (M_r, 42,000) and can be converted to an active proteinase (M_r, 28,000) by reducing agents or proteolytic enzymes (20, 26). Proteinase appears to have the ability to cause general tissue destruction, but its mechanism of action and role in pathogenicity are not clear (7). It is a strong immunogen (28), and Gerlach et al. have pro-

vided evidence that the proteinase precursor or zymogen is identical to erythrogenic toxin B (6).

The proteinase gene from *S. pyogenes* C203S has been cloned in bacteriophage lambda and ex-presses both the precursor and active proteinase forms (Fig. 5). A 407-bp internal fragment, whose deduced amino acid sequence was identical to the sequence reported by Tai et al. (26), was obtained by PCR amplification. Use of this probe in colony hybridization experiments with over 500 clinical strains of group A streptococci showed that all strains possessed the proteinase gene.

Erythrogenic Toxins

Erythrogenic toxins have been described by Dick and Dick as all those substances that cause a skin reaction, regardless of their biological ac-tivity or biochemical nature (4). The erythrogenic toxins have been variously known as scarlet fever toxin, Dick toxin, streptococcal pyrogenic exo-toxin, streptococcal exotoxin, blastogens, and mi-togens; for historical reasons we continue to use the original designation of erythrogenic toxin. Er-ythrogenic toxins have a variety of biological ef-fects, including skin reactivity, pyrogenicity, mi-togenicity, alteration of antibody production and reticuloendothelial clearance function, enhance-ment of endotoxin shock, and cardiotoxicity (1, 7, 32).

Three erythrogenic toxins, A, B, and C, have been described. The type A toxin has been shown to be specified by a gene found in a bacteriophage (15, 33), and when infected into nontoxigenic sus-ceptible hosts it can result in phage conversion to the lysogenic state and the production of ex-tracellular erythrogenic toxin A. The gene spec-

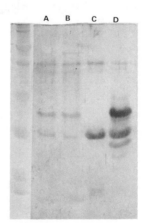

FIG. 5. Western immunoblot of proteinase precursor and proteinase, using antibody to proteinase for detec-tion. Lanes: A and B, proteins from two phage lambda recombinant clones; C, proteinase from strain C203S; D, proteinase precursor and proteinase from strain C203S.

ifying streptococcal erythrogenic toxin A (*speA*) has been sequenced and shown to specify a protein of 221 amino acids (25,787 Da) preceded by a 30-amino-acid signal peptide (34). Erythrogenic toxin A has been shown to have a high degree of homology with staphylococcal enterotoxins B and C1 (34). The gene specifying the type B exotoxin (*speB*) has recently been cloned and specifies a protein of M_r 29,300 (3). The gene specifying the type C exotoxin (*speC*) has been cloned and shown to be part of a bacteriophage. *speC* specifies a mature protein of 208 amino acids (24,354 Da) that is preceded by a 27-amino-acid signal peptide (8). The *speC*-encoded protein was shown to have homology with *speA* gene product.

The role of erythrogenic toxins in effecting streptococcal diseases has been of great interest, and the presence of both *speA* and *speC* on mobile genetic elements, such as bacteriophages, prompted us to examine the association of these genes with clinical strains of group A streptococci related to particular diseases. The knowledge of *speA* and *speC* sequences, as well as the information that the proteinase precursor was identical to erythrogenic toxin B, allowed us to construct specific probes for each of the three exotoxins. Over 500 clinical isolates, obtained from all over the world, were analyzed by colony lift hybridizations with the specific probes under high-stringency conditions. The majority of strains were classified as general group A strains and consisted of strains obtained from patients with tonsilitis, impetigo, cellulitis, pyoderma, abscess, or acute glomerulonephritis. Two other groups of strains were obtained from patients with scarlet fever and rheumatic fever. The results from this analysis are presented in Table 1. *speB* is found in all strains tested, and this information is consistent with the knowledge that all strains elaborate a proteinase. *speC* is present in about 50% of the three groups of strains. However, *speA* is found in only 15% of general strains but in 51 and 52%, respectively, of rheumatic fever- and scarlet fever-associated strains. It would therefore appear that the presence of *speA* is more likely to be associated with severe forms of streptococcal infections, implicating it as an important factor in the virulence or pathogenicity of group A streptococcal infections.

Further analysis of *spe* molecular epidemiology data showed that *speA* was more frequently associated with M1, M3, and M49 strains (35), whereas *speC* was more frequently associated with M2, M4, M6, and M12 strains. Of particular interest here is the high association of *speA* with M1 strains, strains recently associated with a high degree of virulence and mortality.

Discussion

The role of various extracellular products in virulence or in the pathogenesis of group A streptococcal disease is still not clearly defined despite the large amount of information available concerning many of their biological properties and effects in the initiation of tissue lesions (7). The information on the extracellular genes and their proteins discussed here will be supplemented by information about additional genes and products in the near future. Further information on regulation of expression and role in virulence or pathogenicity will be of great importance in understanding the nature of streptococcal disease.

Considerable progress has been made in the genetic dissection of the group A streptococcal chromosome with identification of individual genes and tools to locate their positions on physical maps. Techniques are also available not only for the manipulation and alteration of genes by recombinant DNA technology but also for their reintroduction into cells by electrotransformation. Additionally, the application of DNA hybridization studies and PCR analysis to the identification of organisms and molecular epidemiology will contribute significantly to our understanding of the biology and ecology of this organism. We are now in a position to respond to the challenge of the late Lewis Wannamaker to "apply many of the fascinating tricks produced in the laboratory to epidemiological and clinical studies aimed at determining what really happens in nature" (31).

This work was supported by Public Health Service grant AI19304 from the National Institutes of Health.

ADDENDUM IN PROOF

The nucleotide sequence of streptococcal pyrogenic exotoxin type B (*speB*) and relationship between the toxin and the streptococcal proteinase precursor have recently been reported (A. R. Hauser and P. M. Schlievert, *Infect. Immun.* **172:**4536–4542, 1990).

TABLE 1. Frequency of *speA*, *speB*, and *speC* genes among clinical isolates of group A streptococci

Gene	% Occurrence (no. of isolates)		
	General group A	Scarlet fever	Rheumatic fever
speA	14.9 (47)	52.6 (80)	51.1 (23)
speB	100 (315)	100 (152)	100 (45)
speC	50.8 (160)	48.7 (74)	46.7 (21)

LITERATURE CITED

1. **Alouf, J. E.** 1980. Streptococcal toxins (streptolysin O, streptolysin S, erythrogenic toxin). *Pharmacol. Ther.* **11:**661–717.
2. **Bhakdi, S., J. Tranum-Jensen, and A. Sziegoleit.** 1985. Mechanism of membrane damage by streptolysin O. *Infect. Immun.* **47:**52–60.
3. **Bohach, G. G., A. R. Hauser, and P. M. Schlievert.** 1988. Cloning of the gene, *speB*, for streptococcal pyrogenic ex-

otoxin type B in *Escherichia coli*. *Infect. Immun.* **56**:1665–1667.

4. **Dick, G. F., and G. H. Dick.** 1924. A skin test for susceptibility to scarlet fever. *J. Am. Med. Assoc.* **82**:265–266.

5. **Duran-Reynals, F.** 1942. Tissue permeability and spreading factors in infection. *Bacteriol. Rev.* **6**:197–252.

6. **Gerlach, D., H. Knoll, W. Köhler, J. H. Ozegowski, and V. Hribalova.** 1983. Isolation and characterization of erythrogenic toxins. V. Communication: identity of erythrogenic toxin type B and streptococcal proteinase precursor. *Zentralbl. Bakteriol. Parasitenkd. Infektionskr. Hyg. Abt. 1 Orig. Reihe A* **225**:221–233.

7. **Ginsburg, I.** 1972. Mechanisms of cell and tissue injury induced by group A streptococci: relation to poststreptococcal sequelae. *J. Infect. Dis.* **126**:294–340.

8. **Goshorn, S. C., and P. M. Schlievert.** 1988. Nucleotide sequence of streptococcal pyrogenic exotoxin type C. *Infect. Immun.* **56**:2518–2520.

9. **Goshorn, S. C., and P. M. Schlievert.** 1989. Bacteriophage association of streptococcal pyrogenic exotoxin type C. *J. Bacteriol.* **171**:3068–3073.

10. **Holm, S. E.** 1988. The pathogenesis of acute poststreptococcal glomerulonephritis in new lights. *Acta Pathol. Microbiol. Immunol. Scand.* **96**:189–193.

11. **Holm, S. E., A. M. Bergholm, and K. H. Johnston.** 1988. A streptococcal plasminogen activator in the focus of infection and in the kidneys during the initial phase of experimental streptococcal glomerulonephritis in new lights. *Acta Pathol. Microbiol. Immunol. Scand.* **96**:1097–1108.

12. **Huang, T. T., H. Malke, and J. J. Ferretti.** 1989. Heterogeneity of the streptokinase gene in group A streptococci. *Infect. Immun.* **57**:502–506.

13. **Huang, T. T., H. Malke, and J. J. Ferretti.** 1989. The streptokinase gene of group A streptococci: cloning, expression in *Escherichia coli*, and sequence analysis. *Mol. Microbiol.* **3**:197–205.

14. **Hynes, W. L., and J. J. Ferretti.** 1989. Sequence analysis and expression in *Escherichia coli* of the hyaluronidase gene of *Streptococcus pyogenes* bacteriophage H4489A. *Infect. Immun.* **57**:533–539.

15. **Johnson, L. P., and P. M. Schlievert.** 1984. Group A streptococcal phage T12 carries the structural gene for pyrogenic exotoxin type A. *Mol. Gen. Genet.* **194**:52–56.

16. **Johnston, K. H., and J. B. Zabriskie.** 1986. Purification and partial characterization of the nephritis strain-associated protein from *Streptococcus pyogenes*, group A. *J. Exp. Med.* **163**:697–711.

17. **Kehoe, M., and K. N. Timmis.** 1984. Cloning and expression in *Escherichia coli* of the streptolysin O determinant from *Streptococcus pyogenes*: characterization of the cloned streptolysin O determinant and demonstration of the absence of substantial homology with determinants of other thiol-activated toxins. *Infect. Immun.* **43**:804–810.

18. **Kehoe, M. A., A. L. Miller, J. A. Walker, and G. J. Boulnois.** 1987. Nucleotide sequence of the streptolysin O (SLO) gene: structural homologies between SLO and other membrane damaging thiol-activated toxins. *Infect. Immun.* **55**:3228–3232.

19. **Kohler, W.** 1963. *Die Serologie des Rheumatismus und der Streptokokkeninfektionen*, vol. 3. Aufl. J. J. Bart Verlag, Leipzig, Germany.

20. **Liu, T.-Y., and S. D. Elliott.** 1971. Streptococcal proteinase, p. 609–639. *In* P. D. Boyer (ed.), *The Enzymes*, vol. 3. Academic Press, Inc., New York.

21. **Malke, H., and J. J. Ferretti.** 1984. Streptokinase: cloning, expression, and excretion by *Escherichia coli*. *Proc. Natl. Acad. Sci. USA* **81**:3557–3561.

22. **Malke, H., B. Roe, and J. J. Ferretti.** 1985. Nucleotide sequence of the streptokinase gene from *Streptococcus equisimilis* H46A. *Gene* **34**:357–362.

23. **Maxted, W. R.** 1952. Enhancement of streptococcal bacteriophage lysis by hyaluronidase. *Nature* (London) **170**:1020–1021.

24. **McLean, D.** 1941. The capsulation of streptococci and its relation to diffusion factor (hyaluronidase). *J. Pathol. Bacteriol.* **53**:13–27.

25. **Pinkney, M., E. Beachey, and M. Kehoe.** 1989. The thiol-activated toxin streptolysin O does not require a thiol group for cytolytic activity. *Infect. Immun.* **57**:2553–2558.

26. **Tai, J. Y., A. A. Kortt, T.-Y. Liu, and S. D. Elliot.** 1976. Primary structure of streptococcal proteinase. III. Isolation of cyanogen bromide peptides: complete covalent structure of the polypeptide chain. *J. Biol. Chem.* **251**:1955–1959.

27. **Tillet, W. S., L. B. Edwards, and R. L. Garner.** 1934. Fibrinolytic activity of hemolytic streptococci: development of resistance of fibrinolysis following acute hemolytic streptococcus infections. *J. Clin. Invest.* **13**:47–78.

28. **Todd, E. W.** 1947. A study of the inhibition of streptococcal proteinase by sera of normal and immune animals and of patients infected with group A hemolytic streptococci. *J. Exp. Med.* **85**:591–606.

29. **Walter, F., and H. Malke.** 1989. Nucleotide sequence of the streptokinase gene from *Streptococcus pyogenes* type 1 strain. *Nucleic Acids Res* **17**:1261.

30. **Walter, F., and H. Malke.** 1989. Nucleotide sequence of the streptokinase gene from a group G *Streptococcus*. *Nucleic Acids Res* **17**:1262.

31. **Wannamaker, L. W.** 1982. Transformation, transduction, and bacteriophages, p. 112–116. In D. Schlessinger (ed.), *Microbiology—1982*. American Society for Microbiology, Washington, D.C.

32. **Watson, D. W., and Y. B. Kim.** 1970. Erythrogenic toxin, p. 173–187. *In* T. C. Montie, S. Kadis, and S. J. Ajl (ed.), *Microbial Toxins*, vol. 3. Academic Press, Inc., New York.

33. **Weeks, C. R., and J. J. Ferretti.** 1984. The gene for type A streptococcal exotoxin (erythrogenic toxin) is located in bacteriophage T12. *Infect. Immun.* **46**:531–536.

34. **Weeks, C. R., and J. J. Ferretti.** 1986. Nucleotide sequence of the type A streptococcal exotoxin (erythrogenic toxin) gene from *Streptococcus pyogenes* bacteriophage T12. *Infect. Immun.* **52**:144–150.

35. **Yu, C. E., and J. J. Ferretti.** 1990. Molecular epidemiologic analysis of the type A streptococcal exotoxin (erythrogenic toxin) gene (*speA*) in clinical *Streptococcus pyogenes* strains. *Infect. Immun.* **57**:3715–3719.

Enterococcus faecalis Hemolysin/Bacteriocin

MICHAEL S. GILMORE

Department of Microbiology and Immunology, University of Oklahoma Health Sciences Center, Oklahoma City, Oklahoma 73190

The *Enterococcus faecalis* hemolysin/bacteriocin was initially termed a pseudo-hemolysin (27) because of the ability to detect hemolytic activity on blood agar containing erythrocytes of certain mammalian species but the inability to detect hemolytic activity in supernatants of liquid cultures. The pseudo-hemolysin is a highly evolved cytolysin that lyses both eukaryotic and prokaryotic cells, including those of *E. faecalis* itself; as a result, it presents a significant hazard to the producing bacterium if the activity is not closely regulated. This stringent regulation of hemolysin/bacteriocin activity is achieved through the interaction of a number of gene products and is therefore of considerable interest to geneticists. Additionally, recent evidence has been presented indicating that production of the hemolysin/bacteriocin by strains of *E. faecalis* correlates with severity of infection. Therefore, the *E. faecalis* hemolysin/bacteriocin is also of interest to medical microbiologists.

Historical Perspective

One of the most significant contributions to the study of streptococcal hemolysins was the landmark work of Todd (27) in 1934. Although composing only a small part of this study, a number of important observations were made on the nature of the hemolysin produced by group D streptococci. Among these observations, as cited above, Todd noted that the hemolysin was not detectable in filtrates derived from cultures grown in standard media used for streptococcal cultivation. However, a " . . . satisfactory haemolytic filtrate . . . " using " . . . horse flesh infusion . . . " buffered to pH 8.0 was obtained (27). Using these filtrates, Todd determined that the hemolysin was labile at pH 6.6 when incubated at 37°C for 30 min. However, if the filtrates were adjusted to pH 8.0, measurable hemolytic activity was detected after 1 h in boiling water. Further, it was noted that the hemolysin was oxygen stable, lysed erythrocytes slowly, was not neutralized by antistreptolysin O antiserum, and was nonantigenic. Kobayashi (19) subsequently observed that erythrocytes from different mammalian species exhibited different degrees of sensitivity to the enterococcal hemolysin; human, horse, cow, and rabbit erythrocytes were the most sensitive, and

sheep and goat erythrocytes were comparatively resistant.

In contrast to the observations of Todd, Irwin and Seeley (17) in 1958 observed that the group D streptococcal hemolysin exhibited a half-life of approximately 12 min when shaken. Furthermore, no hemolytic activity could be detected when culture supernatants at pH 7.0 were heated to 96°C for 1 min. On the basis of these observations, Irwin concluded that the hemolysin under study differed from Todd's. In retrospect, most of the observed differences could be attributable to differences in techniques applied. Like Todd, Irwin noted that the enterococcal hemolysin in culture supernatants was only transiently detectable during growth and required much greater quantities of culture fluid for detection than did streptolysin S (17). Additionally, Irwin observed that the rate of hemolysis could be enhanced by inclusion of certain salts and that media supplemented with large amounts of DNA promoted enhanced hemolysin production.

In 1949, Sherwood et al. (25) reported that five of eight beta-hemolytic group D streptococci studied produced "antibiotic substances"; however, a direct connection between hemolysin production and bacteriocin activity was not noted. Stark (26) observed in 1960 that each of 16 beta-hemolytic strains of *Streptococcus faecalis* subsp. *zymogenes* (including 2 type strains and 14 clinical isolates) produced an antibiotic active against some strains of *S. pyogenes*, all strains of *S. pneumoniae*, and all strains of *Clostridium aedematiens*, *C. welchii*, and *C. septicum* but not against gram-negative bacteria, *C. tetani*, or *C. bifermentans*. It was further observed that the antibiotic activity in culture fluids was destroyed by heating to 60°C for 1 h.

In a study of 99 enterococcal isolates, Brock et al. (3) observed that over 50% produced some bacteriocin activity, including all strains of the hemolytic *S. faecalis* subsp. *zymogenes*. The *S. faecalis* subsp. *zymogenes* bacteriocin was observed to have a particularly broad spectrum of action, inhibiting growth of all gram-positive strains tested (except for *Bacillus polymyxa* [and *B. subtilis*, which was only partially sensitive]) but none of the gram-negative bacteria tested. The bacteriocin was resistant to *S. liquefaciens* protease but sensitive to chloroform. Hemolysin secreted into the surrounding agar could be heated to 80°C for 10 to 20 min without detectable loss of activity. Interestingly, hemolytic isolates of the group D streptococcus *S. durans* were not observed to be bacteriocinogenic.

In 1963, Brock and Davie (2) reported on the probable identity of the hemolysin and bacteriocin activities produced by *S. faecalis* subsp. *zymogenes*. Mutants derived by repeated exposure to UV radiation were observed to simultaneously lose the ability to produce substances lytic to blood cells and bacteria. Moreover, an *S. faecalis* subsp. *zymogenes* variant that produced low levels of hemolysin and bacteriocin was restored to proficiency in production of both by similar exposure to UV light. Hemolytic and bacteriolytic activities were only transiently detectable in culture supernatants, and peaks for both activities coincided with mid-log phase. These activities were rapidly lost by incubation of culture supernatants at 37°C but, as mentioned above, were stable to heat when secreted in an agar matrix. Additionally, both activities were inhibited by lecithin. Brock and Davie (2) postulated that this hemolysin/bacteriocin contributes to disease by disrupting resident bacteria and combating host defenses. They observed that the hemolysin/bacteriocin-producing strain displaced sensitive *S. faecalis* in coculture, even when initially present in 100-fold-lower concentration.

These investigators later conducted studies to determine the mechanism of action and nature of resistance of the producing cell to the lytic effects of the hemolysin/bacteriocin. It was observed that spontaneous mutants of *E. faecalis* exhibiting intermediate resistance could be identified by plating them in soft agar on a lawn of hemolysin/bacteriocin-producing cells (6). Hemolysin/bacteriocin-producing strains, spontaneous intermediate-resistance mutants, and wild-type sensitive *E. faecalis* demonstrated a gradient of sensitivity to the hemolysin/bacteriocin both as intact cells and as spheroplasts. Moreover, the same patterns of resistance were observed when these spheroplasts were exposed to streptolysin S. These findings suggested that the cell wall played little if any role in sensitivity or resistance of the *E. faecalis* hemolysin/bacteriocin and that the mechanism of action was similar for the hemolysin/bacteriocin and streptolysin S. However, since all cells were resistant to the lytic effects of streptolysin S when possessing intact cell walls, it was suggested that the hemolysin/bacteriocin of *E. faecalis* may be smaller. In further characterizing the mechanism of resistance, Davie and Brock (7) characterized the inhibition of hemolysin/bacteriocin activity that occurred in late-log-phase liquid cultures. The inhibitor was observed to be cell associated until after mid-log phase, and it bound DEAE-cellulose, suggesting

that it was negatively charged. In the presence of the inhibiting substance, the lysin exhibited a half-life of 5 min at 37 or 46°C. However, when the inhibitor was removed from solution by adsorption to DEAE-cellulose added to the culture medium, the lysin had a half-life of 2 h at 40°C. An inhibiting D-alanyl ribitol teichoic acid was isolated from hemolysin/bacteriocin-producing strains as well as the intermediate-resistance mutant but not from sensitive strains. It was therefore concluded that resistance of hemolysin/bacteriocin-producing *E. faecalis* resulted from the modification of teichoic acids.

The Bicomponent Nature of the *E. faecalis* Cytolysin

Basinger and Jackson (1) further characterized the physical properties and lysis kinetics of the *E. faecalis* hemolysin/bacteriocin (or, more generally, cytolysin). These investigators observed that cell membranes were effective in inhibiting target cell lysis. Moreover, it was found that bacterial membranes were about 3 orders of magnitude more effective than erythrocyte ghosts. Target cell lysis was characterized by a lag with essentially no lysis followed by a sigmoidal lysis curve. Species-related erythrocyte sensitivity affected primarily the length of the lag phase, as did varying the concentration of lysin. Sensitive *E. faecalis* were found to become resistant to the lysin with increasing age of the culture. In a finding similar to those of previous reports, a D-alanyl glycerol teichoic acid from an *S. pyogenes* strain was observed to inhibit lysis. DNase was also found to inhibit lysis, but not as a result of enzymatic activity since inhibition occurred in the presence of EDTA. The authors cited unpublished results in advancing the concept that the lysin consisted of two separate components.

Results suggesting that the *E. faecalis* lysin consisted of two dissimilar components were published in 1969 by Granato and Jackson (11). Noncytolytic mutants of the cytolysin-producing strain X-14 were derived by two cycles of exposure to nitrosoguanidine. It was observed that hemolysis occurred in zones between certain pairs of noncytolytic mutants. Characterization of the kinetics of interaction between culture supernatants of these two classes of mutants revealed that hemolysis was linearly dependent on the concentration of one component, but a hyperbolic optimum in the concentration of the other component existed, either side of which hemolytic activity was reduced. On the basis of these observations, Granato and Jackson (11) termed the former component L (since it behaved as a precursor of the lytic substance) and the latter component A (since it appeared to have an activator role). It was further noted that component L was small and entered a Bio-Gel P-30 column, whereas component A appeared to be large and was essentially excluded from Bio-Gel P-200. We have observed that the active lysin passes through ultrafiltration membranes having a nominal molecular size cutoff of 8,000 Da (unpublished observations).

These investigators subsequently further characterized components A and L. Component A was purified by gel filtration and ion-exchange chromatography and finally eluted from a nondenaturing gel (12). It was found by gel electrophoresis to have a molecular size of 27,000 Da and exhibited a net negative charge at neutral pH. Exclusion of component A from Bio-Gel P-200 was therefore thought to result from aggregation. Component L was similarly purified but was observed to bind DEAE considerably less avidly (13). Component L exhibited a molecular size of 11,000 Da. No phosphatase, hexose, or lipid was associated with either component.

Further research suggested that exposure of cytolysin-producing E. faecalis to sublethal levels of bacitracin, vancomycin, D-cycloserine, or phosphonomycin resulted in rapid loss of cytolytic activity (29). It was later reported that this inhibition resulted from production of phosphatidylserine (J. M. Werth and R. W. Jackson, Abstr. Annu. Meet. Am. Soc. Microbiol. 1973, P25, p. 145).

Cytolysin Genes Are Typically Encoded by Plasmids

Transfer of E. faecalis cytolysin genes was reported in the early 1970s (8, 18, 28). Jacob et al. (18) observed that strains JH1 and JH3 were cured of the hemolysin/bacteriocin phenotype at rates of about 0.5% after overnight culture at 37 and 45°C, indicating plasmid linkage. They therefore mixed JH1 or JH3 with plasmid-free recipients in broth culture. Cytolytic exconjugants were recoverable within 4 h at a rate of 10^{-1} to 10^{-2} per donor. Bacteriocin resistance was observed to transfer with the cytolysin determinant, and both correlated with the acquisition of a 58-kb plasmid. This plasmid could be further transferred to plasmid-free recipients, demonstrating that no specific host factors were required. Noncytolytic derivatives of JH1 that maintained the plasmid and exhibited resistance to the bacteriocin were observed at a relatively high frequency. It was suggested that these plasmids had suffered a mutation in one or both genes encoding the cytolysin components. Unlike that reported by Granato and Jackson (11), no complementation was observed between any of these mutants, suggesting either that production of both components was simultaneously affected or that all mutants represented a single complementation class. One noncytolytic mutant was observed to maintain bacteriocin resistance in the absence of detectable extrachromosomal DNA.

Dunny and Clewell (8) made similar observations on the transmissible nature of the cytolysin determinant. It was observed that recipients were killed during prolonged incubation of the mating mixtures which enriched the population for exconjugants that had acquired the cytolysin plasmid. It was further noted that the large, transmissible cytolysin-encoding plasmid was capable of mobilizing a smaller antibiotic resistance plasmid which itself was incapable of transfer. Clewell et al. (4) observed that transposon integration into one region of pAD1 that included EcoRI fragments F, H, and D resulted in a noncytolytic phenotype. Similar observations were made for Tn3701 insertions into EcoRI fragments E and G of the cytolysin plasmid pIP964 (23).

LeBlanc et al. (22) demonstrated that cytolysin plasmids harbored by E. faecalis isolates of diverse origin shared extensive nucleotide sequence identity, including the region shown to be associated with cytolysin expression. These plasmids included pAD1 and pJH2 (the latter studied by Jacob et al. [18]). It has also recently been demonstrated that pAD1 and pX98 (a plasmid harbored by a strain observed to be phenotypically identical to X-14 [1]) share extensive identity through the cytolysin determinant (18a). Cytolysin plasmids compose a single plasmid incompatibility group (5).

Association between Cytolysin Phenotype and Bacterial Virulence

Ike et al. (15) demonstrated that the 50% lethal dose of E. faecalis in a mouse intraperitoneal model correlated with the cytolytic phenotype. Cytolytic E. faecalis were about an order of magnitude more toxic than isogenic strains harboring a transposon-inactivated pAD1 cytolysin determinant. Moreover, toxicity was observed to double again when an isogenic strain harboring a transposon insertion that elevated plasmid copy number and increased cytolysin expression was tested. We have recently made similar observations with use of a physiological model for endophthalmitis (B. D. Jett, Abstr. 3rd Int. Am. Soc. Microbiol. Conf. Streptococcal Genet., B/27, 1990; unpublished data). Isogenic strains differing only in the location of the Tn917 insertion within or adjacent to the cytolysin determinant of plasmid pAD1 were tested. Cytolytic strains caused a significantly more rapid and severe endophthalmitis than did those harboring a transposon-inactivated cytolysin determinant, as measured by electroretinography and slit lamp biomicroscopy. Histopathologically, little retinal architecture could be discerned in thin sections from animals infected for 72 h with 10^1 to 10^4 cytolytic E. faecalis. In contrast, animals infected for a similar period with identical numbers of noncytolytic transposon insertion mutants retained approximately 24% of retinal function as measured by electroretinography.

Ike et al. (16) presented epidemiological evidence that cytolysin synthesis correlated with enterococcal disease. In this study it was observed that 60% of *E. faecalis* isolates from parenteral infections were hemolytic, compared with 17% for strains isolated from feces of healthy volunteers. Moreover, hemolysin production was associated with a high frequency of antibiotic resistance, although the two were not genetically linked. It was observed that a significant proportion of the hemolysin determinants did not transfer in conjugation experiments, suggesting that this trait may also be chromsomally borne or carried by nonconjugative plasmids. We have made similar observations in retrospective studies of clinical *E. faecalis* isolates from the University of Wisconsin Hospital and Clinics and the Veterans Hospital, Oklahoma City (M. M. Huycke, C. A. Spiegel, and M. S. Gilmore, submitted for publication). In these studies, approximately 40% of blood culture isolates were hemolytic, and nearly all of these were resistant to aminoglycosides. It was observed that the hemolytic phenotype was associated with an acutely terminal outcome. The outcome of the infection was independent of the mode of treatment. These observations suggest that the cytolysin plays a role in the pathogenesis of *E. faecalis* disease and that the poor clinical outcome was not the result of antibiotic resistance.

It has recently been observed that the *E. faecalis* cytolysin inhibits growth of a number of oral streptococci, including those known to contribute to dental caries formation as well as those thought to play a protective role in colonizing the gingival crevice (Jett and Gilmore, in press). Since *E. faecalis* colonizes the gingival crevice and since gram-negative bacteria are uniformly resistant to its effects, it was speculated that cytolytic *E. faecalis* may precipitate a shift in gingival crevice colonization pattern from predominantly gram-positive flora to one containing predominantly gram-negative anaerobes, as is seen in periodontal disease (Jett and Gilmore, in press).

Analysis of the Cytolysin Determinant

In a collaborative study, the cytolysin determinant of pAD1 was extensively mutagenized by Tn917 insertion (14). The location of Tn917 insertion in nonhemolytic pAD1 derivatives was mapped and found to include insertions in *Eco*RI fragments F and H, the fragment F-proximal portion of fragment C, and the fragment H-proximal portion of fragment D (Fig. 1). By making cross-streaks of the various mutants, two complementation classes were observed: those that produced only component L and those that produced exclusively component A. No insertion mutations were found that abrogated expression of both components. Mutants expressing only component A possessed transposon insertions over an approximately 5-kb span of the cytolysin determinant, in sharp contrast to the small molecular size (11,000 Da) reported for component L. A single insertion was observed to affect component A expression. Importantly, this component A-deficient mutant was also observed to be sensitive to lysis by the activated cytolysin, demonstrating that the activator and immunity were linked, if not activities of the same gene product. The lack of observed polar effects of transposon insertion on downstream functions suggested that both components L and A could be independently transcribed. On the basis of this localization, the cytolysin determinant was reconstituted from restriction fragments in *Escherichia coli*, in which expression of the hemolytic phenotype was observed. Deletion and complementation analysis revealed that the component A gene was confined to approximately 1.5 kb of *Eco*RI fragment D near the fragment D-fragment I junction (Fig. 1; M. C. Booth, D. A. Morales, and M. S. Gilmore, *Abstr. 3rd Int. Am. Soc. Microbiol. Conf. Streptococcal Genet.*, B/26, 1990; unpublished data). Both components were individually expressed in *E. coli*.

Nucleotide sequence analysis of the approximately 8-kb region of pAD1 observed to be required for cytolysin expression was initiated with the relatively small portion found to encode component A. A single reading frame was found to be contained within the region of *Eco*RI fragment D shown to be necessary and sufficient for component A expression (14; Booth et al., *Abstr. 3rd Int. Am. Soc. Microbiol. Conf. Streptococcal Genet.*, 1990). Analysis of the inferred amino acid sequence for component A revealed a protein highly related to the subtilisin class of serine proteases (Booth et al., *Abstr. 3rd Int. Am. Soc. Microbiol. Conf. Streptococcal Genet.*, 1990). Identical or conserved amino acids occur at 56% of the amino acid positions, with identities clustered around the amino acids that constitute the catalytic triad and the oxyanion hole. Evidence supporting the inferred proteolytic function of component A was obtained by demonstrating that component A is inhibited by diisopropylfluorophosphate and other inhibitors of serine proteases. Alignment of the component A amino acid sequence with that for subtilisin BPN' suggests that component A is initially synthesized with a conventional signal peptide and is secreted as a zymogen.

We have observed that component A is transiently associated with the cell wall, has an isoelectric point of 4.5, and avidly binds DEAE-cellulose (Booth et al., *Abstr. 3rd Int. Am. Soc. Microbiol. Conf. Streptococcal Genet.*, 1990). It

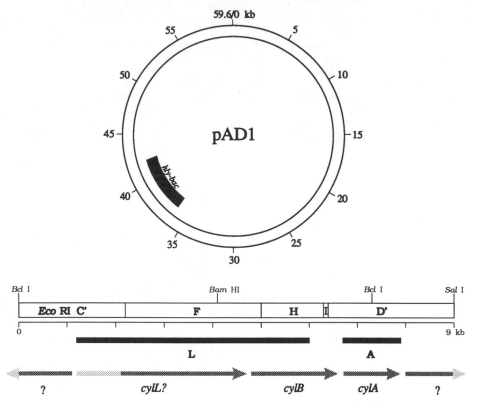

FIG. 1. Genetic organization of the pAD1 hemolysin/bacteriocin determinant. (Top) Schematic representation of pAD1 and location of the hemolysin/bacteriocin determinant (4). (Bottom) Expanded view of the hemolysin/bacteriocin determinant and flanking regions. The cleavage sites that define *Eco*RI fragments F, H, and I are indicated, as are sites for *Bcl*I, *Bam*HI, and *Sal*I recognition. Regions of the determinant associated with expression of components L and A (as determined by transposon insertional mutagenesis [14]) are indicated in black. Deletion analysis and nucleotide sequence determination localized the component A-encoding reading frame (*cylA*). An auxiliary function (CylB) required for component L expression is encoded immediately 5' to *cylA*. An open reading frame 5' to *cylB* originates in the *Eco*RI F fragment-proximal region of *Eco*RI fragment C. Insufficient nucleotide sequence and supporting biological data exist to unambiguously ascribe a function to this reading frame. This reading frame, designated *cylL?*, may encode a very large primary translation product precursor to component L. Alternatively, the component L structural gene may occur immediately 5' to this large reading frame (light shaded area) in an area of the determinant rich in secondary structure that remains to be sequenced. The hemolysin/bacteriocin determinant is flanked by reading frames of unknown function (?) that appear to be unrelated to hemolysin/bacteriocin expression, since Tn917 insertions in these regions (14) and frameshift mutations introduced in vitro (data not shown) into these reading frames do not detectably affect the levels of hemolysin/baceriocin secreted or immunity to its lytic effect.

was therefore of interest to determine whether the acidic side chains contribute to the observed transient cell association. The carbon skeleton of subtilisin BPN', which has been characterized at the angstrom level, was used as a starting point for modeling component A. Subtilisin BPN' surface residues were substituted for those predicted by alignment to be equivalent in position on component A. Negatively charged surface residues compensated by adjacent positively charged residues were masked, and the structure was examined for localized patches of negative charge. All uncompensated negative charges were in fact localized to one predicted face of the protein,

suggesting that charge may contribute to component A association and orientation on the cell wall.

A second open reading frame initiates immediately 3' to the component A reading frame and reads through the *Sal*I recognition site (Booth et al., *Abstr. 3rd Int. Am. Soc. Microbiol. Conf. Streptococcal Genet.*, 1990). Its relationship to cytolysin expression or immunity is unknown. This reading frame is not necessary for cytolysin or component A expression in *E. coli*, and transposon insertion into this reading frame (e.g., pAM9054 [14]) does not detectably affect cytolysin expression or immunity in *E. faecalis*. However, the lack of intergenic space suggests some role associated with the biology of the cytolysin for this extended reading frame.

Immediately 5' to the component A gene (into the region mapped by deletion and transposon insertion analysis to be related to component L expression) is a third open reading frame (9a; R. A. Segarra and M. S. Gilmore, *Abstr. 3rd Int. Am. Soc. Microbiol. Conf. Streptococcal Genet.*, B/15, 1990). This open reading frame encodes a protein of 714 amino acids. Comparison of the inferred amino acid sequence with sequences contained in the NBRF data base revealed that this component L-associated function was highly related to the HlyB protein of the *E. coli* α-hemolysin operon. Conserved or identical amino acids occur at 70% of the positions throughout the length of this reading frame, which is termed *cylB*. HlyB has been shown to form a transport channel through the *E. coli* cytoplasmic membrane, specific for transport of the *E. coli* α-hemolysin (10, 20). The observed similarity between HlyB and CylB suggested a similar role in secretion of component L. To demonstrate that CylB was essential for component L expression and that observed transposon insertional inactivation and associated L-deficient phenotype did not result from potential polar effects on transcription or translation of downstream genes, an insertion mutation was introduced into the *cylB* reading frame. The *cylB* gene was cleaved within the unique *Afl*III recognition sequence and the cohesive ends were filled, resulting in the addition of four nucleotides. An 8-bp linker encoding a *Cla*I recognition sequence was inserted between the blunt ends, resulting in the addition of 12 nucleotides of known sequence. This addition added four codons to the *cylB* reading frame inserting Met, His, Arg, and Ser into the nonpolar central region of CylB. When the mutagenized *cylB* reading frame was reinserted into the cloned cytolysin determinant, the resulting phenotype was L⁻ A⁺. The hemolytic phenotype was restored by introducing the cloned wild-type *cylB* gene on a compatible vector in *trans*. Interestingly, this lesion was not complemented by HlyB expressed in a similar arrangement.

The C-terminal region most highly conserved between CylB and HlyB also shares homology with a number of other prokaryotic and eukaryotic ATP-binding transport proteins (Segarra and Gilmore, *Abstr. 3rd Int. Am. Soc. Microbiol. Conf. Streptococcal Genet.*, 1990; Gilmore et al., submitted). These include the multidrug resistance and cystic fibrosis proteins of eukaryotic cells and MalK and related proteins in prokaryotic cells. A region of these proteins is similar to the domain of adenylate kinase, characterized at high resolution, which binds the α-phosphoryl group of ATP and is shared by *ras* p21 GTPase

(9). However, all of the transport proteins related to CylB possess a serine residue at a position that in *ras* p21 abrogates nucleotide hydrolysis but not binding and results in oncogenic transformation of the cell (10, 24). This finding suggests that members of the family of ATP-binding transport proteins are incapable of hydrolyzing nucleotides except when conformational changes are induced by binding of the transport substrate. Nucleotide hydrolysis would therefore be predicted to occur simultaneously with substrate transport. The eukaryotic multidrug resistance protein (P-glycoprotein) and the cystic fibrosis protein contain tandem duplications of *cylB*-like reading frames. This information further suggests that CylB functions in the membrane as a dimer or multimer of even number.

Immediately 5' to the *cylB* reading frame is a fourth gene that would be predicted to encode a protein of molecular size greater than 100 kDa and consumes most of the remaining region of the determinant known to be associated with expression of component L. It is unknown whether this region encodes a very large precursor for the relatively small component L, and if so what additional functions might be associated with it. On the basis of analogy with the known function of HlyB, component L would be predicted to contain a recognition sequence for CylB-mediated transport, potentially at its C terminus. However, such a sequence as it occurs on the transport substrate for HlyB (HlyA) consists of only about 50 amino acids (21). Speculatively, additional roles for such a large precursor may include autoregulated shutoff of component L expression under conditions in which ATP levels required for secretion are low, thereby preventing intracellular accumulation of component L. No extensive identity between this open reading frame and sequences within the NBRF data base has been found. Approximately 800 bp of the cytolysin determinant remains to be sequenced, which would be sufficient to separately encode an additional small protein such as component L.

Conclusion

The *E. faecalis* cytolysin is encoded by a cluster of genes on pAD1, some of which encode auxiliary functions. Expression of the cytolysin may be regulated by the relative timing of expression of component A and L genes, regulation of component L secretion by ATP availability, or regulation of the release of component A from the cell wall. Because of the ease of detection of cytolytic activity on a solid agar matrix, compared with the difficulty of detecting cytolytic activity in liquid culture, it appears that expression of the cytolysin was optimized through evolution for growth in solid or semisolid media. Component A appears to contribute substantially to immunity to the cytolysin as well as play an activator role. This speculation is supported by several observations; an excess of component A inhibits hemolysis, component A deficiency results in sensitivity to the lysin, and transposon insertions into component A flanking reading frames do not affect cell sensitivity to the lysin. Since component A appears to be a serine protease, related structurally to subtilisin, the dual activator and immunity roles for component A may be explained in a larger hypothetical scheme for expression of the cytolysin determinant as follows. Component L may be transcribed and translated prior to component A (since they appear to be transcribed from separate promoters) or simultaneously and separately secreted by different mechanisms (CylB transport for component L or conventional signal peptide-mediated secretion for component A). Component L would then freely and, because of its small size, rapidly diffuse through the surrounding solid or semisolid medium, whereas component A is transiently retained at the cell wall. Expression and secretion of component L would cease when intracellular ATP levels drop below a threshold value required for CylB-mediated transport, possibly through autogenous feedback regulation. Component A would slowly be released into the surrounding medium and begin activating component L by limited proteolysis at relatively high component L/component A ratios. Component A, because of its delayed release from the cell wall, would remain in relative excess in proximity to the nascent *E. faecalis* colony, potentially cleaving component L at a low-affinity binding site, resulting in its inactivation and hence depletion of active lysin from the local environment. In liquid culture, an imbalance in this diffusion-dependent activation occurs, resulting in rapid hydrolysis of component L to the inactive proteolytic breakdown product. This notion is supported by the observed short half-life of the active lysin in culture fluids at 37°C but its temperature stability when rapidly heated to 98°C or when plates are heated to 80°C.

We are in the process of testing this model for expression and activation of the *E. faecalis* cytolysin. These studies should provide information of interest to geneticists, to medical microbiologists, and potentially to scientists engaged in such diverse research fields as cystic fibrosis and the mechanisms by which cancerous cells achieve resistance to multiple cytotoxic agents.

LITERATURE CITED

1. **Basinger, S. F., and R. W. Jackson.** 1968. Bacteriocin (hemolysin) of *Streptococcus zymogenes. J. Bacteriol.* **96:**1895–1902.
2. **Brock, T. D., and J. M. Davie.** 1963. Probable identity of a group D hemolysin with a bacteriocine. *J. Bacteriol.* **86:**708–712.
3. **Brock, T. D., B. Peacher, and D. Pierson.** 1963. Survey of the bacteriocines of enterococci. *J. Bacteriol.* **86:**702–707.
4. **Clewell, D. B., P. K. Tomich, M. C. Gawron-Burke, A. E. Franke, Y. Yagi, and F. Y. An.** 1982. Mapping of *Streptococcus faecalis* plasmids pAD1 and pAD2 and studies relating to transposition of Tn*917. J. Bacteriol.* **152:**1220–1230.
5. **Colmar, I., and T. Horaud.** 1987. *Enterococcus faecalis* hemolysin-bacteriocin plasmids belong to the same incompatibility group. *Appl. Environ. Microbiol.* **53:**567–570.
6. **Davie, J. M., and T. D. Brock.** 1966. Action of streptolysin S, the group D hemolysin, and phospholipase C on whole cells and spheroplasts. *J. Bacteriol.* **91:**595–600.
7. **Davie, J. M., and T. D. Brock.** 1966. Effect of teichoic acid on resistance to the membrane-lytic agent of *Streptococcus zymogenes. J. Bacteriol.* **92:**1623–1631.
8. **Dunny, G. M., and D. B. Clewell.** 1975. Transmissible toxin (hemolysin) plasmid in *Streptococcus faecalis* and its mobilization of a noninfectious drug resistance plasmid. *J. Bacteriol.* **124:**784–790.
9. **Fry, D. C., S. A. Kuby, and A. S. Mildvan.** 1988. ATP-binding site of adenylate kinase: mechanistic implications of its homology with *ras*-encoded p21, F1-ATPase, and other nucleotide-binding proteins. *Proc. Natl. Acad. Sci. USA* **83:**907–911.
9a. **Gilmore, M. S., R. A. Segarra, and M. C. Booth.** 1990. An HlyB-type function is required for expression of the *Enterococcus faecalis* hemolysin/bacteriocin. *Infect. Immun.* **58:**3914–3923.
10. **Goebel, W., and J. Hedgpeth.** 1982. Cloning and functional characterization of the plasmid-encoded hemolysin determinant of *Escherichia coli. J. Bacteriol.* **151:**1290–1298.
11. **Granato, P. A., and R. W. Jackson.** 1969. Bicomponent nature of lysin from *Streptococcus zymogenes. J. Bacteriol.* **100:**865–868.
12. **Granato, P. A., and R. W. Jackson.** 1971. Characterization of the A component of *Streptococcus zymogenes* lysin. *J. Bacteriol.* **107:**551–556.
13. **Granato, P. A., and R. W. Jackson.** 1971. Purification and characterization of the L component of *Streptococcus zymogenes* lysin. *J. Bacteriol.* **108:**804–808.
14. **Ike, Y., D. B. Clewell, R. A. Segarra, and M. S. Gilmore.** 1990. Genetic analysis of the pAD1 hemolysin/bacteriocin determinant in *Enterococcus faecalis*: Tn*917* insertional mutagenesis and cloning. *J. Bacteriol.* **172:**155–163.
15. **Ike, Y., H. Hashimoto, and D. B. Clewell.** 1984. Hemolysin of *Streptococcus faecalis* subspecies *zymogenes* contributes to virulence in mice. *Infect. Immun.* **45:**528–530.
16. **Ike, Y., H. Hashimoto, and D. B. Clewell.** 1987. High incidence of hemolysin production by *Enterococcus* (*Streptococcus*) *faecalis* strains associated with human parenteral infections. *J. Clin. Microbiol.* **25:**1524–1528.
17. **Irwin, J., and H. W. Seeley.** 1958. Titration and partial characterization of a soluble hemolysin of a group D *Streptococcus. J. Bacteriol.* **76:**29–35.
18. **Jacob, A. E., G. J. Douglas, and S. J. Hobbs.** 1975. Self-transferable plasmids determining the hemolysin and bacteriocin of *Streptococcus faecalis* var. *zymogenes. J. Bacteriol.* **121:**863–872.
18a. **Jett, B. D., and M. S. Gilmore.** 1990. The growth-inhibitory effect of the *Enterococcus faecalis* bacteriocin encoded by pAD1 extends to the oral streptococci. *J. Dent. Res.* **69:**1640–1645.
19. **Kobayashi, R.** 1940. Studies concerning hemolytic streptococci: typing of human hemolytic streptococci and their relation to diseases and their distribution on mucous membranes. *Kitasato Arch. Exp. Med.* **17:**218–241.
20. **Koronakis, V., E. Koronakis, and C. Hughes.** 1988. Comparison of the haemolysin secretion protein HlyB from *Proteus vulgaris* and *Escherichia coli*; site-directed mutagenesis causing impairment of export function. *Mol. Gen. Genet.* **213:**551–555.
21. **Koronakis, V., E. Koronakis, and C. Hughes.** 1989. Iso-

lation and analysis of the C-terminal signal directing export of *Escherichia coli* hemolysin protein across both bacterial membranes. *EMBO J.* **8**:595–605.

22. **LeBlanc, D. J., L. N. Lee, D. B. Clewell, and D. Behnke.** 1983. Broad geographical distribution of a cytotoxin gene mediating beta-hemolysis and bacteriocin activity among *Streptococcus faecalis* strains. *Infect. Immun.* **40**:1015–1022.

23. **Le Bouguénec, C., T. Horaud, C. Geoffroy, and J. E. Alouf.** 1988. Insertional inactivation by Tn*3701* of pIP*964* hemolysin expression in *Enterococcus faecalis*. *FEMS Microbiology Let.* **49**:455–458.

24. **Seeburg, P. H., W. W. Colby, D. J. Capon, D. V. Goeddel, and A. D. Levinson.** 1984. Biological properties of human *c-Ha-ras1* genes mutated at codon 12. *Nature* (London) **312**:71–75.

25. **Sherwood, N. P., B. E. Russell, A. R. Jay, and K. Bowman.** 1949. Studies on streptococci. III. New antibiotic substances produced by beta hemolytic streptococci. *J. Infect. Dis.* **84**:88–91.

26. **Stark, J. M.** 1960. Antibiotic activity of haemolytic enterococci. *Lancet* **i**:733–734.

27. **Todd, E. W.** 1934. A comparative serological study of streptolysins derived from human and from animal infections, with notes on pneumococcal haemolysin, tetanolysin and staphylococcus toxin. *J. Pathol. Bacteriol.* **39**:299–321.

28. **Tomura, T., T. Hirano, T. Ito, and M. Yoshioka.** 1973. Transmission of bacteriocinogenicity by conjugation in group D streptococci. *Jpn. J. Microbiol.* **17**:445–452.

29. **Werth, J. M., and R. W. Jackson.** 1971. Effects of various drugs on hemolytic activity of *Streptococcus zymogenes*. *J. Bacteriol.* **108**:844–848.

C Proteins of Group B Streptococci

JAMES L. MICHEL,[1] LAWRENCE C. MADOFF,[1,2] DAVID E. KLING,[1] DENNIS L. KASPER,[1,2] AND FREDERICK M. AUSUBEL[3]

Channing Laboratory, Department of Medicine, Brigham and Women's Hospital,[1] *and Division of Infectious Diseases, Beth Israel Hospital,*[2] *Department of Medicine, Harvard Medical School, Boston, Massachusetts 02115; and Department of Molecular Biology, Massachusetts General Hospital, Boston, Massachusetts 02114*[3]

Streptococcus agalactiae (group B streptococci [GBS]) is the leading cause of neonatal sepsis and meningitis in the United States (1, 13). Despite advances in understanding the pathogenesis, diagnosis, and management of GBS infections, the incidence of such infections has not changed over the last 15 years (11). The type-specific capsule of GBS has been extensively studied and is an important determinant for both virulence and immunity (28). Recent genetic studies by Rubens et al. (21), Wessels et al. (28), and others have focused on the role of capsule in pathogenesis of GBS. The protein antigens of GBS are less well understood, and genetic studies on a number of these antigens are currently being pursued by us and by workers in several other laboratories (9, 24; J. L. Michel, L. C. Madoff, D. E. Kling, D. L. Kasper, and F. M. Ausubel, *Infect. Immun.*, in press; P. Ferrieri and G. Lindahl, personal communications). We have concentrated on cloning and characterizing the surface-associated C-protein antigens of GBS. The C proteins are thought to play a role in virulence and immunity (12) and have been proposed as a component of a conjugate vaccine against GBS (17). Some biochemical and immunological properties of these proteins have been described (12). However, lack of knowledge on the number, sizes, and biological activities of C proteins has hampered research.

Classification of the C Proteins

GBS are classified by their immunogenic type-specific polysaccharide capsule. Evidence for a new GBS antigen was noted by Wilkinson and Moody (30), who observed a serological cross-reactivity between type Ia and type Ib strains. They found that antisera to capsular type Ib strains cross-reacted with some type Ia strains. Wilkinson and Eagon (29) identified and partially purified two serologically reactive protein determinants. One fraction was susceptible to both trypsin and pepsin (termed trypsin sensitive [TS]); the other fraction was susceptible to pepsin but not trypsin (trypsin resistant [TR]).

Type Ia GBS strains containing these protein antigens were designated type Ic (29), and the TS and TR antigens were called Ibc proteins (16). In 1984 a consensus was reached at an international symposium to categorize GBS by the type-specific polysaccharide capsule and call the Ibc protein antigens C proteins (14). Therefore, capsular type Ia strains carrying the C-protein antigens, formerly classified as type Ic, are now designated type Ia/c.

C proteins are found on all type Ib, on many type Ia and II, and possibly on type IV strains. However, these proteins are uncommon in strains with the type III capsular polysaccharide. Johnson and Ferrieri (15) tested 785 clinical isolates of GBS and found that 462 (59%) expressed the TR or TS antigen (or both) of the C proteins. Using a two-stage radioimmunoassay, Brady et al. (8) looked at the antigenic determinant identified by the C-protein typing sera used by the Centers for Disease Control. They identified four unique antigens, two of which correspond to the C proteins TR and TS. C-protein-positive strains of GBS can carry one or more of the individual C-protein antigens. However, the number of individual genes or unique polypeptides that define the C-protein antigens is still not known.

Biochemical Characterization of the C Proteins

Bevanger and Maeland (5) raised antisera to type Ib and type Ia/c strains and identified two C-protein antigens, termed α and β. The α antigen corresponds to the trypsin-resistant protein, and the β antigen is trypsin sensitive. Techniques for extracting and purifying the C proteins have yielded variable results. These proteins have been extracted from whole cells by using hot HCl (6, 22, 29) and detergents such as sodium dodecyl sulfate (SDS) (23) and Triton X-100 (22) and more recently by enzymatic extraction using mutanolysin (16a). C proteins have been identified on the surface of GBS and in the cytoplasm of some strains (10). In addition, C-protein antigens can be isolated from the GBS culture supernatants (25). Bevanger and Iversen (4) used hot HCl to extract these antigens, which were partially purified by ion-exchange chromatography and isoelectric focusing. The molecular sizes reported for the α antigen range from 14 to more than 190 kDa (4, 9, 16a, 25; Michel et al., in press); there is presently no agreement as to the size of the α antigen. The molecular sizes and the biochemical and immunological properties of C proteins, in-

TABLE 1. Properties of C proteins of group B streptococci

Protein	Molecular size (Da)	Protease susceptibility (5, 29; Michel et al., in press)		Binding to IgA	Ability to elicit protective immunity
		Pepsin sensitive	Trypsin sensitive		
C protein	14,000 (25), 20,000–130,000 (22)				+ (6, 16, 25)
α Antigen	30,000–160,000 (16a), 40,000–116,000 (Michel et al., in press), 75,000 (4), 190,000 (9)	+	−	No (9; Michel et al., in press)	+ / − (6; Michel et al., in press)
β Antigen	23,000–"several hundred thousand" (4), 38,000–145,000 (7), 110,000–130,000 (9, 16a, 23; Michel et al., in press)	+	+	Yes (7, 9, 10, 22; Michel et al., in press)	+ (6; Michel et al., in press)

cluding the α and β antigens, are summarized in Table 1.

Bevanger and Iversen (4) initially described the molecular size of the β antigen as several hundred thousand daltons, with subunits of 70, 45, and 23 kDa that were shown to have antigenic activity. Russell-Jones and Gotschlich (22) used HCl and Triton X-100 extractions and analyzed the size of the protein antigens on Western immunoblots of SDS-polyacrylamide gel electrophoresis (PAGE). Probing with both polyclonal rabbit typing antisera raised against whole organisms and mouse monoclonal antibodies raised against partially purified C proteins, they identified numerous bands ranging in molecular size from 20 to 130 kDa. Brady and Boyle (7) described a heterologous family of β antigens ranging up to 145 kDa. They found low-molecular-size forms of the β antigen at 55, 53, and 38 kDa. Cleat and Timmis (9) isolated a clone of GBS DNA expressed in *Escherichia coli* that produces a single protein species with β-antigen properties and a molecular size of 130 kDa. However, multiple-molecular-size species were seen on Western blots of GBS probed with murine monoclonal antibodies to the C proteins (P. Ferrieri, A. E. Flores, and J. Nelson, in press).

Protective C-Protein Antigens

In 1975, Lancefield et al. (16) raised antibodies in rabbits to Formalin-killed GBS expressing C proteins. They showed that these antibodies provide passive protection to mice challenged with strains of GBS that carry the C proteins. The passive protection seen in the mouse model did not extend to strains that do not express the C proteins. The ability of the C proteins to elicit protective immunity is a key property defining their biological function. This finding suggested

that the C proteins may have a role in natural immunity to GBS infection.

Bevanger (3) used an enzyme-linked immunosorbent assay to screen human sera for antibodies to the C proteins. He found that antibodies to the β antigen are present in a large proportion of the healthy population. Convalescent sera from an elderly patient following septicemia with an α-antigen-producing strain showed a rise in the titer of antibodies to the α antigen. Bevanger and Naess (6) raised antiserum in rabbits to the partially purified α and β antigens. Using a passive protection mouse assay, they found that animals challenged with strains carrying both the α and β antigens were protected by antiserum to the β antigen but not by antiserum directed against the α antigen. Antiserum against each of the antigens protected against challenge with GBS carrying the homologous antigen. However, mice were not protected when challenged with strains carrying only the heterologous antigen. In addition, antisera to the C proteins did not protect animals challenged with strains that do not express C proteins.

Valtonen et al. (25) isolated C proteins from the supernatants of GBS cultures. They identified a 14-kDa protein that elicits protective immunity in the mouse model. Antisera to this 14-kDa GBS protein cross-reacted with a number of larger proteins. The immunological cross-reactivity of this small C protein suggests that it may represent shared epitopes, possibly from a gene family. Alternatively, this protein may represent the secreted or degradation product of a larger C-protein precursor. The 14-kDa supernatant protein is the smallest C protein species found to carry a protective epitope.

It has been shown that antibody to the type-specific capsule of GBS elicits protective immunity that is clinically important (1). However, the role of immunity to the C proteins of GBS in preventing or modulating disease in humans is not known. Bevanger (3) and others (Ferrieri et al., in press) have identified antibodies to the C proteins in human sera. In addition, profiles of antibodies to the C proteins show a high level of antibody in mothers colonized with GBS that express the C proteins. These high levels of antibody to the C proteins are also found in paired sera from their newborn infants (Ferrieri, personal communication).

IgA FcA Binding of C Proteins

In 1984, Russell-Jones et al. (23) reported that a 130-kDa, detergent-extractable C protein binds to the Fc region of human serum immunoglobulin A (IgA) and secretory IgA. This protein represents the trypsin-sensitive β antigen. Smaller (53- and 55-kDa) β antigens that bind the IgA Fc region, as well as a 38-kDa β antigen that does not, have been identified (7). Multiple-size species of β antigen have been found that bind to human IgA, suggesting that there may be more than one IgA binding domain (Ferrieri, personal communication). The presumptive β-antigen clone of Cleat and Timmis (9) expressed a single 130-kDa protein in *E. coli* that bound to IgA on immunoblots.

There are several hypotheses regarding the potential roles of human IgA Fc binding on virulence and immunity. By binding IgA to the cell surface, GBS might be masking other cell surface antigens or blocking binding of other immunoglobulins. Presenting IgA Fab on the surface of GBS may also stimulate production of an anti-idiotype antibody that could bind up specific IgA directed at GBS and prevent IgA from inactivating or opsonizing GBS. The bound IgA would not be available for opsonization of GBS. It is possible that binding IgA Fc might block phagocytosis. The C proteins and other GBS proteins do not appear to bind to either IgG or IgM.

Role in Pathogenesis

As described above, the C proteins of GBS have been defined by their ability to elicit protective antibodies, and the β antigen has been shown to bind the Fc region of human IgA. Several studies have looked at other potential roles for the C proteins in virulence. Payne and Ferrieri (18) studied the C proteins from type II strains of GBS. Expression of C proteins appeared to be one factor contributing to resistance to opsonization of type II GBS. In a subsequent study, they found that C-protein-carrying strains of GBS appeared to resist intracellular killing by phagocytes (19). Genetic approaches to studying virulence properties of the C proteins are discussed below under Future Work.

Genetic Analysis

In 1987, Cleat and Timmis (9) reported the cloning and expression of GBS DNA in *E. coli*. They constructed a gene bank in bacteriophage λ, and immunoblots of recombinants were probed with polyclonal antibodies raised by Bevanger to partially purified α and β antigens (2). Two clones were identified that cross-reacted with the antisera used to screen the library, and they were subcloned into a plasmid vector. One subclone, pPHC10, expressed a 190-kDa protein that cross-reacted with anti-α antiserum. Another subclone, pPHC33, expressed a 130-kDa protein that bound human IgA on immunoblots and cross-reacted with anti-β antiserum. They did not determine whether the gene products expressed by these clones were susceptible to degradation by trypsin or pepsin or whether the cloned antigens were able to elicit protective immunity.

To initiate a genetic analysis of the C proteins of GBS, we prepared a recombinant library of partial restriction endonuclease-cleaved chromosomal DNA from the A909 (type I a/c) strain of GBS in a modified pUC12 plasmid vector (Michel et al., in press). The clones were screened for expression of GBS proteins with antisera prepared against mutanolysin-extracted surface-associated proteins of GBS thought to include the C proteins. These antisera conferred passive protection in a mouse virulence assay against a heterologous capsule type of GBS that expressed the C proteins (16a). Thirty-five positive clones were isolated from the 25,000 *E. coli* transformants screened. Twenty-four positive clones were divided into two distinctive groups by screening with antisera on Western blots and by restriction endonuclease mapping. By Southern blot hybridization, there is no homology between these two groups of clones. Two clones, pJMS1 and pJMS23, that are representative of each group were chosen for further study.

Clone pJMS23 expressed gene products that were immunologically and biochemically similar to the native C-protein α antigen of GBS (16a; Michel et al., in press). The cloned proteins bound to the anti-α antibodies of Bevanger (2). Antibodies to the cloned gene products bound native C-protein α antigen on Western blots and elicited passive protective immunity in a mouse lethality model. Native α antigens bound by antibody to the clone were susceptible to degradation by pepsin but not trypsin. This clone expressed a polymorphic gene product on Western blots when probed with both polyclonal antisera and two monoclonal antibodies.

The cloned proteins produced by pJMS23 showed a broad range of molecular sizes, from 40 to 116 kDa, with a regularly repeating ladder at approximately 8-kDa intervals. These polymorphic gene products were also seen on Western blots of the native GBS C-protein antigens expressed by strain A909 and probed with both polyclonal and monoclonal antibodies. Restriction endonuclease mapping identified an intragenic region with repeating sequences that could be associated with the phenotypic differences in the sizes of the C-protein α antigens of GBS. The restriction map and sizes of the protein products of pJMS23 are different from those of the presumptive α-antigen pPHC10 clone of Cleat and Timmis (9). Clone pPHC10 may express a second α antigen, another C-protein antigen, or a GBS antigen that is not a C protein. Cleat and Timmis (9) did not determine whether pPHC10 expressed an α antigen that was trypsin resistant or was able to elicit protective immunity.

Clone pJMS1 expressed a gene product that was immunologically and biochemically similar to the native C-protein β antigen of GBS (16a; Michel et al., in press). Clone pJMS1 expressed a major protein of 110 kDa that bound to the anti-β-antigen typing sera of Bevanger. However, SDS-PAGE of native A909 proteins demonstrated a 130-kDa protein that reacted with antiserum to pJMS1 on Western blots. On a Western blot using antisera against pJMS1, the native protein was degraded by treatment with both trypsin and pepsin. Both the native A909 protein and the cloned gene product bound to human IgA. Restriction endonuclease mapping of pJMS1 and study of expression by a family of Tn5 transposon mutants indicated that the cloned gene is transcribed from a streptococcal promoter and that at least one IgA binding region is located in the amino portion of the protein.

Rabbit antiserum raised against a lysate of the pJMS1 clone protected mice against lethal challenge with GBS strains that express C proteins but not against challenge with C-protein-negative GBS strains. Therefore, pJMS1 and pJMS23 define two different C-protein genes in GBS whose gene products encode unique protective epitopes.

The gene product of clone pJMS1 may be a truncated version of the pPHC33 clone of Cleat and Timmis (9). Both cloned gene products bind IgA. However, there are several differences in the restriction endonuclease maps between these clones. The molecular size of the pPHC33 product is 130 kDa, which is close to the size of β antigen reported by other investigators (23). In addition, antibody raised against pJMS1 bound a native GBS A909 protein of 130 kDa on Western blots. The numerous molecular sizes reported for the C-protein β antigen may represent variability in gene size or expression, multiple β-antigen genes, or posttranscriptional modification or degradation of the β antigen.

Future Work

The C proteins of GBS are incompletely characterized antigens that have a role in immunity to experimental GBS infection. The availability of clones for some of these genes should open up a variety of genetic techniques for use in characterizing the distribution, regulation, expression, structure and function, and biological role of each of these proteins. Since the C proteins are protective antigens in experimental infections, they may play a protective role in natural immunity to human diseases (3, 6, 12, 16). The protective epitopes of the cloned gene products could be mapped and studied for use in a conjugate vaccine against GBS.

Pairs of isogenic strains of GBS that express only α or β or neither of the antigens would be useful in studying specific differences in virulence related to the C proteins. Wanger and Dunny (26) originally proposed the use of the conjugative transposon Tn916 to study the pathogenesis of GBS. This transposon and a derivative, Tn916ΔE, have been used to create isogenic strains blocking the regulation, expression, or synthesis of putative virulence factors (20, 21, 27, 28). Transposon mutagenesis of the C-protein genes should allow for a more complete genetic and molecular analysis of the role of the C proteins in pathogenesis.

This research was supported by Public Health Service grants AI28500, AI00981, and AI23339 from the National Institute of Allergy and Infectious Diseases, by a grant from Hoechst AG to the Massachusetts General Hospital, and by the William Randolph Hearst Fund at Harvard Medical School. J.L.M. is a recipient of the Lederle Young Investigator Award in Vaccine Development of the Infectious Diseases Society of America.

LITERATURE CITED

1. **Baker, C. J., M. A. Rench, M. S. Edwards, R. J. Carpenter, B. M. Hays, and D. L. Kasper.** 1988. Immunization of pregnant women with a polysaccharide vaccine of group B *Streptococcus*. *N. Engl. J. Med.* **319:**1180–1185.
2. **Bevanger, L.** 1983. Ibc proteins as serotype markers of group B streptococci. *Acta Pathol. Microbiol. Immunol. Scand. Sect. B* **91:**231–234.
3. **Bevanger, L.** 1985. The Ibc proteins of group B streptococci: isolation of the α and β antigens by immunosorbent chromatography and test for human serum antibodies against the two antigens. *Acta Pathol. Microbiol. Immunol. Scand. Sect. B* **93:**113–119.
4. **Bevanger, L., and O.-J. Iversen.** 1981. The Ibc protein fraction of group B streptococci: characterization of protein antigens extracted by HCl. *Acta Pathol. Microbiol. Scand. Sect. B* **89:**205–209.
5. **Bevanger, L., and J. A. Maeland.** 1979. Complete and incomplete Ibc protein fraction in group B streptococci. *Acta Pathol. Microbiol. Scand. Sect. B* **87:**51–54.
6. **Bevanger, L., and A. I. Naess.** 1985. Mouse-protective antibodies against the Ibc proteins of group B streptococci. *Acta Pathol. Microbiol. Immunol. Scand. Sect. B* **93:**121–124.
7. **Brady, L. J., and M. D. P. Boyle.** 1989. Identification of nonimmunoglobulin A-Fc-binding forms and low-molecular-weight secreted forms of the group B streptococcal β antigen. *Infect. Immun.* **57:**1573–1581.
8. **Brady, L. J., U. D. Daphtary, E. M. Ayoub, and M. D. P. Boyle.** 1988. Two novel antigens associated with group B streptococci identified by a rapid two-stage radioimmunoassay. *J. Infect. Dis.* **158:**965–972.
9. **Cleat, P. H., and K. N. Timmis.** 1987. Cloning and expres-

sion in *Escherichia coli* of the Ibc protein genes of group B streptococci: binding of the human immunoglobulin A to the beta antigen. *Infect. Immun.* **55**:1151–1155.

10. **Coleman, S. E., L. J. Brady, and M. D. P. Boyle.** 1990. Colloidal gold immunolabeling of immunoglobulin-binding sites and β antigen in group B streptococci. *Infect. Immun.* **58**:332–340.

11. **Dillon, H. C., Jr., S. Khare, and B. M. Gray.** 1987. Group B streptococcal carriage and disease: a 6-year prospective study. *J. Pediatr.* **110**:31–36.

12. **Ferrieri, P.** 1988. Surface-localized protein antigens of group B streptococci. *Rev. Infect. Dis.* **10**:S363–S366.

13. **Ferrieri, P.** 1990. Neonatal susceptibility and immunity to major bacterial pathogens. *Rev. Infect. Dis.* **12**:S394–S399.

14. **Henricksen, J. P., P. Ferrieri, J. Jelinkova, W. Koehler, and W. R. Maxted.** 1984. Nomenclature of antigens of group B streptococci. *Int. J. Syst. Bacteriol.* **34**:500.

15. **Johnson, D. R., and P. Ferrieri.** 1984. Group B streptococcal Ibc protein antigen: distribution of two determinants of wild-type strains of common serotypes. *J. Clin. Microbiol.* **19**:506–510.

16. **Lancefield, R. C., M. McCarthy, and W. N. Everly.** 1975. Multiple mouse-protective antibodies directed against group B streptococci. *J. Exp. Med.* **142**:165–179.

16a.**Madoff, L. C., J. L. Michel, and D. L. Kasper.** 1991. A monoclonal antibody identifies a protective C-protein alpha-antigen epitope in group B streptococci. *Infect. Immun.* **59**:204–210.

17. **Michel, J. L.** 1990. Group B streptococcal infections: an update. *Infect. Dis. Pract.* **13**:1–12.

18. **Payne, N. R., and P. Ferrieri.** 1985. The relation of the Ibc protein antigen to the opsonization differences between strains of type II group B streptococci. *J. Infect. Dis.* **151**:672–681.

19. **Payne, N. R., Y. Kim, and P. Ferrieri.** 1987. Effects of differences in antibody and complement requirements on phagocytic uptake and intracellular killing of "c" protein-positive and -negative strains of type II group B streptococci. *Infect. Immun.* **55**:1243–1251.

20. **Rubens, C. E., and L. M. Heggen.** 1988. Tn916ΔE: a Tn916 transposon derivative expressing erythromycin resistance. *Plasmid* **20**:137–142.

21. **Rubens, C. E., M. R. Wessels, L. M. Heggen, and D. L. Kasper.** 1987. Transposon mutagenesis of type III group B *Streptococcus*: correlation of capsule expression with virulence. *Proc. Natl. Acad. Sci. USA* **84**:7208–7212.

22. **Russell-Jones, G. J., and E. C. Gotschlich.** 1984. Identification of protein antigens of group B streptococci, with special reference to the Ibc antigens. *J. Exp. Med.* **160**:1476–1484.

23. **Russell-Jones, G. J., E. C. Gotschlich, and M. S. Blake.** 1984. A surface receptor specific for human IgA on group B streptococci possessing the Ibc protein antigen. *J. Exp. Med.* **160**:1467–1475.

24. **Schneewind, O., F. Karlheinz, and R. Lutticken.** 1988. Cloning and expression of the CAMP factor of group B streptococci in *Escherichia coli*. *Infect. Immun.* **56**:2174–2179.

25. **Valtonen, M. V., D. L. Kasper, and N. J. Levy.** 1986. Isolation of a C (Ibc) protein from group B *Streptococcus* which elicits mouse protective antibody. *Microb. Pathog.* **1**:191–204.

26. **Wanger, A. R., and G. M. Dunny.** 1985. Development of a system for genetic and molecular analysis of *Streptococcus agalactiae*. *Res. Vet. Sci.* **38**:202–208.

27. **Weiser, J. N., and C. E. Rubens.** 1987. Transposon mutagenesis of group B *Streptococcus* beta-hemolysin biosynthesis. *Infect. Immun.* **55**:2314–2316.

28. **Wessels, M. R., C. E. Rubens, V.-J. Benedi, and D. L. Kasper.** 1989. Definition of a bacterial virulence factor: sialylation of the group B streptococcal capsule. *Proc. Natl. Acad. Sci. USA* **86**:8983–8987.

29. **Wilkinson, H. W., and R. G. Eagon.** 1971. Type-specific antigens of group B type Ic streptococci. *Infect. Immun.* **4**:596–604.

30. **Wilkinson, H. W., and M. D. Moody.** 1969. Serological relationships of type I antigens of group B streptococci. *J. Bacteriol.* **97**:629–634.

Type III Capsule and Virulence of Group B Streptococci

MICHAEL R. WESSELS,[1,2] VICENTE-JAVIER BENEDÍ,[1†] DENNIS L. KASPER,[1,2]
LAURA M. HEGGEN,[3] AND CRAIG E. RUBENS[3]

*Channing Laboratory, Brigham and Women's Hospital,[1] and Division of Infectious Diseases, Beth Israel
Hospital,[2] Harvard Medical School, Boston, Massachusetts 02215; and Division of Infectious Diseases,
Department of Pediatrics, Children's Hospital and Medical Center, University of Washington,
Seattle, Washington 98105[3]*

Group B streptococci (GBS) constitute a major cause of bacterial sepsis and meningitis among human neonates born in the United States and Western Europe and may be emerging as significant neonatal pathogens in developing countries as well. It is estimated that GBS strains are responsible for 10,000 to 15,000 cases of invasive infection in neonates in the United States alone (1, 2, 6). Despite advances in early diagnosis and treatment, neonatal sepsis due to GBS continues to carry a mortality rate of 15 to 20% (3). In addition, survivors of GBS meningitis have a 30 to 50% incidence of long-term neurologic sequelae (9, 21). The increasing recognition over the past two decades of GBS as an important pathogen for human infants has generated renewed interest in defining the bacterial and host factors important in virulence of GBS and in the immune response to GBS infection.

Particular attention has focused on the capsular polysaccharide as the predominant surface antigen of the organisms. In a modification of the system originally developed by Rebecca Lancefield, GBS strains are serotyped on the basis of antigenic differences in their capsular polysaccharides and the presence or absence of serologically defined C proteins (12, 15, 24). While GBS isolated from nonhuman sources often lack a serologically detectable capsule, a large majority of strains associated with neonatal infection belong to one of four major capsular serotypes, Ia, Ib, II, or III. The capsular polysaccharide forms the outermost layer around the exterior of the bacterial cell, superficial to the cell wall. The capsule is distinct from the cell wall-associated group B carbohydrate. The group B polysaccharide, in contrast to the type-specific capsule, is present on all GBS strains and is the basis for serogrouping of the organisms into Lancefield's group B. Early studies by Lancefield and coworkers showed that antibodies raised in rabbits against whole GBS organisms protected mice against challenge with strains of homologous capsular type, demonstrating the central role of the capsular polysaccharide as a protective antigen

(16, 17). Studies in the 1970s by Baker and Kasper demonstrated that cord blood of human infants with type III GBS sepsis uniformly had low or undetectable levels of antibodies directed against the type III capsule, suggesting that a deficiency of anticapsular antibody was a key factor in susceptibility of human neonates to GBS disease (4).

Immunochemistry of the Type III Capsular Polysaccharide

Type III strains account for roughly two-thirds of cases of GBS neonatal disease. In addition, while all of the GBS serotypes may result in meningitis, meningitis occurs in 80 to 90% of infants with sepsis due to type III strains, suggesting that the presence of the type III capsule may be particularly associated with meningeal invasiveness. Like the other GBS capsular polysaccharides, the type III capsule is a high-molecular-weight polymer of galactose, glucose, N-acetylglucosamine, and N-acetylneuraminic acid (sialic acid) arranged in a regular repeating structure (22). The backbone of the type III polysaccharide is a linear polymer with a trisaccharide repeating unit of $[\rightarrow 4)$-β-D-Glc$p$-$(1 \rightarrow 6)$β-D-Glc$p$NAc$(1 \rightarrow 3)$-β-D-Gal$p$-$(1 \rightarrow]$. Attached to each GlcNAc residue is a disaccharide side chain of galactose and a terminal residue of sialic acid (Fig. 1). Studies by Jennings, Kasper, and colleagues showed that the sialic acid residues of the polysaccharide were important to the immunodeterminant recognized by protective anticapsular antibodies (13, 14, 22). Evidence from immunochemical and nuclear magnetic resonance studies suggested that the carboxylate group of the sialic acid residues exerted conformational control over the epitope of the type III polysaccharide. Removal of the sialic acid residues produced a core polysaccharide that was antigenically incomplete in comparison with the native antigen. Seroepidemiologic studies of human neonates with GBS infection indicated that the presence of antibodies directed against the sialylated (native), and not the desialylated (core), polysaccharide was correlated with protection against GBS infection (4, 14).

In addition to influencing the epitope recognized by anticapsular antibodies, sialic acid ap-

†Present address: Department of Microbiology, University of the Balearic Islands, Palma de Mallorca, Spain.

A

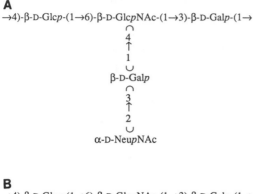

FIG. 1. Repeating unit structures of the GBS type III native (A) and core (B) polysaccharides (22).

pears to affect the capacity of the organisms to activate complement. Terminal sialic acid residues are present on the bacterial capsules of several species of pathogenic bacteria, including capsular types B and C of *Neisseria meningitidis* and *Escherichia coli*, type K1. The presence of sialic acid residues on the surface of a particle may inhibit activation of the alternative pathway of complement (10, 18). Sialic acid increases the affinity of C3b for the regulatory protein factor H, leading to inactivation of bound C3b by the C3b inactivator protein, factor I, thereby interrupting the alternative pathway amplification loop and preventing efficient deposition of opsonic complement fragments on the particle surface. Edwards et al. (7, 8) showed that the presence of sialic acid residues on the capsule of type III GBS prevented direct activation of the alternative pathway of complement. If the sialic acid residues on the type III polysaccharide were removed by treating the organisms with neuraminidase, the desialylated bacterial cells supported complement activation and were efficiently phagocytosed (7, 8). Thus, capsule sialylation may be an important mechanism by which GBS evade host defense. These observations have provided indirect evidence of the importance of the sialylated capsular polysaccharide in pathogenesis of type III GBS infection. However, any general inferences from the results of these studies are limited by the fact that they involve comparisons of chemically modified or incompletely characterized strains which may differ in aspects other than degree of encapsulation or sialic acid content.

Role of the Type III Capsule in Virulence: Studies of Capsule Mutants

To study more directly the role of the type III capsule in pathogenesis of GBS infections, we have used transposon mutagenesis to create isogenic mutants altered in capsule expression from a wild-type strain of type III GBS. Tn916 is a 16.4-kb transposon, carrying a tetracycline resistance marker, that is capable of self-conjugation among a variety of gram-positive organisms (11). Tn916 was transferred from the high-frequency donor strain *Enterococcus faecalis* CG110 to a recipient strain of type III GBS by filter mating (20). A spontaneous streptomycin- and rifampin-resistant mutant of wild-type GBS strain COH31 had been previously selected by plating a broth culture on rifampin- and streptomycin-containing medium. The streptomycin-resistant recipient GBS strain was designated COH31r/s. Following filter mating, transconjugants were selected on tetracycline-streptomycin medium and screened for deficient capsule expression by a colony immunoblot assay using rabbit antiserum specific for the native (sialylated) type III polysaccharide. Eleven transconjugants failed to react with type III GBS antiserum, compatible with complete absence of immunoreactive capsular polysaccharide. Southern hybridization analysis of restriction digests of genomic DNA from these putative unencapsulated mutants revealed from one to four Tn916 insertions. One of the mutant strains, COH31-15, which contained a single Tn916 insertion, was selected for further study. The mutant and parent strains were compared by immunoelectron microscopy. Glutaraldehyde-fixed cells of each strain were incubated with type III GBS antiserum and then with ferritin-conjugated anti-rabbit immunoglobulin G. Electron microscopy revealed a distinct capsular layer decorated with ferritin particles on the parent strain COH31r/s, while mutant strain COH31-15 showed minimal antibody binding and no visible capsular material (Fig. 2). Capsular extracts were prepared from both strains and assayed for presence of immunoreactive type III polysaccharide by enzyme-linked immunosorbent assay (ELISA) inhibition. For these assays, purified type III GBS polysaccharide was coated onto ELISA wells, and then the capsular extracts were added at various concentrations before addition of type III antiserum. The parent strain extract produced dose-dependent inhibition of the type III ELISA, while the COH31-15 extract failed to inhibit. To exclude the possibility that COH31-15 produced an antigenically altered capsule, rabbits were immunized with Formalin-fixed organisms of strain COH31-15 by the same protocol used to produce type-specific antiserum against encapsulated strains of GBS. This rabbit antiserum was used in immunoelectron microscopy experiments, as

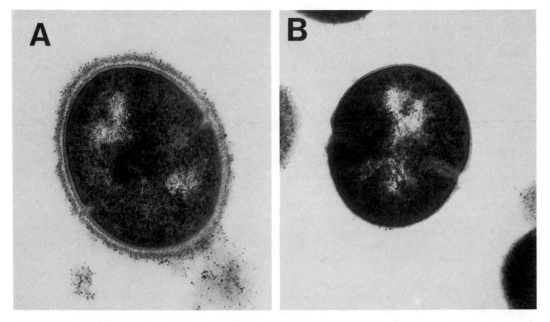

FIG. 2. Electron micrographs of wild-type GBS type III strain COH31r/s (A) and unencapsulated mutant strain COH31-15 (B), incubated with type III GBS antiserum and then with ferritin-conjugated goat anti-rabbit immunoglobulin G (original magnification, ×63,000). Antiserum was previously absorbed with cells of the poorly encapsulated type Ia variant strain 090R to remove antibodies directed against the common group B polysaccharide (20).

described above, and showed binding to the cell wall of COH31-15 only, with no suggestion of a capsule. Thus, a single transposon insertion in the chromosome of a type III strain of GBS completely prevented capsule expression. The effect of loss of encapsulation on virulence was assessed in a neonatal rat model of lethal GBS infection. The 50% lethal dose for mutant strain COH31-15 was 4.1×10^7 CFU/g of body weight, compared with 4.1×10^5 for the parent strain COH31r/s (20, 23).

Importance of Sialic Acid to the Virulence Function of the Type III Capsule

Because of the evidence implicating sialic acid as a critical structural feature in the interaction of the type III capsular polysaccharide with the host immune system, we were interested in deriving not only mutants devoid of capsule expression but also strains deficient only in capsular sialic acid. The desialylated, or core, type III GBS polysaccharide has a repeating unit structure identical to that of type 14 pneumococci, permitting us to use type 14 pneumococcal antiserum to screen for such mutants (23). Colonies failing to react with type III GBS antiserum were tested for immunoreactivity with type 14 pneumococcal antiserum in the same type of immunoblot assay. A single transconjugant, COH31-21, failed to react with type III GBS antiserum, but did react

with type 14 pneumococcal antiserum, consistent with the expression on this strain of the asialo, or core, form of the type III polysaccharide. Strain COH31-21 had two Tn916 insertions. Treatment of strain COH31-21 with neuraminidase released no detectable sialic acid from the surface of the organisms. Immunoelectron microscopy using type III GBS antiserum showed no immunoreactive capsule on COH31-21. However, similar experiments using type 14 pneumococcal antiserum demonstrated a clear capsular layer on COH31-21 but no reaction with the parent strain COH31r/s or with the unencapsulated mutant COH31-15. The capsular polysaccharide of strain COH31-21 was extracted from the organisms with mutanolysin and then purified by ion-exchange and gel filtration chromatography. When tested against type 14 pneumococcal antiserum in Ouchterlony gel diffusion experiments, the COH31-21 polysaccharide formed a line of identity with chemically desialylated type III GBS polysaccharide and with authentic type 14 pneumococcal polysaccharide, while native type III GBS polysaccharide produced no reaction with this serum. To assess the effect of loss of capsular sialic acid on virulence, strain COH31-21 was tested in the neonatal rat model and was found to have a 50% lethal dose of 3.2×10^7 CFU/g, similar to that of the unencapsulated mutant COH31-15 and approximately

2 orders of magnitude greater than that of the wild-type parent strain (23).

Studies of Capsule Mutants Derived from a Highly Encapsulated Type III Strain

These studies of transposon mutants of strain COH31 provided direct evidence that the type III GBS capsular polysaccharide functions as a virulence factor for the organism and that the presence of sialic acid residues as side chain termini of the polysaccharide was a key component of the virulence function. While the COH31 mutants provided a useful model to confirm these basic hypotheses about the role of the capsule, the model was limited by the fact that the parent strain COH31r/s is itself relatively avirulent, having a 50% lethal dose in neonatal rats at least 2 orders of magnitude higher than that of most type III strains isolated from infants with sepsis or meningitis. The latter strains are resistant to phagocytic killing in serum by human peripheral blood leukocytes in the absence of anticapsular antibody. In contrast, strain COH31r/s is readily killed in vitro under these conditions. Strain COH31r/s was selected for our initial studies because it is sensitive to tetracycline, a relatively rare occurrence among GBS isolates (90 to 95% are resistant), permitting the use of the tetracycline resistance marker on Tn916 for selection of mutants.

Because strain COH31 is not entirely representative of invasive type III GBS strains, we have developed another set of mutant strains in the background of a type III strain isolated from an infant with meningitis. This strain, COH1, is typical of clinical isolates of type III GBS in terms of virulence in animals and resistance to phagocytic killing in vitro. As COH1 is tetracycline resistant, a derivative of Tn916 was developed by Rubens and Heggen in which the tetracycline resistance gene was replaced by an erythromycin resistance marker to create a new transposon designated Tn916ΔE (19). Tn916ΔE was transferred to COH1 from a donor strain of *Enterococcus faecalis* and mutants with altered capsule expression were selected by immunoblot assay as described above (see also Rubens et al., this volume).

Screening of Tn916ΔE transconjugants derived from COH1 yielded several unencapsulated mutants, of which a single insertion mutant, COH1-13, was selected for further characterization. As in the case of transposon mutagenesis of COH31, only one mutant of COH1 expressed the asialo, or core, form of the type III polysaccharide. The asialo mutant, COH1-11, like strain COH31-21, had two transposon insertions detected by Southern hybridization analysis. Interestingly, one of the transposon insertions in COH1-11 appeared

to be in the same restriction fragment as one of the insertions in COH31-21, suggesting that interruption of gene expression at this chromosomal site was responsible for the asialo phenotype in both mutants. A 30-kb region on the COH1 chromosome has been mapped which includes the transposon insertion sites of COH1-13 and the site common to COH1-11 and COH31-21. A more detailed description of the mapping studies and of cloning and expression of the genes from this region is included elsewhere in this volume. The unencapsulated and asialo phenotypes of strains COH1-13 and COH1-11, respectively, were confirmed by ELISA inhibition assays, as described above, in which capsular extracts from each strain were assayed for immunoreactive polysaccharide in ELISA by using either type III GBS polysaccharide or type 14 pneumococcal polysaccharide as the coating antigen.

The Type III Capsule in Resistance to Phagocytosis

Immune defense against GBS, like that against other gram-positive bacteria, depends on clearance of the organisms by host phagocytes. Type III GBS isolates from infants with invasive GBS disease are generally not killed by peripheral blood leukocytes in the presence of nonimmune serum. Antibodies directed against the type III capsular polysaccharide, in concert with serum complement, effectively opsonize the organisms for phagocytic killing in vitro (5, 8). Strain COH1 is resistant to killing in vitro by human peripheral blood leukocytes in the presence of nonimmune serum. Addition of specific anticapsular antibody to the assay results in efficient phagocytosis and killing of the organisms (>90% killed in 1 h). These observations are consistent with the seroepidemiologic studies of infected infants, which showed that those infants who had preexisting maternal antibodies against the type III polysaccharide were protected against type III GBS disease. However, it is not possible to determine from these data whether the capsule itself is responsible for resistance of the organisms to phagocytosis or whether it simply serves as a target for opsonic antibodies. Derivation of acapsular and asialo mutants from the virulent strain of type III GBS COH1 allowed us to assess directly the role of the type III capsule in the ability of the organism to resist phagocytosis. Each strain was incubated with human peripheral blood leukocytes and 10% normal human serum. Quantitative cultures were performed at time zero and after 1 h of incubation at 37°C. The parent strain COH1 showed no reduction in CFU under these conditions, while both the unencapsulated mutant COH1-13 and the asialo mutant COH1-11 were efficiently phagocytosed and killed.

That the poorly encapsulated wild-type strain COH31 is susceptible to phagocytic killing in the absence of specific antibody suggests that not only the presence or absence of capsule but also the degree of encapsulation is a critical determinant of the antiphagocytic capacity of the capsular polysaccharide. This interpretation is consistent with observations in our laboratory and by others that the amount of capsule expressed by type III GBS strains is correlated both with virulence in human infections and in infections of experimental animals and with the degree of resistance of the strains to phagocytosis. While the capsule clearly plays an important role in defending highly encapsulated strains from phagocytic killing, it is less clear what other virulence function(s) the capsule may serve. Strain COH31, though sensitive to phagocytosis in vitro, became less virulent in animals after loss of capsule or capsular sialic acid. Further studies will be required to determine whether the organisms may increase capsule expression in vivo or whether the presence of even a thin capsular layer may confer virulence through some other mechanism.

We conclude from these studies that the type III capsular polysaccharide is a key surface structure in the interactions of the bacterium with the host immune system and that the capsule constitutes an important virulence factor for GBS. These results provide the most direct demonstration to date that the type III capsule of GBS protects the organism from phagocytic killing and that the presence of sialic acid residues as side chain termini of the capsule is an essential feature of this antiphagocytic function. Further investigations using the combined experimental approaches of immunochemistry and molecular genetics promise to define in greater detail the molecular mechanisms involved in the interactions of this important human pathogen with the host immune system.

This work was supported by Public Health Service grants AI28040, AI23339, and AI22498 from the National Institutes of Health and by a grant-in-aid from the American Heart Association.

We thank Thomas J. DiCesare and April Blodgett for expert technical assistance.

LITERATURE CITED

1. **Anthony, B. F., and D. M. Okada.** 1977. The emergence of group B streptococci in infections of the newborn infant. *Annu. Rev. Med.* **28:**355–369.
2. **Baker, C. J.** 1980. Group B streptococcal infections. *Adv. Intern. Med.* **25:**475–501.
3. **Baker, C. J.** 1986. Neonatal sepsis: an overview. *In* A. Morell and U. E. Nydegger (ed.), *Clinical Use of Intravenous Immunoglobulins.* Academic Press, London.
4. **Baker, C. J., and D. L. Kasper.** 1976. Correlation of maternal antibody deficiency with susceptibility to neonatal group B streptococcal infection. *N. Engl. J. Med.* **294:**753–756.
5. **Baltimore, R. S., D. L. Kasper, C. J. Baker, and D. K. Goroff.** 1977. Antigenic specificity of opsonophagocytic antibodies in rabbit anti-sera to group B streptococci. *J. Immunol.* **118:**673–678.
6. **Dillon, H. C., Jr., S. Khare, and B. M. Gray.** 1987. Group B streptococcal carriage and disease: a 6-year prospective study. *J. Pediatr.* **110:**31–36.
7. **Edwards, M. S., D. L. Kasper, H. J. Jennings, C. J. Baker, and A. Nicholson-Weller.** 1982. Capsular sialic acid prevents activation of the alternative complement pathway by type III, group B streptococci. *J. Immunol.* **128:**1278–1283.
8. **Edwards, M. S., A Nicholson-Weller, C. J. Baker, and D. L. Kasper.** 1980. The role of specific antibody in alternative pathway-mediated opsonophagocytosis of type III, group B *Streptococcus. J. Exp. Med.* **151:**1275–1287.
9. **Edwards, M. S., M. A. Rench, A. A. Haffar, M. A. Murphy, M. M. Desmond, and C. J. Baker.** 1985. Long-term sequelae of group B streptococcal meningitis in infants. *J. Pediatr.* **106:**717–722.
10. **Fearon, D. T.** 1978. Regulation by membrane sialic acid of β1H-dependent decay-dissociation of amplification C3 convertase of the alternative complement pathway. *Proc. Natl. Acad. Sci. USA* **75:**1971–1975.
11. **Gawron-Burke, C., and D. B. Clewell.** 1982. A transposon in *Streptococcus faecalis* with fertility properties. *Nature* (London) **300:**281–284.
12. **Henrichsen, J., P. Ferrieri, J. Jelinkova, W. Köhler, and W. R. Maxted.** 1984. Nomenclature of antigens of group B streptococci. *Int. J. Syst. Bacteriol.* **34:**500.
13. **Jennings, H. J., C. Lugowski, and D. L. Kasper.** 1981. Conformational aspects critical to the immunospecificity of the type III group B streptococcal polysaccharide. Biochemistry **20:**4511–4518.
14. **Kasper, D. L., C. J. Baker, R. S. Baltimore, J. H. Crabb, G. Schiffman, and H. J. Jennings.** 1979. Immunodeterminant specificity of human immunity to type III group B *Streptococcus. J. Exp. Med.* **149:**327–339.
15. **Lancefield, R. C.** 1934. A serological differentiation of specific types of bovine hemolytic streptococci (group B). *J. Exp. Med.* **59:**441–458.
16. **Lancefield, R. C.** 1938. Two serological types of group B hemolytic streptococci with related, but not identical, type-specific substances. *J. Exp. Med.* **67:**25–40.
17. **Lancefield, R. C., M. McCarty, and W. N. Everly.** 1975. Multiple mouse-protective antibodies directed against group B streptococci. Special reference to antibodies effective against protein antigens. *J. Exp. Med.* **142:**165–179.
18. **Pangburn, M. K., and H. J. Müller-Eberhard.** 1978. Complement C3 convertase: cell surface restriction of β1H control and generation of restriction on neuraminidase treated cells. *Proc. Natl. Acad. Sci. USA* **75:**2416–2420.
19. **Rubens, C. E., and L. M. Heggen.** 1988. Tn916ΔE: a Tn916 derivative expressing erythromycin resistance. *Plasmid* **20:**137–142.
20. **Rubens, C. E., M. R. Wessels, L. M. Heggen, and D. L. Kasper.** 1987. Transposon mutagenesis of group B streptococcal type III capsular polysaccharide: correlation of capsule expression with virulence. *Proc. Natl. Acad. Sci. USA* **84:**7208–7212.
21. **Wald, E. R., I. Bergman, H. G. Taylor, D. Chiponis, C. Porter, and K. Kubek.** 1986. Long-term outcome of group B streptococcal meningitis. *Pediatrics* **77:**217–221.
22. **Wessels, M. R., V. Pozsgay, D. L. Kasper, and H. J. Jennings.** 1987. Structure and immunochemistry of an oligosaccharide repeating unit of the capsular polysaccharide of type III group B *Streptococcus. J. Biol. Chem.* **262:**8262–8267.
23. **Wessels, M. R., C. E. Rubens, V.-J. Benedi, and D. L. Kasper.** 1989. Definition of a bacterial virulence factor: sialylation of the group B streptococcal capsule. *Proc. Natl. Acad. Sci. USA* **86:**8983–8987.
24. **Wilkinson, H. W., and R. G. Eagon.** 1971. Type-specific antigens of group B type Ic streptococci. *Infect. Immun.* **4:**596–604.

Genetics and Characterization of Group B Streptococcal Pigment

DAVID E. WENNERSTROM,[1] LINDA N. LEE,[2] ADAM G. BASEMAN,[2] DONALD J. LeBLANC,[2] CARL E. CERNIGLIA,[3] AND KAREN M. TROTTER[4]

Department of Microbiology and Immunology, University of Arkansas for Medical Sciences, Little Rock, Arkansas 72205[1]; Department of Microbiology, University of Texas Health Science Center at San Antonio, San Antonio, Texas 78284[2]; Microbiology Division, National Center for Toxicological Research, Jefferson, Arkansas 72079[3]; and James Baker Institute, Cornell University, Ithaca, New York 14853[4]

Group B streptococci (GBS) are beta-hemolytic gram-positive cocci that contain the Lancefield group B carbohydrate antigen (5). Six serotypes, designated Ia, Ib, Ic, II, III, and IV, have been identified on the basis of carbohydrate antigens and the C (formerly Ibc) protein antigen. This medically important group of streptococci is unique in that beta-hemolytic strains of all serotypes are capable of producing a pigment during growth on laboratory media. The color of the pigment varies from yellow to red, depending on the strain of GBS and the conditions of its cultivation. Generally, pigment production is enhanced and colonies appear dark orange or red following anaerobic cultivation on media containing starch and horse serum. Under these conditions, pigment is expressed by 99.5% of beta-hemolytic GBS, allowing it to be used for the presumptive identification of these organisms cultured from clinical specimens (2, 4, 9).

The nature and functions of pigment have not been determined because it remains associated with the cell membrane, is extracted with membrane components, or becomes tightly bound to the starch used to elicit its production (7, 11). Because of these limitations, pigment solubilized by starch or sonication of GBS has been characterized only by its UV-visible light absorption spectrum. The spectrum contains multiple peaks in the region between 430 and 520 nm, suggesting that the pigment is a bacterial carotenoid (7, 10). By measuring absorption, Tapsall has observed that GBS produce pigment during all phases of growth and that the spectral characteristics of pigment change during growth and according to the association of pigment with a carrier molecule such as starch or albumin. He has suggested that pigment color may result from the presence of a mixture of compounds (11). Because the production and color of pigment also vary according to growth conditions, the results obtained by Tapsall also do not rule out the possibility that GBS contain multiple genes for pigmentation and that their expression is under complex regulation.

Several observations suggest that pigment and β-hemolysin are closely linked properties of GBS.

Thus, pigment and hemolysin were both lost in spontaneous mutants or in variants isolated after nitrosoguanidine mutagenesis (6, 14).

The role of pigment in pathogenesis of GBS infection of the newborn has been suggested by two studies. R. A. Nemergut and K. Merritt (*Abstr. Annu. Meet. Am. Soc. Microbiol. 1983*, B-32, p. 28) reported that addition of solubilized pigment to phagocytes inhibited the oxidative killing of engulfed GBS. Additionally, Wennerstrom et al. (14) observed that mice infected with nitrosoguanidine-derived nonhemolytic (Hly⁻) pigment-deficient (Pig⁻) GBS survived much longer than mice infected with Hly⁺ Pig⁺ organisms. However, expression of hemolysin and pigment were both altered, and the nitrosoguanidine mutagenesis may have affected other genes. Therefore, a genetic approach has been taken to target the gene(s) encoding pigment to study the trait of pigmentation, its linkage with hemolysin, and its role in pathogenesis or early-onset infection of the newborn.

Cloning of the GBS Gene(s) Encoding Pigment Production

The genetic determinant(s) for pigment production was targeted on the GBS chromosome by transposon insertional mutagenesis with Tn*925::917*, a derivative of the conjugative transposon Tn*925* (encoding resistance to tetracycline) with an insertion of Tn*917* located about 3 kb from one end (1). *Enterococcus faecalis* INY275, containing a chromosomal insertion of Tn*925::917*, was used as the transposon donor in filter matings with a Hly⁺ Pig⁺ GBS serotype Ib recipient strain, H36B, previously selected for spontaneous mutations to rifampin and fusidic acid resistance. GBS transconjugants that had received Tn*925::917* were obtained on selective plates containing rifampin, fusidic acid, and tetracycline. Approximately 16,000 transconjugants were tested for hemolytic activity on sheep blood agar plates. Two did not express any hemolysin activity (Hly⁻) and three were hyperhemolytic (Hly⁺⁺), exhibiting larger zones of hemolysis. When tested for pigment production by streaking onto starch-containing medium (4), the Hly⁻ transconjugants failed to produce pigment (Pig⁻) and the Hly⁺⁺ transconjugants were hyperpigmented (Pig⁺⁺). Chromosomal DNA from each of these transconjugants was digested with HindIII and used to prepare Southern blots, which were hybridized with the Tn925-containing plasmid pCF10 (1). The results indicated that all but one of the transconjugants altered in both hemolysin and pigment production contained a single insertion of the transposon, providing support for the earlier suggestion that hemolysin and pigment production are closely linked traits in GBS (6, 14).

Alteration of hemolysin activity in the serotype III GBS strain COH31C has also been reported in transconjugants containing insertions of Tn916 (13). Four such isolates were obtained from C. Rubens and were tested for pigment production as described above. Two that were Hly⁻ were also Pig⁻, and two that were Hly⁺⁺ were also Pig⁺⁺, indicating that these two traits are linked in serotype III GBS as well.

One of the Hly⁻ Pig⁻ H36B transconjugants, containing a single insertion of Tn925::917, was chosen for cloning of the transposon-targeted chromosomal DNA. Tn925::917, like Tn916, contains no sites for cleavage by EcoRI. Therefore, an EcoRI fragment containing the transposon and flanking GBS chromosomal DNA was cloned from the transconjugant into the EcoRI site of pVA891, in Escherichia coli, by a strategy similar to that used by Yamamoto et al. (15) to clone Tn916 and flanking DNA from pAD1. The experiment yielded a single transformant colony on medium containing tetracycline and erythromycin. The clone produced a bright yellow pigment and contained a recombinant plasmid consisting of pVA891 and a 7-kb EcoRI fragment composed entirely of GBS DNA. Presumably, the transposon had excised precisely from the cloned GBS DNA, resulting in reactivation of a gene(s) encoding pigment production (3). The E. coli transformant, which released the water-soluble pigment into the culture medium, was not hemolytic when grown on sheep blood agar.

Characterization of the Cloned 7-kb GBS DNA Fragment

The 7-kb EcoRI fragment was subcloned, in E. coli, into the EcoRI site of the E. coli-Streptococcus shuttle vector pDL276 (L. N. Lee, D. J. LeBlanc, and G. M. Dunny, Abstr. Annu. Meet. Am. Soc. Microbiol. 1989, H-214, p. 205). Recombinant plasmid DNA was isolated from the E. coli clone and used to transform, by electroporation, a Hly⁻ E. faecalis strain, OG1RF (1). A transformant selected for further study was nonhemolytic on sheep blood agar. However, hemolytic activity was expressed when this trans-

formant was grown in Todd-Hewitt broth supplemented with 1% Tween 80, which can be used to stimulate production and release of hemolysin by GBS. An E. faecalis OG1RF transformant containing pDL276 did not express any hemolytic activity when grown under identical conditions. Thus, the cloned 7-kb fragment of GBS DNA also encodes hemolysin.

The original plasmid vector used to clone the 7-kb fragment and the recombinant plasmid containing this fragment were used to transform the E. coli maxicell strain CSR603, and the proteins encoded by each plasmid were determined. Three proteins were attributed to the vector, and six additional proteins, ranging in size from 22 to 56 kDa, were encoded by the 7-kb GBS-derived fragment. Which of these proteins are required for pigment production has not been determined. A series of deletion derivatives of the 7-kb fragment has been constructed in preparation for such a determination. The 7-kb fragment was first subcloned into the EcoRI site of pBR322 in both orientations relative to the vector (pLL46 and pLL48; Fig. 1; only the GBS DNA is shown). Digestion of pLL46 (in pBR322) with BamHI and XbaI, religation, and transformation yielded pLL58. Similarly, NdeI-digested pLL48 (one site each in the 7-kb fragment and pBR322) provided pLL68, and digestion of pLL68 with EcoRI and XbaI provided a clone containing pLL70. Although each of the E. coli recombinant clones containing a deletion derivative of the original 7-kb fragment produced pigment, the only one producing nearly as much pigment as those containing the entire 7-kb fragment harbored pLL68. These subclones will be used to identify the proteins required for pigment production as well as any regulatory elements encoded by the 7-kb fragment.

Expression and Characterization of Pigment Made by E. coli

In contrast to GBS, E. coli did not produce pigment throughout growth. E. coli containing the cloned 7-kb fragment achieved about three doublings in LB broth before pigment could be detected in the medium by measurement of its fluorescence. Once pigment production was initiated, it proceeded at a constant rate throughout growth. However, the initial appearance of pigment was associated with a sudden decrease in the growth rate of the organism. When cells producing pigment were observed by phase-contrast microscopy, the culture contained some transparent cells that possessed inclusions (Fig. 2). The inclusions were fluorescent when the cells were exposed to UV light, indicating that intracellular pigment was associated with the inclusion granules.

Production and release of a water-soluble GBS

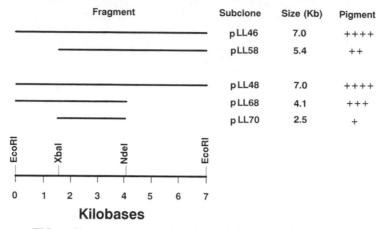

FIG. 1. Pigment production by subcloned fragments of Pig⁺ DNA.

pigment by the recombinant *E. coli* has made possible a more complete purification and characterization of pigment than heretofore possible. Pigment in LB broth eluted as a single peak when purified isocratically by semipreparative high-pressure liquid chromatography using 40% methanol in water. Rechromatography with 35% acetonitrile in water revealed the presence of isomeric forms of the pigment. The chromatographic behavior of pigment was very similar to that of water-insoluble carotenoids extracted from bacteria (8).

Purified pigment exhibited UV-visible light absorption maxima at 270, 350, and 450 nm and emitted fluorescence at 520 nm. Thus, the purified pigment appears to be different from that produced by GBS in both solubility and absorption maxima, although its fluorescence emission is identical to absorbance of GBS pigment at the longest wavelength (10). Further comparison awaits extraction and purification of pigment made by GBS. Pigment made by *E. coli* from the subcloned fragments (Fig. 1) was not altered in its chromatographic properties, suggesting that the entire pigment is encoded by 2.5 kb or less. Nevertheless, pigment characterized by electron impact and fast-atom bombardment-mass spectrometry supplemented by proton nuclear mag-

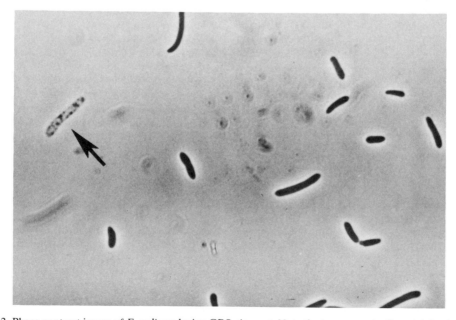

FIG. 2. Phase-contrast image of *E. coli* producing GBS pigment. Note the transparent cell containing inclusions (arrow).

netic resonance spectroscopy appears to be a complex molecule that contains an aromatic moiety, multiple CH_2 and CH_3 groups, and an odd number of nitrogen atoms. A mass loss of 307 detected by fast-atom bombardment-mass spectrometry is consistent with the presence of a tripeptide. Further, pigment tests positive for the presence of protein. These results indicate that pigment is very different from the usual carotenoids obtained from bacteria (8, 12).

The possible role of the purified GBS pigment on phagocytes is of interest because of the importance of phagocytosis in infection. Preincubation of thioglycolate-stimulated mouse peritoneal macrophages for 90 min with 15 μg of pigment in RPMI 1640 medium without serum inhibited the uptake of a radiolabeled Hly⁻ Pig⁻ transconjugant of GBS strain H36B. Pigment had no effect on the number of peritoneal cells recovered after the assay period (90 min). The effect of pigment was not observed when 10% serum was included in the medium. Determination of the significance of these results and the basis for the effect of serum await further investigation.

D.E.W. gratefully acknowledges the guidance and gracious hospitality of Gary Dunny, in whose laboratory the cloning work was performed. Craig Rubens kindly provided the isolates of type III GBS.

LITERATURE CITED

1. **Christie, P. J., R. Z. Korman, S. A. Zahler, J. C. Adsit, and G. M. Dunny.** 1987. Two conjugation systems associated with *Streptococcus faecalis* plasmid pCF10: identification of a conjugative transposon that transfers between *S. faecalis* and *Bacillus subtilis. J. Bacteriol.* **169:**2529–2536.
2. **de la Rosa, M., R. Villareal, D. Vega, C. Miranda, and A. Martinezbrocal.** 1983. Granada medium for detection and identification of group B streptococci. *J. Clin. Microbiol.* **18:**779–785.
3. **Gawron-Burke, C., and D. B. Clewell.** 1984. Regeneration of insertionally inactivated streptococcal DNA fragments after excision of transposon Tn916 in *Escherichia coli*: strategy for targeting and cloning of genes from gram-positive bacteria. *J. Bacteriol.* **159:**214–221.
4. **Islam, A. K. M. S.** 1977. Rapid recognition of group B streptococci. *Lancet* **i:**256–257.
5. **Lancefield, R. C.** 1933. A serological differentiation of human and other groups of hemolytic streptococci. *J. Exp. Med.* **57:**571–595.
6. **Lancefield, R. C.** 1934. Loss of the properties of hemolysin and pigment formation without change in immunological specificity in a strain of *Streptococcus haemolyticus. J. Exp. Med.* **59:**459–469.
7. **Merritt, K., and N. J. Jacobs.** 1978. Characterization and incidence of pigment production by human clinical group B streptococci. *J. Clin. Microbiol.* **8:**105–107.
8. **Nelis, H. J., and A. P. De Leenheer.** 1989. Profiling and quantitation of bacterial carotenoids by liquid chromatography and photodiode array detection. *Appl. Environ. Microbiol.* **55:**3065–3071.
9. **Noble, M. A., J. M. Bent, and A. B. West.** 1983. Detection and identification of group B streptococci by use of pigment production. *J. Clin. Pathol.* **36:**350–352.
10. **Tapsall, J. W.** 1986. Pigment production by Lancefield group B streptococci (*Streptococcus agalactiae*). *J. Med. Microbiol.* **21:**75–81.
11. **Tapsall, J. W.** 1987. Relationship between pigment production and haemolysin formation by Lancefield group B streptococci. *J. Med. Microbiol.* **24:**83–87.
12. **Taylor, R. F.** 1984. Bacterial triterpenoids. *Microbiol. Rev.* **48:**181–198.
13. **Weiser, J. N., and C. E. Rubens.** 1987. Transposon mutagenesis of group B streptococcus beta-hemolysin biosynthesis. *Infect. Immun.* **55:**2314–2316.
14. **Wennerstrom, D. E., J. C. Tsaihong, and J. T. Crawford.** 1985. Evaluation of the role of hemolysin and pigment in pathogenesis of early-onset group B streptococcal infection, p. 155–156. *In* Y. Kimura, S. Kotami, and Y. Shiokawa (ed.), *Recent Advances in Streptococci and Streptococcal Diseases*. Reedbooks, Bracknell, Berkshire, England.
15. **Yamamoto, M., J. M. Jones, E. Senghas, C. Gawron-Burke, and D. B. Clewell.** 1987. Generation of Tn5 insertions in streptococcal conjugative transposon Tn916. *Appl. Environ. Microbiol.* **53:**1069–1072.

Cloning of Immunoglobulin G Fc Receptor Protein of *Streptococcus pyogenes* Type M48

V. GOLUBKOV, L. NESTERCHUK, A. DUKHIN, V. KATEROV, A. SUVOROV, AND A. TOTOLIAN

Institute of Experimental Medicine, Leningrad, USSR

Group A streptococcal Fc receptors (FcRA) belong to the type II immunoglobulin G (IgG) Fc receptors of gram-positive cocci (4, 7). They are localized on the microbial cell surface and exhibit nonimmune binding through the Fc part of the IgG molecule (2, 5). Although FcRA protein differs from M protein in function (1), significant homology between their respective genes (*emm* and *fcrA*) was found. The high degree of sequence similarity between M and FcRA proteins suggests that their physicochemical properties are also similar. By cloning and sequencing FcRA protein genes from different M serotypes, the biological and evolutionary relationship between the *fcrA* and *emm* genes can be better understood. This report discusses the cloning of the FcRA48 protein gene.

The bacterial strains and plasmids used are described in Table 1.

Both chromosomal (*Streptococcus pyogenes*) and plasmid DNAs were prepared as described previously (3, 6). A genomic library was constructed as described by Maniatis et al. (6). Recombinant phages were screened for *fcrA48* by using as a probe pDH56, which contains almost the complete sequence of the *fcrA76* gene (4).

Expression was detected by colony blot with the peroxidase-antiperoxidase (PAP) system (Fig. 1). The PAP system was prepared ex tempore by mixing antiperoxidase rabbit serum with a concentrated solution of horseradish peroxidase. An immune complex of one IgG molecule with two molecules of peroxidase is able to find FcRA protein.

A library of *S. pyogenes* genes was constructed in the λL47.1 replacement vector. For this pur-

pose, *Sau*3A partially digested DNA from strain 1/64 was ligated with *Bam*HI fragments of vector DNA, packed into the phage particles, and plated on *Escherichia coli* Q359. Among 2×10^4 recombinant particles, only 10 clones expressed FcRA. Expression was detected by colony blot with PAP detection. In comparison with the second-antibody technique, FcRA detection with the PAP system has some advantages. First, it is well known that most rabbit antisera contain anti-*E. coli* antibodies which must be absorbed to obtain specific sera. The PAP system avoids this problem. Second, the PAP system is more sensitive, because the binding constant of PAP-IgG to FcRA is 30 times higher than IgG binding via the Fab binding site (A. Dukhin, personal communication).

The 10 phage clones selected hybridized with [^{32}P]DNA of pDH56 harboring *fcrA76* (kindly provided by P. Cleary, University of Minnesota, Minneapolis). Phage clone λ7-2 was chosen for further study. It contained an insertion (17.8 kb) that consisted of six *Hin*dIII fragments of 5.6, 2.9, 2.7, 2.3, 2.2, and 2.0 kb. This DNA insert was partially digested with *Hin*dIII and subcloned into pBR322. Among the recombinant plasmids obtained, only one (pGC22) expressed FcRA protein. Plasmid pGC22 contained an insertion of 4.9 kb which included two of the above *Hin*dIII fragments carried by λ7-2 DNA (2.7 and 2.2 kb). The subcloned fragment of pGC22 was mapped by using nine endonucleases: *Bam*HI, *Bgl*II, *Eco*RI, *Hin*dIII, *Hpa*I, *Hpa*II, *Pst*I, *Pvu*II, and *Sal*I. There were no *Bam*HI, *Eco*RI, or *Sal*I restriction sites (Fig. 2). The two *Hin*dIII fragments from pGC22 were cloned into pBR322 and

TABLE 1. Strains and plasmids

Strain or plasmid	Properties	Reference
Strain		
E. coli HB101	Recipient in transformation	6
E. coli Q359	Indicator for recombinant phage clones	6
S. pyogenes 1/64	M48; source of chromosomal DNA	World Health Organization Reference Laboratory, Prague
Plasmid		
pBR322	4.36 kb; Ap^r Tc^r	6
pUC8	2.7 kb; Ap^r	6
pDH56	4.5 kb; contains *fcrA76*	3

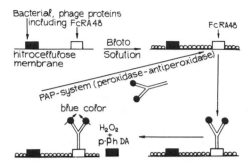

FIG. 1. PAP system for detection of FcRA proteins.

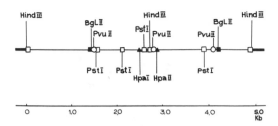

FIG. 3. Map of plasmid pGC22 containing the *fcrA* gene.

pUC8, respectively. These constructs (pGC210 and pGC211) did not express FcRA protein. Plasmid pGC301, which contained a *Bgl*II fragment (2.7 kb) internal to the insert, did, however, express FcRA48 protein (Fig. 3). This fragment probably contains the complete *fcrA48* gene because Western immunoblot analysis revealed that the streptococcal and the cloned FcRA proteins have the same electrophoretic mobility (data not shown).

The levels of FcRA expression in the *E. coli* subclones harboring pGC22 or GC301 were similar. Moreover, expression was similar in *E. coli*(pGC301₁) and *E. coli*(pGC301₂). These plasmids differed from each other by the inversion of

the *Bgl*II fragment in pUC8. Therefore, expression of the *fcrA48* gene is probably under the control of its own promoter. The failure of plasmids pGC210 and pGC211 to produce FcRA48 suggests that *Hin*dIII restriction inactivated the *fcrA48* gene.

From other digests, the map shown in Fig. 3 was developed. *fcrA48* contains *Pst*I, *Hin*dIII, and *Pvu*II in an orientation similar to that of *fcrA76*; location of the *Hpa*I and *Hpa*II sites is different (4; P. Cleary, personal communication). In contrast to our results, cleavage of the *Hin*dIII site of *fcrA76* in pDH91 and pDH56 has no influence on the level of FcRA76 protein expression (4).

From these data, we conclude that *fcrA76* and *fcrA48* differ at the nucleotide level. It is possible that this difference arose by intergenic (between *emm* and *fcrA*) or intragenic (between repeat sequences in the receptor gene) recombination.

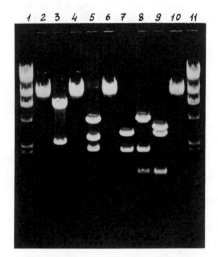

FIG. 2. Agarose (0.8%) gel electrophoresis of pGC22 digested with *Bam*HI (lane 2), *Bgl*II (lane 3), *Eco*RI (lane 4), *Hin*dIII (lane 5), *Hpa*I (lane 6), *Hpa*II (lane 7), *Pst*I (lane 8), *Pvu*II (lane 9), and *Sal*I (lane 10). Lanes 1 and 11 contain λ DNA digested with *Hin*dIII.

LITERATURE CITED

1. **Christensen, P., P. Grabb, R. Grabb, G. Samuelsson, C. Shalen, and M. Swensson.** 1979. Demonstration of non identity between the Fc receptor for human IgG from group A streptococci type 15 and M protein, peptidoglycan and the group specific carbohydrate. *Acta Pathol. Microbiol. Scand. Sect. C* **87:**257–261.
2. **Glinn, L., and M. Stuard.** 1983. *Structure and Function of Antibody.* Mir Publishing House, Moscow.
3. **Heath, D. G., and P. P. Cleary.** 1987. Cloning and expression of the gene for immunoglobulin G Fc receptor protein from a group A streptococcus. *Infect. Immun.* **55:**1233–1238.
4. **Heath, D. G., and P. P. Cleary.** 1989. Fc-receptor and M-protein genes of group A streptococci are products of gene duplication. *Proc. Natl. Acad. Sci. USA* **86:**4741–4745.
5. **Kronvall, G.** 1973. A surface component in group A, C and G streptococci with non-immune reactivity of immunoglobulin G. *J. Immunol.* **111:**1401–1406.
6. **Maniatis, T., E. Fritsch, and J. Sambrook.** 1982. *Molecular Cloning: a Laboratory Manual.* Cold Spring Harbor Laboratory, Cold Spring Harbor, N.Y.
7. **Myhre, E. B., and G. Kronvall.** 1982. Immunoglobulin specificities of defined types of streptococcal Ig receptors, p. 209–210. *In* S. E. Holm and P. Christensen (ed.), *Basic Concepts of Streptococci and Streptococcal Diseases.* Reedbooks Ltd., Chertsey, England.

C5a Peptidase Gene from Group B Streptococci

ALEXANDER N. SUVOROV,[1] P. PATRICK CLEARY,[2] AND PATRICIA FERRIERI[2]

University of Minnesota Medical School, Minneapolis, Minnesota, 55455,[2] and Institute of Experimental Medicine, Leningrad, USSR[1]

C5a is known to be a major chemoattractant generated in serum after activation of complement. It stimulates smooth muscle contraction and increases vascular permeability and release of histamine from mast cells and basophils. Furthermore, C5a is a potent stimulant for phagocytic leukocytes, inducing a burst of metabolic activity, release of toxic superoxide radicals, and release of lysosomal enzymes (10). All of these features contribute substantially to host defenses against invading microorganisms. Group A streptococci (GAS) can express a cell surface factor, C5a peptidase (SCP), which can inactivate complement-derived chemotactic activity (11). The mechanism of action of SCP is based on the ability to specifically cleave C5a (10). GAS cells that express SCP have been shown to retard the recruitment of polymorphonuclear leukocytes (5).

The GAS SCP (*scpA*) gene was cloned, sequenced, and shown to have limited homology with *Bacillus* subtilisins (1, 2). More recently, group B streptococci (GBS) were found to express an inhibitor of C5a activity (4). Preliminary experiments suggested that the SCPs from GBS and GAS (SCP B and SCP A, respectively) were antigenically distinct. To further explore the relationship between SCP B and SCP A at the molecular level, we cloned and sequenced a segment of the SCP B gene (*scpB*).

On the basis of the SCP A sequence, we designed oligonucleotide primers in order to amplify the *scpB* gene by using the polymerase chain reaction (PCR) technique. Primers for DNA amplification corresponded to the previously published sequence of *scpA* (2). For ease of cloning, the PCR products *Hin*dIII and *Eco*RI restriction sites were included on the 5′ ends of the primers (Fig. 1). Oligonucleotide synthesis and DNA sequencing were carried out by the Microchemical Facility laboratories (University of Minnesota, Health Sciences Center).

GBS chromosomal DNA (strain 78-471 serotype II/c) was used as a template. Amplification was carried out over a total of 30 cycles with an annealing-temperature ramp from 37 to 45°C. A Perkin-Elmer Cetus thermal cycler and *Taq*I DNA polymerase (Perkin-Elmer Cetus, Norwalk, Conn.) were used for amplification. We decided to amplify the GBS chromosomal region that corresponds to an internal region of the *scpA*

gene. This region of the gene codes for the putative catalytic domains as determined from homology to several *Bacillus subtilis* serine proteases (2) that contain the charge transfer residues Asp-130 and His-194.

Electrophoretic analysis of the PCR-amplified product is presented in Fig. 2. A major amplification band of 1,200 bp corresponded in size to that obtained from the *scpA* template (1,145 bp). To assess the reproducibility of this result, we also examined a GAS M49 strain and another GBS type II/c strain (78-334). When we used these other GBS and GAS chromosomal DNAs as templates, 1,200-bp amplified fragments were produced. PCR without template did not yield these fragments, demonstrating the purity of the PCR components.

The amplified GBS DNA fragments and plasmid pTT45, which contains a DNA insert internal to *scpA*, were labeled with digoxygenin by using the Genius kit (Boehringer Mannheim Biochemicals, Indianapolis, Ind.) and used to probe pTT45, GBS, and GAS amplified fragments (Fig. 2). The GBS PCR product shared homology with the cloned *scpA* gene (pTT45; Fig. 2, lane 3) as well as with the *scpA* gene fragment amplified from chromosomal DNA (lane 1). Only the largest and most predominant of the GBS amplified fragments hybridized to the pTT45 probe (lane 4).

To clone the *scpB* sequence, the amplified fragment was extracted from the agarose gel and ligated to pUC18 DNA digested with *Hin*dIII and *Eco*RI. One of the recombinant plasmids obtained, pC5a1, contained the correct 1,200-bp *scpB Hin*dIII-*Eco*RI insert (Fig. 1).

The *scpB* gene fragment was subcloned from the plasmid into M13mp18 and -mp19 vectors in order to determine its sequence. From phage M13 subclones, 400 bp from both ends of the amplified *scpB* fragment were sequenced by the dideoxy-chain termination method of Sanger et al. (8), using a model ABI 370 DNA sequencing apparatus (Applied Biosystems, Foster City, Calif.). The sequence analysis and comparison were performed by using programs in the Molecular Biology Resource software package. The sequence revealed that this part of the *scpB* gene was highly homologous (97 and 98%) to *scpA* (Fig. 3). Moreover, the predicted amino acid sequence derived

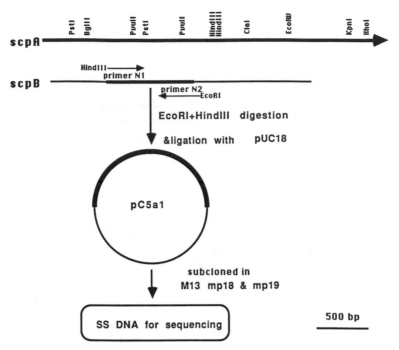

FIG. 1. Physical maps of *scpA* and the corresponding amplified *scpB* gene fragment. The arrow on the *scpA* gene shows the direction of transcription. Primers N1 and N2 represent reverse and direct oligonucleotide primers used for PCR amplification.

from the *scpB* sequence contained Asp-130 and His-194, which are considered to be catalytic centers of SCP A (2). To determine whether the cloned *scpB* sequence was universally present in GBS, DNAs from a variety of strains were ana-

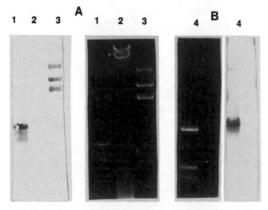

FIG. 2. Cross-hybridization between *scpA* and *scpB*. (A) Results of Southern hybridization of *scpA* with an immunolabeled *scpB* amplified gene fragment used as a probe. (B) Hybridization of the *scpB* gene fragment with the *scpA* gene used as a probe. Lanes: 1, *scpA* gene fragment amplified by using GAS (M49) chromosomal DNA as a template; 2, λ DASHII digested with *Eco*RI; 3, pTT45 (plasmid containing an internal part of *scpA*); 4, *scpB* amplified gene fragment chromosomal DNA (strain 78-471 was used as a template).

lyzed by Southern hybridization using capillary transfer of *Eco*RI-digested genomic DNA. The 17 GBS strains listed in Table 1 were tested in Southern hybridization experiments for the presence of the *scpB* gene, using plasmids pTT45 and pC5a1 as probes. All 17 strains hybridized to both probes, giving one reactive chromosomal fragment in each strain. The fragment size, however, varied from strain to strain.

The high degree of homology between *scpA* and the cloned *scpB* fragment was surprising in light of the previously reported antigenic difference between the two proteins (4). This could be explained by the fact that the *scpB* sequenced fragment represents only one-third of the whole *scpA* gene. Moreover, the primers used in PCR were expected to amplify a segment of the gene that was predicted to encode the catalytic domain of the peptidase. This region of SCP A was discovered to be also highly conserved in the *B. subtilis* subtilisins (2) and protease III from lactococci (de Vos et al., this volume). Our data indicate a close evolutionary relationship between the *scp* genes of GAS and GBS. Other segments of the *scpB* gene may be considerably different in sequence from *scpA*. This question will be resolved by cloning and sequencing of the entire *scpB* gene.

The expression of SCP by GBS and GAS, both known to colonize mucosal tissues, strengthens

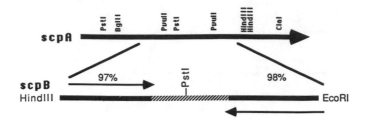

FIG. 3. Strategy for sequencing of the *scpB* gene fragment. Arrows represent the extent and direction of the DNA sequenced by the method of Sanger et al. (8). ▬, *scpA* and sequenced *scpB* DNA; ▨, unsequenced *scpB* DNA; ➤, direction of sequencing.

the idea that this protein may have a role in the virulence mechanisms of these pathogens. Wexler et al. (11) suggested that SCP could enhance the process of colonization by initially retarding the influx of polymorphonuclear leukocytes to the site of infection, which would allow the bacteria time to alter their microenvironment for optimal expansion of the colonizing clone. Thus, the SCP B protein could likewise improve the capacity of GBS to colonize the vaginal mucosa. It may or may not contribute to further invasion of deep tissues or infection of infants. Other surface-localized proteins of GBS have been described (3), and one of these, the c protein, contributes to resistance to opsonophagocytic killing (6) and also possesses immunoglobulin A Fc receptor activity (7). In some aspects, c protein resembles the M protein of GAS (Cleary et al., this volume) M-protein gene expression was found to be co-regulated with expression of the *scpA* and immunoglobulin G Fc receptor genes. It may be possible that the *scpB* gene is regulated in a similar way. The complex interaction of the GBS surface-expressed proteins and their roles in virulence require further study.

LITERATURE CITED

1. **Chen, C. C., and P. P. Cleary.** 1989. Cloning and expression of the streptococcal C5a peptidase gene in *Escherichia coli*: linkage to the type 12 M protein gene. *Infect. Immun.* **57:**1740–1745.
2. **Chen, C. C., and P. P. Cleary.** 1989. Complete nucleotide sequence of the streptococcal C5a peptidase gene of Streptococcus pyogenes. *J. Biol. Chem.* **265:**3161–3167.
3. **Ferrieri, P.** 1988. Surface-localized protein antigens of group B streptococci. *Rev. Infect. Dis.* **10:**S363–S366.
4. **Hill, H. R., J. F. Bohnsack, E. Z. Morris, N. H. Augustine, C. J. Parker, P. P. Cleary, and J. T. Wu.** 1988. Group B streptococci inhibit the chemotactic activity of the fifth component of complement. *J. Immunol.* **141:**3551–3556.
5. **O'Connor, S. P., and P. P. Cleary.** 1987. *In vivo Streptococcus pyogenes* C5a peptidase activity: analysis using transposon- and nitrosoguanidine-induced mutants. *J. Infect. Dis.* **156:**495–503.
6. **Payne, N. R., Y. Kim, and P. Ferrieri.** 1987. Effect of differences in antibody and complement requirements on phagocytic uptake and intercellular killing of "c" protein-positive and -negative strains of type II group B streptococci. *Infect. Immun.* **55:**1243–1251.
7. **Russell-Jones, G. J., E. C. Gotschlich, and M. S. Blake.** 1984. A surface receptor specific for human IgA on Group B streptococci possessing the Ibc protein antigen. *J. Exp. Med.* **160:**1467–1475.
8. **Sanger, F., S. Nicklen, and A. R. Coulson.** 1977. DNA sequencing with chain-terminating inhibitors. *Proc. Natl. Acad. Sci. USA* **74:**5463–5467.
9. **Simpson, W. J., D. La Penta, and P. P. Cleary.** 1990. Coregulation of type-12 M protein and streptococcal C5a peptidase genes in group A streptococci: evidence for a virulence regulon controlled by *virR* locus. *J. Bacteriol.* **172:**696–700.
10. **Wexler, D. E., D. E. Chenoweth, and P. P. Cleary.** 1985. Mechanism of action of the group A streptococcal C5a inactivator. *Proc. Natl. Acad. Sci. USA* **82:**8144–8148.
11. **Wexler, D. E., R. D. Nelson, and P. P. Cleary.** 1983. Human neutrophil chemotactic response to group A streptococci: bacteria-mediated interference with complement-derived chemotactic factors. *Infect. Immun.* **39:**239–246.

TABLE 1. GBS strains used

Strain	Serotype	Source
78-471	II/c	Umbilical cord
74-660	II/c	Blood
76-040	Ib/c	CSF[a]
76-043	III/R4	CSF
78-334	II/c	Ear canal
79-240	III	Blood
80-015	Ia/c	Blood
80-285	III	CSF
80-351	III	CSF
80-426	NT/c	Vagina
80-481	Ib/c	CSF
81-017	III	Blood
81-418	NT/c/R1	Vagina
81-440	II	Blood
82-316	II	Vagina
82-454	NT/c	Ear canal
82-518	III	Blood

[a]CSF, Cerebrospinal fluid.

Genetic Analysis of *Streptococcus pyogenes* M49 Chromosomal DNA

ALEXANDER N. SUVOROV[1] AND JOSEPH J. FERRETTI[2]

University of Oklahoma, Oklahoma City, Oklahoma 73190,[2] and Institute of Experimental Medicine, Leningrad, USSR[1]

Pulsed-field gel electrophoresis (PFGE) is a powerful approach for the separation of large DNA fragments. During the last 2 years, numerous publications have reported on the analysis of different bacterial genomes. Estimation of genomic size and construction of physical and genetic maps of the bacterial genome have been possible for microorganisms such as *Haemophilus parainfluenzae* (5), *Clostridium perfringens* (8), and *Streptococcus mutans* (7). However, little is known about the organization of the group A streptococcal chromosome. The only linkage data reported have involved the transduction of markers associated with antibiotic resistance (9) and *virR*-derived regulation of the M-protein gene, the streptococcal C5a peptidase gene, and the immunoglobulin G Fc receptor gene (Cleary et al., this volume). In recent years, a number of genes have been cloned from this organism. In combination with studies aimed at establishing a physical map of the group A streptococcal chromosome, we used available gene probes to identify the locations of these genes on separated DNA fragments. In this work, we used PFGE for analysis of the group A streptococcal genome. Another aim of the project was to locate several genes thought to be associated with virulence or pathogenicity on the PFGE DNA fragments.

Streptococcus pyogenes NZ131 type M49 DNA was chosen for PFGE analysis. Streptococcal cells were mixed with low-melting-point agarose, and the mixture was allowed to solidify in a 1-cm² mold. Following solidification, the plug from this mold was incubated for 18 h at 37°C in lysis solution (50 U of mutanolysin per ml, 10 mM Tris [pH 7.5], 1 M NaCl, 0.5% sarcosyl, 1 mg of lysozyme per ml) and then deproteinized in 0.5 M EDTA–0.5% sarcosyl–proteinase K (2 mg/ml) for 48 h (24 h at 60°C and 24 h at 37°C). After being washed in TE buffer, the plugs were cut into five sections, and group A streptococcal chromosomal DNA in each section was digested for 4 to 6 h with various restriction enzymes in buffers recommended by manufacturers. The restriction enzymes included *Sma*I, *Apa*I, *Hin*dIII, *Sfi*I, *Eco*RI, *Not*I, *Ksp*I, *Sac*I, and *Ecl*I. Electrophoresis was performed in 0.5× TBE. DNA samples were subjected to contour-clamped homogeneous electric field electropho-resis for 24 to 48 h at 120 to 180 V with a ramp from 6 to 20 s.

Since *S. pyogenes* has a low G+C content (36 to 38%), enzymes that recognize GC sequences were first tested for the ability to cleave the DNA into a low and separable number of fragments (Table 1). With use of conventional electrophoresis of restriction fragments after digestion with *Sma*I, *Apa*I, *Ksp*I, *Ecl*I, *Sfi*I, *Not*I, and *Sac*I, the separation of fragments was difficult to discern. However, the use of PFGE made it possible to resolve restriction fragments larger than 50 kb in size. In terms of fragment size and separation of fragments, *Apa*I, *Ksp*I, and *Sma*I gave the best results. Fragment sizes were determined by comparing the mobilities of a ladder of known fragment sizes (λ DNA concatemers). *Apa*I, *Sma*I, and *Ksp*I cleaved M49 DNA into 25, 15, and 18 fragments, respectively; by addition of these fragment sizes, the molecular size of the M49 group A streptococcal chromosome was estimated to be 1,900 kb (Table 1).

DNA hybridization with known group A streptococcal gene probes was performed with PFGE DNA fragments that had been transferred to nylon filters by the Southern procedure. Gene probes included streptokinase (*ska*) (6), streptococcal erythrogenic toxin A (*speA*) (10), hyaluronidase phage (*hylP*) (2), and proteinase (*pro*) from the University of Oklahoma collection. The M protein 48 probe (*emm48*) was from the Institute of Experimental Medicine, Leningrad, USSR. Streptococcal acid glycoprotein (*sagp*) (3) and streptococcal pyrogenic exotoxin C (*speC*) (1) were oligonucleotide probes synthesized from published sequences, and the streptolysin O (*slo*) (4) probe was obtained by the polymerase chain reaction. Primers for DNA amplification corresponded to the previously published *slo* sequence (4). All DNA probes were labeled by using the Genius labeling and detection kit (Boehringer Mannheim, Indianapolis, Ind.).

Results of hybridization of these probes to the various fragments are summarized in Table 2. The *emm* gene, known to be closely linked to the C5a peptidase and Fc receptor genes (Cleary et al., this volume), is located on the same *Apa*I, *Ksp*I, and *Sma*I fragments as the proteinase (erythrogenic toxin B) gene. The *slo* and *ska* genes were

TABLE 1. Molecular sizes of restriction fragments after PFGE

No.	Size (kb)		
	KspI	SmaI	ApaI
1	450 ± 70	500 ± 80	250 ± 30
2	250 ± 30	250 ± 30	200 ± 25
3	190 ± 20	245 ± 30	170 ± 20
4	150 ± 15	220 ± 25	150 ± 15
5	145 ± 15	195 ± 20	110 ± 10
6	110 ± 10	120 ± 15	100 ± 10
7	90 ± 10	100 ± 10	90 ± 10
8	85 ± 10	65 ± 10	85 ± 10
9	80 ± 10	60 ± 10	80 ± 10
10	75 ± 10	50 ± 8	80 ± 10
11	70 ± 10	45 ± 5	75 ± 10
12	60 ± 5	30 ± 5	70 ± 10
13	55 ± 5	25 ± 3	65 ± 5
14	50 ± 5	10 ± 2	60 ± 5
15	23 ± 3	6 ± 1	55 ± 5
16	20 ± 3		50 ± 5
17	15 ± 2		30 ± 3
18	10 ± 2		25 ± 2
19			17 ± 2
20, 21, 23–25			14 ± 2, 9, 8, 6, 3
Total	1,928 ± 232	1,921 ± 254	1,802 ± 210

TABLE 2. Molecular sizes of group A streptococcal DNA restriction fragments hybridizing with immunolabeled DNA probes

Probe	Size (kb)			
	ApaI	KspI	SacI	SmaI
emm12 or emm48	150	100	9, 17	195
hylP	150, 170	23	9	220
speA	100	190	9	
pro(speB)	150	100		195
slo	25	150	17	100
ska	25, 100	60	9	100
speC			8	
sagp	250		8	

found on the same 100-kb SmaI fragment but on different ApaI and KspI fragments (Table 2).

LITERATURE CITED

1. **Goshorn, S. C., and P. M. Schlievert.** 1988. Nucleotide sequence of streptococcal pyrogenic exotoxin type C. *Infect. Immun.* **56:**2518–2520.
2. **Hynes, W. L., and J. J. Ferretti.** 1989. Sequence analysis and expression in *Escherichia coli* of the hyaluronidase gene of *Streptococcus pyogenes* bacteriophage H4489A. *Infect. Immun.* **57:**533–539.
3. **Kanaoka, M., C. Kawanaka, T. Negoro, Y. Fukita, K. Taya, and H. Agui.** 1987. Cloning and expression of antitumor glycoprotein gene of Streptococcus pyogenes Su in Escherichia coli. *Agric. Biol. Chem.* **51:**2641–2648.
4. **Kehoe, M. A., L. Miller, J. A. Walker, and G. J. Boulnois.** 1987. Nucleotide sequence of the streptolysin O (SLO) gene: structural homologies between SLO and other membrane-damaging thiol-activated toxins. *Infect. Immun.* **55:**3228–3232.
5. **Kuac, L., and S. H. Goodgal.** 1989. The size and physical map of the chromosome of *Haemophilus parainfluenzae*. *Gene* **83:**377–380.
6. **Malke, H., B. Roe, and J. J. Ferretti.** 1985. Nucleotide sequence of streptokinase gene from *Streptococcus equisimilis* H46A. *Gene* **34:**357–362.
7. **Smith, C. L., and G. Condemine.** 1990. New approaches for physical mapping of small genomes. *J. Bacteriol.* **172:**1167–1172.
8. **Tudor, J., L. Marri, P. J. Piggot, and L. Daneo-Moore.** 1990. Size of the *Streptococcus mutans* GS-5 chromosome as determined by pulsed-field gel electrophoresis. *Infect. Immun.* **58:**838–840.
9. **Wannamaker, L. W.** 1982. Transformation, transduction, and bacteriophages, p. 112–116. *In* D. Schlessinger (ed.), *Microbiology—1982*. American Society for Microbiology, Washington, D.C.
10. **Weeks, C. R., and J. J. Ferretti.** 1986. Nucleotide sequence of the type A streptococcal exotoxin (erythrogenic toxin) gene from *Streptococcus pyogenes* bacteriophage T12. *Infect. Immun.* **52:**144–150.

VI. Molecular Biology of Oral Streptococci

Replication Functions of pVA380-1

DONALD J. LeBLANC AND LINDA N. LEE

*Department of Microbiology, The University of Texas Health Science Center at San Antonio,
San Antonio, Texas 78284*

The replication properties of several plasmids of gram-negative bacterial origin, particularly *Escherichia coli*, have been studied for many years (7). Detailed analysis of plasmid replication in gram-positive bacteria has been limited, for the most part, to a group of replicons originally isolated from staphylococci (15). The majority of these latter plasmids replicate via a rolling circle mechanism (6). The detailed information obtained from these studies, relative to nucleotide base sequence, incompatibility, regulation of plasmid copy number, control of replication initiation, and plasmid host range, has made possible the construction of recombinant DNA cloning vectors with a variety of uses (16). Plasmids of streptococcal origin also have been used as cloning vectors for the past 10 years (1, 12). These plasmids have been used to clone genes encoding selectable traits directly in streptococcal species and to construct *E. coli-Streptococcus* shuttle vectors (16). The usefulness of these vectors remains somewhat limited, however, in part because of the paucity of information available on the replication properties of streptococcal plasmids. In fact, the replication of only two plasmids of streptococcal origin, pLS1 (8), originally from a strain of *Streptococcus agalactiae* (pMV158 [2]), and pSH71, from a *Lactococcus lactis* strain (5), has been studied in any significant detail. One plasmid that has been used extensively in streptococcal recombinant DNA technology is pVA380-1 (16). We have initiated studies on the replication functions of this plasmid, with the hope of improving its usefulness as a genetic tool.

pVA380-1—Historical Perspective

pVA380-1 was first described in a strain of *Streptococcus ferus* (13). It is a 4.2-kb cryptic plasmid that is maintained in *Streptococcus sanguis* (Challis) at 15 to 25 copies per genome equivalent. The original plasmid was used to clone several plasmid-mediated antibiotic resistance determinants directly in the Challis strain of *S. sanguis* (9, 11). Chromosomal resistance deter-

minants from different strains of *Streptococcus mutans* were also cloned in *S. sanguis* by using pVA380-1 as both a vector and a resident helper plasmid (10, 17). A number of these cloning experiments provided streptococcal vectors with selectable or insertionally inactivatable resistance markers. At the First International Streptococcal Genetics Conference, in 1981, at approximately the same time that the National Aeronautics and Space Administration's first space shuttle was being launched, Macrina and co-workers (12) described the construction of the first *E. coli-Streptococcus* shuttle vector. This vector, pVA838, contained pVA380-1 as the streptococcal replicon and pACYC184 (3) as the *E. coli* replicon. It was 9.2 kb in size, possessed seven different restriction endonuclease sites suitable for cloning, and contained two antibiotic resistance markers selectable in *E. coli*, chloramphenicol and erythromycin, the latter of which was also expressed in streptococci.

Delineation of the pVA380-1 Basic Replicon

The term "basic replicon" has been defined as the smallest contiguous segment of plasmid DNA that is able to replicate with the same properties as the original wild-type plasmid (14). The first goal in the analysis of pVA380-1 replication was to establish the basic replicon portion of the plasmid. This was achieved primarily by the cloning and subcloning of plasmid-mediated antibiotic resistance traits in *S. sanguis* (Fig. 1). Early cloning experiments indicated that the single *Eco*RI, *Hin*dIII, *Ava*I, and *Cla*I sites on the plasmid could be interrupted without affecting replication (9, 11). This information was extended by cloning into pVA380-1 DNA that had been doubly digested with the combinations *Hin*dIII-*Eco*RI, *Hin*dIII-*Ava*I, and *Ava*I-*Hpa*II. Thus, in the construction of pVA736 and pVA738 (11) and pVA797 (16), it was shown that nearly 2 kb of plasmid DNA was not required for replication. A streptococcal spectinomycin resistance gene (L. N. Lee, J. M. Inamine, and D. J. LeBlanc,

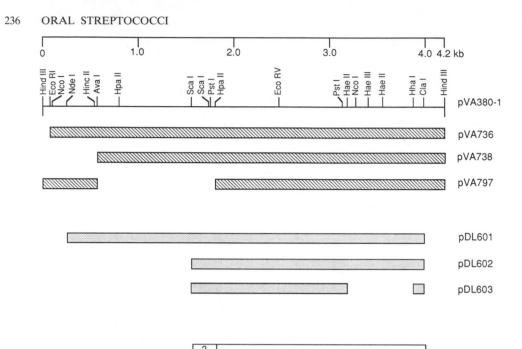

FIG. 1. Delineation of the pVA380-1 basic replicon. The top line illustrates the kilobase coordinates of pVA380-1; the next line represents the restriction endonuclease map of pVA380-1; the remaining lines represent derivatives of pVA380-1 able to replicate in *S. sanguis* (only the pVA380-1 portion of the respective derivative is shown). pVA736 and pVA738 each contained a segment of pVA1 encoding resistance to erythromycin (11); pVA797 contained a segment of pIP501 encoding resistance to chloramphenicol (16); pDL601 and pDL602 contained a 1.1-kb fragment of *E. faecalis* plasmid DNA encoding resistance to spectinomycin (see text). pDL603 was derived from pDL602 by digestion with *Hha*I, religation, and transformation of *S. sanguis* with selection for spectinomycin resistance.

Plasmid **16**:230, 1986) originally cloned in *S. sanguis* on a derivative of pVA380-1 was subcloned into pVA380-1 by using the combinations *Eco*RI-*Cla*I and *Sca*I-*Cla*I, which resulted in the construction of pDL601 and pDL602, respectively. In the construction of pVA797, Macrina and coworkers (described in reference 16) were able to delete all of the DNA between the *Ava*I site and the *Hpa*II site on pVA380-1. However, we were unable to delete DNA beyond the first of two *Sca*I sites from the left of the map (Fig. 1). Thus, the basic replicon has been defined as the 2.5-kb segment of pVA380-1 represented in pDL602, with a question mark between the left-most *Sca*I site and the *Hpa*II site. An attempt to shorten the basic replicon involved digestion of pDL602 with *Hha*I, which also cleaves at *Hae*II sites. The digested DNA was religated and used to transform *S. sanguis*. The deletion derivative pDL603 was obtained, which appeared to delineate further the basic replicon. However, whereas native pVA380-1 as well as pDL602 were maintained stably by *S. sanguis* under nonselective conditions, pDL603 was lost from more than 50% of the population within 20 generations. Thus, since

pDL603 did not replicate with the same property of stability as the wild-type plasmid, the 2.5 kb of contiguous pVA380-1 DNA present in pDL602 was retained as the basic replicon. Visually on agarose gels, pDL602 was maintained at approximately the same copy number as pVA380-1, and it exhibited incompatibility with other pVA380-1 derivatives. All subsequent experiments on the replication of pVA380-1 were initiated with molecules containing only the 2.5-kb basic replicon portion of the plasmid.

Analysis of the pVA380-1 Basic Replicon

Strains of *S. sanguis*, *S. mutans*, *Streptococcus pyogenes*, *Enterococcus faecalis*, and *L. lactis* were able to support the replication of pDL602 or of other pVA380-1 derivatives containing only the 2.5-kb basic replicon. Derivatives containing insertions of the *aphA3* kanamycin resistance gene from pJH1 (9) at the *Eco*RV site or at the *Nco*I site (Fig. 2) transformed and replicated in *S. sanguis*. It was not possible to insert this gene into the *Hpa*II site, a result consistent with other data indicating that the DNA between the first *Sca*I site on the basic replicon and the *Hpa*II site

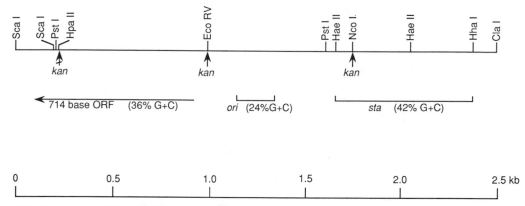

FIG. 2. Analysis of the pVA380-1 basic replicon. The top line represents a restriction endonuclease map of the 2.5-kb basic replicon. Vertical arrows refer to sites of insertion of the *aphA3* (*kan*) gene encoding resistance to kanamycin (see text). Designations below the map refer to the ORF, *ori* (putative origin of replication), and *sta* (stability) regions of the basic replicon, as determined from the nucleotide base sequence (ORF, *ori*, and *sta*) and deletion analysis (*sta*) of the basic replicon.

is necessary for replication. These results suggest that the successful insertion, by Evans et al. (described in reference 16), of a portion of pIP501 in place of the DNA between the *Ava*I and *Hpa*II sites (Fig. 1; pVA797) may reflect a peculiarity of this particular hybrid.

The nucleotide base sequence of the 2.5-kb basic replicon was determined. The G+C content of the entire sequence was 37%. Only one open reading frame (ORF) of significant size, 714 bp (Fig. 2), was found, encoding a protein with a predicted size of 29 kDa. A 200-bp region was particularly A+T-rich and has been temporarily designated the origin of replication; however, there are no firm data to support this assignment. For reasons that are not immediately obvious, the G+C content of the region associated with plasmid stability, as defined by pDL603 (Fig. 1), was considerably higher than in any other region of the replicon. Since the basic replicon could be interrupted at the *Eco*RV site, the DNA between this site and the *Cla*I site, plus the spectinomycin resistance gene of pDL602, was subcloned in both orientations onto a derivative of the *E. coli* vector, pUC19 (20), in which the ampicillin resistance gene had been inactivated. *S. sanguis* could not be transformed with either chimera. Attempts were made to clone the ORF region, on a streptococcal plasmid compatible with pVA380-1, with the intention of providing a putative replication protein in *trans*. To date, it has not been possible to obtain such an ORF-containing hybrid molecule.

Transposon Mutagenesis of the Basic Replicon

Since attempts to supply the ORF function in *trans* by the approach described above were unsuccessful, transposon mutagenesis was used to

establish the ORF requirement for pVA380-1 replication. An *E. coli-Streptococcus* shuttle vector, pDL276 (L. N. Lee, D. J. LeBlanc, and G. M. Dunny, *Abstr. Annu. Meet. Am. Soc. Microbiol. 1989*, H-214, p. 205) was chosen as the starting point for these experiments. The basic features of pDL276 include the pUC origin of replication, the *aphA3* gene that encodes resistance to kanamycin in *E. coli* as well as streptococci, the pVA380-1 basic replicon, and the pUC19 mcs flanked by two strong transcriptional terminators first described by Chen and Morrison (4). Since Tn5, the transposon chosen for mutagenesis, encodes resistance to kanamycin, the *aphA3* gene of pDL276 was replaced with the spectinomycin resistance gene from pDL602, which is also expressed in *E. coli* and streptococci. The new shuttle vector, pDL278, was similar to pDL276 in that both molecules transformed and replicated stably in strains of *S. sanguis*, *S. mutans*, *E. faecalis*, *L. lactis*, and *E. coli*. When purified from a transformant of the Challis strain of *S. sanguis*, either plasmid transformed the plasmid-free Challis strain at a frequency of ~10^{-3}/CFU. The two shuttle vectors differed in size, 6.9 kb (pDL276) and 6.6 kb (pDL278), and in the resistance trait encoded. The new vector also differed from its progenitor in one very advantageous way. When isolated from an *E. coli* transformant, pDL276 transformed the Challis strain at a very low frequency, 10^{-7}/CFU, whereas pDL278 DNA isolated from *E. coli* transformed the Challis strain with a frequency of 10^{-4}/CFU. These results provided an unanticipated advantage for the Tn5 mutagenesis studies.

pDL278 was mutagenized with Tn5 in *E. coli* as described by Yamamoto et al. (19), and the locations of individual transposon insertions were mapped. Approximately 40 insertion mutants contained Tn5 in the pVA380-1 basic replicon segment of the shuttle vector and were tested for their ability to transform and replicate in *S. sanguis*. All insertions within and upstream of the ORF, but to the left of the *Eco*RV site (Fig. 2), prevented replication in *S. sanguis*. No insertions were obtained within the putative origin of replication.

Two of the Tn5 insertion mutants were examined for their ability to transform *E. faecalis* by electroporation. Tn5-8, with Tn5 inserted just to the left of the *Cla*I site (Fig. 2), transformed and replicated stably in the transformants. The second insertion mutant, Tn5-30, with the transposon within the ORF between the *Pst*I and *Hpa*II sites, failed to transform either *S. sanguis* or *E. faecalis*. These results suggested an experiment that might provide information relative to the *trans* activity of the ORF function. The recombination-defective *E. faecalis* strain UV202 (18), containing the kanamycin resistance shuttle vector pDL276, was transformed with pDL278::Tn5-8 and pDL278::Tn5-30 by electroporation. In this way, the resident plasmid, pDL276, would provide the ORF function in *trans*, and because the strain is Rec⁻, recombination between resident and transforming plasmids would be prevented. Both of the mutant plasmids produced transformant colonies on selection plates containing spectinomycin or spectinomycin plus kanamycin. Plasmid DNA was obtained from transformants from each of the four selection plates and was digested with *Eco*RI, which cleaves pDL276 and the pDL278::Tn5 derivatives only once. All transformants selected for resistance to both antibiotics retained both the 6.9-kb pDL276 plasmid and the transforming 12-kb pDL278::Tn5 mutant plasmid. The pDL278::Tn5-8 transformants, selected on spectinomycin alone, contained only the 12-kb transforming plasmid, presumably because of incompatibility between pDL276 and pDL278. However, the isolates transformed with pDL278::Tn5-30, in which the ORF had been interrupted, retained both pDL276 and the transforming plasmid following selection with spectinomycin alone. This result was interpreted as a reflection of the requirement for the ORF function supplied in *trans* by pDL276 in order for pDL278::Tn5-30 to replicate.

Conclusions

pVA380-1 is a relatively broad-host-range, intermediate-copy-number plasmid that is able to replicate stably in at least six streptococcal species. The basic replicon of pVA380-1 is contained within a 2.5-kb contiguous segment of DNA. Deletion analysis of the basic replicon revealed a region associated with plasmid stability that may represent a partition function. The nucleotide base sequence revealed the presence of only one significant ORF of 714 bp and a 200-bp putative origin region with an A+T content of 76%. Tn5 insertional mutagenesis was used to provide data suggesting a requirement for the ORF region in replication and that the ORF function can be provided in *trans*.

LITERATURE CITED

1. **Behnke, D., M. S. Gilmore, and J. J. Ferretti.** 1982. pGB301 vector plasmid family and its use for molecular cloning in streptococci, p. 239–242. *In* D. Schlessinger (ed.), *Microbiology—1982.* American Society for Microbiology, Washington, D.C.
2. **Burdett, V.** 1980. Identification of tetracycline-resistant R-plasmids in *Streptococcus agalactiae* (group B). *Antimicrob. Agents Chemother.* **18:**753–760.
3. **Chang, A. C. Y., and S. N. Cohen.** 1978. Construction and characterization of amplifiable multicopy DNA cloning vehicles from the p15A cryptic miniplasmid. *J. Bacteriol.* **134:**1141–1156.
4. **Chen, J., and D. A. Morrison.** 1987. Cloning of *Streptococcus pneumoniae* DNA fragments in *Escherichia coli* requires vectors protected by strong transcriptional terminators. *Gene* **55:**179–187.
5. **de Vos, W.** 1987. Gene cloning and expression in lactic streptococci. *FEMS Microbiol. Rev.* **46:**281–295.
6. **Gruss, A., and S. D. Ehrlich.** 1989. The family of highly interrelated single-stranded deoxyribonucleic acid plasmids. *Microbiol. Rev.* **53:**231–241.
7. **Kues, U., and U. Stahl.** 1989. Replication of plasmids in gram-negative bacteria. *Microbiol. Rev.* **53:**491–516.
8. **Lacks, S., P. Lopez, B. Greenberg, and M. Espinosa.** 1986. Identification and analysis of genes for tetracycline resistance and replication functions in the broad-host-range plasmid pLS1. *J. Mol. Biol.* **192:**753–765.
9. **LeBlanc, D. J., J. M. Inamine, and L. N. Lee.** 1986. Broad geographical distribution of homologous erythromycin, kanamycin, and streptomycin resistance determinants among group D streptococci of human and animal origin. *Antimicrob. Agents Chemother.* **29:**549–555.
10. **LeBlanc, D. J., L. N. Lee, B. M. Titmas, C. J. Smith, and F. C. Tenover.** 1988. Nucleotide sequence analysis of tetracycline resistance gene *tetO* from *Streptococcus mutans* DL5. *J. Bacteriol.* **170:**3618–3626.
11. **Macrina, F. L., K. R. Jones, and P. H. Wood.** 1980. Chimeric streptococcal plasmids and their use as molecular cloning vehicles in *Streptococcus sanguis* (Challis). *J. Bacteriol.* **143:**1425–1435.
12. **Macrina, F. L., J. A. Tobian, R. P. Evans, and K. R. Jones.** 1982. Molecular cloning strategies for the *Streptococcus sanguis* host-vector system, p. 234–238. *In* D. Schlessinger (ed.), *Microbiology—1982.* American Society for Microbiology, Washington, D.C.
13. **Macrina, F. L., P. H. Wood, and K. R. Jones.** 1980. Genetic transformation of *Streptococcus sanguis* (Challis) with cryptic plasmids from *Streptococcus ferus. Infect. Immun.* **28:**692–699.
14. **Nordstrom, K.** 1985. Control of plasmid replication: theoretical considerations and practical solutions, p. 189–214. *In* D. R. Helinski, S. N. Cohen, D. B. Clewell, D. A. Jackson, A. Hollaender, L. Hager, S. Kaplan, J. Konisky, and C. M. Wilson (ed.), *Plasmids in Bacteria.* Plenum Press, New York.
15. **Novick, R. P.** 1989. Staphylococcal plasmids and their replication. *Annu. Rev. Microbiol.* **43:**537–565.
16. **Pouwels, P. H., B. E. EngerValk, and W. J. Bramar.** 1985. Cloning vectors: a laboratory manual. Elsevier Science Publishers, Amsterdam.
17. **Tobian, J. A., and F. L. Macrina.** 1982. Helper plasmid cloning in *Streptococcus sanguis*: cloning of a tetracycline

resistance determinant from the *Streptococcus mutans* chromosome. *J. Bacteriol.* **152:**215–222.

18. **Yagi, Y., and D. B. Clewell.** 1980. Recombination-deficient mutant of *Streptococcus faecalis. J. Bacteriol.* **143:**966–970.

19. **Yamamoto, M., J. M. Jones, E. Senghas, C. Gawron-Burke, and D. B. Clewell.** 1987. Generation of Tn5 inser-

tions in the streptococcal conjugative transposon Tn*916. Appl. Environ. Microbiol.* **53:**1069–1072.

20. **Yanisch-Perron, C., J. Vieira, and J. Messing.** 1985. Improved M13 phage cloning vectors and host strains: nucleotide sequences of the M13mp18 and pUC19 vectors. *Gene* **33:**103–119.

Molecular Structure of Fimbriae-Associated Genes of *Streptococcus sanguis*

P. FIVES-TAYLOR, J. C. FENNO, E. HOLDEN, L. LINEHAN, L. OLIGINO, AND M. VOLANSKY

Department of Microbiology and Molecular Genetics, University of Vermont, Burlington, Vermont 05405

Streptococcus sanguis is believed to play a key role in the initiation of plaque development as a primary colonizer of the tooth surface (1, 13, 14) and in the accumulation of dental plaque by coaggregating with other bacterial species (2).

The species *S. sanguis* is ill defined. Several investigators suggest that these organisms belong to at least four distinct genetic groups (3, 11), with as little as 30% DNA homology between any two groups. All strains appear to have in common properties required for the colonization of the tooth surface and detectable surface structures. The latter include fimbriae (relatively long and having a distinct width along their length) and fibrils (short and width unable to be measured because they aggregate in the stain) (10).

Our laboratory has focused on adhesion mechanisms associated with *S. sanguis* FW213, a strain of genetic group IV according to Coykendall's scheme (3). The organism has numerous long, peritrichous fimbriae (3.5 nm) and short, peritrichous fibrils (5). Fimbriae play a major role in the adhesion of FW213 to saliva-coated hydroxyapatite (SHAP) (5). Nonadherent mutants isolated strictly on the inability to adhere to SHAP were fimbria negative (8). Fab fragments of monoclonal antibodies that bind specifically to fimbriae block the adhesion of FW213 to SHAP by greater than 90% (4).

In contrast to the extensive body of knowledge on fimbriae of gram-negative organisms, very little is known about the fimbriae of gram-positive organisms. One reason for this paucity of knowledge is the inability to dissociate intact fimbriae into subunits. It has been suggested that the subunits are covalently linked, since treatment of intact fimbriae with hydrophobic and ionic disrupting agents alone or in combination fails to dissociate them (S. Fachon-Kalweit, Ph.D. dissertation, University of Vermont, Burlington, 1985). Therefore, strategies were developed to identify and characterize putative subunits and regulatory genes involved in fimbrial expression by cloning the respective genes into *Escherichia coli*.

Cloning and Sequencing of a Fimbriae-Related Gene

A 6-kb *Eco*RI fragment of DNA encoding an FW213 antigen was cloned in *E. coli* (7). The nucleotide sequence of a 927-bp open reading frame (ORF), *fimA*, coding for a protein with an apparent size of 36 kDa was determined (6). The location of *fimA* within the 6-kb fragment was confirmed by insertional mutagenesis within the ORF, which resulted in loss of antigenicity. A signal sequence of 29 amino acids, with a cleavage site at an alanine residue, is suggested. Except for the signal sequence, *fimA* has little or no homology to any known sequenced gene.

VT321, a nonadhesive mutant of FW213 (8), was electrotransformed with a construct containing the *E. coli-S. sanguis* shuttle vector pDL276 (3a) and the 6-kb *Eco*RI FW213 fragment. A transformant, maintaining the construct as a plasmid, expressed the 36-kDa antigen and had approximately 79% of the adhesive capacity of the wild type (data not shown).

A 2.9-kb fragment of DNA encoding a 36-kDa protein (SsaB) was cloned in *E. coli* from *S. sanguis* 12, an organism with peritrichous fibrils only (9). SsaB mediates the binding of *S. sanguis* 12 to SHAP. A comparison between the deduced amino acid compositions of FimA and SsaB (N. Ganeshkumar, Ph.D. dissertation, University of British Columbia, Vancouver, British Columbia, Canada, 1988) suggests that the two molecules are very similar, indicating a common origin. SsaB does not appear to be associated with the fibrils of *S. sanguis* 12. Antiserum raised against the purified fibrils does not react with purified SsaB. Furthermore, anti-SsaB serum does not react with purified fibrils. Yet when *S. sanguis* 12 was reacted with immunogold-labeled anti-SsaB, a halo at a point removed from the cell was noted. FimA is associated with the fimbriae of FW213. Perhaps SsaB is associated with the fibrils of *S. sanguis* 12 and the methods used to purify the fibrils removed the SsaB antigen as well.

A DNA fragment internal to *fimA* was used to probe *S. sanguis* strains from each of the genetic groups and several clinical isolates in Southern hybridization experiments. All strains of *S. sanguis* tested contained at least one copy of *fimA* (data not shown). Thus, this gene encodes an adhesin universal among *S. sanguis* strains. Kolenbrander and Andersen (12) identified a 38-kDa protein in *S. gordonii* (*S. sanguis*) implicated in coaggregation reactions. This protein also was found in all *S. sanguis* strains tested. The relative mobilities of the two proteins are similar, and it would be extremely interesting if they proved to be related.

Analysis of DNA Regions Flanking *fimA*

A segment spanning 4 kb of the 6-kb *Eco*RI fragment was sequenced. Four ORFs were identified, including the one associated with *fimA* (Fig. 1). All ORFs were preceded by typical ribosome binding sites. Only 11 bp separated the stop codon of ORF 1 and the start codon of *fimA*, and no transcriptional terminators were identified between them. The amino acid sequence deduced from ORF 1 predicts an extremely hydrophobic protein of 29.9 kDa. ORF 2 coded for a predicted protein of 16 kDa and was contained within ORF 1. ORF 3, downstream of *fimA*, coded for a predicted protein of 20 kDa. The *E. coli* clone isolated by Ganeshkumar et al. (9) also contained DNA coding for a 20-kDa protein.

Total cell RNA was isolated from FW213 and subjected to Northern (RNA) blot hybridization (Fig. 2). A probe, an internal 660-bp fragment of *fimA*, hybridized to a single transcript of 3.2 kb. This result suggested that *fimA* was transcribed as part of a polycistronic message. Initial attempts to transform nonadherent mutants to wild type with the cloned 6-kb *Eco*RI FW213 fragment were not successful when the entire plasmid (vector plus insert) recombined into the chromosome. One explanation for this result is that Campbell insertion of the plasmid via the homologous insert led to disruption of the polycistronic message and premature termination of the required proteins.

Tn*phoA* Mutagenesis of the 6-kb *Eco*RI Fragment

Tn*phoA* insertions within the 6-kb *Eco*RI fragment in *E. coli* were used in an attempt to identify additional genes upstream of *fimA* transcribed in the same polycistronic message. Disruption of expression of *fimA* occurred only when the insertions were within the *fimA* reading frame. At least four transcripts hybridized to the internal

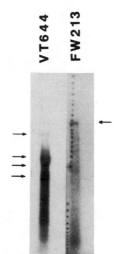

FIG. 2. Northern blot hybridization of cellular RNA from *S. sanguis* FW213 and the *E. coli* clone containing the 6-kb *Eco*RI FW213 fragment (VT644). Arrows point to bands that hybridized with the *fimA* probe.

fimA fragment when total cellular RNA from the *E. coli* clone carrying the 6-kb *Eco*RI FW213 fragment was probed by Northern blot (Fig. 2). The different transcript sizes may be due to variable readthrough of streptococcal terminators by *E. coli*, or *E. coli* may recognize promoters not recognized by *S. sanguis*. One of the *E. coli* transcripts was 1.2 kb, suggesting that the putative promoters just upstream of *fimA* (6) are functional, at least in *E. coli*.

Unexpectedly, Tn*phoA* insertions in *fimA* led to overexpression of the 20-kDa protein whose coding region was downstream from *fimA*. This effect was not an enhancer effect of the transposon, since it occurred with the transposon inserted in either orientation. Perhaps *fimA* regulates the expression of the 20-kDa protein, or perhaps the 36- and 20-kDa proteins interact. Further experiments are needed to determine the nature of the overproduction of the 20-kDa protein.

Insertion of Tn*phoA* directly downstream of the putative signal sequence-coding region of *fimA* resulted in the expression of alkaline phosphatase activity, thus confirming the FimA signal sequence as functional in *E. coli*. The 36-kDa protein was found in the cytoplasm and in the supernatant but not in the *E. coli* periplasmic space. Western immunoblot analysis of the 36-kDa protein revealed a doublet of bands at 36 and 34 kDa, respectively. The secreted form is the "clipped" lower band at 34 kDa.

Western Blot Analysis of Nonadherent Mutants of FW213

Concentrated extracts of FW213 lysed cells (LE) and of surface components isolated by

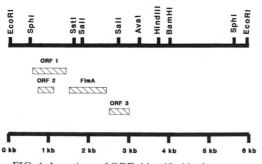

FIG. 1. Locations of ORFs identified in the sequence of the 6-kb FW213 *Eco*RI fragment. All ORFs are transcribed from left to right.

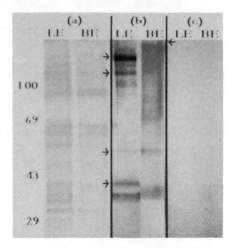

FIG. 3. Western blot analysis of LE and BE of FW213 nonadhesive mutants. (a) Total protein stained with amido black; (b) total protein with fimbria-specific antiserum; (c) total protein probed with preimmune antiserum.

shearing the cells in a Sorvall Omnimixer (BE) were probed with fimbriae-specific antiserum by Western blot (Fig. 3). LE contained a series of fimbriae-specific proteins at 36, 53, 125, and 160 kDa, although the 53-kDa protein was found in very low concentrations. The same proteins were found in BE except that the 53-kDa protein was

in much higher concentrations, and the larger forms ran as a smear from 200 to 80 kDa. Sodium dodecyl sulfate-polyacrylamide gel electrophoresis of BE, with subsequent periodic acid-Schiff reagent staining, was positive only for the 200-kDa protein, suggesting that this protein was glycosylated. Monoclonal antibody F51 (4) reacts only with the 200-kDa protein in BE and the 160-kDa protein in LE (E. Holden, M.S. thesis, University of Vermont, Burlington, 1990). Treatment with proteases of nitrocellulose strips containing electrophoresed BE did not affect the binding of monoclonal antibody F51. However, incubation with periodate at levels as low as 5 to 20 mM completely inhibited it. Since low concentrations of periodate inactivate carbohydrates, these data are consistent with the hypothesis that the 200-kDa protein was a glycoprotein. Further experiments utilizing endoglycosidases are required to confirm these results.

Nonadherent mutants of FW213 were isolated by the use of chemical mutagens, particularly ethyl methanesulfonate (8). All mutants were independent isolates and were mutagenized in a manner that favored single-hit kinetics. The results from Western blot analyses of LE of these mutants probed with fimbriae-specific antiserum are summarized in Table 1. The most striking feature is that all of the mutants lacked the 36-kDa protein. These data are consistent with the observation that the cloned 36-kDa protein in *E.*

TABLE 1. Summary of Western blot analysis of extracts of lysed cells of nonadherent mutants

Phenotypic group[a]	Strain	Protein band[b]			
		36 kDa	53 kDa	125 kDa	160 kDa
	FW213	−	+	+	+
I	VT321	−	+	−	−
	VT325	−	−	−	−
	VT377	−	+	+	+
	VT378	−	+	−	+
	VT361	−	+	−	+
	VT360	−	−	−	−
II	VT344	−	−	−	−
	VT342	−	−	−	−
III	VT345	−	+	+	+
	VT324	−	+ +	−	+
V	VT379	−	+	+	+
	VT380	−	+	−	+
VI	VT343	−	+	+	+
VII	VT346	−	+	+	+
	VT367	−	−	−	−
	VT357	−	−	−	−

[a] From reference 8.
[b] +, Band present; −, band absent; + +, greater than wild-type amounts.

coli may be an adhesin. All mutants that expressed any of the larger fimbrial proteins always expressed the 53-kDa protein as well. These data suggest the 53-kDa protein may be the structural subunit of the fimbriae. Interestingly, Ganeshkumar et al. (9) isolated a clone in *E. coli* that expressed a 50-kDa protein that reacted with antifibril serum. Perhaps these two proteins are also related.

The mutants were placed into three groups on the basis of their reactions with fimbriae-specific antiserum. Six mutants did not make any of the fimbria-specific antigens (group A). Two of the mutants made only the 53-kDa protein, and one of these mutants, VT324, overproduced it (group B). Eight of the mutants made both the 53- and 160-kDa proteins (group C). However, the 160-kDa protein was no longer reactive with monoclonal antibody F51 in three of the group C mutants, indicating that an epitope change had occurred.

Summary

The expression of fimbriae in FW213 appears to be a complex process. A number of specific antigens are involved, but no confirmed role has been assigned to any of them as yet. The 36-kDa protein appears to be an adhesin. We propose that the 53-kDa protein is the structural subunit of the fimbriae. Additional experiments are required to determine the functions of the ORFs and the nature and extent of the polycistronic message.

This investigation was supported by Public Health Service grant RO1-DE05606 from the National Institute of Dental Research.

LITERATURE CITED

1. Carlsson, J., H. Grahnen, G. Jonsson, and S. Wikner. 1970. Establishment of *S. sanguis* in the mouths of infants. *Arch. Oral Biol.* **15**:1143–1148.

2. Cisar, J. O., M. J. Brennan, and A. L. Sandberg. 1985. Lectin-specific interaction of *Actinomyces* fimbriae with oral streptococci, p. 159–163. *In* S. E. Mergenhagen and B. Rosan (ed.), *Molecular Basis of Oral Microbial Adhesion*. American Society for Microbiology, Washington, D.C.

3. Coykendall, A. L. 1989. Classification and identification of viridans streptococci. *Clin. Microbiol. Rev.* **2**:315–328.

3a. Dunny, G. M., L. N. Lee, and D. J. LeBlanc. 1991. Improved electroporation and cloning vector system for gram-positive bacteria. *Appl. Environ. Microbiol.* **57**:1194–1201.

4. Elder, B. L., and P. Fives-Taylor. 1986. Characterization of monoclonal antibodies specific for adhesion: isolation of an adhesin of *Streptococcus sanguis* FW213. *Infect. Immun.* **54**:421–427.

5. Fachon-Kalweit, S., B. Elder, and P. Fives-Taylor. 1985. Antibodies that bind to fimbriae block adhesion of *Streptococcus sanguis* to saliva-coated hydroxyapatite. *Infect. Immun.* **48**:617–624.

6. Fenno, J. C., D. J. LeBlanc, and P. Fives-Taylor. 1989. Nucleotide sequence analysis of a type 1 fimbrial gene of *Streptococcus sanguis* FW213. *Infect. Immun.* **57**:3527–3533.

7. Fives-Taylor, P. M., F. L. Macrina, T. J. Pritchard, and S. S. Peene. 1987. Expression of *Streptococcus sanguis* antigens in *Escherichia coli*: cloning of a structural gene for adhesion fimbriae. *Infect. Immun.* **55**:123–128.

8. Fives-Taylor, P., and D. Thompson. 1985. Surface properties of *Streptococcus sanguis* FW213 mutants nonadherent to saliva-coated hydroxyapatite. *Infect. Immun.* **47**:752–759.

9. Ganeshkumar, N., M. Song, and B. McBride. 1988. Cloning of a *Streptococcus sanguis* adhesin which mediates binding to saliva-coated hydroxyapatite. *Infect. Immun.* **56**:1150–1157.

10. Handley, P. S., P. L. Carter, J. E. Wyatt, and L. M. Hesketh. 1985. Surface structures (peritrichous fibrils and tufts of fibrils) found on *Streptococcus sanguis* strains may be related to their ability to coaggregate with other oral genera. *Infect. Immun.* **47**:217–227.

11. Killian, M., M. Mikkelsen, and J. Henrichsen. 1989. Taxonomic study of viridans streptococci: description of *Streptococcus gordonii* sp. nov. and emended descriptions of *Streptococcus sanguis* (White and Niven, 1946), *Streptococcus oralis* (Bridge and Sneath, 1982), and *Streptococcus mitis* (Andrews and Horder, 1906). *Int. J. Syst. Bacteriol.* **39**:471–484.

12. Kolenbrander, P. E., and R. N. Andersen. 1990. Characterization of *Streptococcus gordonii* (*S. sanguis*) PK488 adhesin-mediated coaggregation with *Actinomyces naeslundii* PK606. *Infect. Immun.* **58**:3064–3073.

13. Nyvad, B., and M. Kilian. 1987. Microbiology of the early colonization of human enamel and root surfaces in vivo. *Scand. J. Dent. Res.* **95**:369–380.

14. Socransky, S. S., S. D. Manganiello, D. Propas, Y. Oram, and J. van Houte. 1977. Bacteriological studies of developing supragingival dental plaque. *J. Periodont. Res.* **12**:90–106.

Binding Protein-Dependent Transport System in *Streptococcus mutans*

R. R. B. RUSSELL,[1] J. ADUSE-OPOKU,[1] L. TAO,[2] AND J. J. FERRETTI[2]

Hunterian Dental Research Unit, London Hospital Medical College, London, E1 2AD, United Kingdom,[1] and Department of Microbiology, University of Oklahoma, Oklahoma City, Oklahoma 73190[2]

In the oral streptococci, the mechanisms by which sugars are transported is of interest because of the contribution of fermentable carbohydrates to the production in dental plaque of acids, which can cause enamel demineralization and result in dental caries. Sucrose has attracted most interest because of its dual role as a substrate for extracellular glucosyltransferases (GTF) and fructosyltransferases (FTF), which form polymers important in the formation and maintenance of plaque, and as a metabolic substrate which enters glycolytic pathways, leading to acid production. However, other sugars in the diet also contribute to the activity of plaque bacteria, and there is increasing awareness that sugars released from host glycoproteins such as salivary mucins provide the primary nutrient for bacteria in the oral cavity (18). *Streptococcus mutans* is the bacterium most closely implicated as a cause of dental caries; much has been ascertained about its saccharolytic capabilities, though the complexity of the many pathways present in this organism has complicated physiological and biochemical studies. Consequently, a number of laboratories have used molecular genetic approaches in their investigations, and we describe here the results obtained from extensive nucleotide sequencing of the region of the *S. mutans* chromosome surrounding the *gtfA* gene. This region specifies three enzymes involved in sugar metabolism, four genes involved in transport, and a regulatory protein.

The Sucrose Phosphorylase Gene, *gtfA*

We have described a procedure for screening a gene bank of fragments of the *S. mutans* chromosome cloned in bacteriophage lambda for expression of enzymes that utilize sucrose; this procedure relies on plating the phage and host *Escherichia coli* on medium with sucrose as the sole carbon source (16). By this method, we found recombinant plaques that expressed GTF or FTF activity and accumulated polymer above the area of lysis. However, the majority of sucrose-degrading recombinants did not accumulate polymer but did show cross-feeding of the surrounding *E. coli* (which can utilize monosaccharides but not sucrose), and these recombinants were found to carry either the *scrB* gene for su-

crose-6-phosphate hydrolase or the *gtfA* gene. The latter gene encodes the enzyme sucrose phosphorylase, which has sucrose as its only known substrate and transfers the glucose moiety to P_i to form glucose-1-phosphate, fructose being the other product of the reaction (17). Since the *gtfA* enzyme has the characteristics of an intracellular enzyme, it becomes necessary to explain the means by which sucrose becomes available inside the cell. Sequence data indicated that another open reading frame began immediately downstream of *gtfA* and probably cotranscribed with it (7), so interest was focused on this region, where others had previously concluded that proteins were encoded (5, 14).

The Dextran Glucosidase Gene, *dexB*

Burne et al. (5) reported the existence of a gene linked to *gtfA* that encoded an enzyme capable of releasing reducing sugars from high-molecular-weight dextran, i.e., acted as a dextranase. Further work on this gene, designated *dexB*, has shown that it specifies an enzyme that is much more efficient at releasing glucose from short isomaltosaccharides three or four glucose units long than from dextran. This preference for small substrates is consistent with the fact that it is located inside the cell, as deduced from sequencing data (15). It is tempting to propose that this intracellular enzyme (dextran glucosidase) acts in series with the extracellular dextranase which cleaves high-molecular-weight dextran into the shorter oligosaccharides, but this has yet to be demonstrated. It should be noted that *S. mutans* is unable to produce acid from dextran even though it appears to possess the requisite enzymes, and it is possible that the true substrate of the *dexB* enzyme is some other α-glucoside. α-Glucosidases generally have quite wide substrate specificities, and it appears that *dexB* is the only α-glucosidase in *S. mutans*; when *dexB* is inactivated by insertional mutagenesis, all α-glucosidase activity is lost.

The α-Galactosidase Gene, *aga*

Among the collection of recombinant phage that formed cross-feeding plaques on sucrose medium, approximately half of those carrying *gtfA* were found also to show cross-feeding on medium containing raffinose (a trisaccharide of galactose-glucose-fructose). Such an activity could be due to α-galactosidase, invertase, or FTF. Of these three possible activities, only α-galactosidase could be detected in lysates of the recombinants, suggesting the linkage of the gene encoding this enzyme to *gtfA*. The α-galactosidase encoded by this gene (*aga*) is again an intracellular enzyme, with a deduced molecular weight of 82,040. It releases galactose frc n melibiose (galactose-glucose) and from raffinose; in the latter case, sucrose is the other product. The presence of the *aga* enzyme thus provides a likely explanation for the generation of sucrose inside the cell, if raffinose is transported intact across the membrane.

A proposed pathway by which melibiose and raffinose might be metabolized by the *aga* and *gtfA* enzymes is shown in Fig. 1. Note that raffinose will also be utilized by extracellular FTF and perhaps by an invertase, though there is uncertainty as to whether a distinct invertase exists in *S. mutans* (other enzymes, including GTF, FTF, GtfA, and ScrB, can display invertaselike activity).

The deduced amino acid sequence of the α-galactosidase of *S. mutans* showed no obvious homology to the chromosomally determined (*melA*) gene of *E. coli* or to human or yeast α-galactosidases. It did, however, show extensive homology to the product of the *rafA* gene carried by Raf plasmids, which enable *E. coli* to grow on raffinose (3), and also to the two α-galactosidases from *Bacillus stearothermophilus* (8; R. Mattes, personal communication). The apparent relationship between the *E. coli* plasmid gene and those from a streptococcus and a bacillus suggests that its origin may have been in a gram-positive organism.

Transport Genes

Nucleotide sequencing of the region between *aga*, *gtfA*, and *dexB* has revealed the existence of further open reading frames (Fig. 2). The deduced amino acid sequences of the proteins encoded by these genes were compared with known sequences in data banks, and the unexpected result was obtained: all showed strong homology to components of binding-protein-dependent transport systems of gram-negative bacteria. Such systems commonly consist of a substrate-binding protein, located in the periplasm, one or two integral membrane proteins, and a cytoplasmic component that binds ATP and is believed to energize the uptake mechanism (2).

Immediately downstream of *aga*, and separated from it by only 14 bp, is a gene specifying a protein the first 29 amino acids of which conform to the pattern found in the signal peptides of other extracellular streptococcal proteins. The deduced amino acid sequence shows a strong homology to that of the maltose-binding protein of *E. coli* (22.8% identical amino acids in a 333-amino-acid overlap). The deduced amino acid sequences of the next two gene products also have homology to *E. coli* maltose uptake system components, namely, the inner membrane proteins encoded by *malF* and *malG*, and to the corresponding proteins of the *ugp* operon, which is involved in uptake of *sn*-glycerol-3-phosphate (13). These similarities are reflected in closely similar hydropathy profiles with seven transmembrane domains. The final component expected to complete the transport system is an ATP-binding protein; the region between *gtfA* and *dexB* encodes such a product, which shows strong homology to other known ATP-binding proteins from a wide range of organisms, possessing all of the universally conserved residues (12).

The finding of a binding-protein-dependent uptake system was unexpected, since until recently it was thought that such systems were confined to gram-negative bacteria, in which the binding protein is located in the periplasmic space. However, Gilson et al. (9) recently observed that the product of the *S. pneumoniae malX* gene, known to be somehow involved in transport, had homology to the *E. coli* maltose-binding protein and have also deduced the existence of another binding protein locus (*amiA*) in *S. pneumoniae*. The *ami* genes have now been characterized in more detail by Alloing et al. (1). In addition to these

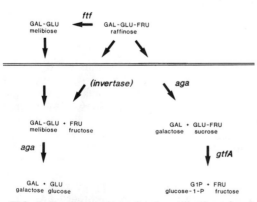

FIG. 1. Possible pathway for the utilization of melibiose and raffinose in *S. mutans*, showing involvement of the enzymes specified by genes *ftf* (fructosyltransferase), *aga* (α-galactosidase), and *gtfA* (sucrose phosphorylase).

FIG. 2. Arrangement of genes on the *S. mutans* chromosome. *R*, Activator gene; *aga*, α-galactosidase; 1, binding protein; 2 and 3, integral membrane components; *gtfA*, sucrose phosphorylase; 4, ATP-binding protein; *dexB*, dextran glucosidase.

examples in the streptococci, Dudler et al. (6) have reported the presence in *Mycoplasma hyorhinis* of a group of genes specifying the several components of a binding-protein-dependent system, of function as yet unknown, and preliminary mention has been made of such uptake systems in *Mycobacterium* and *Bacillus* species (11). One of the puzzles to be solved concerns the location of the binding protein component, which is confined by the outer membrane in gram-negative bacteria but would seem likely to be lost to the surroundings of a streptococcus once secreted. It has been proposed that the binding proteins remain cell associated by virtue of being lipoproteins, the lipid moiety being attached to a cysteine residue at the N terminus of the mature binding protein after removal of the signal peptide. The sequence of the *S. mutans* binding protein in the proposed cleavage region is Leu-Ala-Ala-Cys, which fits closely to the consensus sequence in known bacterial lipoproteins (19).

Function of the Transport System

Two clues exist to the nature of the substrate transported by the group of genes identified: the presence of *aga* and *gtfA* genes suggested melibiose and/or raffinose as possibilities, and Barletta and Curtiss (4) observed that inactivation of *gtfA* resulted in *S. mutans* mutants unable to ferment melibiose. This result was difficult to interpret in the light of what is known about the function of the *gtfA* enzyme but can readily be explained now it is known that the gene for the ATP-binding protein is separated from *gtfA* by only 14 nucleotides and cotranscribed with *gtfA*. We have now demonstrated that inactivation of *gtfA* by insertion of an antibiotic resistance marker prevents the uptake of [^{3}H]melibiose and acid production from melibiose. Insertional inactivation of any of the other transport component genes or of *aga* also results in mutants unable to ferment melibiose, while acid production from raffinose is reduced (presumably residual activity is due to the action of FTF). Inactivation of *dexB* or any of the structural genes upstream of it result in loss of α-glucosidase activity and hence of the ability to ferment isomaltose or isomaltotriose. It remains to be discovered whether the transport of other substrates, or other physiological activities, is affected by the mutations.

Regulation

The polar effect of mutations in *aga* on the transport genes and of upstream mutations on *dexB* suggested that all three genes encoding enzymes, as well as the four transport system genes, were organized in an operonlike arrangement and cotranscribed. The entire region appears to be under the control of a gene that is divergently transcribed from a point immediately upstream of *aga*: expression of *aga* requires the presence of this gene (unless *aga* is cloned under the control of another promoter, such as that of *lacZ*), which has all the properties of a positive effector (activator) gene. The deduced amino acid sequence of this activator shows substantial homology to a family of DNA-binding proteins, of which the best known is the *E. coli araC* gene product (10), several of which are also known to be divergently transcribed from the genes they control. It is interesting to note that the gene to which the *S. mutans* activator shows greatest similarity is the *melR* gene of *E. coli* (20), though the α-galactosidase and transport genes in the two organisms are completely different. While melibiose can induce α-galactosidase in *S. mutans*, raffinose is a much better inducer; it therefore seems likely that raffinose is transported into the cell by a mechanism other than that described here and that raffinose or a product derived from it (maybe intracellular melibiose) interacts with the activator protein to promote transcription of the operon.

Conclusion

There is evidence for a strong evolutionary conservation of sugar transport systems across taxonomic boundaries; for example, components of the phosphotransferase system from gram-positive and gram-negative bacteria show a high degree of homology, and homologies have been reported between proton symport transport systems of prokaryotic and eukaryotic cells. The finding of a binding-protein-dependent uptake system in *S. mutans* thus broadens the range of species in which such systems are found. In *S. mutans*, it functions alongside phosphotransferase and proton motive force-driven systems, and it seems probable that other examples of such uptake systems remain to be recognized in this and other gram-positive bacteria. It will also be necessary to reevaluate some of the literature on sugar transport in *S. mutans*, taking note of the possibility that previously unrecognized mechanisms may have been functioning.

While our present data indicate that the group of genes we have identified serve to transport melibiose and to metabolize melibiose and raffinose, it is important to bear in mind that other α-galactosides may be encountered in nature. Similarly, we cannot yet be sure of the physiological role of the *dexB* enzyme—does it really act on isomaltosaccharides or on some other α-glucosides? At present we are faced with the enigma of why this enzyme should be under the same regulatory control as the other genes identified. Further exploration of the physiological capabilities of wild-type and defined mutant strains should help to clarify the function of this and other sugar metabolic pathways in *S. mutans*.

LITERATURE CITED

1. **Alloing, G., M.-C. Trombe, and J.-P. Claverys.** 1990. The *ami* locus of the gram-positive bacterium *Streptococcus pneumoniae* is similar to binding protein-dependent transport operons of gram-negative bacteria. *Mol. Microbiol.* **4:**633–644.

2. **Ames, G. F. L.** 1988. Structure and mechanism of bacterial periplasmic transport systems. *Bioenerg. Biomembr.* **20:**1–18.

3. **Aslanidis, C., K. Schmid, and R. Schmitt.** 1989. Nucleotide sequences and operon structure of plasmid-borne genes mediating uptake and utilization of raffinose in *Escherichia coli. J. Bacteriol.* **171:**6753–6763.

4. **Barletta, R. G., and R. Curtiss III.** 1989. Impairment of melibiose utilization in *Streptococcus mutans* serotype c *gtfA* mutants. *Infect. Immun.* **57:**992–995.

5. **Burne, R. A., B. Rubinfeld, W. H. Bowen, and R. E. Yasbin.** 1986. Tight genetic linkage of a glucosyltransferase and dextranase of *Streptococcus mutans* GS-5. *J. Dent. Res.* **65:**1392–1401.

6. **Dudler, R., C. Schmidhauser, W. Parish, E. H. Wettenhall, and T. Schmidt.** 1988. A mycoplasma high-affinity transport system and the in vitro invasiveness of mouse sarcoma cells. *EMBO J.* **7:**3963–3970.

7. **Ferretti, J. J., T.-T. Huang, and R. R. B. Russell.** 1988. Sequence analysis of the glucosyltransferase A gene (*gtfA*) from *Streptococcus mutans* Ingbritt. *Infect. Immun.* **56:**1585–1588.

8. **Ganter, C., A. Bock, P. Buckel, and R. Mattes.** 1988. Production of thermostable, recombinant α-galactosidase suitable for raffinose elimination from sugar beet syrup. *J. Biotechnol.* **8:**301–310.

9. **Gilson, E., G. Alloing, T. Schmidt, J. P. Claverys, R. Dudler, and M. Hofnung.** 1988. Evidence for high affinity binding-protein dependent transport systems in Gram-positive bacteria and in Mycoplasma. *EMBO J.* **7:**3371–3374.

10. **Henikoff, S., J. C. Wallace, and J. P. Brown.** 1990. Finding protein similarities with nucleotide sequence databases. *Methods Enzymol.* **183:**111–132.

11. **Higgins, C. F., M. P. Gallagher, S. C. Hyde, M. L. Mimmack, and S. R. Pearce.** 1990. Periplasmic binding protein-dependent transport systems: the membrane-associated components. *Philos. Trans. R. Soc. London Ser. B* **326:**353–365.

12. **Higgins, C. F., M. P. Gallagher, M. L. Mimmack, and S. R. Pearce.** 1988. A family of closely related ATP-binding subunits from prokaryotic and eukaryotic cells. *BioEssays* **8:**111–116.

13. **Overduin, P., W. Boos, and J. Tommassen.** 1988. Nucleotide sequence of the *ugp* genes of *Escherichia coli* K-12: homology to the maltose system. *Mol. Microbiol.* **2:**767–775.

14. **Pucci, M. J., and F. L. Macrina.** 1986. Molecular organization of the *gtfA* gene of *Streptococcus mutans* LM7. *Infect. Immun.* **54:**77–84.

15. **Russell, R. R. B., and J. J. Ferretti.** 1990. Nucleotide sequence of the dextran glucosidase (*dexB*) gene of *Streptococcus mutans. J. Gen. Microbiol.* **136:**803–810.

16. **Russell, R. R. B., P. Morrissey, and G. Dougan.** 1985. Cloning of sucrase genes from *Streptococcus mutans* in bacteriophage lambda. *FEMS Microbiol. Lett.* **30:**37–41.

17. **Russell, R. R. B., H. Mukasa, A. Shimamura, and J. J. Ferretti.** 1988. *Streptococcus mutans gtfA* gene specifies sucrose phosphorylase. *Infect. Immun.* **56:**2763–2765.

18. **Smith, K., and D. Beighton.** 1986. The effect of the availability of diet on exoglycosidases in the supragingival plaque of monkeys. *J. Dent. Res.* **65:**1349–1352.

19. **von Heijne, G.** 1989. The structure of signal peptides from bacterial lipoproteins. *Protein Eng.* **2:**531–534.

20. **Webster, C., L. Gardner, and S. Busby.** 1989. The *Escherichia coli melR* gene encodes a DNA-binding protein with affinity for specific sequences located in the melibiose-operon regulatory region. *Gene* **83:**207–213.

Genetics of the Phosphoenolpyruvate-Dependent Sucrose Phosphotransferase System in *Streptococcus mutans* V403

FRANCIS L. MACRINA,[1] KEVIN R. JONES,[1] R. DWAYNE LUNSFORD,[1] CARL-ALFRED ALPERT,[2] AND BRUCE M. CHASSY[2†]

Department of Microbiology and Immunology, Virginia Commonwealth University, Richmond, Virginia 23298-0678,[1] and Laboratory of Microbiology and Immunology, National Institute of Dental Research, Bethesda, Maryland 20205[2]

The transport of sucrose into cariogenic *Streptococcus mutans* cells appears to be primarily governed by a phosphoenolpyruvate-dependent sucrose phosphotransferase (PTS) system (10, 13). Although initial observations suggested that there were multiple sucrose transport systems in *S. mutans* (12), recent evidence indicates that a secondary sucrose transport system in strain GS-5 is actually a high-affinity trehalose uptake system (5). Sucrose is the prime substrate for the synthesis of exopolymers which are crucial to *S. mutans* adherence to the tooth surface. Moreover, sucrose transport into the cell results in acid production, which contributes to cariogenicity by demineralizing the enamel tooth surface. The high affinity of the sucrose PTS for its substrate (apparent K_m, 0.07 mM) as compared with the exopolymer-synthesizing glucosyltransferases and fructosyltransferases (apparent K_m, 1 to 10 mM) suggests the possibility of an extracellular competition for dietary sucrose by these various enzyme and transport systems. The sucrose PTS of *S. mutans* consists of the membrane-associated, sugar-specific enzyme II ($EII^{sucrose}$, encoded by the *scrA* gene) as well as cytoplasmic HPr and enzyme I components (4). Translocation of sucrose into the cell is accompanied by phosphorylation of the sucrose to form sucrose-6-phosphate. An intracellular sucrase activity, sucrose-6-phosphate hydrolase, then cleaves the sucrose-6-phosphate into glucose-6-phosphate and fructose. Lunsford and Macrina (3) cloned the gene for sucrose-6-phosphate hydrolase (*scrB*) from *S. mutans* GS-5 in *Escherichia coli*. Using the technique of allelic exchange, we then (3) constructed a *scrB* mutant of *S. mutans* V403. As expected, this mutant was sensitive to sucrose when grown on complex media containing sucrose at concentrations of 0.05 mM or more. This phenotype is believed to be the result of accumulation of intracellular sucrose-6-phosphate, which is toxic to cells (14). We isolated a number of sucrose-resistant revertants from the *scrB* mutant strain by selection on sucrose-containing complex medium (brain heart infusion agar plus 0.5% sucrose). We predicted that one class of resistant mutant would contain second-site lesions that eliminated another component of the sucrose PTS system (e.g., EII). Such revertants would be of interest since they would now be defective in their ability to transport sucrose into the cell by the normal means, i.e., PTS transport. One such mutant obtained in this fashion was found to lack the ability to transport sucrose into the cell via the PTS system. Genetic analysis of this strain revealed it to represent a class of mutant that contained a second genomic alteration very near the *scrB* locus. In this report, we detail the nature of this genomic rearrangement and conclude it to be the result of a duplicative transposition event into the *scrA* gene by an apparently novel insertion sequence (IS) element that is found in multiple copies on the *S. mutans* V403 chromosome.

Isolation of V1356 and Related Mutants

Methods for DNA preparation and enzymology, genetic transformation, agarose gel electrophoresis, and Southern blotting were as previously described (3). Oligonucleotide synthesis was performed by using an Applied Biosystems ABI 380A instrument. Polymerase chain reactions (PCR) were performed by using a Perkin-Elmer Cetus DNA Thermal Cycler according to published methods (7).

The *scrB* gene from *S. mutans* initially was isolated as part of a 2.9-kb DNA fragment cloned into pBR322 (3). This recombinant plasmid (pVA1343) conferred raffinose fermentation in *E. coli* by virtue of the sucrase activity of the *scrB* product. The *scrB* gene was localized and its orientation was determined by deletion mapping and the analysis of pVA1343-derived, truncated polypeptides synthesized in *E. coli* minicells (3). A defective *scrB* sequence was prepared by the insertion of a 2-kb DNA fragment carrying the *ermAM* gene (which confers erythromycin resistance). This construct, prepared in *E. coli* by standard recombinant DNA methodology, was linearized and used to transform *S. mutans* V403 to erythromycin resistance. Southern blot anal-

†Present address: ABL 103, University of Illinois, Urbana, IL 61801.

ysis of erythromycin-resistant transformants revealed them to carry a defective copy of the *scrB* locus bearing an internal insertion corresponding to the *ermAM* fragment. Such transformants exhibited a sucrose-sensitive phenotype consistent with the accumulation of intracellular sucrose-6-phosphate due to the *scrB* defect. The mutant constructed in this fashion which was used in these studies was designated V1355.

S. mutans V1355 was grown in brain heart infusion broth to stationary phase, and approximately 10^8 cells were plated on brain heart infusion agar containing 0.5% sucrose. Mutants able to grow on this medium (Suc^r) were detected at a frequency of approximately 3×10^{-7}. Mutants from several independently performed experiments were kept for study. In this work we have focused on one such mutant, designated V1356. We expected that some Suc^r mutations might reside in other genes of the sucrose PTS system (4). Moreover, we were aware of the common genetic linkage of cognate PTS genes in other systems (4). Therefore, we carried out simple transformational genetic linkage studies to determine whether the secondary mutation conferring Suc^r in V1356 was linked to the *ermAM* locus, which served to mark the *scrB* gene. Four independently performed experiments revealed that the Suc^r phenotype cotransformed with Em^r in about 94% of colonies examined (100 to 500 colonies per experiment). The tight genetic linkage of these markers indicated that the mutation conferring sucrose resistance was very close to the *scrB* locus on the *S. mutans* V403 chromosome.

Physical Characterization of V1356

These data were consistent with the genetic map prepared by Sato et al. (9), using both restriction enzyme site placement and nucleotide sequencing of the *S. mutans* GS-5 chromosome region bearing the *scrB* gene. This region, found on a cloned 6.3-kb *Eco*RI restriction fragment, appeared indistinguishable from the corresponding 6.3-kb *Eco*RI fragment isolated from *S. mutans* V403 (3; R. D. Lunsford, Ph.D. dissertation, Virginia Commonwealth University, Richmond, 1986). A map of this fragment, adapted from the work of Sato et al. (9), is illustrated in Fig. 1 and displays selected restriction enzyme sites as well

as the location and orientation of the *scrB* (sucrose-6-phosphate hydrolase) and *scrA* (EII^{sucrose}) genes. The nucleotide sequence of *scrB* was determined by Sato and Kuramitsu (8). Additional nucleotide sequence analysis by the Kuramitsu group (9) revealed the gene (*scrA*) for the sucrose-specific EII to be tightly linked to *scrB*. The *scrA* gene was located immediately upstream from *scrB*, with the genes being transcribed in a divergent fashion. The DNA sequences corresponding to the start codons of these two genes were separated by approximately 200 bp. These mapping data taken together with our linkage analyses provided compelling evidence to support the notion that the secondary mutation in V1356 was in the *scrA* gene.

Physical Analysis of V1355 and V1356 DNA

Next we physically analyzed the region around the *scrB-scrA* locus in V1356 and its parent V1355. Whole-cell DNA was prepared from both strains, cleaved with restriction endonuclease, and subjected to agarose gel electrophoresis. The gel was blotted to nitrocellulose and was hybridized with nick-translated ^{32}P-labeled probe DNA. For use as a probe, the 3-kb *Eco*RI-*Bam*HI fragment containing all of *scrA* and a small portion of *scrB* was purified from a recombinant plasmid (pVA1404) carrying the 6.3-kb *Eco*RI fragment cloned from V403 (Fig. 1). The probe fragment corresponded to the DNA sequence spanning coordinates 3.3 to 6.3 kb in Fig. 1. An autoradiogram of this Southern blot is presented in Fig. 2. Lanes A and B contained *Eco*RI-cleaved chromosomal DNAs from strains V1355 and V1356, respectively. The 8.1-kb fragment seen in lane A corresponded to the 6.3-kb *Eco*RI fragment seen in Fig. 1 into which a 1.8-kb fragment carrying the *ermAM* gene had been inserted within the *scrB* gene (3). The analogous hybridizing signal seen in V1356 (lane B) showed an additional increase in size, migrating at 9.3 kb. Additional genetic mapping (data not shown) coupled with Southern blot analysis indicated that the original 8.1-kb *Eco*RI fragment present in V1355 contained an insertion of approximately 1.2 kb in size. Similar results were obtained when a different cleavage strategy was used with these two strains. Namely, when V1355 (lane C) and V1356 (lane D) were simultaneously digested with

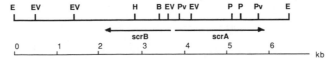

FIG. 1. Map of the *S. mutans* sucrose PTS region. Restriction enzyme sites: E, *Eco*RI; H, *Hin*dIII; B, *Bam*HI; Pv, *Pvu*II; EV, *Eco*RV; P, *Pst*I. Genetic distance in kilobase pairs is illustrated below the linear map. The relative positions of the *scrB* gene (sucrose-6-phosphate hydrolase) and the *scrA* gene (EII^{sucrose}) are noted below the map. Arrows indicate direction of transcription.

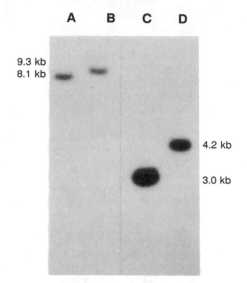

FIG. 2. Analysis of the *scrA-scrB* region in *S. mutans* V1355 and V1356. Total cellular DNA from strains V1355 and V1356 was cleaved with restriction endonuclease(s), electrophoresed through 0.8% agarose, and blotted to nitrocellulose. Southern blots were probed with nick-translated ^{32}P-labeled probe DNA consisting of the 3-kp *Eco*RI-*Bam*HI fragment carrying all of *scrA* and part of *scrB*. A photograph of the resultant autoradiogram is shown. Sizes of the DNA fragments, shown at the left and right, were determined by using bacteriophage lambda *Hin*dIII size reference standards. Lanes: A, V1355 DNA cleaved with *Eco*RI; B, V1356 DNA cleaved with *Eco*RI; C, V1355 DNA cleaved with *Eco*RI and *Bam*HI; D, V1356 DNA cleaved with *Eco*RI and *Bam*HI.

*Eco*RI and *Bam*HI, a size difference was noted between the corresponding signals seen in these two strains (3.0 versus 4.2 kb). These results again suggested that an insertion into the *scrA-scrB* region had occurred. In this case, however, the location of the insert could be assigned to the *Eco*RI-*Bam*HI fragment defined between coordinates 3.3 and 6.3 kb on the map seen in Fig. 1. Detailed mapping indicated the insert fell between the *Eco*RV and the first *Pst*I site (ca. coordinate 5.0 kb) in this region of the map (Fig. 1). Therefore, the insert occurred within the open reading frame of the *scrA* gene.

Characterization of the Insert Found in the V1356 *scr* Region

To characterize the presumed insertion mutation in the *scrA* gene, we attempted to isolate the relevant DNA fragment by gene cloning. Repeated attempts to capture this region by using a variety of restriction fragments as well as a number of plasmid and phage gene cloning vectors in *E. coli* failed. From this work we concluded that

this region has unusual properties that precluded its stable maintenance in *E. coli* host strains. The nature of this problem continues to remain unexplained.

We were able to amplify this region of the V1356 chromosome by using PCR. Since the entire nucleotide sequence of the *scrA* region of the *S. mutans* chromosome has been determined for strain GS-5 (9), we used this information to design two oligonucleotide primers that could be used to amplify across the region where we believed the insert to occur. The first primer encompassed the *Eco*RV site seen at ca. 4 kb in Fig. 1. Its sequence was 5'-CT<u>GATATC</u>TTTGT-ACCAA-3', the *Eco*RV site indicated by underlining. The oligonucleotide used to prime synthesis on the opposing strand had the sequence 3'-CGATCAC<u>GACGTC</u>GTTAA-5' and encompassed the *Pst*I site (also noted by underlining) located at ca. 5 kb on the map shown in Fig. 1. These primers were used to amplify V1355 and V1356 DNA, and the resulting products were analyzed by agarose gel electrophoresis. The V1355 amplification reaction revealed a major product of about 0.9 kb, while the V1356 reaction yielded a major product of about 2.1 kb. These data confirmed that an insert of approximately 1.2 kb in size existed within the *scrA* reading frame, as had been predicted from the physical mapping data. We proceeded to purify the 2.1-kb fragment obtained from the V1356 amplification reaction. Restriction mapping of this purified fragment revealed the inserted sequence to contain unique *Hin*dIII and *Pvu*II sites. Moreover, one end of the 1.2-kb insert was within 200 to 300 bp of the *Eco*RV site contained within the *scrA* gene (Fig. 1).

On the basis of mapping data, fragment size calculations, and DNA nucleotide sequence data, we concluded that the V1356 PCR-amplified fragment contained about 0.9 kb of internal *scrA* sequence flanking the insertion. We proceeded to use the V1356 PCR-amplified fragment as a probe to explore the presence of the inserted DNA detected in the *scrA* gene. The PCR fragment was ^{32}P labeled by nick translation and used as a probe against *Eco*RI-*Bam*HI-digested V403, V1355, and V1356 chromosomal DNAs. A photograph of the resultant autoradiogram is shown in Fig. 3. Six hybridizing *Eco*RI-*Bam*HI fragments were seen in V403 and V1355. These ranged in size from 5.3 to 1.7 kb. One of the six fragments (3.0 kb) corresponded to the *Eco*RI-*Bam*HI fragment carrying the *scrA* gene (arrow in Fig. 3). Hybridization to V1356 DNA also revealed six fragments; however, the 3.0-kb *scrA* fragment was missing in this strain, and a new 4.2-kb fragment appeared that hybridized to the PCR-amplified fragment.

Our interpretation of these data is as follows.

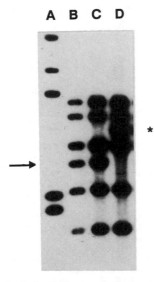

FIG. 3. Analysis of *S. mutans* strains by using an amplified sequence found in the *scrA* gene of V1356. Total cellular DNA was cleaved with *Eco*RI and *Bam*HI, electrophoresed through 0.8% agarose, blotted to nitrocellulose, and probed with a ^{32}P-labeled PCR fragment corresponding to the *Eco*RV-*Pst*I fragment internal to *scrA* (see Fig. 1). This fragment was amplified from V1356 DNA. It contains the entire insertion mutation flanked by *scrA* sequences internal to the reading frame of this gene. A photograph of the autoradiogram is shown. Lanes: A, ^{32}P-labeled bacteriophage lambda *Hin*dIII linear size reference fragments (sizes [top to bottom]: 23.1, 9.4, 6.5, 4.3 [faint signal], 2.3, and 2.0 kb); B to D, *Eco*RI-*Bam*HI-cleaved whole-cell DNAs from V403, V1355, and V1356, respectively. Signals seen in V403 (lane B) and V1355 (lane C) correspond to DNA fragments of 5.3, 4.5, 3.4, 3.0, 2.5, and 1.7 kb. The 3.0-kb fragment represents the *Eco*RI-*Bam*HI fragment that carries *scrA* (arrow; see Fig. 1). Since the PCR-amplified fragment carries *scrA* sequences, it serves as an internal hybridization control. Lane D contains similarly analyzed DNA from V1356 and reveals the 3.0-kb fragment to be missing, with the concomitant appearance of a novel 4.3-kb fragment (asterisk).

Since the 3.0-kb *Eco*RI-*Bam*HI *scrA* fragment by itself hybridized only to a single 3.0-kb fragment in similarly cleaved V403 and V1355 DNAs (data not shown), the multiple signals seen in lanes B and C of Fig. 3 must be the result of the inserted DNA present in the *scrA* subfragment. No *Eco*RI or *Bam*HI sites were found to exist in the insert DNA; therefore, the results of this experiment revealed that five copies of this sequence are present on the chromosomes of V403 and V1355 (sizes of 5.3, 4.5, 3.4, 2.5, and 1.7 kb). The sixth signal (3.0 kb) is assumed to be the unaltered *scrA*-containing *Eco*RI-*Bam*HI fragment. Thus, there are repeated copies of this sequence present on

the chromosome of V403, and this appears to be the case for V1355 as well. The appearance of a novel 4.2-kb fragment in V1356 along with the disappearance of the 3.0-kb *scrA*-containing *Eco*RI-*Bam*HI fragment is consistent with the following interpretation: the sequence being detected with the PCR-amplified DNA probe is able to move by replicative transposition from one of its sites on the chromosome into the *scrA* gene sequence.

PTS-Mediated Uptake of Sucrose in V1355 and V1356

The genotypes of *S. mutans* V1355 and V1356 give rise to the following predictions. First, the V1355 strain should transport sucrose into the cell by the PTS system. Although the *scrB* lesion destroys the cell's ability to use the phosphorylated sugar, the sucrose PTS transport system itself is left intact. Second, the V1356 strain should be unable to transport sucrose into the cell via the PTS system, because it is missing the essential EII^{sucrose} component as a result of the insertion in the *scrA* gene. To test this hypothesis, both strains were tested for sucrose PTS transport activity by using the cell growth and assay procedures described by Thompson and Chassy (15) (Fig. 4). V1355 demonstrated rapid accumulation of radiolabeled sucrose under these assay conditions. V1356, on the other hand, displayed only baseline activity over the time course of the assay period. We conclude that the insertion mutation in the *scrA* gene of V1356 indeed results in an inactivation of the sucrose PTS in this strain.

Discussion and Conclusions

The work described above typifies the investigational power that recombinant DNA methodology has brought to bear on the study of the

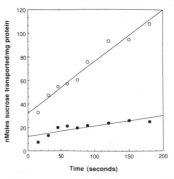

FIG. 4. PTS sucrose uptake by *S. mutans* V1355 and V1356. Cells were prepared as described by Thompson and Chassy (15). [^{14}C]sucrose uptake was monitored by collecting cell samples by filtration at the time points indicated. Radioactivity in such samples was measured by liquid scintillation spectrometry. Symbols: ○, V1355 cells; ●, V1356 cells.

cariogenic streptococci as well as other pathogenic microorganisms. Although there was a significant extant literature on the biochemistry of the sucrose PTS in *S. mutans* (1, 10–14), the genetic basis of this system began to unfold only after the sucrose-6-phosphate gene was cloned (3). This led to the construction of mutants by use of the technique of allelic exchange (2). This precise mutagenesis technique requires that the gene in question be cloned in whole or in part so that it can be characterized and inactivated in vitro. Mutants obtained by this technique then can be further analyzed biochemically and genetically in order to investigate the contribution of the gene to the system being studied. Of equal importance, cloned genes immediately provide the means to determine nucleotide sequence analysis, such as was the case with the *scrB* gene (8). Our genetic analysis of a *scrB* mutant (sucrose sensitive) led to the discovery that a second-site mutation conferring sucrose resistance mapped very near the *scrB* locus. These findings dovetailed well with the extended nucleotide sequence findings of Sato et al. (9), which established the juxtaposition of the *scrB* and *scrA* genes. It should be noted that sequencing of the regions adjacent to the *scrB* gene was commenced by this group after they demonstrated that transposon mutagenesis of that region effectively destroyed PTS-mediated uptake of sucrose in *S. mutans*. This finding substantiated the notion that the *scrB* gene might be linked to other PTS loci and ultimately led to the discovery of the *scrA* gene.

Our findings are remarkable in that they have uncovered a novel mechanism involved in the genetic events that relieve the sucrose sensitivity imposed by the lack of a functional *scrB* gene product. The repeated nature of the inserted DNA sequence found in the *scrA* gene provides strong evidence that it is a mobile genetic element. In addition, we have found that it is possible to isolate mutants similar to V1356 in which this sequence is inserted in the *scrA* gene at a site different from that observed in the V1356 strain (data not shown). Such sequences are indistinguishable from the one found in the *scrA* gene of V1356 in terms of size, restriction sites, and DNA hybridization to the PCR-generated probe discussed above. It is our hypothesis that this element is an IS element, and we are continuing to explore this possibility. Preliminary results indicate this sequence is not found widely in laboratory strains or fresh clinical isolates of *S. mutans*. An IS was recently discovered and characterized in a group B streptococcal strain (6), but to our knowledge IS elements have not been described in any other of the human pathogenic streptococci. The molecular comparison of the repeated sequence found in *S. mutans* V403 with the IS element found in the group B streptococcus is presently under way.

This work was supported by Public Health Service grants DE04224 and DE09035 from the National Institute of Dental Research to F.L.M.

The assistance of Arunsri Brown and Steve Reuther is gratefully acknowledged. We thank Cindy Munro for critical reading of the manuscript.

LITERATURE CITED

1. **Chassy, B. M., and E. V. Porter.** 1979. Initial characterization of sucrose-6-phosphate hydrolase from *Streptococcus mutans* and its apparent identity with intracellular invertase. *Biochem. Biophys. Res. Commun.* **89:**307–314.
2. **Dertzbaugh, M. T., and F. L. Macrina.** 1989. Molecular genetic approaches to the study of oral microflora, p. 18–32. *In* H. M. Myers (ed.), *New Biotechnology in Oral Research.* S. Karger, Basel.
3. **Lunsford, R. D., and F. L. Macrina.** 1986. Molecular cloning and characterization of *scrB*, the structural gene for the *Streptococcus mutans* phosphoenolpyruvate-dependent phosphotransferase system sucrose-6-phosphate hydrolase. *J. Bacteriol.* **166:**426–434.
4. **Postma, P. W., and J. W. Lengeler.** 1985. Phosphoenolpyruvate:carbohydrate phosphotransferase system of bacteria. *Microbiol. Rev.* **49:**232–269.
5. **Poy, F., and G. R. Jacobson.** 1990. Evidence that a low-affinity sucrose phosphotransferase activity in *Streptococcus mutans* GS-5 is a high-affinity trehalose uptake system. *Infect. Immun.* **58:**1479–1480.
6. **Rubens, C. E., L. M. Heggen, and J. M. Kuypers.** 1989. IS*861*, a group B streptococcal insertion sequence related to IS*150* and IS*3* of *Escherichia coli. J. Bacteriol.* **171:**5531–5535.
7. **Saiki, R. K., D. H. Gelfand, S. Stoffel, S. J. Scharf, R. Higuchi, G. T. Horn, K. B. Mullis, and H. A. Ehrlich.** 1988. Primer-directed enzymatic amplification of DNA with a thermostable DNA polymerase. *Science* **239:**487–491.
8. **Sato, Y., and H. K. Kuramitsu.** 1988. Sequence analysis of the *Streptococcus mutans scrB* gene. *Infect. Immun.* **56:**1956–1960.
9. **Sato, Y., F. Poy, G. R. Jacobson, and H. K. Kuramitsu.** 1989. Characterization and sequence analysis of the *scrA* gene encoding system II^{Scr} of the *Streptococcus mutans* phosphoenolpyruvate-dependent sucrose phosphotransferase system. *J. Bacteriol.* **171:**263–271.
10. **Slee, A. M., and J. M. Tanzer.** 1979. Phosphoenolpyruvate-dependent sucrose phosphotransferase activity in *Streptococcus mutans* NCTC 10449. *Infect. Immun.* **24:**821–828.
11. **Slee, A. M., and J. M. Tanzer.** 1979. Phosphoenolpyruvate-dependent sucrose phosphotransferase activity in five serotypes of *Streptococcus mutans. Infect. Immun.* **26:**783–786.
12. **Slee, A. M., and J. M. Tanzer.** 1982. Sucrose transport by *Streptococcus mutans*—evidence for multiple transport systems. *Biochim. Biophys. Acta* **692:**415–424.
13. **St. Martin, E. J., and C. L. Wittenberger.** 1979. Characterization of a phosphoenolpyruvate-dependent sucrose phosphotransferase system in *Streptococcus mutans. Infect. Immun.* **24:**865–868.
14. **St. Martin, E. J., and C. L. Wittenberger.** 1979. Regulation and function of sucrose 6-phosphate hydrolase in *Streptococcus mutans. Infect. Immun.* **26:**487–491.
15. **Thompson, J., and B. M. Chassy.** 1981. Uptake and metabolism of sucrose by *Streptococcus lactis. J. Bacteriol.* **147:**543–551.
16. **Ueda, S., and H. K. Kuramitsu.** 1988. Molecular basis for the spontaneous generation of colonization-defective mutants of *Streptococcus mutans. Mol. Microbiol.* **2:**135–140.

Genetic Analysis of *Streptococcus mutans* Glucosyltransferases

H. K. KURAMITSU, C. KATO, H. JIN, AND K. FUKUSHIMA[†]

Department of Pediatric Dentistry, University of Texas Health Science Center,
San Antonio, Texas 78284

Background

The central role of *Streptococcus mutans* in the development of human dental caries has been well documented (5). In addition, the conversion of dietary sucrose to water-insoluble glucan has been demonstrated to be an important cariogenic property of these organisms. These polymers are synthesized by the concerted action of extracellular glucosyltransferases (GTFs) elaborated by *S. mutans* and related mutans streptococci. Our laboratory has demonstrated that one human cariogenic strain of *S. mutans*, GS-5, produces three distinct GTFs (1, 3, 4). The genes encoding the two GTFs responsible for insoluble glucan synthesis, *gtfB* and *gtfC*, are tandemly arranged on the GS-5 chromosome and share extensive amino acid sequence similarity. However, the *gtfD* gene expressing GTF-S, which synthesizes primarily water-soluble glucan, is not closely linked to the other two genes (7a).

Utilizing *gtfB* gene probes, two laboratories have recently demonstrated that most human oral isolates of *S. mutans* also appear to harbor two *gtf* genes involved in insoluble glucan synthesis (J. S. Chia et al., submitted for publication; Y. Sato, personal communication). However, polymorphism of the *gtfBC* gene cluster is also apparent in a few isolates. The functional significance of these differences still remains to be determined.

The three GS-5 *gtf* genes have been recently sequenced and found to share extensive sequence similarity (4a, 8, 9). This sequence information is being used to investigate several aspects of GTF physiology: the mechanism of action of the enzymes, their respective roles in sucrose-dependent colonization of tooth surfaces, the regulation of their expression, GTF secretion through bacterial membranes, and expression of the *gtf* genes in heterologous oral streptococci.

In Vivo Testing of Insertionally Inactivated *gtf* Genes

Mutants defective in the *gtfB*, -*C*, and -*D* genes have been constructed following insertional inactivation (1, 3, 4). Since *S. mutans* GS-5 is not highly active in the rat caries model system, these

mutations have been transferred to the more cariogenic UA101 strain following transformation. In conjunction with W. Bowen (University of Rochester, Rochester, N.Y.), these latter strains are currently under investigation in rats. The results so far indicate that inactivation of the *gtfD* gene coding for the GTF-S enzyme, as well as the *ftf* gene coding for fructosyltransferase activity, do not affect the ability of strain UA101 to produce smooth-surface caries when rats are fed a sucrose-rich diet.

Mechanism of GTF activity. Mooser et al. (G. Mooser, R. J. Paxton, S. A. Hefta, J. E. Shively, and T. Lee, *J. Dent. Res.* **69**[special issue]:325, abstr. 1729, 1990) have recently identified sucrose-binding peptides from the GTFs from *Streptococcus sobrinus* 6715 (Fig. 1). Essentially identical peptides exist within the structures of the three *S. mutans* GTFs. To confirm the identity of the Asp residue actually involved in sucrose binding, site-directed mutagenesis of the putative sucrose-binding Asp residue (Asp-450) of GTF-I (insoluble glucan synthesizing) from strain GS-5 was carried out. Conversion of this residue to either Thr, Asn, or Glu resulted in complete inactivation of the resulting enzyme. The mutated enzymes displayed negligible GTF and sucrase activities. Therefore, these results confirm the identity of the Asp residue that is involved in covalently binding sucrose to the enzyme. Binding of the second substrate, glucan, to the parental and mutant enzymes is under investigation.

Recent biochemical (7) as well as genetic (2) approaches have also suggested that the GTFs are composed of two functionally distinct domains: a sucrose binding domain and a carboxyl-terminal glucan binding domain. To define the glucan binding domain of the GS-5 GTF-I enzyme, deletion analysis of the enzyme was initiated. This enzyme also contains six similar direct repeating units of 65 amino acids each at the carboxyl terminus. Deletion of all six repeating units following exonuclease III treatment of plasmid pTSU5 containing the *gtfB* gene resulted in an enzyme with no detectable GTF or sucrase activity (Fig. 2). Moreover, the presence of at least two repeating units was necessary for significant GTF activity. In contrast, significant sucrase (sucrose binding and hydrolysis) activity was present when at least a portion of one repeat unit was

[†]Present address: Department of Bacteriology, Nihon University School of Dentistry, Matsudo 271, Japan.

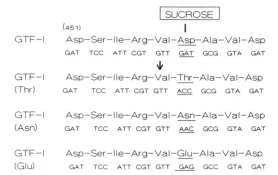

FIG. 1. Site-directed mutagenesis of the Asp residue of *S. mutans* GS-5 involved in sucrose binding. Synthetic oligonucleotides were used to construct the indicated amino acid substitutions.

present. The differential effects of the deletions on both sucrase and GTF activities are consistent with the presence of two distinct functional domains within the GTF enzymes. In addition, these results are consistent with previous suggestions that the carboxyl-terminal domain of the GTFs is important for glucan binding.

Secretion of GTFs. Isolation of the GS-5 *gtf* genes in *Escherichia coli* has revealed that the extracellular proteins expressed by these genes are not exported through the cytoplasmic membranes of this gram-negative organism. To determine the molecular basis for this behavior, two different approaches have been undertaken. Alkaline phosphatase gene fusions with the *gtfB* gene were initially constructed by using Tn*phoA* (6). Fusions in different regions of the gene were isolated and analyzed for their cellular distribution in *E. coli*. These results indicated that the signal sequence of the *gtfB* gene was functional in *E. coli* and allowed transport of the fusion

proteins. However, the extent of transport was dependent on the fusion junction points. This finding indicated that certain combinations of GTF-I and alkaline phosphatase sequences were transported more efficiently than others. However, fusions within the carboxyl-terminal direct repeating units of the GTF-I enzyme were not secreted into the periplasmic space of *E. coli*. As suggested below, this may be due to the presence of secretion-inhibiting sequences in this region of the enzyme.

Another approach to examining the inability of the GTFs to be exported in *E. coli* involved the use of deletion mutants generated from the functional domain analysis described above. These mutants were examined for secretion through the *E. coli* membrane by determining either GTF activity or GTF protein concentration following Western immunoblotting. These results indicated that the presence of the direct repeating units at the carboxyl terminus of the GTF-I protein inhibited secretion. Deletion of this region resulted in inactive enzymes that could be transported into the periplasmic space of *E. coli*. Therefore, the use of both deletion analysis and *phoA* fusions suggests that the carboxyl-terminal sequences of the GTF proteins may be involved in preventing secretion through the *E. coli* membrane. Interestingly, these sequences do not prevent secretion through gram-positive membranes (*S. mutans*, *Streptococcus sanguis*, or *Streptococcus milleri*).

Regulation of GTF expression. The presence of tandem *gtfB* and *-C* genes on the GS-5 chromosome suggests that the two genes may constitute an operon and could be translated from a single polycistronic mRNA. However, previous nucleotide sequencing did suggest the presence of both a weak transcription terminator and a promoter between the two genes (8). To examine

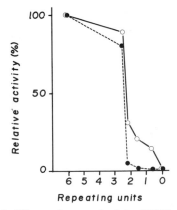

FIG. 2. Effects of removal of the GTF-I direct repeating units on enzymatic activity. Symbols: ○, Sucrase activity; ●, GTF activity.

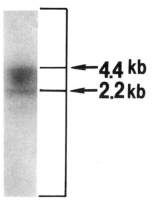

FIG. 3. Northern blot analysis of mRNA corresponding to the *gtfBC* gene cluster. *S. mutans* GS-5 RNA was isolated and probed with the 1.6-kb *Bam*HI fragment from the *gtfB* gene (1).

TABLE 1. Transfer of *S. mutans gtf* genes into *S. milleri*

Gene transferred	GTF activity	Plasmid[a]	Colonization[b]
gtfB	+ + +	None	−
gtfC	+	Yes	+
gtfD	+ + +	None	−

[a]Detected on agarose gels.
[b]Sucrose-dependent colonization of glass surfaces.

the regulation of expression of the *gtfB* and -*C* genes, Northern (RNA) blot analysis of *S. mutans* GS-5 and *E. coli* clones containing the *gtf* genes was carried out. In this assay, no detectable polycistronic mRNA corresponding to both the *gtfB* and -*C* genes could be detected (Fig. 3). In contrast, mRNA transcripts that corresponded to either the *gtfB* or -*C* gene were readily identified in both organisms. Expression of each gene in *S. milleri* (K. Fukushima and H. K. Kuramitsu, unpublished data) suggested that the *gtfB* promoter is much stronger than that of the *gtfC* gene in both oral streptococci and *E. coli*. Therefore, the 4.2-kb mRNA band detected primarily represents the *gtfB* transcript.

Heterologous expression of the *gtf* genes. To further examine the role of each *gtf* gene in *S. mutans* cariogenicity, each gene was introduced into a streptococcus-*E. coli* shuttle plasmid (pVA838, pTS-A, or pTS150) and transformed into competent *S. milleri* cells. However, only the *gtfC* gene could be recovered on a plasmid in the transformants. With use of either the *gtfB* or -*D* gene, GTF expression was detected only when the gene was integrated into the host chromosome (Table 1). This occurred spontaneously in one *gtfB* transformant and following incorporation of random *S. milleri* chromosomal fragments into a shuttle plasmid containing the *gtfD* gene prior to transformation. These results suggest that *S. milleri*, unlike *S. sanguis* and *S. mutans*, cannot tolerate high-level expression of the extracellular proteins. However, since expression of the *gtfC* gene is under the control of a relatively weak promoter (GTF activity in these transformants is still much lower than that from the single-copy *gtfB* gene integrated into the *S. milleri* chromosome), this gene can be recovered apparently intact on a plasmid in these organisms. It will be

of interest to examine the molecular basis for this property of *S. milleri*.

Use of the *S. milleri* transformants also revealed that transformants harboring only the *gtfC* gene can adhere to hard surfaces in the presence of sucrose (Table 1). In contrast, transformants expressing either the *gtfB* or -*D* gene did not colonize in vitro. However, it is not yet known whether these differences will also be expressed in vivo. In addition, since *S. milleri* does not express glucan-binding protein activity, it is likely that such activity is not required for sucrose-dependent colonization of smooth surfaces in vitro. These *S. milleri* constructs will be examined in a rat model system to confirm the roles of the individual *gtf* genes in sucrose-dependent colonization of tooth surfaces.

This investigation was supported in part by Public Health Service grant DE03258 from National Instititutes of Health.

LITERATURE CITED

1. **Aoki, H., T. Shiroza, M. Hayakawa, S. Sato, and H. K. Kuramitsu.** 1986. Cloning of a *Streptococcus mutans* gene coding for insoluble glucan synthesis. *Infect. Immun.* **53**:587–594.
2. **Ferretti, J. J., M. L. Gilpin, and R. R. B. Russell.** 1986. Nucleotide sequence of the glucosyltransferase gene from *Streptococcus sobrinus* MFe28. *J. Bacteriol.* **169**:4271–4278.
3. **Hanada, N., and H. K. Kuramitsu.** 1988. Isolation and characterization of the *Streptococcus mutans gtfC* gene, coding for synthesis of both soluble and insoluble glucans. *Infect. Immun.* **56**:1999–2005.
4. **Hanada, N., and H. K. Kuramitsu.** 1989. Isolation and characterization of the *Streptococcus mutans gtfD* gene, coding for primer-dependent soluble glucan synthesis. *Infect. Immun.* **57**:2079–2085.
4a. **Honda, O., C. Kato, and H. K. Kuramitsu.** 1990. Nucleotide sequence of the *Streptococcus mutans gtfD* gene encoding the glucosyltransferase-S enzyme. *J. Gen. Microbiol.* **136**:2099–2105.
5. **Loesche, W. J.** 1986. Role of *Streptococcus mutans* in human dental decay. *Microbiol. Rev.* **50**:353–380.
6. **Manoil, C., and J. Beckwith.** 1985. TnphoA: a transposon probe for protein export signals. *Proc. Natl. Acad. Sci. USA* **82**:8129–8133.
7. **Mooser, G., and C. Wong.** 1988. Isolation of a glucan-binding domain of glucosyltransferase (1,6-α-glucan synthase) from *Streptococcus sobrinus*. *Infect. Immun.* **56**:880–884.
7a. **Perry, D., and H. K. Kuramitsu.** 1990. Linkage of sucrose-metabolizing genes in *Streptococcus mutans*. *Infect. Immun.* **58**:3462–3464.
8. **Shiroza, T., S. Ueda, and H. K. Kuramitsu.** 1987. Sequence analysis of the *gtfB* gene from *Streptococcus mutans*. *J. Bacteriol.* **169**:4262–4270.
9. **Ueda, S., T. Shiroza, and H. K. Kuramitsu.** 1988. Sequence analysis of the *gtfC* gene from *Streptococcus mutans* GS-5. *Gene* **69**:101–109.

Genetic and Physiological Studies of Variants of *Streptococcus mutans* GS-5 That Produce Nonmucoid Colonies on Sucrose-Containing Media

S. SUN, L. DANEO-MOORE, L. MARRI,[†] M. HANTMAN, AND G. D. SHOCKMAN

Department of Microbiology and Immunology, Temple University School of Medicine, Philadelphia, Pennsylvania 19140

The synthesis of extracellular, water-insoluble glucan from sucrose via the action of glucosyltransferases (GTFs) has been considered to be a major pathogenic factor for *Streptococcus mutans* and closely related species. In turn, the presence of extracellular insoluble glucan is thought to be responsible for the hard, small, mucoid colonies with volcanolike collapsed centers on sucrose-containing agar such as mitis-salivarius agar (MSA). The presence of nonmucoid (NM) colonies of *S. mutans* on sucrose-containing media is considered to indicate the presence of variants that no longer produce extracellular insoluble glucan and therefore possess a defect in their GTF activity system. Variants that produce such colonies occur spontaneously in cultures of *S. mutans* GS-5 (3).

The production of extracellular insoluble glucan, as expressed by mucoid, wild-type colonies on MSA, is a complex process that involves synthesis of the relevant GTF activity, its secretion by the producing bacterium, its binding to the surface of the bacterial cell, its catalytic activity resulting in hydrolysis of sucrose coupled with polymerization of glucose units into linear strands, addition of glucosyl branches to the linear strands, and binding of the branched glucan to the bacterial surface. At least some of these enzyme activities are apparently provided by large-molecular-size proteins (ca. 150 to 170 kDa).

Dissection of this complex process is further complicated by the presence in *S. mutans* GS-5 of two tandemly arranged genes, *gtfB* and *gtfC*, that contain regions of extensive homology (7) and can readily recombine (6). When cloned in a recombination-proficient, but not in a *recA*, *Escherichia coli* host, *gtfB-gtfC* recombinants contained deletions ranging from about 3.7 to 6 kb in size (6). A spontaneous NM mutant of *S. mutans* GS-5, SP2, contained a 4.7-kb deletion in the *gtfB-gtfC* region (6), suggesting a direct relationship between expression of these two genes and colony phenotype.

In an attempt to sort out this process, strains that produced NM colonies on sucrose-containing media were isolated and partially characterized.

Selection and Enrichment for NM Mutants

Two separate and independent reagents were used for selection and enrichment for NM mutants. The first method utilized the growth characteristics of *S. mutans* GS-5 in sucrose-containing medium (Fig. 1, tube 1). In Todd-Hewitt broth (THB) plus 2% sucrose, cells of *S. mutans* GS-5 clumped and fell to the bottom of the tube, leaving a lightly turbid to nearly clear supernatant. Thus, the 16-L series NM mutants were obtained after 15 serial passages of the top 0.5 ml of 10-ml cultures after overnight growth of GS-5 in this medium. The turbid supernatant of the last transfer contained cells that produced only NM colonies on MSA. The S series of mutants were selected from NM colonies on MSA after plating of the top of the supernatant after only three cycles of overnight growth in THB plus 2% sucrose. This supernatant contained cells that produced mucoid and NM colonies at a ratio of about 1 to 1. The third series of NM strains was obtained by taking advantage of the ability of an anti-GTF monoclonal antibody, MAb 16B2, to agglutinate cells of *S. mutans* GS-5. Cells from an overnight culture grown in THB plus 2.0% glucose at 37°C were suspended in phosphate-buffered saline containing a 1:10 dilution of MAb 16B2. After incubation for 17 h at 4°C, bacteria in the apparently clear supernatant were harvested and regrown in THB plus glucose, and the process was repeated three more times. Sixteen NM colonies were obtained from the supernatant of the fourth passage. In separate experiments, the frequency of unselected spontaneous NM mutants was found to be about 10^{-4}.

In addition, a B series of NM transformants was obtained by using DNA from one of the L series NM isolates, L7, which possessed both a streptomycin resistance (Sm^r) and a tetracycline resistance (Tet^r on Tn916) determinant to transform competent *S. mutans* GS-5 to Sm^r or Tet^r. NM colonies were obtained at frequencies of 9×10^{-2} among Sm^r transformants and 11×10^{-2} among Tet^r transformants. In separate ex-

---

[†]Present address: Università di Siena, Siena, Italy.

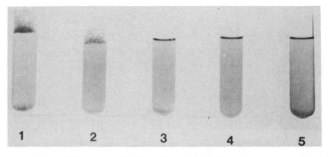

FIG. 1. Growth characteristics of *S. mutans* GS-5 and four NM variants in sucrose-containing medium. Cultures were grown overnight at 37°C in THB containing 2% sucrose. Tubes were inoculated with *S. mutans* GS-5 (tube 1), strain L7 (tube 2), strain B2 (tube 3), strain B4 (tube 4), and strain B6 (tube 5). Note that strains L7 and B4 (tubes 2 and 4) grew as uniformly suspended cells, in contrast to the flocculent aggregated growth of the parent GS-5 strain. Strains B2 and B6 (tubes 3 and 5) grew in an intermediate fashion, showing some aggregation and settlement of cells to the bottom of the tubes.

periments using DNA from other L series mutants, the frequency of NM colonies among Sm^r transformants ranged from 0.13 to 14%. The frequency of transformation to either Sm^r or Tet^r was about 10^{-4}. Sixty-seven Sm^r Tet^s transformants obtained by using L7 DNA were stored for further study.

Unexpected Range of Phenotypic Properties of Mutants All Producing NM Colonies on MSA

Growth in THB plus 2% sucrose. As expected, when examined for growth characteristics in sucrose-containing medium, some of the NM strains grew as homogeneously turbid suspensions, as illustrated in Fig. 1 by the growth of strain L7 in tube 2 and of strain B4 in tube 4. Other NM strains such as the transformants B2 and B6 continued to show some aggregative behavior more typical of the parent GS-5 strain (tube 1). Since these two strains presumably resulted from transformations which utilized DNA of L7, a strain that did not display the aggregating property (tube 2), the difference in phenotype between L7 and B2 or B6 was of interest.

Reactivity of NM strains with MAb 16B2. MAb 16B2 is one of several MAbs that react with GTF of serotype c strains of *S. mutans* (Go et al., unpublished data). For example, in Western immunoblots, MAb 16B2 reacts with two high-molecular-weight proteins that are capable of producing a band of white precipitate in the gels after incubation in a sucrose-containing buffer (4). When tested in a colony immunoblot assay system, as expected, all 16 of the A series NM strains tested and all 20 of the S series NM strains tested reacted much more weakly with MAb 16B2 than did the GS-5 parent strain. The L series NM strains and the B series NM transformants showed a greater variety of reactivities in colony blots with MAb 16B2. Of the 18 L strains tested by colony blots, 11 reacted more weakly than the

parent strain, while 7, including strain L7, reacted at least as strongly as the parent strain. Of the 68 strains resulting from the transformation with DNA of strain L7, only 4 reacted weakly with MAb 16B2, 62 reacted with the same or slightly greater intensity than did the parent strain, and 2 strains, B1 and B2, reacted with a greater intensity than did the parent strain.

Analysis of the DNA of NM Strains by Southern Blotting and Hybridization with pKS5

The pKS5 probe has homology with a 1.7-kb *Bam*HI fragment found in both *gtfB* and *gtfC*, with the region spanning *gtfB* and *gtfC*, and with the carboxyl terminus of *gtfC* (Fig. 2A and B). We used this probe on *Sph*I-, *Eco*RI-, *Pst*I-, and *Pst*I-*Bam*HI-cut DNA to examine the sizes of the deletions in two L series, three A series, three S series, and three B series NM strains.

Southern blot analysis of the *Pst*I digests of the DNA from eight spontaneous NM strains all showed the presence of an 8.9-kb fragment that hybridized with pKS5 (Fig. 3, lanes 2 to 9). In a separate experiment (data not shown), Southern blot hybridizations with pKS5 to DNA from three B series NM transformants (strains B2, B4, and B6) also showed the presence of the same 8.9-kb *Pst*I fragment. In contrast to the NM strains, *S. mutans* GS-5 DNA contained two fragments of 7.2 and 6.4 kb that hybridized with pKS5 (Fig. 3, lane 1). These two pKS5-hybridizing fragments were expected from the published restriction map of GS-5 DNA (7; see also Fig. 2).

The difference in restriction enzyme fragments between the parent and all of the NM strains examined could be accounted for by a 4.7-kb deletion which would include the *Pst*I site that is present in the *gtfC* gene but is absent in the *gtfB* gene (Fig. 2). This could have occurred via a recombination between these two highly homologous but not identical genes producing a fused

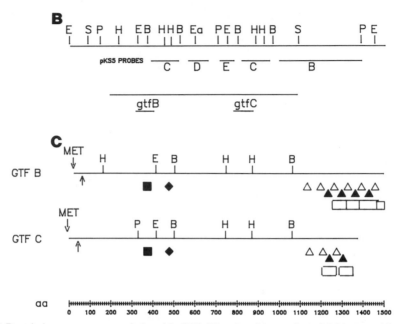

FIG. 2. (A) Restriction enzyme map of plasmid pKS5. The plasmid contains a 9.3-kb recombined fragment of GS-5 chromosomal DNA which includes sequences from both *gtfB* and *gtfC*. Following digestion with restriction enzymes *Pst*I (P) and *Bam*HI (B), five fragments (A to E) are produced, which were separated and isolated from agarose gels and used as hybridization probes. (B) Restriction enzyme map of the *gtfB-gtfC* region of the *S. mutans* GS-5 chromosome, reconstructed from published data (7). Each of the pKS5 fragments was hybridized to a *Pst*I-*Bam*HI double digest of GS-5 chromosomal DNA, and the homologous regions are indicated. (C) Alignment of the *gtfB* and *gtfC* regions. The solid square indicates the epitope reported by Dertzbaugh and Macrina (2). The antiserum that recognizes this 15-amino-acid sequence can inhibit GTF activity. The diamond represents a seven-amino-acid sequence reported by Russell and Ferretti (5) to share homology to sequences found in several glucan-hydrolyzing enzymes. The open and solid triangles represent the A and C repeats, respectively, found in both GTFs and glucan-binding protein (1). The direct repeats (open squares) described by Ueda et al. (7) are also shown. The up arrow is the putative signal sequence cleavage site. Restriction sties: B, *Bam*HI; E, *Eco*RI; Ea, *Eag*I; H, *Hin*dIII; P, *Pst*I; S, *Sph*I.

hybrid gene. Consistent with this interpretation are the results of Southern blot analyses of *Sph*I DNA digests. *Sph*I cleavage produced a 10.5-kb pKS5-hybridizing fragment from GS-5 DNA and a 5.8-kb pKS5-positive fragment from the DNA of all NM strains tested (data not shown). Additional evidence favoring the presence of a 4.7-kb deletion in the *gtfB-gtfC* gene system in the NM strains was the absence of a 4.6-kb pKS5-positive DNA fragment in *Eco*RI digests of three NM B series transformants (compare lanes 3 to 5 with lane 2 in Fig. 4). The same pKS5 hybridization pattern was observed with *Eco*RI digests of DNA from several spontaneous NM strains (data not shown). Lastly, *Pst*I and *Bam*HI double digests of GS-5 DNA showed the presence of three fragments at 5.1, 2.7, and 1.7 kb that hy-

bridized strongly with pKS5 (Fig. 4, lane 6), whereas only two of these fragments, at 5.1 and 1.7 kb, were seen in similar double digests of DNA from three of the NM transformants (Fig. 4, lanes 7 to 9). DNA of eight spontaneous NM mutants also showed the same restriction pattern as the three NM transformants (data not shown).

All of the results briefly summarized above are consistent with the presence in the NM mutants of a fused gene product consisting of an amino-terminal end derived from *gtfB* of at least 1,000 bp and a carboxyl-terminal end derived from *gtfC*. The fused gene would remain under the control of the original *gtfB* promoter.

The tandem nature of *gtfB* and *gtfC* plus the presence of long stretches of homology, if not identity, provide nearly ideal circumstances for

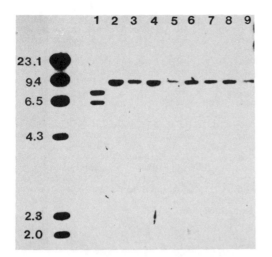

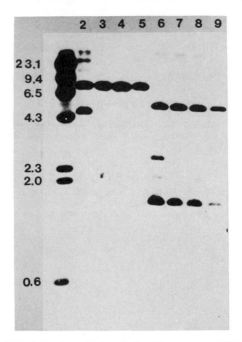

FIG. 3. Southern blot analysis of *Pst*I-digested chromosomal DNA of GS-5 (lane 1) and of strains A1, A2, A3, S1, S2, S3, L7, and L8 (lanes 2 to 9). On the extreme left are *Hin*dIII-digested lambda markers. DNA fragments separated by agarose gel electrophoresis were transferred to nitrocellulose membrane by using the Vacublot transfer system (American Bionetics, Emeryville, Calif.). Hybridization was carried out with plasmid pKS5 labeled with [α-³²P]dCTP according to the Standard Multiprime DNA labeling protocol (Amersham). pKS5 was a gift from R. A. Burne (University of Rochester, Rochester, N.Y.). It contains a 9.3-kb recombined fragment of *gtfB* and *gtfC* (Fig. 2A). DNA from GS-5 shows the presence of 7.2- and 6.4-kb hybridizing fragments, whereas all of the other strains tested show the presence of a single 8.9-kb hybridizing fragment.

FIG. 4. Southern blot analysis of chromosomal DNA digested with *Eco*RI (lanes 2 to 5) or *Pst*I plus *Bam*HI (lanes 6 to 9). DNAs from *S. mutans* GS-5 and derived strains B2, B4, and B6 were hybridized to an [α-³²P]dCTP-labeled pKS5 probe. *Eco*RI-digested GS-5 DNA (lane 2) generated 4.6- and 7.8-kb hybridizing fragments, whereas *Eco*RI-digested DNA from B2, B4, and B6 (lanes 3 to 5, respectively) shows only the presence of a 7.6-kb hybridizing fragment. A *Pst*I-plus-*Bam*HI double digest of DNA from GS-5 (lane 6) shows the presence of three strongly hybridizing fragments of 5.1, 2.7, and 1.7 kb. The 1.7-kb *Bam*HI fragment from *gtfB* and from *gtfC* would comigrate, and an expected 0.6-kb *Bam*HI-*Pst*I fragment appears faintly but cannot be seen on this exposure. Digests of DNA from B2, B4, and B6 (lanes 7 to 9, respectively) show only the presence of 5.1- and 1.7-kb hybridizing fragments.

the frequent occurrence of homologous recombinations accompanied by deletions of stretches of DNA (6). It is of considerable interest that all of the NM derivatives of *S. mutans* GS-5 examined so far contain a 4.7-kb deletion in this region. This deletion appears to be present constantly, despite significant differences in the phenotypes of the NM strains. Presumably, a small shift in the exact spot at which a recombination event occurs could result in a major change in coding, amino acid sequence, and function. Thus, we initiated determinations of the nucleotide (and amino acid) sequence of this region of the DNA of selected NM strains. The 4,703 bases between the *Eco*RI site (or, for that matter, the *Bam*HI site) in *gtfB* and the corresponding *Eco*RI (or *Bam*HI site) in *gtfC* in the chromosome of *S. mutans* GS-5 is consistent with the 4.7-kb deletion observed in our studies and in the SP2 strain (6).

Preliminary sequence data have indicated that the fused *gtfB*-*gtfC* genes present in various NM strains do differ in exact sites of recombination (S. Sun et al., unpublished data). For example, the *gtfB*-*gtfC* region of the DNA of NM strain

L7 appears to consist of the *gtfB* sequence to just past the first *Bam*HI site, followed by the *gtfC* sequence. The same DNA region of the B4 NM transformant consists of a longer stretch of *gtfB* bases before it switches to the *gtfC* sequence. Thus, the strain B4 sequence may not be due to transformation by L7 DNA, although the L7 DNA could have been integrated into the GS-5 DNA before a spontaneous recombinational event.

The recombinational events that occurred in strains B2 and B6 appear to be even more complex. The fused gene produced appears to result from more than one recombinational event and consists of *gtfB* regions followed by *gtfC*, *gtfB*, and then *gtfC* sequences.

Our interpretations are based on comparison between the genetic and physiological changes observed. However, in at least some instances, secondary mutations cannot be ruled out. At a minimum, the strains obtained present a tool for the study of the various functional domains of the GTFs of *S. mutans*.

We thank R. A. Burne for providing plasmid pKS5.

This work was supported by Public Health Service research grants DE-03487 and DE-05180 from the National Institutes of Health and by Biomedical Research Support grant 2 S07 RR05417 to Temple University.

LITERATURE CITED

1. **Banas, J. A., R. R. B. Russell, and J. J. Ferretti.** 1990. Sequence analysis of the gene for the glucan-binding protein of *Streptococcus mutans* Ingbritt. *Infect. Immun.* **58:**667–673.

2. **Dertzbaugh, M. T., and F. L. Macrina.** 1990. Inhibition of *Streptococcus mutans* glucosyltransferase activity by antiserum to a subsequence peptide. *Infect. Immun.* **58:**1509–1513.

3. **Perry, D., L. M. Wondrack, and H. K. Kuramitsu.** 1983. Genetic transformation of putative cariogenic properties in *Streptococcus mutans. Infect. Immun.* **41:**722–727.

4. **Russell, R. R. B.** 1979. Use of Triton X-100 to overcome the inhibition of fructosyltransferase by SDS. *Anal. Biochem.* **97:**173–175.

5. **Russell, R. R. B., and J. J. Ferretti.** 1990. Nucleotide sequence of the dextran glucosidase (*dexB*) gene of *Streptococcus mutans. J. Gen. Microbiol.* **136:**803–810.

6. **Ueda, S., and H. K. Kuramitsu.** 1988. Molecular basis for the spontaneous generation of colonization-defective mutants of *Streptococcus mutans. Mol. Microbiol.* **2:**135–140.

7. **Ueda, S., T. Shiroza, and H. K. Kuramitsu.** 1988. Sequence analysis of the *gtfC* gene from *Streptococcus mutans* GS-5. *Gene* **69:**100–109.

Use of Operon Fusions To Study Gene Expression in *Streptococcus mutans*

MICHAEL C. HUDSON[†] AND ROY CURTISS III

Department of Biology, Washington University, St. Louis, Missouri 63130

Most genes encode products for which no easy assay exists. For this reason, transcriptional gene fusions that join the promoters of genes of interest to the *Escherichia coli* lactose operon have been shown to be important tools in the study of gene regulation (22). Although many of the methods used for constructing *lac* fusions are sufficiently general to be suitable for any target gene from *E. coli*, other genes from this bacterium have also been used and are in certain cases superior to *lac* (22). Other genes used for fusions would include *galK* (13), *npt* (26), *phoA* (12), and *cat* (5). Unfortunately, the majority of fusion approaches rely on the molecular genetics of *E. coli*, and only sporadically has the technique involved other organisms and their genes.

We chose to use *cat* fusions as a tool to study the regulation of gene expression in *Streptococcus mutans*. The in vitro assay for chloramphenicol acetyltransferase (CAT) is both rapid and simple, and the *cat* gene product can be assayed with specificity and great sensitivity (18). Apart from the ability to assay CAT easily, this fusion system has other advantages. If necessary, the CAT enzyme can be purified to homogeneity from crude extracts in one step by affinity chromatography (28). Additionally, very few organisms produce CAT, which alleviates the problem of background enzymatic activities (19). This should eventually allow an investigation into *S. mutans* gene expression in the background of the eukaryotic host.

To illustrate the utility of the *cat* fusion approach to study *S. mutans* gene expression, two important *S. mutans* genes were chosen. Both the glucosyltransferase B and fructosyltransferase genes (*gtfB* and *ftf*, respectively) encode products that contribute to the virulence of *S. mutans* (16, 17), and studies were undertaken to investigate the regulation of expression of these determinants. In the case of the *gtfB* gene, sequence analysis of the nearby *gtfC* determinant conducted by Ueda et al. (25) suggests that *gtfB*, *gtfC*, and a small downstream coding region are located within a single operon transcribed from the *gtfB* promoter. Operon fusions were constructed between the glucosyltransferase B/C (*gtfB/C*) operon promoter and a promoterless *cat*

[†]Present address: Department of Biology, University of North Carolina at Charlotte, Charlotte, NC 28223.

determinant, and the fusion was returned to the *S. mutans* genome (7). For fructosyltransferase, Shiroza and Kuramitsu (20) report several open reading frames near the *ftf* coding region and an inverted repeat structure upstream of the *ftf* determinant. A *cat* operon fusion was generated to the promoter of the *ftf* coding region to examine the potential regulation of expression of *ftf* (7).

Construction of the *S. mutans* Fusion Strains SMS101 and SMS102

Figures 1 and 2 show the predicted structures of the relevant regions of the genomes of *S. mutans* fusion strains SMS101 (*ftf-cat*) and SMS102 (*gtfB/C-cat*) as well as those of the parental strain UA130 (serotype c; from P. Caufield) (14). Importantly, the fusion strains contain a single-copy operon fusion as well as a tandem wild-type copy of the gene whose regulatory sequences direct *cat*. Construction of the fusion vectors pYA2810 and pYA2808 has been previously described (7). Briefly, pYA2810 was generated by isolating the *S. mutans ftf* promoter from pTS102 (20) and ligating the fragment into the *Sma*I site of the promoter-cloning vehicle pMH109 (8). This *ftf-cat* operon fusion was then subcloned as a *Bam*HI fragment into pVA891 (11). The vector pYA2808 was constructed similarly, using the *S. mutans gtfB/C* regulatory region found on plasmid pSU20 (21).

CAT Specific Activities as a Measure of *S. mutans ftf* and *gtfB/C* Expression

Studies were undertaken to investigate the regulation of expression of the *gtfB/C* operon and *ftf* genes (7). CAT specific activities were used to monitor expression, with values expressed as nanomoles of chloramphenicol acetylated per minute per milligram of total protein. Strains SMS101 and SMS102 were grown anaerobically overnight in defined (FMC) medium (24) supplemented with 1% glucose; 3 ml of each culture was then separately inoculated into 30 ml of fresh prewarmed FMC medium plus 1% glucose. Strains were then incubated statically at 37°C for 2 h. These cultures were designated as the zero time points. Samples (1 ml) of the zero time cultures were inoculated into separate tubes containing 10 ml of either FMC medium with 1% glucose or FMC medium with 1% sucrose (prewarmed). The

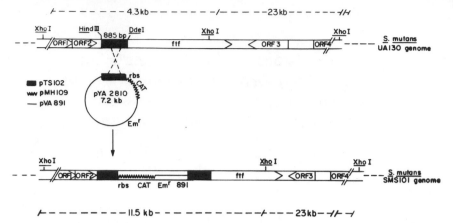

FIG. 1. Introduction of the integration vector pYA2810 (*ftf-cat*) into the *S. mutans* genome by Campbell-type insertion (7). The parental chromosome structure is shown above the plasmid; the genome of the fusion strain SMS101 is shown below. ORF, Open reading frame; rbs, *cat* gene ribosome binding site; Em^r, erythromycin resistance. Arrows depict the direction of transcription.

strains were incubated as static cultures at 37°C for various times. Cultures removed at various time points were harvested by centrifugation, and bacterial pellets were transferred to microfuge tubes containing one-third volume of glass beads (B. Braun Melsungen AG). Cells were disrupted in a Braun homogenizer at maximum speed for 8 min with intermittent CO_2 cooling, and the lysates were centrifuged to remove glass beads and cellular debris. The resulting supernatant fluids were stored at 0°C prior to being assayed for both CAT activity (spectrophotometic method [18]) and total protein content (1).

The results of this analysis using static cultures are presented in Fig. 3 and 4. The amount of total protein obtained was the same for either glucose-

or sucrose-grown cells. Both strains SMS101 and SMS102 demonstrated a nearly steady level of CAT specific activity over time when grown in FMC medium with glucose as the carbohydrate. This same steady level of CAT specific activity over time was also seen in FMC medium with fructose as the carbohydrate (data not shown). However, both fusion strains demonstrated an initial induction of expression when sucrose (relative to glucose or fructose) was the sole carbohydrate, followed by a rapid decline in CAT specific activity after about 1 h.

Stability of CAT in *S. mutans*

Fusion strain SMS101 was used to determine the half-life of CAT in *S. mutans*. CAT specific

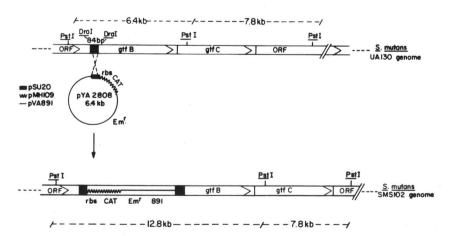

FIG. 2. Introduction of the integration vector pYA2808 (*gtfB/C-cat*) into the *S. mutans* genome by Campbell-type insertion (7). The parental chromosome structure is shown above the plasmid; the genome of the fusion strain SMS102 is shown below. ORF, Open reading frame; rbs, *cat* gene ribosome binding site; Em^r, erythromycin resistance. Arrows depict the direction of transscription.

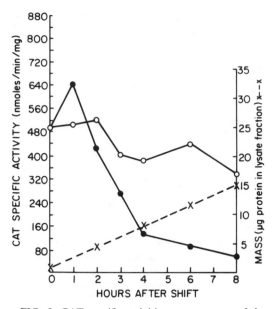

FIG. 3. CAT specific activities as a measure of the expression of the *S. mutans ftf* gene (7). Symbols: ●, growth in sucrose-containing media; ○, growth in glucose-containing media. Hours after shift designates the time following transfer of cells to FMC medium plus 1% glucose or 1% sucrose. Values represent the averages of three independent experiments.

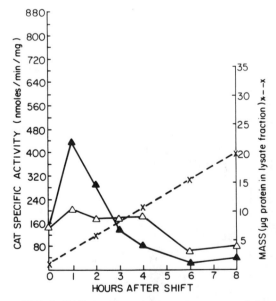

FIG. 4. CAT specific activities as a measure of the expression of the *S. mutans gtfB/C* operon (7). Symbols: ▲, growth in sucrose-containing media; △, growth in glucose-containing media. Hours after shift designates the time following transfer of cells to FMC medium plus 1% glucose or 1% sucrose. Values represent the averages of three independent experiments.

activities were determined as described above except that rifampin was used to block *S. mutans* protein synthesis. Specifically, 1-ml samples of the zero time culture (see above) were inoculated into separate tubes containing 10 ml of FMC medium with 1% sucrose (prewarmed). The tubes were incubated statically at 37°C for various times. After 1 h, rifampin (100 μg/ml) was added to the remaining tubes and incubation was continued. Cell growth and protein synthesis were completely blocked after rifampin addition. CAT protein amounts were determined on replicate cultures and compared with the initial value obtained after growth for 1 h in FMC with sucrose. After the amount of CAT was plotted versus time, the CAT enzyme was determined to have a half-life of 2.9 h in *S. mutans*.

Influence of Other Factors on *gtfB/C* and *ftf* Expression

Anaerobiosis. Since the experiments described above were done with static aerobic cultures, we wanted to determine whether strictly anaerobic conditions might affect expression of the two determinants. We made use of the Oxyrase enzyme system (Oxyrase, Inc., Ashland, Ohio), which has been used to cultivate anaerobic microorganisms (4). The Oxyrase enzyme system was used as recommended by the supplier, and determinations of CAT specific activities were made following incorporation of Oxyrase into all growth media. The results of this analysis were indistinguishable from those presented in Fig. 3 and 4 for static aerobic cultures.

pH and NaCl concentration. For the experiments with static cultures described above, cells of the fusion strains were grown in FMC medium at pH 7.4. Expression studies were repeated using FMC medium adjusted to both pH 5 and pH 9. No difference in expression of *gtfB/C* or *ftf* was observed from that presented in Fig. 3 and 4. Similarly, no effect on expression of either determinant was observed when the NaCl concentration ranged from 5 mM to as high as 100 mM in FMC medium.

Polymers. The effect of fructose polymers on expression of *ftf* was examined. When either inulin [β-(2,1)-linked fructan] or levan [β-(2,6)-linked fructan] was added to FMC medium at 0.5% final concentration, no effect on *ftf* expression was observed. Higher inulin concentrations (>1% final) partially inhibited the sucrose induction of *ftf* expression, although the physiological significance of this observation is unclear.

Gene Expression in a Saliva-Coated Hydroxyapatite Model System

Saliva-coated hydroxyapatite was used to investigate gene expression by *S. mutans* bound to artificial tooth pellicles (7). CAT specific activities were again used to monitor expression of the *gtfB/C* operon and *ftf* genes, with results expressed as nanomoles of chloramphenicol acetylated per minute per 2×10^8 ^{3}H cpm. *S. mutans* fusion strains were grown anaerobically overnight in 40 ml of FMC medium containing 0.2% glucose, 0.01% sucrose, and 10 µCi of [^{3}H]thymidine (*methyl-^{3}H*; Dupont) per ml. Cells were harvested by centrifugation and suspended in 4 ml of KCl buffer (3) containing 10 µCi of [^{3}H]thymidine per ml. Strains were then passed through 27-gauge needles 15 to 20 times to break up streptococcal chains. The bacterial suspensions were mixed with saliva-coated beads that were prepared as follows. Spheroidal hydroxyapatite beads (200 mg) were placed in 1.5-ml microfuge tubes and washed with 7 to 10 volumes of distilled water to remove very small particles. Beads were then equilibrated in KCl buffer overnight. Following equilibration, beads were incubated overnight with either 1 ml of clarified whole saliva (isolated from a single subject, heat treated at 60°C for 30 min, and containing 0.04% sodium azide) or KCl buffer containing sodium azide. Tubes were incubated at room temperature with inversion at a rate of 10 times per min. The beads were then washed two times with KCl buffer in the absence of sodium azide prior to the addition of 1 ml of the bacterial suspensions. Mixtures of bacteria and beads were rotated at room temperature for 3 h and then permitted to stand for 60 s. Unadsorbed bacteria were gently pipetted to separate tubes, and all bacteria were washed extensively to remove free [^{3}H]thymidine. Following the washes, pellets were suspended in KCl buffer, glass beads were added, and the cells were disrupted in a Braun homogenizer for 8 min with intermittent CO_2 cooling. After removal of debris, supernatant fluids were assayed for both CAT activity and ^{3}H counts (EcoLite [+] scintillation solution [ICN Biomedicals, Inc.]; Beckman LS-3150T scintillation counter).

Binding to saliva-coated hydroxyapatite had no effect on expression of the *ftf* determinant, since cells bound to saliva-coated beads expressed CAT at essentially the same level as did unbound organisms (Fig. 5). There did seem to be some down-regulation of *ftf* expression by cells bound to buffer-incubated beads. The significance of this observation is currently unclear. Expression of the *gtfB/C* operon appeared to be increased in bound cells. CAT specific activity directed by the *gtfB/C* operon promoter was approximately two-fold higher when cells of the fusion strain were bound to saliva-coated beads (compared with the corresponding activity in unbound cells), although the standard errors of the mean were rather large. A similar but less pronounced effect was seen with organisms bound to beads that were incubated in buffer alone.

Induction of *gtfB/C*

Since sucrose was shown to induce *gtfB/C* expression, saliva-coated hydroxyapatite experi-

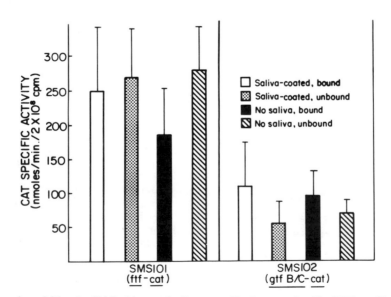

FIG. 5. Expression of *ftf* and *gtfB/C* with use of saliva-coated hydroxyapatite (7). CAT specific activities were used to monitor the expression of the *ftf* and *gtfB/C* determinants. Values represent the averages of three independent experiments. Standard errors of the means are indicated.

ments were repeated without the presence of sucrose during cell growth. Importantly, the same binding effect on *gtfB/C* expression is also seen without the presence of sucrose during growth of the fusion strain. This result suggests that the slight induction of *gtfB/C* expression is a result of binding to artificial pellicles and does not represent a prebinding induction.

Conclusions

The use of *cat* fusions has enabled several conclusions to be drawn regarding the regulation of expression of the *gtfB/C* operon and *ftf* genes.

(i) The fructosyltransferase gene is more highly expressed than the glucosyltransferase operon under all conditions tested. These findings using operon fusions are consistent with those of Wenham et al. (27), who studied the production of extracellular sucrose-metabolizing enzymes by *S. mutans* Ingbritt. These investigators found the most active enzyme to be fructosyltransferase, implying that the *cat* fusion approach can indeed yield insights into expression.

(ii) Expression of both the *gtfB/C* operon and *ftf* genes is induced by sucrose (relative to glucose or fructose), followed by a rapid decline in expression over time. This may indicate that *S. mutans* has evolved some mechanism(s) for shutting off the formation of extracellular polymers from sucrose once a sufficient amount of either glucan or fructan is formed. It has indeed been observed that during growth on sucrose, *S. mutans* assimilates only 2 to 5% of sucrose carbon as extracellular polymers (2, 23).

(iii) There is a greater induction of expression of the *gtfB/C* operon than of *ftf* by sucrose.

(iv) Expression of the *gtfB/C* operon is somewhat increased in cells bound to either saliva-treated or untreated hydroxyapatite. This induction of expression could be explained by the notion that the product ultimately encoded by *gtfB/C*, insoluble glucan, is intimately involved in the firmer attachment of *S. mutans* to tooth surfaces (6). Initial binding could, in some way, signal *S. mutans* to produce the needed glucan.

The use of *cat* fusions has proved to be invaluable in studying the expression of *S. mutans* genes. In this regard, *cat* fusions have now been made to *scrB* (10), *asd* (9), and *gtfA* (15). The expression of these *S. mutans* genes is currently being examined.

We express our gratitude to Howard K. Kuramitsu and Francis L. Macrina for strains and plasmids.

This research was supported by National Research Service award F32-DE05493 to M.C.H. and by Public Health Service grant DE06801 from the National Institutes of Health.

LITERATURE CITED

1. **Bradford, M. M.** 1976. A rapid and sensitive method for the quantitation of microgram quantities of protein utilizing the principle of protein-dye binding. *Anal. Biochem.* **72:**248–254.

2. **Chassy, B. M., J. R. Beall, R. M. Bielawski, E. V. Porter, and J. A. Donkersloot.** 1976. Occurrence and distribution of sucrose-metabolizing enzymes in oral streptococci. *Infect. Immun.* **14:**408–415.

3. **Clark, W. B., L. L. Bammann, and R. J. Gibbons.** 1978. Comparative estimates of bacterial affinities and adsorption sites on hydroxyapatite surfaces. *Infect. Immun.* **19:**846–853.

4. **Crow, W. D., R. Machanoff, and H. I. Adler.** 1985. Isolation of anaerobes using an oxygen reducing membrane fraction: experiments with acetone butanol producing organisms. *J. Microbiol. Methods* **4:**133.

5. **Guan, C., B. Wanner, and H. Inouye.** 1983. Analysis of regulation of *phoB* expression using a *phoB-cat* fusion. *J. Bacteriol.* **156:**710–717.

6. **Hamada, S., and H. D. Slade.** 1980. Biology, immunology, and cariogenicity of *Streptococcus mutans*. *Microbiol. Rev.* **44:**331–384.

7. **Hudson, M. C., and R. Curtiss III.** 1990. Regulation of expression of *Streptococcus mutans* genes important to virulence. *Infect. Immun.* **58:**464–470.

8. **Hudson, M. C., and G. C. Stewart.** 1986. Differential utilization of *Staphylococcus aureus* promoter sequences by *Escherichia coli* and *Bacillus subtilis*. *Gene* **48:**93–100.

9. **Jagusztyn-Krynicka, E. K., M. Smorawinska, and R. Curtiss III.** 1982. Expression of *Streptococcus mutans* aspartate-semialdehyde dehydrogenase gene cloned into plasmid pBR322. *J. Gen. Microbiol.* **128:**1135–1145.

10. **Lunsford, R. D., and F. L. Macrina.** 1986. Molecular cloning and characterization of *scrB*, the structural gene for the *Streptococcus mutans* phosphoenolpyruvate-dependent sucrose phosphotransferase system sucrose-6-phosphate hydrolase. *J. Bacteriol.* **166:**426–434.

11. **Macrina, F. L., R. P. Evans, J. A. Tobian, D. L. Hartley, D. B. Clewell, and K. R. Jones.** 1983. Novel shuttle plasmid vehicles for *Escherichia-Streptococcus* transgeneric cloning. *Gene* **25:**145–150.

12. **Manoil, C., and J. Beckwith.** 1985. Tn*phoA*: a transposon probe for protein export signals. *Proc. Natl. Acad. Sci. USA* **82:**8129–8133.

13. **McKenney, K., H. Shimatake, D. Court, U. Schmeissner, C. Brady, and M. Rosenberg.** 1981. A system to study promoter and terminator signals recognized by *Escherichia coli* RNA polymerase, p. 384–415. *In* J. S. Chirikjian and T. S. Papas (ed.), *Gene Amplification and Analysis*. Elsevier Science Publishing, Inc., New York.

14. **Murchison, H. H., J. F. Barrett, G. A. Cardineau, and R. Curtiss III.** 1986. Transformation of *Streptococcus mutans* with chromosomal and shuttle plasmid (pYA629) DNAs. *Infect. Immun.* **54:**273–282.

15. **Robeson, J. P., R. G. Barletta, and R. Curtiss III.** 1983. Expression of a *Streptococcus mutans* glucosyltransferase gene in *Escherichia coli*. *J. Bacteriol.* **153:**211–221.

16. **Sato, S., S. Ueda, and H. K. Kuramitsu.** 1987. Construction of *Streptococcus mutans* glucosyltransferase mutants utilizing a cloned gene fragment. *FEMS Microbiol. Lett.* **48:**207–210.

17. **Schroeder, V. A., S. M. Michalek, and F. L. Macrina.** 1989. Biochemical characterization and evaluation of virulence of fructosyltransferase-deficient mutant of *Streptococcus mutans* V403. *Infect. Immun.* **57:**3560–3569.

18. **Shaw, W. V.** 1979. Chloramphenicol acetyltransferase from chloramphenicol-resistant bacteria. *Methods Enzymol.* **43:**737–755.

19. **Shaw, W. V., L. C. Packman, B. D. Burleigh, A. Dell, H. R. Morris, and B. S. Harltey.** 1979. Primary structure of a chloramphenicol acetyltransferase specified by R plasmids. *Nature* (London) **282:**870–872.

20. **Shiroza, T., and H. K. Kuramitsu.** 1988. Sequence analysis of the *Streptococcus mutans* fructosyltransferase gene and flanking regions. *J. Bacteriol.* **170:**810–816.

21. **Shiroza, T., S. Ueda, and H. K. Kuramitsu.** 1987. Sequence analysis of the *gtfB* gene from *Streptococcus mutans*. *J. Bacteriol.* **169:**4263–4270.

22. **Silhavy, T. J., and J. R. Beckwith.** 1985. Uses of *lac* fusions for the study of biological problems. *Microbiol. Rev.* **49:**398–418.

23. **Tanzer, J. M.** 1972. Studies on the fate of the glucosyl moiety of sucrose metabolized by *Streptococcus mutans*. *J. Dent. Res.* **51:**415–423.

24. **Terleckyj, B., N. P. Willett, and G. D. Shockman.** 1975. Growth of several cariogenic strains of oral streptococci in a chemically defined medium. *Infect. Immun.* **11:**649–655.

25. **Ueda, S., T. Shiroza, and H. K. Kuramitsu.** 1988. Sequence analysis of the *gtfC* gene from *Streptococcus mutans* GS-5. *Gene* **69:**101–109.

26. **Van den Broeck, G., M. P. Timko, A. P. Kausch, A. R. Cashmore, M. Van Montagu, and L. Herrera-Estrella.** 1985. Targeting of a foreign protein to chloroplasts by fusion to the transit peptide from the small subunit of ribulose 1,5-biphosphate carboxylase. *Nature* (London) **313:**358–363.

27. **Wenham, D. G., T. D. Hennessey, and J. A. Cole.** 1979. Regulation of glucosyl- and fructosyltransferase synthesis by continuous cultures of *Streptococcus mutans*. *J. Gen. Microbiol.* **114:**117–124.

28. **Zaidenzaig, Y., and W. V. Shaw.** 1976. Affinity and hydrophobic chromatography of three variants of chloramphenicol acetyltransferases specified by R factors in *Escherichia coli*. *FEBS Lett.* **62:**266–271.

Characterization of the *iga* Gene Encoding Immunoglobulin A Protease in *Streptococcus sanguis*

JOANNE V. GILBERT, ANDREW G. PLAUT, AND ANDREW WRIGHT

Department of Medicine, Tufts-New England Medical Center Hospital, and Department of Molecular Biology and Microbiology, Tufts University School of Medicine, Boston, Massachusetts 02111

Immunoglobulin A (IgA) proteases are extracellular enzymes of pathogenic bacteria, first clearly identified among streptococci in the human oral cavity (19). Numerous other enzymes satisfying the criteria for this group of endopeptidases have since been found among human bacterial pathogens in the genera *Neisseria* (20), *Haemophilus* (11, 15), *Clostridium* (5), and *Bacteroides* (16). In all cases, the initial sites of entry and colonization of these bacteria are the mucosal surfaces of the human oral cavity, gut, respiratory tree, and genital tract, sites at which IgA is the dominant antibody isotype. Interest in how these unusual enzymes are involved in the infectious process has been sustained by their unique properties, including a high degree of specificity for human IgA as substrate, the invariant cleavage of hinge region peptide bonds to which proline contributes the carboxyl group (shown in Fig. 1), the large size of these enzymes (exceeding 100 kDa even after removal of a large segment of the precursor forms), and the existence of natural inhibitors of their activity in the form of secretory IgA antibody in human milk and other relevant mucosal secretions (6).

The IgA proteases of gram-negative organisms have been studied in detail. The *iga* genes encoding the enzyme in *Neisseria gonorrhoeae* type 1 and type 2, *Haemophilus influenzae* type 1 and type 2, and *Neisseria meningitidis* type 1 have been cloned into *Escherichia coli* and characterized (2, 7, 8, 13, 21, 22). Deletion and insertion mutations have provided insights into the secretion mechanism from the bacterial cell, and the generation of hybrid enzyme proteins has localized the region of the enzyme that determines subtrate specificity. Also, nucleotide and deduced amino acid sequences have allowed comparison of the various proteases to one another and have provided information required for correct classification of their catalytic mechanisms. It has been shown in the case of the *N. gonorrhoeae* IgA protease (21) that secretion involves the autoproteolytic processing of a larger precursor having four domains: an amino-terminal signal peptide, a catalytic domain that eventually becomes the mature enzyme, an alpha region of uncertain function, and a carboxy-terminal helper segment that provides a pore through which the enzyme is transported across the outer membrane. Because the IgA proteases of the gram-negative genera *Haemophilus* and *Neisseria* have a consensus sequence GDSGS characteristic of serine-type proteases and are inhibited by peptide boronic acid analogs of the hinge region substrate (1), these enzymes have been tentatively classified as serine proteases. Although the *Neisseria* and *Haemophilus* enzymes appear to function quite similarly, the IgA proteases as a group exhibit marked differences in catalytic mechanism. The *Bacteroides melaninogenicus* enzyme was shown by Mortensen and Kilian (16) to be a thiol-activated protease, and the streptococcal and clostridial (5, 17) IgA proteases were considered metalloproteases on the basis of their inhibition by metal chelators. The divergence in catalytic mechanism among enzymes that share substrate specificity for human IgA reinforces the likelihood that these enzymes participate in the colonization or infections involving these diverse bacterial pathogens.

Studies of the streptococcal IgA proteases have already provided valuable insights into their possible role in oral health. Reinholdt and Kilian (23) have shown that streptococcal IgA proteases impair the capacity of secretory IgA to carry out one of its documented functions, inhibition of adherence of streptococci to saliva-coated hydroxyapatite. In a series of detailed studies, they also showed that Fab fragments remaining bound to streptococcal surface antigens did not block binding of these cells to hydroxyapatite and possibly even enhanced it. Kilian and co-workers have recently used IgA protease as one of the biochemical and serological criteria to reclassify viridans streptococci (12). As newly classified, all *Streptococcus sanguis* strains (biovars 1 to 4) and *Streptococcus oralis* (formerly called *Streptococcus mitior*) produce IgA proteases. This is of particular interest because these species are dominant in the initiation of dental plaque, the precursor to caries formation.

We have recently identified the *iga* gene encoding IgA protease in *S. sanguis* ATCC 10556 (19). This strain, a human isolate, was included in the recent reclassification of oral streptococci and found to be biovar 1. The *iga* gene was cloned into *E. coli*, in which it encoded an active protease

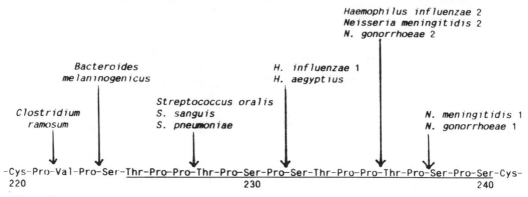

FIG. 1. Cleavage sites of the IgA proteases of various bacterial pathogens. The sequence is the hinge region of the human IgA1 heavy chain. As shown, all known IgA proteases cleave to the carboxy-terminal site of proline residues.

of approximately 200 kDa having the same substrate specificity as the wild-type enzyme; 75% of this activity was localized to the periplasmic space. With use of two DNA probes more than 2 kb in length from within the cloned *iga* gene, there was no detectable homology, under stringent conditions, between the *S. sanguis iga* gene and the chromosomal DNA of bacteria that also produce IgA proteases. These organisms included *H. influenzae* and *N. gonorrhoeae*, *Clostridium ramosum*, *B. melaninogenicus*, and *Streptococcus pneumoniae*. These probes also failed to hybridize with chromosomal DNA of IgA protease-negative streptococcal strains (e.g., *Streptococcus mutans*) and of strain Challis, an organism earlier classified as *S. sanguis* but now classified as *Streptococcus gordonii*.

To further characterize the *S. sanguis* IgA protease, we have sequenced the entire *iga* gene by the dideoxy-chain termination method (24) on M13 clones from plasmid pJG1 (6a). The nucleotide and deduced amino acid sequences (Fig. 2) reveal an open reading frame (ORF) consisting of 5,655 bp, at least a part of which encodes active IgA protease. Translation of the *iga* gene in *E. coli* begins 652 bp from the beginning of the ORF, as determined by matching the deduced amino acid sequence to the amino terminus of the enzyme protein recovered from the periplasm of *E. coli*. The first residue in the protein is methionine encoded by GTG, which establishes this as the amino terminus because GTG, which typically encodes valine, encodes methionine only when initiating a polypeptide chain (14). Knowing that the methionine is the authentic initiating amino acid also excludes the possibility that the *E. coli iga* gene product has undergone posttranslational processing at its amino-terminal end. In addition, no persuasive signal sequence was found at the amino terminus, leaving unexplained the peri-

plasmic localization of the enzyme. The deduced streptococcal IgA protease amino acid sequence shares no significant homology with sequences of the serine-type IgA proteases of *Haemophilus* and *Neisseria* species or of any other protein in the GenBank and NBRF data banks.

The deduced protein sequence also reveals a lengthy tandem repeat beginning 139 amino acids from the amino terminus and consisting of 10 segments of 20 residues each. Although not identical, these segments are highly homologous. After insertion of two gaps (consisting of single residues) to maximize homology, the amino acids at six positions in all of the repeats are invariant, and in 10 of the 20 positions there is at least 90% homology. Analysis for hydrophobicity and potential secondary structure indicates that the repeated region is hydrophobic despite a content of 28 glutamic acids and a single lysine residue and, because of the numerous proline and glycine residues, is unlikely to have a helical structure. This stretch has no discernible homology with tandemly repeated regions in other streptococcal proteins, such as the immunoglobulin-binding proteins, the antiphagocytic M protein of group A strains, and the glucosyltransferase enzymes of cariogenic streptococci (3, 4, 9, 25, 26).

The deduced amino acid sequence of the *S. sanguis* protease has provided important insights into the classification of this enzyme because it contains the pentapeptide HEMTH beginning at amino acid 1284. This sequence corresponds to a consensus zinc binding signature HExxH (x is variable) found in many metalloproteases in both prokaryotes and eukaryotes and allows tentative classification of the *S. sanguis* enzyme as a metalloprotease (10), consistent with earlier findings that the catalytic activity is inhibited by the metal chelators. Supportive evidence that this region is essential for activity has been obtained by site-

FIG. 2. Nucleotide and deduced amino acid sequences of the *S. sanguis* 10556 *iga* gene. The start of transcription is indicated as +1, and putative promoter regions are indicated as −10 and −35. SD is the proposed ribosome binding site. MEDKEALNQN is the amino terminus of the enzyme protein encoded by pJG1 in *E. coli*, as determined by microsequencing. The boxed amino acids HEMTH, and E 20 residues away, are elements of a proposed zinc binding site of the *S. sanguis* enzyme.

```
1201  GCCATCGTTGAGCCTGAACAAATTGAACCGGAGATTGGAGGTGTCCAATCCGGTGCGATAGTAGAACTGAGCAAGTAGACACAGGAACTCAAGCAGGCGCCGTA  1320
177    A  I  V  E  P  E  Q  I  E  P  E  I  G  G  V  Q  S  G  A  I | V  E  P  E  Q  V  T  P  L  P  E  Y  T  G  T  Q  A  G  A  V   216

1321  GTGTCACCCGAACAAGTAGCTCCATTGCCAGAATACACAGGTACACAGTCTGGGGCCATAGTTGAACCGGCTCAAGTTACTCCATTGCCGGAGTACACAGGCGTCCAATCTGGGGCCATA  1440
217    V  S  P  E  Q  V  A  P  L  P  E  Y  T  G  T  Q  S  G  A  I | V  E  P  A  Q  V  T  P  L  P  E  Y  T  G  V  Q  S  G  A  I   256

1441  GTAAAACCCGCGCAAGTCAAGTTACTCCGTTGCCAGAGTATACGGGCACAGAGTCAATCTGGGGCCATAGTTGAACCTACGTCTCCAGCAGGAGCAATC  1560
257    V  K  P  A  Q  V  T  P  L  P  E  Y  T  G  T  Q  S  G  A  I | V  E  P  E  Q  V  T  P  S  P  E  Y  T  G  V  Q  A  G  A  I   296

1561  GTTGAACCTGAACAAGTGGCTTCATTACCTGAATACACAGGATCTCAAGCTGGAGCGATTGTTGAGCCTGAGCAGGAGCTCTCCGCAAGAATACTGAAACATGAACCTGCAGCG  1680
297    V  E  P  E  Q  V  A  S  L  P  E  Y  T  G  S  Q  A  G  A  I | V  E  P  E  Q  V  E  P  P  Q  E  Y  T  G  N  I  E  P  A  A   336

1681  CCAGAGGCGGAAAATCCTGAAAAAAGCTCAAGAGACCAGAAGCAGAAGCAAGAACCAGAAAGAATCGAGTTAAGAAATGTATGTGATGTAGAGCTCTAGCTGATGGA  1800
337    P  E  A  E  N  P  T  E  K  A  Q  E  P  K  E  Q  K  Q  E  P  E  K  N  I  E  L  R  N  V  S  D  V  E  L  Y  S  L  A  D  G   376

1801  AAATACAAACAGCACGTTCTCTAGATGCCATTCCAAGCAATCAAGAGAATTATTCGTGAAAGTAAAATCTTCTAAATTAAAGATGTTTTCTTGCCGATTCTTCAATAGTCGATAGC  1920
377    K  Y  K  Q  H  V  S  L  D  A  I  P  S  N  Q  E  N  Y  F  V  K  V  K  S  S  K  F  K  D  V  F  L  P  I  S  S  I  V  D  S   416

1921  ACAAAAAGATGTCAGCCGGTTTATAAATTACGGCAAGTGCTGAAAAATTAAAACGAGACCTCAACAATAAGTACGAGACAATTTACTTTTATCTGCTAAAAAGGCAGAGAGAGAA  2040
417    T  K  D  G  Q  P  V  Y  K  I  T  A  S  A  E  K  L  K  Q  D  V  N  N  K  Y  E  D  N  F  T  F  Y  L  A  K  K  A  E  R  E   456

2041  GTCACAAACTTCACTTCCTTAGTAACTTGGTTCAAGCTATAAATAAACAATCTCAATGGAACTATTATTTAGCTGCTAGTCTGAATGCCAACGAGGTCGAACTAGAAAATGGTGCTAGC  2160
457    V  T  N  F  T  S  F  S  N  L  V  Q  A  I  N  N  N  L  N  G  T  Y  Y  L  A  A  S  L  N  A  N  E  V  E  L  E  N  G  A  S   496

2161  AGTTATATAAAGGGTAGATTTACTGGTAAGCTCTTTGGCAGCAAAGACGGGAAAAAATTATGCTATTTATAAATTTGAAGAAACCTTATTTGACACCATTGAGCGCTGCTACTGTAGAAAAT  2280
497    S  Y  I  K  G  R  F  T  G  K  L  F  G  S  K  D  G  K  N  Y  A  I  Y  N  L  K  K  P  L  F  D  T  L  S  A  A  T  V  E  N   536

2281  CTGACTCTTAAAGATGTGAATATCTCAGGAAAAACTGATATTGGGGCCCTTGCAAATGAAGCCAATAATGCACAAGGATTAACAATGTCCATGTAGACGGTGTTCTGGCTGGCAACGT  2400
537    L  T  L  K  D  V  N  I  S  G  K  T  D  I  G  A  L  A  N  E  A  N  N  A  T  R  I  N  N  H  V  D  G  V  L  A  G  E  R   576

2401  GGCATTGGTGGCTTGGTGTTGGAAGGCTGATAATTCTAATAGTAGTTTCAACTCCTATGAAACGAAGCACCATACAATATCCGGAGGATTAGTAGGC  2520
577    G  I  G  G  L  V  W  K  A  D  N  S  K  I  S  N  S  S  F  K  G  R  I  V  N  S  Y  E  T  K  A  P  Y  N  I  G  G  L  V  G   616
```

FIG. 2—Continued.

```
2521  CAACTGACTGGCATCAATGGCATTGGTTGATAAAGTCAAAAGCTACAATTACCATCTCGTCAAATGCGGATAGTAGTCAAAAGCGGGGTCTTGTTGAGAAAGATGCG  2640
617    Q  L  T  G  I  N  A  L  V  D  K  S  K  A  T  I  T  S  S  N  A  D  S  T  N  Q  T  V  G  G  L  A  G  L  V  E  K  D  A   656

2641  CTTATCAGCAATAGTTATGCCGAGGCAACATTAATAATGTGAAACGCTTGGAAGTGTTGCTGGCTGGCTACTGTGGGATAGAGATTCTAGCGAAGAGAGACATGCTGGAAGA  2760
657    L  I  S  N  S  Y  A  E  G  N  I  N  N  V  K  R  F  G  S  V  A  G  V  A  G  Y  L  W  D  R  D  S  S  E  E  R  H  A  G  R   696

2761  TTGCATAATGTTCTTAGTGATATCAATGTTATGAACGGAGATCGGATTAGTGGTTATCACTATCGAGGAATGAGGATAACTGACTCATATAGCAACAAAGACAACAGAGTCTACAAAGTG  2880
697    L  H  N  V  L  S  D  I  N  V  M  N  G  N  A  I  S  G  Y  H  Y  R  G  M  R  I  T  D  S  Y  S  N  K  D  N  R  V  Y  K  V   736

2881  ACTCTTGAAAAGGATGAGGTTGTCACCAAGGAATCTCGAAGAGAGGGACAATCCTTGATGTTTCTCAAATCCAAGTAAGAAATCTGAAATTAACTCTCTTTCTGCACCGCAAAGTC  3000
737    T  L  E  K  D  E  V  V  T  K  E  S  L  E  E  R  G  T  I  L  D  V  S  Q  I  A  S  K  K  S  E  I  N  S  L  S  A  P  K  V   776

3001  GAAACCTTGCTGACTAGCACTAATAAAGAAGTGATTTTCTAAGGTTAAAGACTATCAAGCCAGTCGAGCTTTAGCATATAAGAATATTGAAAAATTGCTGCCGTTTATAATAAGCCA  3120
777    E  T  L  L  T  S  T  N  K  E  S  D  F  S  K  V  K  D  Y  Q  A  S  R  A  L  A  Y  K  N  I  E  K  L  L  P  F  Y  N  K  A   816

3121  ACCATAGTCAAATACGGTAATCTAGTAAAAGAAGATAGCACCTTGTATGAAAAGAAAATCTTATCTGCAGTCATGATGAAGGATAATGAAGTGATCACAGATATCGCTTCGCATAAGAG  3240
817    T  I  V  K  Y  G  N  L  V  K  E  D  S  T  L  Y  E  K  E  I  L  S  A  V  M  M  K  D  N  E  V  I  T  D  I  A  S  H  K  E   856

3241  GCAGCTAATAAGCTCTTGATTCATTATAAAGATCATTCATCTGAAAAGCTAGATCTCACTACCAATCTGACTTTAGTAAATTAGCGGAATATCGTGTGGGCGATCAGGTCTCATCTAT  3360
857    A  A  N  K  L  L  I  H  Y  K  D  H  S  S  E  K  L  D  L  T  Y  Q  S  D  F  S  K  L  A  E  Y  R  V  G  D  T  G  L  I  Y   896

3361  ACGCCAAATCAATTCTTGCAAAATCATAGTTCAATGAAGTTTGCCTGATTGAAAGAAGTTGCGATTATCAGTCAGAGCTATCAGAAACACCCTAGGTATTTCTTCAGGCCTT  3480
897    T  P  N  Q  F  L  Q  N  H  S  S  I  V  N  E  V  L  P  D  L  K  A  V  D  Y  Q  S  E  A  I  R  N  T  L  G  I  S  S  G  V   936

3481  TCACTGACGGAATTATTACTTAGAGAAGCAGTTGCCAAACGAGTTTGCCAAAATCAATTAGCAACACATTGGAAAAACTGTTGTCTGCTGCAGTCAGTGAAAATCAAACGATTAAT  3600
937    S  L  T  E  L  Y  L  E  E  Q  F  A  K  T  K  E  N  L  A  N  T  L  E  K  L  L  S  A  D  A  V  I  A  S  E  N  Q  T  I  N   976

3601  GGTTATGTCGTTGATAAAAATCAAACGCAATAAGGAGGCCTTGCTTCTTAGAGTTGACCTTATTAGAGCGCTGTGTGACAACTTTAACTATGGTGATGTCAAAGACTTAGTCATGTAT  3720
977    G  Y  V  V  D  K  I  K  R  N  K  E  A  L  L  L  G  L  T  Y  L  E  R  W  Y  N  F  N  Y  G  D  V  N  V  K  D  L  V  M  Y   1016

3721  CACATGGATTTCTTTGGTAAGGGCAATGTGTCACCGCTAGACACCATCATTGAATTAGGTAAATCTGGCTTTAACAATCTTCTGGCCAAGAACAACCTAGATGCCTATAACATCAGCCTT  3840
1017   H  M  D  F  F  G  K  G  N  V  S  P  L  D  T  I  I  E  L  G  K  S  G  F  N  N  L  L  A  K  N  N  V  D  A  Y  N  I  S  L   1056
```

FIG. 2—Continued.

```
3841  GCAAACAATAATGCAACAAAAGATTGTTCAGCACGCTTGCCAATTACCGAGAAGTATTTTACCAAACAAACAAATAATCAATGTTTAAAGAGCAACCAAGGCTTATATAGTTGAA  3960
1057  A  N  N  A  T  K  D  L  F  S  T  L  A  N  Y  R  E  V  F  L  P  N  K  T  N  N  Q  W  F  K  E  Q  T  K  A  Y  I  V  E     1096

3961  GAAAAATCAGCTATTGATGAGGTCAGGGGTCAAACAAGAGAGCAGGGTCTGGCAGCAAATACCTCTATCGGGGTGTATGATCGAATTACCAGTGACACTTGGAAATACCGAAATATGTCCTCCT  4080
1097  E  K  S  A  I  D  E  V  R  V  K  Q  E  Q  A  G  S  K  Y  S  I  G  V  Y  D  R  I  T  S  D  T  W  K  Y  R  N  M  V  L  P    1136

4081  TTGCTGACAATGCTGAAAGATCAGTCTTTGTCATCAGATATATCCAGTCTTGGTTTTGGTGCCTATGACAGATATAGAAATAATGAGACAGAGCAGGGCAGAACTTAATAAGTT  4200
1137  L  L  T  M  P  E  R  S  V  F  V  I  S  T  I  S  S  L  G  F  G  A  Y  D  R  Y  R  N  N  E  H  R  A  G  A  E  L  N  K  F    1176

4201  GTTGAGGATAATGCCCAAGAAACCGCAAAACGTCAACAGCGAGATCATTATGATTATTGGTATAGAAATATTAGACGACAGGAGAAAGCCGTGAAAAGCCTATCGAAATATCTTGGTCTATGATGCC  4320
1177  V  E  D  N  A  Q  E  T  A  K  R  Q  R  D  H  Y  D  Y  W  Y  R  I  L  D  E  Q  G  R  E  K  L  Y  R  N  I  L  V  Y  D  A    1216

4321  TATAAGTTCGGAGATGATACTACTGTCGACAAGGCTACAGTAGAGGCGCAATTGACAGTTCTAATCACCTATGAAGTACTTCTTGGCCGGTTGAAACAAGGTTGTACAACAATAAG  4440
1217  Y  K  F  G  D  T  T  V  D  K  A  T  V  E  A  Q  F  D  S  S  N  P  A  M  K  Y  F  F  G  P  V  G  N  K  V  V  H  N  K     1256

4441  CATGGAGCTTACGCTACTGGTGATAGTGTTTACTACATGGGCTATCGGATGCTGGACAAGGATGGAGCTATTACCTATCCATGAAATGACGCATGTCATTCTGATAACGAGATCTATTA  4560
1257  H  G  A  Y  A  T  G  D  S  V  Y  Y  M  G  Y  R  M  L  D  K  D  G  A  I  T  Y  T [H  E  M  T] H  D  S  D  N  E  I  Y  L   1296
                                                                                    F  K

4561  GGTGGATATGGAGAAGAAGTGGTCTTGCCCAGAATCTTTTGCCAAGGCTTGTTGCAGCAATGCGACACCGGTCAACTCAATATGACAAG  4680
1297  G  G  Y  G  R  R  S  G  L  G [P  E] F  F  A  K  G  L  L  Q  A  P  D  H  P  D  D  A  T  I  T  V  N  S  I  L  K  Y  D  K  1336

4681  AATGATGCATCTGAAAAATCTCGTTTGCAAGTCTTGGATCCAACTAAACGTTTCCAAAATGCGGATGATCTGAAAAACTATGTTGTAAACATGTTTGATGTCATTTATATGTTGGAGTAC  4800
1337  N  D  A  S  E  K  S  R  L  Q  V  L  D  P  T  K  R  F  Q  N  A  D  D  L  K  N  Y  V  V  H  N  M  F  D  V  I  Y  M  L  E  Y  1376

4801  CTAGAAGGAATGTCAATCGTAAATCGTCTAAACCGTCTGTCCGATGTACAGAAAGTACAAGAAAGTGAATGCTGAGAGAATAAATATGTCCGATGCGGAAATGATGTTTACGCAACCAATGTG  4920
1377  L  E  G  M  S  I  V  N  R  L  S  D  V  Q  K  V  N  A  L  R  K  I  E  N  K  Y  V  R  D  A  D  G  N  D  V  Y  A  T  N  V    1416

4921  ATAAAAAATATTACAATGGCGGATCAGAAACTAAATTCATTCAACAGTCTGATTGAAAATGATATTCTTCAGCGCTGAGTACAAAAATGGCGATGTAGAAAGAAATGGCTACCAT  5040
1417  I  K  N  I  T  M  A  D  A  Q  K  L  N  S  F  N  S  L  I  E  N  D  I  L  S  A  R  E  Y  K  N  G  D  V  E  R  N  G  Y  H    1456

5041  ACGATTAAACTCTTCTCCGATTTATTCTGCTTTAAGCAGTGAAAAAGGAACTCCTGGACGTCGTATGAACTTGCCAGCCAAAGGCTTCGCAGCGATGGT  5160
1457  T  I  K  L  F  S  P  I  Y  S  A  L  S  S  E  K  G  T  P  G  D  L  M  G  R  R  I  A  Y  E  L  L  A  A  K  G  F  K  D  G    1496
```

FIG. 2—Continued.

specific mutagenesis of the HE dipeptide in this sequence, which resulted in complete loss of activity.

Turning to the entire ORF, there are 651 bp between the start of the ORF and the GTG which begins translation in *E. coli*. Clearly translation of this upstream region is not required for the IgA protease activity, because the product of pJG1 is active. Inspection of this upstream region shows three hydrophobic segments near the amino terminus. Interestingly, the hydrophobic region that follows a methionine at bp 43 has features typical of a gram-positive signal sequence. Although the role of this upstream segment is not known, if it is translated in *S. sanguis*, one or more of the hydrophobic regions could contribute to enzyme binding or transport through the *S. sanguis* cell membrane. In beginning to examine the possibility that translation

```
5161  ATGGTTCCTTATATCTCTAATCAATATGAAGATGATGCCAAGCAAACAATCAGTATCTATGGTAAGACTAGAGGTCTGGTAACAGATGACTTGGTTTACGTAAGGTCTTC  5280
1497   M  V  P  Y  I  S  N  Q  Y  E  D  D  A  K  Q  N  G  K  T  I  S  I  Y  G  K  T  R  G  L  V  T  D  D  L  V  L  R  K  V  F   1536

5281  AATGGTCAGTTTAATAATTGGACTGAGTTTAAGAAGGCTAAAATGTATGAAGAAAGAAAGAACAAGTTCGACAGCCTGAACAAGTTCACATTGATGATACAAGACACCATGGACAAGTTAT  5400
1537   N  G  Q  F  N  N  W  T  E  F  K  K  A  M  Y  E  E  R  K  N  K  F  D  S  L  N  K  V  T  F  D  D  T  R  Q  P  W  T  S  Y   1576

5401  GCTACTAAGACTATAAGTACTGTAGAAGAGTTGCAAACCTTGATGGATGAAGCCGTTCTCCAAGATGCAAATGATAATTGGTATTCTTGGAGCGGCTATAAACCAGAATATAACAGTGCT  5520
1577   A  T  K  T  I  S  T  V  E  E  L  Q  T  L  M  D  E  A  V  L  Q  D  A  N  D  N  W  Y  S  W  S  G  Y  K  P  E  Y  N  S  A   1616

5521  GTCCATAAGCTAAAAAAGCAGTCTTCAAAGCTTACCTCGATCAGACTAAAGATTTAGAAAATCAATCTTTGAAAACCAGAGTGATTGGTTCGAGCAGTCTCGCAAAGTTAC  5640
1617   V  H  K  L  K  K  Q  S  S  K  L  T  S  I  R  L  K  I  L  E  N  Q  S  L  K  T  R  S  D  W  F  E  Q  S  N  G  L  Q  S  Y   1656

5641  AAGTCTGAAATGTTAGGGCAGAGCTTGAATTACTCTTAAATTCAGACCTTCTTGCCAGATAAGAAATGACGAACCTAGTGAAAATTAGGTTGTTTTCTATATAAGTCATGCAATTAAAA  5760
1657   K  S  E  M  L  G  Q  S  L  N  Y  S  *   1669

5761  CTGTAGCCAGTCTGCTTCAGGCAGTTGCTTCTGATAGTCATTTATGAGGCTGATGATAAGCAGC  5826
```

FIG. 2—*Continued.*

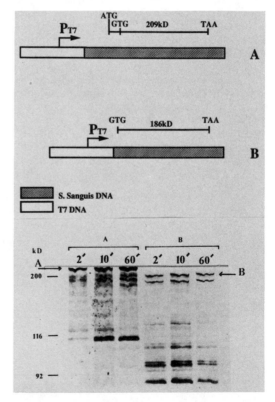

FIG. 3. IgA protease gene expression under control of a T7 promoter. Plasmid A contains the entire ORF, and plasmid B contains the ORF having a 360-bp deletion from the 5' end to remove the possible ATG start and signal sequence. The autoradiograph shows *E. coli* extracts prepared from cells 2, 10, and 60 min after labeling with [^{35}S]methionine. Plasmid A gives rise to a larger *iga* gene product (arrow A) than does plasmid B (arrow B). This indicates that once a promoter is provided, in this case that of T7, the upstream ATG is a suitable translation start site.

begins upstream of the GTG, we expressed the *iga* gene under the control of a T7 promoter. Plasmids (Fig. 3) containing the entire ORF (plasmid A) and the ORF with 360 bp deleted from the 5′ end to remove the possible ATG start and signal sequence (plasmid B) were introduced into *E. coli* K38, from which ^{35}S-labeled cell extracts were examined by sodium dodecyl sulfate-polyacrylamide gel electrophoresis. Plasmids A and B gave rise to proteins of different sizes, with the product of plasmid A being approximately 20 kDa larger than that of plasmid B. Although this finding verifies that once a promoter is provided, the upstream ATG is a suitable translation start site, its participation in the synthesis and secretion of the protease in *S. sanguis* must await analysis of the amino terminus of the *S. sanguis* protein.

Figure 4 shows a map of pJG1 and a schematic diagram depicting the relative positions of the more important features of the *S. sanguis* IgA protease, as discussed above.

The experiments described above clearly do not reveal the details of the secretion mechanisms of the IgA proteases in streptococci as they have been determined for gram-negative IgA proteases. One can predict that among gram-positive bacteria, translocation will be different because these cells have no outer membrane through which a means for transport must be provided. This does not, however, exclude the possibility that structural components of the enzyme could favor interaction with the cell wall during secretion. Available data on extracellular proteases of gram-positive bacteria reveal that all have a pre-pro protein consisting of (i) a signal sequence and (ii) a region ultimately removed by processing to reach the mature form (27).

It will be of considerable interest to determine whether processing of IgA proteases in gram-positive species occurs by an autoproteolytic mechanism similar to that of gram-negative species. There are two potential sites for such cleavage in the enzyme sequence. One (PSP) is upstream of the *E. coli* GTG translation start and lies within the sequence LPSPGE. This is very similar to a region (LPSTGE) thought to be involved in membrane anchoring and subsequent processing of the antiphagocytic M protein of group A streptococci (18). The second possible autocatalytic site (TPSP) is found within the tandem repeat. If the tandem repeat region turns out to be a hot spot for recombination events, it could become a site of antigenic variability, similar to that documented for M protein. In this case, cleavage in the repeat would create antigenic variability in the secreted enzyme. Moreover, because the amino-terminal end of the ORF encodes a hydrophobic domain(s), a portion of the protein could, after processing, be associated with the cell surface. If this view is correct, this protein could itself be antigenically variable. Because of these possibilities, further insights into the structure and secretion mechanisms of the *S. sanguis* protease may substantially advance our understanding of its role in colonization.

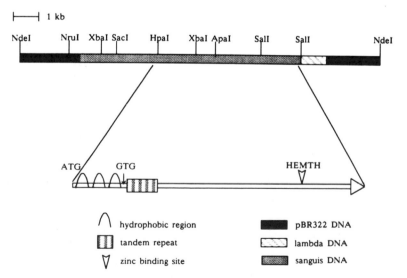

FIG. 4. Partial restriction map of plasmid pJG1 showing the entire *S. sanguis* DNA insert and the extent of the ORF of the *iga* gene. The center diagram of the enzyme protein indicates the extent of the ORF in pJG1 and shows the relationships of the hydrophobic upstream regions, the GTG start site in *E. coli*, the tandem repeat region, and the putative zinc binding site HEMTH to one another.

LITERATURE CITED

1. **Bachovchin, W. W., A. G. Plaut, G. R. Flentke, M. Lynch, and C. A. Kettner.** 1990. Inhibition of IgA1 proteases from *Neisseria gonorrhoeae* and *Hemophilus influenzae* by peptide prolyl boronic acids. *J. Biol. Chem.* **265:**3738–3743.

2. **Bricker, J., M. H. Mulks, A. G. Plaut, E. R. Moxon, and A. Wright.** 1983. IgA1 proteases of *Hemophilus influenzae*: cloning and characterization in *E. coli* K-12. *Proc. Natl. Acad. Sci. USA* **80:**2681–2685.

3. **Fahnestock, S. R., P. Alexander, J. Nagle, and D. Filpula.** 1986. Gene for an immunoglobulin-binding protein from a group G streptococcus. *J. Bacteriol.* **167:**870–880.

4. **Ferretti, J. J., M. L. Gilpin, and R. R. B. Russell.** 1987. Nucleotide sequence of a glucosyltransferase gene from *Streptococcus sobrinus* MFe28. *J. Bacteriol.* **169:**4271–4278.

5. **Fujiyama, Y., M. Iwaki, K. Hodohara, S. Hosoda, and K. Kobayashi.** 1986. The site of cleavage of human alpha chain IgA1 IgA2:A2m(1) allotype paraproteins by the clostridial IgA protease. *Mol. Immunol.* **23:**147–150.

6. **Gilbert, J. V., A. G. Plaut, B. Longmaid, and M. E. Lamm.** 1983. Inhibition of microbial IgA proteases by human secretory IgA and serum. *Mol. Immunol.* **20:**1039–1049.

6a.**Gilbert, J. V., A. G. Plaut, and A. Wright.** 1991. Analysis of the immunoglobulin A protease gene of *Streptococcus sanguis*. *Infect. Immun.* **59:**7–17.

7. **Grundy, F., A. G. Plaut, and A. Wright.** 1987. *Haemophilus influenzae* A1 protease genes: cloning by plasmid integration-excision, comparative analyses, and localization of secretion determinants. *J. Bacteriol.* **169:**4442–4450.

8. **Grundy, F. J., A. G. Plaut, and A. Wright.** 1990. Localization of the cleavage site specificity determinant of *Haemophilus influenzae* immunoglobulin A1 protease genes. *Infect. Immun.* **58:**320–331.

9. **Heath, D. G., and P. P. Cleary.** 1989. Fc-receptor and M protein genes of group A streptococci are products of gene duplication. *Proc. Natl. Acad. Sci. USA* **86:**4741–4745.

10. **Jongeneel, C.-V., J. Bouvier, and A. Bairoch.** 1989. A unique signature identifies a family of zinc-dependent metallopeptidases. *FEBS Lett.* **242:**211–214.

11. **Kilian, M., J. Mestecky, and R. E. Schronhenloher.** 1979. Pathogenic species of the genus *Haemophilus* and *Streptococcus pneumoniae* produce immunoglobulin A2 protease. *Infect. Immun.* **26:**143–149.

12. **Kilian, M., L. Mikkelsen, and J. Henrichsen.** 1989. A taxonomic study of viridans streptococci: description of *Streptococcus gordonii* sp. nov. and amended descriptions of *Streptococcus sanguis* (White and Niven 1946), *Streptococcus oralis* (Bridge and Sneath 1982), and *Streptococcus mitis* (Andrewes and Horder 1906). *Int. J. Syst. Bacteriol.* **39:**471–484.

13. **Koomey, J. M., and S. Falkow.** 1984. Nucleotide sequence homology between the immunoglobulin A1 protease genes of *Neisseria gonorrhoeae*, *Neisseria meningitidis*, and *Haemophilus influenzae*. *Infect. Immun.* **43:**101–107.

14. **Kozak, M.** 1983. Comparison of initiation of protein synthesis in procaryotes, eucaryotes, and organelles. *Microbiol. Rev.* **47:**1–45.

15. **Male, C.** 1979. Immunoglobulin A1 protease production by *Haemophilus influenzae* and *Streptococcus pneumoniae*. *Infect. Immun.* **26:**254–261.

16. **Mortensen, S. B., and M. Kilian.** 1984. Purification and characterization of an immunoglobulin A1 protease from *Bacteroides melaninogenicus*. *Infect. Immun.* **45:**550–557.

17. **Mulks, M. H.** 1985. Microbial IgA proteases, p. 82–104. In I. A. Holder (ed.), *Bacterial Enzymes and Virulence.* CRC Press, Inc., Boca Raton, Fla.

18. **Pancholi, V., and V. A. Fischetti.** 1989. Identification of an endogenous membrane anchor-cleaving enzyme for group A streptococcal M protein. *J. Exp. Med.* **170:**2119–2133.

19. **Plaut, A. G., R. J. Genco, and T. B. Tomasi, Jr.** 1974. Isolation of an enzyme from *Streptococcus sanguis* which specifically cleaves IgA. *J. Immunol.* **113:**289–291.

20. **Plaut, A. G., J. V. Gilbert, M. S. Artenstein, and J. D. Capra.** 1975. *Neisseria gonorrhoeae* and *N. meningitidis*: production of an extracellular enzyme which cleaves human IgA. *Science* **190:**1103–1105.

21. **Pohlner, J., R. Halter, K. Beyreuther, and T. F. Meyer.** 1987. Gene structure and extracellular secretion of *Neisseria gonorrhoeae* IgA protease. *Nature* (London) **325:**458–462.

22. **Poulsen, J., J. Grandt, P. Hjorth, H. C. Thogersen, and K. Kilian.** 1989. Cloning and sequencing of the immunoglobulin A1 protease gene (*iga*) of *Haemophilus influenzae* serotype b. *Infect. Immun.* **57:**3097–3105.

23. **Reinholdt, J., and M. Kilian.** 1987. Interference of IgA protease with the effect of secretory IgA on adherence of oral streptococci to saliva-coated hydroxyapatite. *J. Dent. Res.* **66:**492–497.

24. **Sanger, F., S. Nicklen, and A. R. Coulson.** 1977. DNA sequencing with chain-terminating inhibitors. *Proc. Natl. Acad. Sci. USA* **74:**5643–5647.

25. **Shiroza, T., S. Ueda, and H. K. Kuramitsu.** 1987. Sequence analysis of the *gtfB* gene from *Streptococcus mutans*. *J. Bacteriol.* **169:**4263–4270.

26. **Ueda, S., T. Shiroza, and H. K. Kuramitsu.** 1988. Sequence analysis of the *gtfC* gene from *Streptococcus mutans* GS-5. *Gene* **69:**101–109.

27. **Wandersman, C.** 1989. Secretion, processing and activation of bacterial extracellular proteases. *Mol. Microbiol.* **3:**1825–1831.

Environmental Variables Affecting Arginine Deiminase Expression in Oral Streptococci

ROBERT A. BURNE, DAWN T. PARSONS, AND ROBERT E. MARQUIS

Cariology Center, Department of Dental Research and Department of Microbiology and Immunology, University of Rochester Medical Center, Rochester, New York 14642

Dental caries remains an enormous health problem in this country and abroad (11, 16). The current view of the initiation and progression of dental caries is one of a dynamic process in which dental plaque undergoes repetitive cycles of demineralization and remineralization as a result of glycolytic acidification followed by an alkalinization period. A carious lesion occurs when these cycles are no longer balanced and acid-mediated enamel demineralization predominates because of changes in plaque flora, diet, or other intrinsic and extrinsic factors. Recent evidence supports the notion that metabolism of nitrogenous compounds with generation of basic compounds is a principal mechanism for neutralization of organic acids (2, 15, 17, 22, 23) and likely of major importance in the inhibition of the initiation and progression of dental caries. In particular, studies comparing caries-susceptible and caries-resistant subjects indicate increased production of ammonia, particularly from arginine (2, 17), in plaque by caries-resistant individuals.

Many prokaryotes, including the oral streptococci, can catabolize arginine via the arginine deiminase system (ADS) (1, 7). This three-enzyme pathway (shown below) converts one molecule of arginine to one molecule each of ornithine and carbon dioxide and two molecules of ammonia, with the concomitant production of ATP from ADP.

NCTC 10904 have been isolated and shown to be clustered in the chromosome (4). It seems that these genes could be manipulated to produce strains of oral streptococci able to modulate the cariogenic potential of dental plaque. These strains could be developed either by engineering hyperproducers from organisms that normally possess the system or through the introduction of the genes of the ADS into cariogenic *Streptococcus mutans* (4, 11, 16), which normally lacks the system. In the mouth, bacteria experience radical and abrupt environmental changes. Sugar concentrations change rapidly and radically, as do pH levels, oxygen tension, and fluoride levels. If the ADS were to be active in modulating plaque acidification, then the system would need to function under the extreme conditions found in dental plaque. In this report, we describe the effects of variables commonly encountered in the human mouth on the expression of the ADS and compare the ADS of *S. sanguis* and *Streptococcus rattus*.

Fluoride Effects

Fluoride is a known cariostatic agent that is widely disseminated in drinking water, foods, and oral hygiene products. Its principal cariostatic effect is thought to involve enhanced remineralization, but it can also affect a variety of microbial processes, especially in plaque, where it is con-

$$\text{arginine} \xrightarrow{\text{arginine deiminase (AD)}} \text{citrulline} + NH_3$$

$$\text{citrulline} + P_i \xleftarrow{\text{ornithine carbamyltransferase}} \text{ornithine} + \text{carbamyl phosphate}$$

$$\text{carbamyl phosphate} + ADP \xleftarrow{\text{carbamate kinase}} CO_2 + NH_3 + ATP$$

Arginine can be transported into the cell by an arginine/ornithine antiporter which exchanges arginine (outside) for ornithine (inside) with no expenditure of energy (8). In systems examined to date, coordinate regulation of the ADS enzymes (1, 9, 21), arginine transport (20), and in one case arginine-specific aminopeptidase (13) have been observed.

The genes for the ADS of *Streptococcus sanguis*

centrated to levels as high as 5 μmol/g (wet weight). Fluoride has three known inhibitory effects on oral streptococci which can reduce their cariogenic potential (12). Enolase is inhibited directly (14). Proton expulsion via the F_1F_0 proton-translocating ATPase is inhibited by fluoroaluminate complexes (M. G. Sturr and R. E. Marquis, *Arch. Microbiol.*, in press), and the HF form of fluoride ($pK_a = 3.15$) can diffuse some

TABLE 1. Effects of fluoride on ADS enzyme activity[a]

| Fluoride concn (mM) | % Activity[b] | | | | | |
| | AD | | OTC | | CK | |
	S. sanguis	S. rattus	S. sanguis	S. rattus	S. sanguis	S. rattus
0.0	100	100	100	100	100	100
1.0	102	105	95	97	106	101
5.0	101	107	95	93	105	100
10.0	98	104	94	77	106	101
20.0	104	104	74	59	106	104
50.0	103	107	72	38	100	104
100.0	97	110	56	18	105	104

[a]Reactions were carried out by using permeabilized cells as previously described (5) and contained the indicated concentrations of sodium fluoride. Briefly, AD activity was assayed by the colorimetric citrulline procedure (3, 5). Ornithine carbamyltransferase (OTC) activity was assayed in the direction of citrulline synthesis from ornithine and carbamyl phosphate, and carbamate kinase (CK) activity was determined by measuring ATP synthesized from ADP and carbamyl phosphate in a reaction linked to the reduction of NADP. Cells were preincubated for 5 min with NaF before addition of substrate. NaCl controls for assessing salts effects showed no significant inhibition.

[b]In all assays, the standard error was approximately ±5%.

10^7 times faster than F^- to short-circuit transmembrane proton currents (10).

Fluoride at 1 mM had almost no effect on the enzymes of the ADS in permeabilized cells of S. sanguis NCTC 10904 or S. rattus FA-1 (Table 1). Levels of fluoride above about 10 mM did inhibit ornithine carbamyltransferase, presumably because fluoride-metal complexes can mimic phosphate and affect transfer reactions. In contrast, carbamate kinase and arginine deiminase were highly insensitive, even to 100 mM fluoride.

Fluoride in the growth medium can repress the ADS, as measured by AD enzyme activity (Table 2). S. rattus was not significantly affected by fluoride concentrations as high as 2.0 mM, but 5.0 mM fluoride in the growth medium caused 60% inhibition of production of AD activity. S. sanguis ADS expression was more sensitive to fluoride, with 22% inhibition at 1.0 mM and 65% inhibition at 2.0 mM. At a fluoride level of 5.0 mM, final growth of S. sanguis was reduced and AD expression was 20% that observed for control cultures.

pH Effects

Fluoride is known to upset the acid-base physiology of the oral streptococci (3), and the effects of fluoride on expression of the ADS genes could possibly be related to cytoplasmic acidification. Additionally, a wide variety of components of the oral streptococci are known to be influenced by the growth pH. The results of previous studies indicated that the ADS of the oral streptococci can function at pH values below the minima for growth or glycolysis (18). However, acid environments could affect expression of ADS genes.

Initially, batch cultures buffered with phosphate were used to assess the effects of growth pH on expression of ADS genes for cells grown at a maintained value of approximately 7.0, 6.5, or 6.0. No major differences in specific activities

of ADS enzymes were detected. Cultures were then grown in glucose-limited chemostat cultures in brain heart infusion broth (Difco Laboratories, Detroit, Mich.) supplemented with 0.3% glucose and 1.0% arginine. The dilution rate was 0.1/h (generation time = 6.93 h), and pH was maintained by addition of 1 N solutions of HCl or NaOH. Cultures were allowed to reach steady state over 10 generations prior to sampling for enzyme assays.

The optimal pH value for expression of the ADS was approximately 6.5 for both S. rattus and S. sanguis, as indicated by AD enzyme levels in permeabilized cells. AD activity of S. sanguis grown at pH 5.7 was only 21% that of cells grown at pH 6.0, while for S. rattus AD activity of pH 6.0 cells was about 70% that of pH 6.5 cells. In

TABLE 2. AD levels in cells grown in subinhibitory concentrations of sodium fluoride[a]

| NaF concn (mM) in growth medium | % AD activity[b] | |
	S. sanguis NCTC 10904	S. rattus FA-1
0.0	100 (3.0 U/mg)	100 (9.2 U/mg)
1.0	78	96
2.0	35	93
5.0	21[c]	40

[a]Cultures were grown to early stationary phase in TY medium (6) supplemented with 0.5% galactose and 1.0% arginine and the concentration of sodium fluoride indicated. Cells were harvested and permeabilized by two freeze-thaw cycles followed by toluene treatment, and AD activity was measured by assaying the production of citrulline colorimetrically (3). One unit was defined as the amount of enzyme required to produce 1 μmol of product per h. Normalization was to cell dry weight determined after cells were washed twice in distilled H_2O and dried to constant weight in a drying oven.

[b]Level of AD found in cells grown in the absence of fluoride as 100%. Numbers in parentheses are specific activities per milligram (dry weight) of cells.

[c]Some inhibition of total growth was observed in the S. sanguis sample at 5.0 mM NaF.

chemostat culture, glucose was not repressive because it was a limiting nutrient. Although expression of ADS genes was reduced at lower growth pH values, cells grown in acid media still had the capacity to produce ammonia at low pH. This acid tolerance may be important for any desired anticaries effects associated with ADS activity.

Catabolite Repressibility

In many bacteria, including *Pseudomonas aeruginosa*, *Enterococcus hirae*, and *S. sanguis*, the ADS is repressible by catabolites such as glucose (1, 9, 21). Coordinate repression of all three enzymes of the ADS and catabolic arginine transport occurs in these organisms (20). In *Streptococcus mitis*, it appears that an arginine aminopeptidase is also coordinately regulated with the system (13). For maximal anticaries effect, ADS expression should be insensitive to catabolite repression so that arginolysis concomitant with glycolysis could moderate glycolytic pH drops. Table 3 shows the levels of AD activity of *S. sanguis* and *S. rattus* from the early stationary phase of growth in TY medium (6) with 0, 1, or 2% initial glucose. The ADS of *S. sanguis* proved to be very sensitive to glucose repression (Table 3). Moreover, exponential-phase cells from cultures had only about 5% the AD activity of fully derepressed cells. Similar results were obtained for cells from media with glucose but without supplemental arginine. Expression of the ADS by *S. rattus* was much less sensitive to glucose repression (Table 3). Cells grown in medium with 1% glucose showed no repression, but those grown in 2% glucose showed about 72% repression. The insensitivity of the ADS of *S. rattus* to glucose repression was evident also for cells harvested from cultures in the exponential phase.

Oxygen Effects

Major variations in oxygen tension occur in various regions of the human mouth. A freshly cleaned tooth surface exposed to air would be expected to have a positive redox potential (E_h value). However, as pellicle and plaque form and mature, the combined effects of restricted diffusion and microbial reduction of oxygen would result in a reducing environment and growth of strictly anaerobic as well as facultative bacteria. Although the levels of oxygen in plaque are low, they are not zero, even in subgingival plaque, in which the average, detectable oxygen tension was found to be some 13.3 mm Hg (19). The redox environment in plaque can have important effects on the metabolism of facultative and anaerobic bacteria (5, 24). Moreover, the responses of organisms to changes in oxygen tension or E_h values often involve multiple regulons.

The effects of aeration on the expression of the ADS of *S. sanguis* and *S. rattus* were assessed by growing cultures in TY medium with 0.5% galactose and 1.0% arginine under anaerobic conditions in a GasPak jar (BBL Co., Cockeysville, Md.) or with agitation and aeration at 250 rpm with 25 ml of culture in a 250-ml flask. After overnight cultivation, the AD activity of the aerobically grown culture was about 10 times lower than that of the anaerobically grown culture (Table 3). Aeration had a less repressive effect on *S. rattus*, and the AD level in the aerated culture was only about 75% that of the anaerobic culture.

Discussion

The data presented here indicate that fluoride, acid growth media, glucose, and oxygen all affect expression of ADS genes in *S. sanguis* NCTC 10904 and *S. rattus* FA-1. The overall conclusion is that the system is subject to complex regulatory interactions that need to be considered in designing implantable strains for moderation of plaque cariogenicity through enhanced base production. Moreover, the complex regulatory aspects of the ADS make it an excellent vehicle for study of gene regulation in the oral streptococci. The molecular bases of the differential expression of ADS in response to environmental stimuli and the different sensitivities of *S. sanguis* and *S. rattus* to these stimuli are undoubtedly complex and multifaceted.

Repression of ADS expression by fluoride may be mediated by a variety of systems. Poolman et al. (20) have demonstrated that agents which dissipate ΔpH or $\Delta\Psi$ do not significantly inhibit ADS, but that dicyclohexylcarbodiimide, which inhibits the F_1F_0-ATPase as does fluoride, decreases ADS activity in *Streptococcus lactis*. Presumably, this decrease reflects down-regulation of the ADS by ATP (20, 21). Because of its effects on transmembrane proton currents, fluoride could also affect arginine transport or the phosphoenolpyruvate phosphotransferase system, which mediates catabolite repression of

TABLE 3. Effects of glucose and oxygen on AD expression

Growth condition[a]	% AD sp act	
	S. sanguis	*S. rattus*
0.0% Glucose	100.0	100.0
1.0% Glucose	13.2	102.0
2.0% Glucose	11.5	28.3
Anaerobic[b]	100.0	100.0
Aerated	11.0	75.0

[a] Cells were grown to early stationary phase in TY medium supplemented with 0.5% galactose and 1.0% arginine plus 0, 1.0, or 2.0% glucose without aeration. To assess effects of aeration, cells were cultured in TY medium plus 0.5% galactose and 1.0% arginine as outlined in the text. Cells were harvested and permeabilized, and AD activity was assayed as for Table 1 and expressed as a percentage of the no-glucose control value.

[b] Values represent the percentage of activity from cells grown anaerobically.

the ADS. Therefore, there may be a connection between fluoride effects and pH regulation of the ADS.

Glucose repression of the ADS is quite common, so it is of interest that the *S. rattus* ADS is relatively refractile to repression by this sugar. The molecular basis of this insensitivity of the *S. rattus* ADS to glucose repression could be a peculiarity of ADS regulatory elements, but it might also be a general feature of catabolite-repressible genes of *S. rattus*. It is notable that the ADS enzymes of *Treponema denticola* are not subject to glucose repression at all (C. E. Caldwell, R. A. Burne, D. T. Parsons, and R. E. Marquis, *J. Dent. Res.* **69:**184, 1990), so that the ADS of *S. rattus* is not unique in this respect.

Oxygen could affect transcription of the ADS. However, H_2O_2 produced from oxygen could influence ADS activity or expression. H_2O_2 could direct inactivate ADS enzymes or could perturb the cytoplasmic membrane to disrupt transport of arginine or maintenance of proton motive force. Thus, pH, fluoride, and oxygen could act through a common pathway rather than having distinct regulatory influences on ADS.

A potentially important observation for designing implantable strains is that the ADS of *S. rattus* FA-1 is more highly expressed and less subject to repression, especially by glucose, than the ADS of *S. sanguis* NCTC 10904. Thus, key genetic and physiologic mechanisms for repression and derepression of the ADS may be identified by comparative studies. Clearly, the ADS has properties that make it an attractive candidate for manipulation to control plaque cariogenicity. A more thorough understanding of the molecular events not only will help to dissect general strategies of gene regulation in the streptococci, but also should provide information about molecular ecology of plaque and the potential for manipulating the ADS for modulating plaque cariogenicity. The differences in the effects of the various agents on ADS expression in *S. sanguis* and *S. rattus* should be investigated to determine whether these are characteristics of the ADS of these organisms or whether they simply represent general regulatory differences between these two organisms. Overall, it seems reasonable to draw the conclusion that the *S. rattus* ADS may be better suited to experiments involving manipulation of the ADS to modulate plaque cariogenicity, since these genes are less sensitive to repression by environmental influences. It also seems that the *S. rattus* ADS genes may be more amenable to strategies involving introduction of the system into *S. mutans* (4) than are those of *S. sanguis*, not only because of the *S. rattus* ADS regulatory characteristics but also because *S. rattus* is more closely related to *S. mutans* than is *S. sanguis* (6).

We thank Aida Casiano-Colon for helpful discussions.

This research was supported by Public Health Service award DE2 P50 DE07003 to the Rochester Cariology Center from the National Institute of Dental Research.

LITERATURE CITED

1. **Abdelal, A. T.** 1979. Arginine catabolism by microorganisms. *Annu. Rev. Microbiol.* **83:**139–168.
2. **Abelson, D. C., and I. D. Mandel.** 1981. The effect of saliva on plaque pH *in vivo*. *J. Dent. Res.* **60:**1634–1638.
3. **Bender, G. R., S. Sutton, and R. E. Marquis.** 1986. Acid tolerance, proton permeability, and membrane ATPases of oral streptococci. *Infect. Immun.* **53:**331–338.
4. **Burne, R. A., D. T. Parsons, and R. E. Marquis.** 1989. Cloning and expression in *Escherichia coli* of the genes of the arginine deiminase system of *Streptococcus sanguis*. *Infect. Immun.* **57:**3540–3548.
5. **Condon, S.** 1987. Responses of lactic acid bacteria to oxygen. *FEMS Microbiol. Rev.* **46:**269–280.
6. **Coykendall, A. L.** 1987. Proposal to elevate the subspecies of *Streptococcus mutans* to species based on their molecular composition. *Int. J. Syst. Bacteriol.* **27:**26–30.
7. **Cunin, R., N. Glansdorff, A. Piérard, and V. Stalon.** 1986. Biosynthesis and metabolism of arginine in bacteria. *Microbiol. Rev.* **50:**314–352.
8. **Driessen, A. J., B. Poolman, R. Kiewiet, and W. N. Konings.** 1987. Arginine transport in *Streptococcus lactis* is catalyzed by a cationic exchanger. *Proc. Natl. Acad. Sci. USA* **84:**6093–6097.
9. **Ferro, K. J., G. R. Bender, and R. E. Marquis.** 1983. Coordinately repressible arginine deiminase system in *Streptococcus sanguis*. *Curr. Microbiol.* **9:**145–150.
10. **Gutknecht, J., and A. Walter.** 1981. Hydrofluoride and nitric acid transport through lipid bilayer membranes. *Biochim. Biophy. Acta* **644:**153–156.
11. **Hamada, S., and H. D. Slade.** 1980. Biology, immunology, and cariogenicity of *Streptococcus mutans*. *Microbiol. Rev.* **44:**331–384.
12. **Hamilton, I. R.** 1990. Biochemical effects of fluoride on oral bacteria. *J. Dent. Res.* **69:**660–667.
13. **Hiraoka, B. Y., M. Mogi, K. Fukasawa, and M. Harada.** 1986. Coordinate represssion of arginine aminopeptidase and three enzymes of the arginine deiminase pathway in *Streptococcus mitis*. *Biochem. Int.* **12:**881–887.
14. **Hüther, F.-J., N. Psarros, and H. Duschner.** 1990. Isolation, characterization, and inhibition kinetics of enolase from *Streptococcus rattus* FA-1. *Infect. Immun.* **58:**1043–1047.
15. **Kleinberg, I., J. A. Kanapka, and D. Craw.** 1976. Effect of saliva and salivary factors on the metabolism of the mixed oral flora, p. 433–464. *In* H. M. Stiles, W. J. Loesch, and T. C. O'Brien (ed.), *Microbial Aspects of Dental Caries*. Information Retrieval Inc., Washington, D.C.
16. **Loesch, W. J.** 1986. Role of *Streptococcus mutans* in human dental decay. *Microbiol. Rev.* **50:**353–380.
17. **Margolis, H. C., J. H. Duckworth, and E. C. Moreno.** 1988. Composition and buffer capacity of pooled starved plaque fluid from caries-free and caries susceptible individuals. *J. Dent. Res.* **67:**1476–1482.
18. **Marquis, R. E., G. R. Bender, D. R. Murray, and A. Wong.** 1987. Arginine deiminase system and bacterial adaptation to acid environments. *Appl. Environ. Microbiol.* **53:**198–200.
19. **Mettraux, G. R., F. A. Gusberti, and H. Graf.** 1984. Oxygen tension (pO2) in untreated human periodontal pockets. *J. Periodont.* **55:**516–521.
20. **Poolman, B., A. M. Driessen, and W. N. Konings.** 1987. Regulation of arginine-ornithine exchange and the arginine deiminase pathway in *Streptococcus lactis*. *J. Bacteriol.* **169:**5597–5604.
21. **Simon, J.-P., B. Wargnies, and V. Stalon.** 1982. Control of enzyme synthesis in the arginine deiminase pathway of *Streptococcus faecalis*. *J. Bacteriol.* **150:**1085–1090.

22. **Wijeyeweera, R. L., and I. Kleinberg.** 1989. Acid-base pH curves *in vitro* with mixtures of pure cultures of human oral microorganisms. *Arch. Oral Biol.* **34:**55–64.

23. **Wijeyeweera, R. L., and I. Kleinberg.** 1989. Arginolytic and ureolytic activities of pure cultures of human oral bacteria and their effects on the pH response of salivary sediment and dental plaque *in vitro*. *Arch. Oral Biol.* **34:**43–53.

24. **Yamada, T.** 1987. Regulation of glycolysis in streptococci, p. 69–93. *In* J. Reiser and A. Peterkofsky (ed.), *Sugar Transport and Metabolism in Gram-Positive Bacteria.* Ellis Horwood, Ltd., Chichester, England.

Analysis of *Streptococcus mutans* and *Streptococcus downei* Mutants Insertionally Inactivated in the *gbp* and *gtfS* Genes

JEFFREY A. BANAS AND KEETA S. GILMORE

Department of Microbiology and Immunology, University of Oklahoma Health Sciences Center,
Oklahoma City, Oklahoma 73190

The glucosyltransferases (GTFs) of the mutans streptococci and the glucan-binding protein (GBP) of *Streptococcus mutans* are thought to contribute to processes important in the cariogenicity of these organisms (2, 7, 9). The GTFs hydrolyze sucrose and utilize the glucose moiety to synthesize glucans that are involved in adherence and aggregation of *S. mutans* on the tooth surface (3, 8) and that can act as an extracellular carbohydrate source (12). The GBP is believed to play a role in sucrose-dependent adherence and the formation of coherent plaque (2, 9). In many of the mutans streptococci there are more than one type of GTF, which may synthesize soluble (GTF-S) or insoluble glucans or a mixture of the two. Some of the GTF enzymes are primer dependent, and it has been hypothesized that the soluble glucan made by one GTF can act as a primer for the synthesis of the insoluble glucan made by another (5, 11). To begin to evaluate the relative contribution of each type of GTF, and the contributions of both GTF and GBP to virulence, we have inactivated the *gtfS* gene of *Streptococcus downei* and the *gbp* gene of *S. mutans* and examined the properties of the mutants in comparison with the wild types.

The sequences of *S. mutans gbp* (1) and *S. downei gtfS* (4) have been previously determined, and portions of each gene were removed and replaced with a gene encoding erythromycin resistance (Fig. 1). The inactivated genes were subcloned into *Escherichia coli*, and the DNA was isolated, linearized, and introduced back into the host mutans streptococci by electrotransformation. Western immunoblots and Southern hybridizations were performed to confirm that a single inactivated copy of the gene was present in the transformed streptococci.

Several differences between the *S. downei* GTF-S mutant and the wild-type strain were noted. The mutant organisms formed smoother and larger colonies on mitis-salivarius agar and were surrounded by watery pools of glucan. The mutant colonies no longer retained the characteristic hardness or ground-glass appearance of wild-type *S. downei*, and they exhibited reduced adherence to glass (Fig. 2).

These results suggest that GTF-S plays a key role in the synthesis of insoluble glucans although its own product is a short-chain soluble glucan (10). Evidence has recently been obtained indicating the existence of additional genes specifying GTFs which synthesize water-soluble glucan (R. R. B. Russell, personal communication). In the absence of large amounts of insoluble glucan production, substrate availability may promote synthesis of soluble glucans by these other gene products. However, the structure or perhaps the size of the glucan produced by these alternative GTFs appears to differ from that catalyzed by the *gtfS* gene product in that the watery glucan produced is incapable of efficiently priming insoluble glucan synthesis.

Differences between the GBP mutant and wild-type *S. mutans* were less obvious. Colony morphology was similar, and when grown in sucrose-enriched Todd-Hewitt broth into which steel wires were suspended, similar amounts of plaque were deposited on the wires. The plaques formed by the GBP mutant and wild-type *S. mutans* were similar in consistency and adhered to the wires to similar degrees. These results differ from the original reports of a GBP mutant generated by mutagenesis (9). The chemically mutagenized GPB mutant formed a less coherent plaque on wires suspended in broth cultures (9). The most likely explanation for the different results is that the chemically induced mutant contained an additional mutation that affected plaque formation.

Douglas and Russell previously showed that anti-GPB antisera interfered with the in vitro adherence, but not aggregation, of *S. mutans* (2). The adherence properties of the GBP mutant were tested in polystyrene microtiter dishes in a modification of the procedure of Larrimore et al. (6). No difference between mutant and wild type was observed. More sensitive assays capable of detecting subtle changes in adherence are currently being carried out. Both the GBP mutant and wild-type *S. mutans* also showed extensive sucrose-induced aggregation; however, the aggregated mutant cells appeared to be less dense and did not settle as far toward the bottom of the test tubes as did the wild type (Fig. 3). It is possible that the GBP promotes aggregation and that in its absence the cells do not become packed as densely. Alternatively, the non-glucan-binding portion of GBP may bind some other molecule

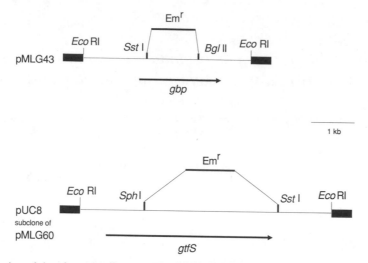

FIG. 1. Inactivation of the *gbp* and *gtfS* genes. The 1.2-kb *Sst*I-*Bgl*II fragment from *gbp* (1) was removed and replaced with the 1.2-kb gene for erythromycin resistance. Likewise, the 3.0-kb *Sph*I-*Sst*I fragment containing the majority of *gtfS* (4) was replaced with the erythromycin resistance gene.

to the cell surface that is utilized by the cell and can contribute to the density of cellular aggregates. At present the GBP has no known function other than binding of glucans. These observations may begin to contribute to the discovery of additional roles for GBP.

We thank Joseph J. Ferretti and Roy R. B. Russell for their support and guidance.

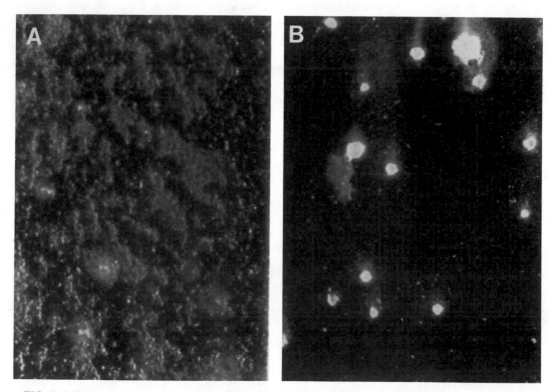

FIG. 2. Adherence to glass of overnight culture of *S. downei* with 5% sucrose. (A) Wild-type *S. downei*; (B) *gtfS* mutant.

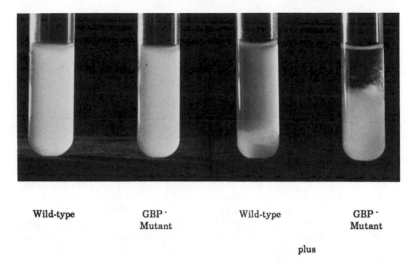

| Wild-type | GBP⁻ Mutant | Wild-type | GBP⁻ Mutant |

plus

3% sucrose

FIG. 3. Sucrose-induced aggregation of wild-type and GBP⁻ *S. mutans* strains. Overnight growth of the strains was resuspended in resting-cell suspensions containing phosphate-buffered saline (two left-most tubes) or phosphate-buffered saline plus 3% sucrose (two right-most tubes). The cells were allowed to aggregate for 5 h at 37°C.

This research was supported by Public Health Service grant DE08191 from the National Institutes of Health. J.A.B. was supported by National Research Service award DE05545 from the National Institute of Dental Research.

LITERATURE CITED

1. **Banas, J. A., R. R. B. Russell, and J. J. Ferretti.** 1990. Sequence analysis of the gene for the glucan-binding protein of *Streptococcus mutans* Ingbritt. *Infect. Immun.* **58:** 667–673.
2. **Douglas, C. W. I., and R. R. B. Russell.** 1982. Effect of specific antisera on adherence properties of the oral bacterium *Streptococcus mutans. Arch. Oral Biol.* **27:**1039–1045.
3. **Gibbons, R. J., and M. Nygaard.** 1968. Synthesis of insoluble dextran and its significance in the formation of gelatinous deposits by plaque-forming streptococci. *Arch. Oral Biol.* **13:**1249–1262.
4. **Gilmore, K. S., R. R. B. Russell, and J. J. Ferretti.** 1990. Analysis of the *Streptococcus downei gtfS* gene, which specifies a glucosyltransferase that synthesizes soluble glucans. *Infect. Immun.* **58:**2452–2458.
5. **Koga, T., S. Sato, M. Inoue, K. Takeuchi, T. Furuta, and S. Hamada.** 1983. Role of primers in glucan synthesis by glucosyltransferases from *Streptococcus mutans* strain OMZ176. *J. Gen. Microbiol.* **129:**751–754.
6. **Larrimore, S., H. Murchison, T. Shiota, S. M. Michalek, and R. Curtiss III.** 1983. In vitro and in vivo complementation of *Streptococcus mutans* mutants defective in adherence. *Infect. Immun.* **42:**558–566.
7. **Loesche, W. J.** 1986. Role of *Streptococcus mutans* in human dental decay. *Microbiol. Rev.* **50:**353–380.
8. **Mukasa, H., and H. D. Slade.** 1973. Mechanism of adherence of *Streptococcus mutans* to smooth surfaces. *Infect. Immun.* **8:**555–562.
9. **Russell, R. R. B., A. C. Donald, and C. W. I. Douglas.** 1983. Fructosyltransferase activity of a glucan-binding protein from *Streptococcus mutans. J. Gen. Microbiol.* **129:** 3243–3250.
10. **Russell, R. R. B., M. L. Gilpin, N. Hanada, Y. Yamashita, Y. Shibata, and T. Takehara.** 1990. Characterization of the product of the *gtfS* gene of *Streptococcus downei*, a primer-independent enzyme synthesizing oligo-isomaltosaccharides. *J. Gen. Microbiol.* **136:**1631–1637.
11. **Shimamura, A., H. Tsumori, and H. Mukasa.** 1983. Three kinds of extracellular glucosyltransferases from *Streptococcus mutans* 6715 (serotype g). *FEBS Lett.* **157:**79–84.
12. **Walker, G. J.** 1978. Dextrans, p. 75–126. *In* D. J. Manners (ed.), *Biochemistry of Carbohydrates*, II, vol. 16. University Park Press, Baltimore.

Phenotypic Effects of Inactivating the Gene Encoding a Cell Surface Binding Protein in *Streptococcus gordonii* Challis

HOWARD F. JENKINSON

Department of Oral Biology and Oral Pathology, University of Otago, Dunedin, New Zealand

Cell surface components of streptococci are recognized as being colonization determinants and virulence factors. Several species of oral streptococci have antigenically related salivary glycoprotein receptors on their cell surfaces (4, 16) which may be important for bacterial growth and survival in the human mouth. However, other colonization factors on the surfaces of oral streptococcal cells are not well characterized. *Streptococcus sanguis* and *S. gordonii* (formerly *S. sanguis*; see reference 17) have perhaps the widest range of adherence and aggregation reactions of all the oral streptococci. Strains of these species interact with protein components in saliva (19, 21, 26), crevicular fluid (3), serum (25), and plasma (2), with platelets (9), and with many other oral bacteria in coaggregations (18). Our work is concerned with identifying and characterizing surface components (mainly proteins) of *S. gordonii* and *S. sanguis* to understand the molecular basis for adherence and aggregation.

Cloning of Genes Encoding Cell Surface Polypeptides

Various physical methods involving blending or agitation have been used to try to remove surface polypeptides from *S. sanguis* (6, 20, 24). Incubation of *S. gordonii* cells with lauroyl sarcosine at room temperature removes peripheral proteins and polysaccharides that are involved in conferring hydrophobicity and coaggregation properties (12). This treatment also removes membrane-associated components, for example a phosphocarrier protein (14). A lauroyl sarcosinate extract of *S. gordonii* Challis was fractionated by anion-exchange high-performance liquid chromatography (14) into about eight polypeptides ranging in molecular mass from 94 to 26 kDa, and antiserum was raised to this preparation in rabbits. On Western immunoblots of cell envelope proteins prepared from *S. gordonii* Challis and separated by sodium dodecyl sulfate (SDS)-polyacrylamide gel electrophoresis (PAGE), the antiserum reacted strongly with four polypeptides and weakly with an additional three polypeptides (Fig. 1, lane 3). The antiserum was used to screen a library of *S. gordonii* Challis DNA prepared in λgt11 (28) for production of streptococcal antigens, using ^{125}I-labeled protein A as a detection reagent. One of several antigen-positive plaques from about 80,000 screened was purified. Restriction enzyme analysis of the DNA obtained from this phage (designated λgt11-cp2) showed an additional *Eco*RI fragment of 1.85 kb in the *lacZ* cloning site (Fig. 2). When *Escherichia coli* Y1089 (28) lysogenic for this recombinant phage was induced by heat shock, a 29-kDa antigenic polypeptide was produced.

Functional Map of the Cloned Fragment

The 1.85-kb *Eco*RI fragment of *S. gordonii* DNA in λgt11-cp2 could not be cloned into either multicopy or low-copy-number plasmid vectors in *E. coli*. The *Pst*I-*Pvu*II fragment (to the right of the *Pst*I site in Fig. 2) was, however, ligated into plasmid pNL9740 (consisting of the erythromycin resistance determinant from pVA736 [22] cloned in the unique *Nde*I site of pUC9 [30]). This plasmid (designated pNL9750) was stable in *E. coli* and did not express an antigenic polypeptide. DNA sequencing from this plasmid, and of numerous smaller fragments of the streptococcal DNA in M13, suggested the presence of two open reading frames (ORFs) within the original 1.85-kb fragment. The first ORF was at the left-hand end, terminating at approximately 100 bp past the second *Hin*dIII site (Fig. 2); the second ORF extended throughout the *Pst*I-*Pvu*II fragment, with promoterlike sequences in an A + T-rich region upstream of the *Pst*I site. These reading frames are indicated in Fig. 2; *sarA* (see below for explanation of nomenclature) encoded the 29-kDa polypeptide detected in induced *E. coli* Y1089 carrying λgt11-cp2.

Insertional Inactivation of the Cloned Gene

The *Pst*I-*Pvu*II fragment in plasmid pNL9750 lay wholly within the transcription unit of the cloned *sarA* gene (Fig. 2). The plasmid therefore was used to inactivate the chromosomal gene in *S. gordonii* by insertion-duplication mutagenesis. So as not to introduce the intact ampicillin resistance determinant into *S. gordonii*, a portion of the determinant was deleted at the unique *Cfr*10I site. The resulting plasmid (pNL9760d1, 5.4 kb) conferred erythromycin resistance but not ampicillin resistance in *E. coli* DH5α. It was transformed into *S. gordonii* Challis with selection for erythromycin resistance (1 μg/ml). Plasmid integration onto the *S. gordonii* ge-

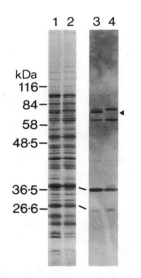

FIG. 1. SDS-PAGE profiles of cell-envelope proteins extracted from wild-type *S. gordonii* Challis (lane 1) and mutant strain d1b (lane 2) and corresponding Western blots (lanes 3 and 4) reacted with surface protein antiserum and peroxidase-linked second antibody. Cells were grown to late exponential phase in TY-glucose medium (15) containing 0.05% Tween 80, harvested by centrifugation, washed, and broken with glass beads (13). Cell envelopes were pelleted by high-speed centrifugation (13), extracted with SDS sample buffer, and subjected to electrophoresis through 10% acrylamide (13). Positions of molecular size markers are indicated. The arrowhead marks the position of the 76-kDa polypeptide (missing in lanes 2 and 4).

nome (within the 1.85-kb *Eco*RI fragment) was confirmed by Southern hybridization of DNA extracted from several transformants with nick-translated pNL9740.

Envelope proteins were extracted from disrupted cells of strain Challis and from erythromycin-resistant transformants and were compared by SDS-PAGE and by reaction of protein blots with *S. gordonii* surface protein antiserum. The Western blot of cell envelope proteins from one representative mutant (designated strain d1b) lacked an antigenic polypeptide of 76 kDa (Fig. 1, lane 4). This deficiency was apparent also in the stained profiles of proteins (Fig. 1, lane 2). There were further differences in the profiles of envelope proteins from the mutant strain, with minor components missing at 39 and 31 kDa and reduction in amounts of bands at 94, 29, and 24 kDa (Fig. 1, lane 2). Antibodies within the original antiserum that reacted with the 29-kDa expression product in *E. coli* were affinity purified from recombinant plaque lifts (28). These antibodies reacted exclusively with the 76-kDa cell envelope polypeptide, showing that the 1.85-kb cloned fragment of *S. gordonii* DNA in λgt11-cp2 coded for a 29-kDa truncated form.

The 76-kDa polypeptide was not present in a cytoplasmic (soluble) fraction prepared from broken cells and was not detectable in extracellular culture fluid except in late stationary phase. Vectorial labeling experiments using ^{125}I and lactoperoxidase (12) showed that the 76-kDa

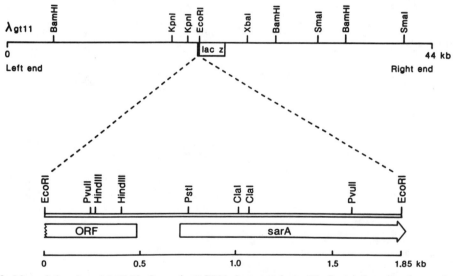

FIG. 2. Map of the cloned 1.85-kb *S. gordonii* DNA fragment in λgt11. Restriction sites for a selection of enzymes recognizing hexanucleotide sequences are shown. Nucleotide sequencing was performed by the Sanger method on a number of overlapping clones in phage M13 and also on both strands of plasmid pNL9750 (see text). The ORF commencing just to the left of the *Pst*I site and extending throughout the cloned fragment encodes a protein approximately one-third the size of complete 76-kDa (SarA) protein in *S. gordonii*.

polypeptide was exposed at the cell surface. The protein was tightly associated with the cell envelope fraction of *S. gordonii*, and the 29-kDa truncated form produced from λgt11-cp2 remained associated with the insoluble (envelope) fraction in *E. coli*.

Phenotypic Effects of Gene Inactivation

Phenotypic properties of strain d1b were compared with those of the wild-type strain Challis carrying pVA736 (strains were grown in medium containing 1 μg of erythromycin per ml). The mutant strain was identical to the wild-type strain in the following properties: hydrophobicity, adherence to saliva-treated hydroxyapatite or buccal epithelial cells, glucosyltransferase production, protease production, and competence for DNA-mediated transformation. However, the mutant cells were affected in their aggregation properties; they showed reduced coaggregation with some *Actinomyces* strains and reduced aggregation in saliva. Serum-mediated agglutination of mutant cells was virtually abolished (Table 1). The reduction in saliva-mediated aggregation was probably related to the immunoglobulin A (IgA) component of saliva. Saliva aggregation titers of mutant and wild-type cells were more similar in heated saliva (heating inactivates IgA), and mutant cells showed reduced aggregation titers in purified IgA from colostrum (Table 1).

TABLE 1. Aggregation reactions of *S. gordonii* Challis and mutant strain d1b

Aggregation with[a]:	Aggregation titer of *S. gordonii*[b]:	
	Wild-type Challis	Mutant d1b
Actinomyces viscosus T14V	32	32
A. viscosus WVU627	32	2
A. naeslundii EF1006	0[c]	0
A. naeslundii W1544	32	4
Whole saliva	64–128	4–8
Heated saliva (80°C, 30 min)	16	2–4
Human serum	1,024–2,048	0–8
Human serum IgG (1 mg/ml)	0	0
Human colostrum IgA (1 mg/ml)	16–32	4

[a]Aggregations were performed in 10 mM Tris hydrochloride (pH 7.2) containing 5 mM CaCl$_2$ in wells of microtiter plates with reciprocal shaking at 20°C for 15 min.

[b]Highest dilution of streptococcal cells (initially 5 × 10^8) still able to fully coaggregate with *Actinomyces* strains (2.5 × 10^8 cells) or the highest dilution of saliva, serum, or protein solution that caused complete bacterial agglutination. Duplicate titers were identical, and any range of titer over three experiments with different batches of cells is indicated.

[c]No aggregation.

Sequence Similarities in N-Terminal Regions of SarA, AmiA, and OppA Polypeptides

The inferred amino acid sequence of the N-terminal 260 amino acids of SarA (76 kDa) showed 69% identity with AmiA from *Streptococcus pneumoniae* (1), and both proteins showed about 30% similarity over 200 residues with OppA (Fig. 3), the periplasmic binding protein component of the oligopeptide permease system in *Salmonella typhimurium* (11). The *ami* locus in *S. pneumoniae* has six ORFs, *amiABCDEF* (1), and the C, D, E, and F proteins exhibit similarity to OppB, -C, -D, and -E components (1). Mutations in *amiC*, *-D*, *-E*, or *-F* cause increased resistance to aminopterin (27) and to methotrexate (29), while mutation in *amiA* does not result in a properly resistant phenotype (1). AmiC$^-$ and AmiE$^-$ mutants of *S. pneumoniae* are unable to incorporate certain oligopeptides, supporting the suggestion that the *ami* operon in *S. pneumoniae* may function as a peptide transport system (1). However, *ami* mutations also have secondary effects, for example on electric transmembrane potential (29). *S. gordonii* mutant strain d1b was tested for growth in defined medium (12) with various di- and tripeptides of leucine, alanine, and proline in place of these free amino acids, but no reduction in growth rate was observed. In TY-glucose medium (15), mutant strain d1b showed slightly increased resistance to methotrexate (MIC, 2 × 10^{-6} M) over that of the wild-type strain Challis (MIC, 5 × 10^{-7} M).

Cell Surface Binding Proteins in Streptococci

It seems likely that the *sarA* gene (so called because inactivation of it affected cell aggregation) in *S. gordonii* is part of an operon analogous to *ami*, although the predicted size of the binding protein AmiA in *S. pneumoniae* (52.3 kDa) is smaller than that of SarA (76 kDa). Further sequencing and mutagenesis of the region surrounding *sarA* in *S. gordonii* is necessary to determine whether *sarA* is part of an *ami*-like operon. The sequence for SarA in Fig. 3 is currently incomplete because of problems encountered with subcloning *sarA* in plasmid and phage M13 vectors. Similar problems were reported with *amiA* (23). AmiA protein is possibly held at the periphery of the cell membrane through a lipoamide or thioether linkage to fatty acid through the N-terminal cysteine (7, 8). Overexpression of the lipoprotein in *E. coli* is thought to disrupt membrane functions (23).

Binding-protein-dependent transport systems have now been described for gram-positive bacteria (10). In streptococci, the *ami* operon in *S. pneumoniae* may serve a peptide transport function, and a binding-protein-dependent uptake system in *S. mutans* transports melibiose (Russell

```
SarA          -KVNPA---A-LWLL--TSLA-CSIL--H-SSFNYIYE-V---DPENL
              * *       * * **    ** **      **   *  *    *** *
AmiA     1    MKKNRVFATAGLVLLAAGVLAACSSSKSSDSSAPKAYGTVYTADPETL
                *    *   * * *    *                            *
OppA     1    MSNITKKSLIAAGILTALIAATPTAADVPAGVQLADKQTLVRNNGSEVQSL


SarA          NYLISSKAATTDLTANLIDGLLENDNYGNLVPSMAEDWTVSKDGLTYTYTL
              **** *   **  * * **** ******* *  *** *********
AmiA    49    DYLISRKNSTTVVTSNGIDGLFTNDNYGNLAPAVAEDTEVSKDGLTYTYKI
              *          *       **  *  *   * *** *   **   *
OppA    52    DPHKIEGVPESNVSRDLFEGLLISDVEGHPSPGVAEKWE-NKDFKVWTFLR


SarA          RKDAKWYTSDGEEYADVQ--DA--GLKYAADNK-ETLY---LVQSSIKGLD
              **   ** ******** *    *    *** *** * *    *   * ***
AmiA   100    RKGVKWFTSDGEEYAEVTAKDFVNGLKHAADKKSEAMY---LAENSVKGLA
               *   **       *** ***     **       *    *
OppA   103    RENAKW-----SDGTPVTAHDFVYSWQRLADPNTASPYASYLQYGHIANID


SarA          DYVNGKTKDFSSVGVKAVDDHTVQYTLNEPESFWNSKTTMGILYPVNEEFL
              **   *   *** ******** * ***** ** ***** *    * * ****
AmiA   148    DYLSGTSTDFSTVGVKAVDDYTLQYTLNQPEPFWNSKLTYSIPWPLNEEFE
               *    *     **** ** *   **  * *        *      *
OppA   149    DIIAGKKPA-TDLGVKALDDHTFEVTLSEPVPYFYKLLVHPSVSPVPKS-A


SarA          KSKGDKFAQSADPTSLLYNGPFLLKSITSKSSIEFAKNPNYWDKDNVHVSD
              ***  **   ************** * *** ** **  **** ***
AmiA   199    TSKGSDFAKPTDPTSLLYNGPFLLKGLTAKSSVEFVKNEQYWDKENVHLDT
               *  *  * *         **  **          * ****
OppA   198    VEKFGD--KWTQPANIVTNGAYKLKNWVVNERIVLERNPQYWDNAKTVIN-


SarA          VKLTYFDGQDQ---------GK-------PGGSVDLQ.....
              * * ** *          *          *  *
AmiA   250    INLAYYDGSDNESLERNFTSGAYSYARLYPTSSNYSKVAEEY......483
                  *    *              *            *   *
OppA   245    -QVTYLPISSEVTDVNRYRSGEIDMTYNNMPIELFQKLKKEI......542
```

FIG. 3. Alignment of the inferred amino acid sequences of the N-terminal regions of SarA and the AmiA precursor (7) and of the AmiA precursor and the OppA precursor from *Salmonella typhimurium* (11). Gaps have been introduced to maximize similarity, and identical residues are marked with asterisks. Arrowheads mark the potential cleavage site of the AmiA precursor (7) and the site of cleavage of the OppA precursor (11).

et al., this volume). SarA is somewhat larger than the binding proteins in these systems; does SarA have a transport function? It seems likely that some binding proteins may serve processes other than transport. The Opp system in *E. coli* is required for recycling of wall components (10), an Opp-like system in mycoplasma is involved in virulence (5), and *Bacillus subtilis* has an *opp* operon in which the binding protein component is found mainly in the cell-free culture fluid and which may be necessary for initiation of sporulation (10). These observations raise the possibility that SarA in *S. gordonii* acts as a receptor for factors that promote streptococcal cell aggregation. However, inactivation of *sarA* expression caused a pleiotropic effect on cell envelope polypeptide composition, which may have influenced ability

of mutant cells to aggregate in the various conditions. This pleiotropic effect could be due to the plasmid insertion affecting expression of neighboring genes or to the aberrant assembly of surface components in the absence of SarA (or to both of these factors). Clearly, expression at the locus containing *sarA* is necessary for functional conformation of the cell surface in *S. gordonii*, and further genetic analysis may reveal a role for this locus in virulence.

I thank R. A. Easingwood for technical assistance and R. D. Cannon for advice and discussions.

This work was supported in part by the Medical Research Council of New Zealand.

LITERATURE CITED

1. **Alloing, G., M.-C. Trombe, and J.-P. Claverys.** 1990. The *ami* locus of the gram-positive bacterium *Streptococcus*

pneumoniae is similar to binding protein-dependent transport operons of gram-negative bacteria. *Mol. Microbiol.* **4:**633–644.

2. **Babu, J. P., W. A. Simpson, H. S. Courtney, and E. H. Beachey.** 1983. Interaction of human plasma fibronectin with cariogenic and noncariogenic oral streptococci. *Infect. Immun.* **41:**162–168.

3. **Cimasoni, G., M. Song, and B. C. McBride.** 1987. Effect of crevicular fluid and lysosomal enzymes on the adherence of streptococci and bacteroides to hydroxyapatite. *Infect. Immun.* **55:**1484–1489.

4. **Demuth, D. R., E. E. Golub, and D. Malamud.** 1990. Streptococcal-host interactions. Structural and functional analysis of a *Streptococcus sanguis* receptor for a human salivary glycoprotein. *J. Biol. Chem.* **265:**7120–7126.

5. **Dudler, R., C. Schmidhauser, R. W. Parish, R. E. H. Wettenhall, and T. Schmidt.** 1988. A mycoplasma high-affinity transport system and the *in vitro* invasiveness of mouse sarcoma cells. *EMBO J.* **7:**3963–3970.

6. **Fachon-Kalweit, S., B. L. Elder, and P. Fives-Taylor.** 1985. Antibodies that bind to fimbriae block adhesion of *Streptococcus sanguis* to saliva-coated hydroxyapatite. *Infect. Immun.* **48:**617–624.

7. **Gilson, E., G. Alloing, T. Schmidt, J.-P. Claverys, R. Dudler, and M. Hofnung.** 1988. Evidence for high affinity binding-protein dependent transport systems in gram-positive bacteria and in *Mycoplasma. EMBO J.* **7:**3971–3974.

8. **Hantke, K., and V. Braun.** 1973. Covalent binding of lipid to protein. Diglyceride and amide-linked fatty acid at the N-terminal end of the murein-lipoprotein of the *Escherichia coli* outer membrane. *Eur. J. Biochem.* **34:**284–296.

9. **Herzberg, M. C., K. Gong, G. D. MacFarlane, P. M. Erickson, A. H. Soberay, P. H. Krebsbach, G. Manjula, K. Schilling, and W. H. Bowen.** 1990. Phenotypic characterization of *Streptococcus sanguis* virulence factors associated with bacterial endocarditis. *Infect. Immun.* **58:**515–522.

10. **Higgins, C. F., M. P. Gallagher, S. C. Hyde, M. L. Mimmack, and S. R. Pearce.** 1990. Periplasmic binding protein-dependent transport systems: the membrane-associated components. *Philos. Trans. R. Soc. London Ser. B* **326:**353–365.

11. **Hiles, I. D., M. P. Gallagher, D. J. Jamieson, and C. F. Higgins.** 1987. Molecular characterization of the oligopeptide permease of *Salmonella typhimurium. J. Mol. Biol.* **195:**125–142.

12. **Jenkinson, H. F.** 1986. Cell-surface proteins of *Streptococcus sanguis* associated with cell hydrophobicity and coaggregation properties. *J. Gen. Microbiol.* **132:**1575–1589.

13. **Jenkinson, H. F.** 1987. Novobiocin-resistant mutants of *Streptococcus sanguis* with reduced cell hydrophobicity and defective in coaggregation. *J. Gen. Microbiol.* **133:**1909–1918.

14. **Jenkinson, H. F.** 1989. Properties of a phosphocarrier protein (HPr) extracted from intact cells of *Streptococcus sanguis. J. Gen. Microbiol.* **135:**3183–3197.

15. **Jenkinson, H. F., H. C. Lala, and M. G. Shepherd.** 1990. Coaggregation of *Streptococcus sanguis* and other streptococci with *Candida albicans. Infect. Immun.* **58:**1429–1436.

16. **Kelly, C., P. Evans, L. Bergmeier, S. F. Lee, A. Progulske-Fox, A. C. Harris, A. Aitken, A. S. Bleiweis, and T. Lehner.** 1989. Sequence analysis of the cloned streptococcal cell surface antigen I/II. *FEBS Lett.* **258:**127–132.

17. **Kilian, M., L. Mikkelsen, and J. Henrichsen.** 1989. Taxonomic study of viridans streptococci: description of *Streptococcus gordonii* sp. nov. and emended descriptions of *Streptococcus sanguis* (White and Niven 1946), *Streptococcus oralis* (Bridge and Sneath 1982), and *Streptococcus mitis* (Andrewes and Horder 1906). *Int. J. Syst. Bacteriol.* **39:**471–484.

18. **Kolenbrander, P. E.** 1989. Surface recognition among oral bacteria: multigeneric coaggregations and their mediators. *Crit. Rev. Microbiol.* **17:**137–159.

19. **Laible, N. J., and G. R. Germaine.** 1982. Adsorption of lysozyme from human whole saliva by *Streptococcus sanguis* 903 and other oral microorganisms. *Infect. Immun.* **36:**148–159.

20. **Lamont, R. J., B. Rosan, C. T. Baker, and G. M. Murphy.** 1988. Characterization of an adhesin antigen of *Streptococcus sanguis* G9B. *Infect. Immun.* **56:**2417–2423.

21. **Liljemark, W. F., C. G. Bloomquist, and J. C. Ofstehage.** 1979. Aggregation and adherence of *Streptococcus sanguis*: role of human salivary immunoglobulin A. *Infect. Immun.* **26:**1104–1110.

22. **Macrina, F. L., K. R. Jones, and P. H. Wood.** 1980. Chimeric streptococcal plasmids and their use as molecular cloning vehicles in *Streptococcus sanguis* (Challis). *J. Bacteriol.* **143:**1425–1435.

23. **Martin, B., G. Alloing, C. Boucraut, and J.-P. Claverys.** 1989. The difficulty in cloning *Streptococcus pneumoniae mal* and *ami* loci in *Escherichia coli*: toxicity of *malX* and *amiA* gene products. *Gene* **80:**227–238.

24. **Morris, E. J., N. Ganeshkumar, M. Song, and B. C. McBride.** 1987. Identification and preliminary characterization of a *Streptococcus sanguis* fibrillar glycoprotein. *J. Bacteriol.* **169:**164–171.

25. **Morris, E. J., and B. C. McBride.** 1983. Aggregation of *Streptococcus sanguis* by a neuraminidase-sensitive component of serum and crevicular fluid. *Infect. Immun.* **42:**1073–1080.

26. **Rundegren, J.** 1986. Calcium-dependent salivary agglutinin with reactivity to various oral bacterial species. *Infect. Immun.* **53:**173–178.

27. **Sicard, A. M., and H. Ephrussi-Taylor.** 1965. Genetic recombination in the DNA-induced transformation of pneumococcus. II. Mapping the *amiA* region. *Genetics* **50:**31–44.

28. **Snyder, M., S. Elledge, D. Sweetser, R. A. Young, and R. W. Davis.** 1987. λgt11: gene isolation with antibody probes and other applications. *Methods Enzymol.* **154:**107–128.

29. **Trombe, M.-C., G. Laneelle, and A. M. Sicard.** 1984. Characterization of a *Streptococcus pneumoniae* mutant with altered electric transmembrane potential. *J. Bacteriol.* **158:**1109–1114.

30. **Vieira, J., and J. Messing.** 1982. The pUC plasmids, an M13mp7-derived system for insertion mutagenesis and sequencing with synthetic universal primers. *Gene* **19:**259–268.

Physical and Genetic Mapping of the *Streptococcus mutans* GS-5 Genome

M. J. HANTMAN,[1] J. J. TUDOR,[2] S. SUN,[1] L. MARRI,[3] P. J. PIGGOT,[1] AND L. DANEO-MOORE[1]

Temple University School of Medicine, Philadelphia, Pennsylvania 19140[1]; Saint Joseph's University, Philadelphia, Pennsylvania 19131[2]; and Università di Siena, Siena, Italy[3]

	Fragment size (kbp)								
Fragment	*Sma*I		*Not*I		*Apa*I		*Rsr*II	
1 A	327	*met-2, met-3, met-4, met-8, thr*	440 [a]	*met-1, met-4, met-5, met-6, met-7*	460		230	
2 B	240	*gtfB, gtfC*	440		248		190	
3 C	230		330	*gtfB, gtfC*	218	*met-1, gtfB, gtfC*	170	
4 D	218	*pro*	282	*gtfA, ftf, scrB*	198	*met-3, met-4, met-8*	168	
5 E	135 [a]	*met-1, lys*	220		172	*gbp, ftf*	150	
6 F	132		130	*gbp*	133	*scrB, gtfA*	133	*gtfA, ftf*
7 G	108	*gtfA, ftf*	124		112	*met-5, met-6, met-7*	124	
8 H	85		83	*met-8*	99		121	*gbp*
9 I	85		33		56		112	
10 J	70	*met-5, scrB*	28		55		90	
11 K	60				54		78	*scrB*
12 L	60				52	*gbp*	75	
13 M	50				48		75	
14 N	50				47		60	
15 O	45 [a]	*met-6, met-7*			46		55	
16 P	45				25		55	
17 Q	35				20		50	
18 R	33				13		30	
19 S	15				7		15	
20 T	13						5	
Total	2,034		2,110		2,062		2,067	

[a]It has not yet been determined which fragment contains the indicated genes.

The calculated genome size is 2,068 kbp. Comigrating bands are predicted by band intensity or transposon insertion. *met-1, met-2, met-3, met-4, met-5, met-6, met-7, met-8, thr, pro*, and *lys* are Tn*916*-generated auxotrophs and were located by hybridization using labeled Tn*916* (J. Procino et al., *Infect. Immun.* **56**:2866–2870, 1988). The other genes were located by hybridization with labeled cloned genes. Sources and references for data are as follows:

gtfA (provided by R. Curtiss III): J. P. Robeson, R. G. Barletta, and R. Curtiss III, *J. Bacteriol.* **153**:211–221, 1983

gtfB (provided by R. Burne; see Sun et al., this volume): H. Aoki, T. Shiroza, M. Hayakawa, S. Sato, and H. K. Kuramitsu, *Infect. Immun.* **53**:387–394, 1986

gtfC (provided by R. Burne): N. Hanada and H. K. Kuramitsu, *Infect. Immun.* **56**:1999–2005, 1988

ftf (provided by H. K. Kuramitsu): S. Sato and H. K. Kuramitsu, *Infect. Immun.* **52**:166–170, 1986

scrB (provided by F. L. Macrina): R. D. Lunsford and F. L. Macrina, *J. Bacteriol.* **166**: 426, 1986

gbp (provided by R. R. B. Russell): R. R. B. Russell, D. Coleman, and G. Dougan, *J. Gen. Microbiol.* **131**:295–299, 1985

Appendix 1

Common Characteristics of the Surface Proteins from Gram-Positive Cocci

V. A. FISCHETTI, V. PANCHOLI, AND O. SCHNEEWIND

The Rockefeller University, New York, New York 10021

At the time of this writing, 19 surface molecules from gram-positive cocci have been cloned and sequenced, 17 of which are published (Table 1). The availability of this information for detailed analysis and comparisons has revealed that these proteins possess certain common features which are unique to these organisms.

C-Terminal Hydrophobic Domain

Starting at the C-terminal end of each of these surface molecules (the part located within the cell), we find that all have a similar arrangement of amino acids. Up to seven charged amino acids, composed of a mixture of both negative and positive charged residues, are found at the C terminus. Immediately N terminal to this short charged region is a segment of 15 to 22 predominantly hydrophobic amino acids (Fig. 1). In all of these proteins, the sequences of the hydrophobic and charged regions are not necessarily the same but the chemical characteristics of the amino acids used to compose them are conserved. Beginning about nine amino acids N terminal to the hydrophobic domain is found a hexapeptide with the consensus sequence LPSTGE that is extremely conserved among all 19 proteins examined (4a). While in some of these proteins, substitutions are seen in positions 3 and 6 of the hexapeptide, positions 1, 2, 4, and 5 are, with one exception (SCP), completely conserved (Table 1 and Fig. 1). This conservation is maintained also at the DNA level (4a). The preservation of this hexapeptide and the high homology within the hydrophobic and charged regions suggest that the method of anchoring these molecules within the bacterial cell is also conserved. In a recent study it was found that M protein from group A streptococci anchored within the streptococcal membrane (22). A detailed analysis addressing the question of the importance of the C-terminal hydrophobic domain and the LPSTGE motif on anchoring M protein (and by analogy all of these surface proteins) is presented by Schneewind et al. in this volume and elsewhere (O. Schneewind and V. A. Fischetti, submitted for publication).

Wall-Associated Region

Continuing toward the N terminus and beginning from the hydrophobic domain (and including the LPSTGE motif) is a wall-associated region that spans about 50 to as much as 125 residues (in the case of FnBP) and found in nearly all of the proteins analyzed. This region is characterized by a high percentage of proline/glycine and threonine/serine residues (Table 2). For some proteins (like FnBP and M protein) the concentration of proline/glycine is significantly higher than that of threonine/serine, while in most others this relationship is either reversed or nearly equal. Because of its proximity to the region responsible for anchoring these molecules in the bacterial membrane, this region is likely located within the peptidoglycan of the cell wall (21). The reason for the presence of these particular amino acids at this location in these surface molecules has not been fully explored. One hypothesis, however, may be that the prolines and glycines, with their ability to initiate bends and turns within proteins, allow this region of the molecule to weave its way through the highly cross-linked peptidoglycan, thus stabilizing the molecule within the cell wall (21). The function of the threonines and serines is not immediately apparent. While these amino acids are commonly used as O-linked glycosylation sites in eukaryotic proteins, such substitutions have not as yet been established in bacteria.

Surface-Exposed Region

As the C-terminal region is characteristically conserved among these 19 surface molecules, the regions exposed on the cell surface are characteristically different. Despite these differences, the molecules appear to fall into two groups, those with repeating sequence blocks and those without (Table 1 and Fig. 2). Streptococcal M protein may be considered the prototype mole-

TABLE 1. Presence of repeat regions and the conserved nature of the LPSTGE motif in surface proteins from gram-positive cocci

Protein[a]	Reference	Repeats	LPSTGE
M6	12	Yes	LPSTGE
M5	17	Yes	LPSTGE
M12	24	Yes	LPSTGE
M24	18	Yes	LPSTGE
M49	10	Yes	LPSTGE
Arp4	6	Yes	LPSTGE
WapA	3	No	LPSTGE
FcRA	11	Yes	LPSTGE
Protein H	8	Yes	LPSTGE
Protein G	20	Yes	LPtTGE
PAc	19	Yes	LPnTGE
Protein A	9	Yes	LPeTGv
Wg2	16	ND[b]	LPkTGE
SpaP	15	Yes	LPnTGE
SCP	2	No	LPtTnd
Sec10	PC[c]	Min[d]	LPqTGE
FnBP	27	Min	LPeTGg
Asc10	PC	No	LPkTGE
T6	26	No	LPSTGs

[a]Abbreviations: M6, M5, M12, M24, and M49, M proteins (*Streptococcus pyogenes*); Arp4, immunoglobulin A (IgA)-binding protein (from an M4 *S. pyogenes*); WapA, wall-associated protein A (*S. mutans*); FcRA, Fc-binding protein from *S. pyogenes*; protein H, human IgG Fc-binding protein (*S. pyogenes*); protein G, IgG-binding protein (group G streptococci); PAc, cell surface protein (*S. mutans*); protein A, IgG-binding protein (*S. aureus*); Wg2, cell wall protease (*S. cremoris*); SpaP, surface protein (*S. mutans*); SCP, streptococcal C5a peptidase; Sec10 and Asc10, surface proteins (*S. faecalis*); FnBP, fibronectin-binding protein (*S. aureus*); T6, surface protein (*S. pyogenes*).

[b]ND, Not determined.

[c]PC, Gary Dunny, Personal communication.

[d]Min, Minimal.

cule for those containing sequence repeats. The M protein is composed of four repeat blocks, each differing in size and sequence (5, 12) (Fig. 2). The A repeats are composed of 14 amino acids each in which the central blocks are identical and the end blocks slightly diverge from the central consensus repeats. The B repeats, composed of 25 amino acids each, are arranged the same as the A repeats. The C repeats, composed of 2.5 blocks of 42 amino acids, are not as identical to each other as are the A and B repeats. There are also four short D repeats which show some homology. These repeat regions make up the central helical rod region of the M molecule because of the high helical potential ascribed to the amino acids found in this region as determined by conformational analysis (5, 23).

Conformational analysis of 12 of the surface molecules by the algorithm of Garnier et al. (7) revealed that those proteins containing repeat sequences were predominantly helical within the region containing the repeat segments (Fig. 3). Conversely, regions and molecules without repeat blocks were composed predominantly of amino acids exhibiting high β-sheet, β-turn, and/or random-coil potential. This is graphically seen in protein PAc, in which the region containing repeats A1-A2-A3 (Fig. 2) was the only segment exhibiting strong helix potential, while the P region exhibited weak helix potential (Fig. 3). Apparently, within most of these molecules, the presence of repeat segments usually predicts the position of a helical domain. Possibly one of the pressures for the maintenance of repeat blocks is

M6	KPNQNKAPMKETKRQ**LPSTGE**TANPFFTAAALTVMATAGVAAVVKRKEEN
M5	-------------------------------------
M19	--------------------AT------------------
M24	-------------------------------------
M30	-------------------------------------
M55	--------------------A--------------------
M49	QANRSRSAMTQQKRT**LPSTGE**TANPFFTAAAATVMVSAGMLALKRKEEN
ARP4	QANRSRSAMTQQKRT**LPSTGE**TANPFFTAAAATVMVSAGMLALKRKEEN
FcRA	RTNTNKAPMAQTKRQ**LPSTGE**ETTNPFFTA...
T6	TVLLETDIPNTKLGE**LPSTGS**IGTYLFKAIGSAAMIGAIGIYIVKRRKA
wapA	TTTSKQVTKQKAKFV**LPSTGE**QAGLLLTTVGLVIVAVAGVYFYRTRR
Protein H	KPNQNKAPMKETKRQ**LPSTGE**TANPFFTAAALTVMATAGVAAVVKRKEEN
Protein G	KKPEAKKDDAKKAET**LPTTGE**GSNPFFTAAALAVMAGAGALAVASKRKED
Protein A	KKQPANHADANKAQA**LPETGE**ENPLIGTTVFGGLSLALGAALLAGRRREL
FnBP	KAVAPTKKPQSKKSE**LPETGG**EESTNKGMLFGGLFSILGLALLRRNKKNHKA
wg2	DTTDRNGQLTSGKGA**LPKTGE**TTERPAFGFLGVIVVILMGVLGLKRKQREE
PAc	KPQSTAYQPSSVQET**LPNTGV**TNNAYMPLLGIIGLVTSFSLLGLKAKKD
Sec10	SEPRKTKQVAKAPES**LPQTGE**QQSIWLTIIGLLMAAGTIKNKKRKKNS
Asc10	PVEPLVVEKASVVPE**LPKTGE**KQNVLLTVVGSLAAMLGLAGLGFKRRRETK
spaP	KPQSTAYQPSSVQKT**LPNTGV**TNNAYMPLLGIIGLVTSFSLLGLKAKKD
SCP	SSKRALATKASDRDQ**LPTTND**KDTNRLHLLKLVMTTFFFGLVAHIFKTKRQKETKK

FIG. 1. Alignment of the C-terminal ends of surface proteins from gram-positive cocci. The sequences are aligned along the LPSTGE motif common to all of the proteins (boxed and shaded). The hydrophobic region found in all of these proteins is also boxed. Sequences are derived from references in Table 1. The sequences of M19, M30, and M55 are from Hollingshead et al. (13). For abbreviations, see Table 1, footnote a.

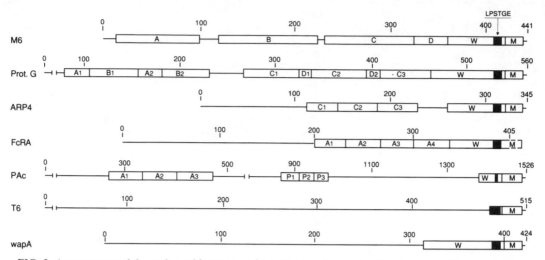

FIG. 2. Arrangement of the amino acid sequences from seven surface molecules of gram-positive cocci. Repeat regions are designated by the letters A, B, C, D, or P according to the analysis of the various authors. The cell wall-associated region (W) is the region found to contain a high concentration of Pro/Gly or Thr/Ser. The black box is the location of the LPSTGE sequence motif, and M signifies the location of the hydrophobic region. Sequences are derived from the references in Table 1. For abbreviations, see Table 1, footnote *a*.

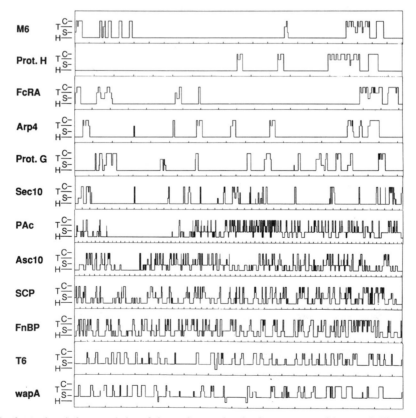

FIG. 3. Conformational characteristics of the surface molecules from gram-positive cocci. The sequences were analyzed by the Garnier-Robson algorithm supplied with the EuGene protein analysis package. The locations within the molecules of regions exhibiting random coil (C), β turn (T), β sheet, and α helix (H) are designated. Sequences were derived from the references in Table 1. For abbreviations, see Table 1, footnote *a*.

TABLE 2. Preponderance of certain amino acids in the cell wall region of surface molecules from gram-positive cocci[a]

Protein	Content (%)	
	Pro/Gly	Thr/Ser
FnBP	32	19
M6	26	14
Protein H	24	18
WapA	10	38
PAc	19	31
T6	16	30
Asc10	13	28
PAc	17	25
FcRA	20	22
SCP	21	20
Sec10	10	18
Arp4	12	16
Protein G	15	13

[a] The region 50 to 125 amino acids N terminal to the C-terminal hydrophobic domain. For abbreviations, see Table 1, footnote a.

to preserve the helix potential within specific regions of these molecules, the presence of which may determine an extended protein structure as has been shown for the M protein (23). An exception to this is found in the Sec10 protein, a predominantly helical molecule with limited repeat segments (Table 2).

Biological Activity

M protein and protein G are among the best-analyzed molecules with respect to specific functional regions. The C-repeat region of the M molecule has recently been found to bind factor H, the control protein of the alternative pathway of complement (V. A. Fischetti et al., unpublished data). It has also been suggested that the B-repeat region of this molecule may be responsible for binding fibrinogen (25). In protein G, distinct repeat regions have been shown to have the specificity for either albumin or immunoglobulin (1). Thus, it is apparent that the repeat regions within some of these molecules are tailored to perform a specific biological function. Recently, it was shown that through continual recombination events the repeat regions of the M protein are able to change in both size and sequence (4, 14). To date, however, intergenic recombination has been suggested but not demonstrated. Perhaps through a dynamic recombination scheme in the presence of a multitude of biological pressures, there is a selection of organisms containing surface proteins having specific functional domains.

LITERATURE CITED

1. **Akerstrom, B., E. Nielsen, and L. Bjorck.** 1987. Definition of IgG- and albumin-binding regions of streptococcal protein G. *J. Biol. Chem.* **262:**13388–13391.

2. **Chen, C. C., and P. P. Cleary.** 1990. Complete nucleotide sequence of the streptococcal C5a peptidase gene of *Streptococcus pyogenes. J. Biol. Chem.* **265:**3161–3167.

3. **Ferretti, J. J., R. R. B. Russell, and M. L. Dao.** 1989. Sequence analysis of the wall-associated protein precursor of *Streptococcus mutans* antigen A. *Mol. Microbiol.* **3:**469–478.

4. **Fischetti, V. A., K. F. Jones, and J. R. Scott.** 1985. Size variation of the M protein in group A streptococci. *J. Exp. Med.* **161:**1384–1401.

4a.**Fischetti, V. A., V. Pancholi, and O. Schneewind.** 1991. Conservation of a hexapeptide sequence in the anchor region of surface proteins from gram-positive cocci. *Mol. Microbiol.* **4:**1603–1605.

5. **Fischetti, V. A., D. A. D. Parry, B. L. Trus, S. K. Hollingshead, J. R. Scott, and B. N. Manjula.** 1988. Conformational characteristics of the complete sequence of group A streptococcal M6 protein. *Proteins Struct. Funct. Genet.* **3:**60–69.

6. **Frithz, E., L.-O. Heden, and G. Lindahl.** 1989. Extensive sequence homology between IgA receptor and M protein in *Streptococcus pyogenes. Mol. Microbiol.* **3:**1111–1119.

7. **Garnier, J., D. J Osguthorpe, and B. Robson.** 1978. Analysis of the accuracy and implications of simple methods for predicting the secondary structure of globular proteins. *J. Mol. Biol.* **120:**97–120.

8. **Gomi, H., T. Hozumi, S. Hattori, C. Tagawa, F. Kishimoto, and L. Bjorck.** 1990. The gene sequence and some properties of protein H. *J. Immunol.* **144:**4046–4052.

9. **Guss, B., M. Uhlen, B. Nilsson, M. Lindberg, J. Sjoquist, and J. Sjodahl.** 1984. Region X, the cell-wall-attachment part of staphylococcal protein A. *Eur. J. Biochem.* **138:**413–420.

10. **Haanes, E. J., and P. P. Cleary.** 1989. Identification of a divergent M protein gene and an M protein-related gene family in serotype 49 *Streptococcus pyogenes. J. Bacteriol.* **171:**6397–6408.

11. **Heath, D. G., and P. P. Cleary.** 1989. Fc-receptor and M protein genes of group A streptococci are products of gene duplication. *Proc. Natl. Acad. Sci. USA* **86:**4741–4745.

12. **Hollingshead, S. K., V. A. Fischetti, and J. R. Scott.** 1986. Complete nucleotide sequence of type 6 M protein of the group A streptococcus: repetitive structure and membrane anchor. *J. Biol. Chem.* **261:**1677–1686.

13. **Hollingshead, S. K., V. A. Fischetti, and J. R. Scott.** 1987. A highly conserved region present in transcripts encoding heterologous M proteins of group A streptococcus. *Infect. Immun.* **55:**3237–3239.

14. **Jones, K. F., S. K. Hollingshead, J. R. Scott, and V. A. Fischetti.** 1988. Spontaneous M6 protein size mutants of group A streptococci display variation in antigenic and opsonogenic epitopes. *Proc. Natl. Acad. Sci. USA* **85:**8271–8275.

15. **Kelly, C., P. Evans, L. Bergmeier, S. F. Lee, H. Fox, A. Progulske, A. C. Harris, A. Aitken, A. S. Bleiweis, and T. Lerner.** 1990. Sequence analysis of a cloned streptococcal surface antigen I/II. *FEBS Lett.* **258:**127–132.

16. **Kok, J., K. J. Leenhouts, A. J. Haandrikman, A. M. Ledeboer, and G. Venema.** 1988. Nucleotide sequence of the cell wall proteinase gene of *Streptococcus cremoris* Wg2. *Appl. Environ. Microbiol.* **54:**231–238.

17. **Miller, L., L. Gray, E. H. Beachey, and M. A. Kehoe.** 1988. Antigenic variation among group A streptococcal M proteins: nucleotide sequence of the serotype 5 M protein gene and its relationship with genes encoding types 6 and 24 M proteins. *J. Biol. Chem.* **263:**5668–5673.

18. **Mouw, A. R., E. H. Beachey, and V. Burdett.** 1988. Molecular evolution of streptococcal M protein: cloning and nucleotide sequence of type 24 M protein gene and relation to other genes of *Streptococcus pyogenes. J. Bacteriol.* **170:**676–684.

19. **Okahashi, N., C. Sasakawa, S. Yoshikawa, S. Hamada, and T. Koga.** 1989. Molecular characterization of a surface protein antigen gene from serotype c *Streptococcus mutans* implicated in dental caries. *Mol. Microbiol.* **3:**673–678.

20. **Olsson, A., M. Eliasson, B. Guss, B. Nilsson, U. Hellman, M. Lindberg, and M. Uhlen.** 1987. Structure and evolution of the repetitive gene encoding streptococcal protein G. *Eur. J. Biochem.* **168:**319–324.

21. **Pancholi, V., and V. A. Fischetti.** 1988. Isolation and characterization of the cell-associated region of group A streptococcal M6 protein. *J. Bacteriol.* **170:**2618–2624.

22. **Pancholi, V., and V. A. Fischetti.** 1989. Identification of an endogeneous membrane anchor-cleaving enzyme for group A streptococcal M protein. *J. Exp. Med.* **170:**2119–2133.

23. **Phillips, G. N., P. F. Flicker, C. Cohen, B. N. Manjula, and V. A. Fischetti.** 1981. Streptococcal M protein: alpha-helical coiled-coil structure and arrangement on the cell surface. *Proc. Natl. Acad. Sci. USA* **78:**4689–4693.

24. **Robbins, J. C., J. G. Spanier, S. J. Jones, W. J. Simpson,** and **P. P. Cleary.** 1987. *Streptococcus pyogenes* type 12 M protein regulation by upstream sequences. *J. Bacteriol.* **169:**5633–5640.

25. **Ryc, M., E. H. Beachey, and E. Whitnack.** 1989. Ultrastructural localization of the fibrinogen-binding domain of streptococcal M protein. *Infect. Immun.* **57:**2397–2404.

26. **Schneewind, O., K. F. Jones, and V. A. Fischetti.** 1990. Sequence and structural characterization of the trypsin-resistant T6 surface protein of group A streptococci. *J. Bacteriol.* **172:**3310–3317.

27. **Signas, C., G. Raucci, K. Jonsson, P. Lindgren, G. M. Anantharamaiah, M. Hook, and M. Lindberg.** 1989. Nucleotide sequence of the gene for fibronectin-binding protein from *Staphylococcus aureus*: use of this peptide sequence in the synthesis of biologically active peptides. *Proc. Natl. Acad. Sci. USA* **86:**699–703.

Appendix 2

Codon Usage Patterns for *Streptococcus pneumoniae* and *Escherichia coli*

BERNARD MARTIN AND JEAN-PIERRE CLAVERYS

*Centre de Recherche de Biochimie et de Génétique Cellulaires du Centre National de la Recherche Scientifique,
Université Paul Sabatier, 31062 Toulouse Cedex, France*

TABLE 1. Codon usage by amino acid for *Streptococcus pneumoniae*[a] and *Escherichia coli*[b]

Amino acid	Codon	Total no. of codons[a]	% 1[a]	% 2[b]	Amino acid	Codon	Total no. of codons[a]	% 1[a]	% 2[b]
Ala	GCT	391	42	26		CTC	120	10	7
	GCC	193	21	21		CTA	98	9	2
	GCA	228	24	22		CTG	108	9	68
	GCG	123	13	31	Lys	AAA	456	57	76
Arg	CGT	227	48	56		AAG	345	43	24
	CGC	71	15	36	Met	ATG	279	100	100
	CGA	46	10	3	Phe	TTT	333	68	37
	CGG	40	9	3		TTC	158	32	63
	AGA	68	15	1	Pro	CCT	171	39	12
	AGG	16	3	1		CCC	31	7	7
Asn	AAT	368	67	26		CCA	191	44	16
	AAC	179	33	74		CCG	46	10	65
Asp	GAT	461	66	46	Ser	TCT	218	27	23
	GAC	238	34	54		TCC	59	7	27
Cys	TGT	25	54	43		TCA	184	23	7
	TGC	21	46	57		TCG	52	7	11
Gln	CAA	305	62	24		AGT	180	23	6
	CAG	190	38	76		AGC	103	13	26
Glu	GAA	527	65	73	Thr	ACT	231	32	25
	GAG	305	35	27		ACC	170	23	50
Gly	GGT	310	41	48		ACA	235	32	7
	GGC	107	14	39		ACG	92	13	18
	GGA	224	30	5	Trp	TGG	143	100	100
	GGG	116	15	8	Tyr	TAT	314	64	40
His	CAT	139	67	37		TAC	178	36	60
	CAC	70	33	63	Val	GTT	308	38	36
Ile	ATT	458	61	36		GTC	177	22	15
	ATC	247	32	61		GTA	161	20	22
	ATA	56	7	3		GTG	170	21	27
Leu	TTA	191	17	7	Stop	TAA	16	57	75
	TTG	419	37	9		TGA	2	7	17
	CTT	211	18	7		TAG	10	36	8

[a]Based on a 31-gene compilation (28 complete genes; see Table 2). Total number of codons = 11,909. All DNA sequences used for this compilation were taken from the EMBL data bank except for the *recP* and *comA* (D. A. Morrison, personal communication), *sulD* (manual entry from reference 10), *epuA* (manual entry from reference 16), *endA* (S. A. Lacks, personal communication), *ami*, *hexB*, and *ung* sequences.
[b]Based on the 52-gene compilation of Alff-Steinberger (1). Total number of codons = 16,351.

TABLE 2. *Streptococcus pneumoniae* genes

From SPDPN1A bases 526 to 1041	From SPMALX bases 1 to 1149
From SPDPN1A bases 1038 to 1499	From SPMALMXP bases 1758 to 3275
From SPDPN2A bases 441 to 1295	From SPMALMXP bases 3301 to 3543
From SPDPN2A bases 1321 to 2091	From AMI bases 256 to 1737
From SPDPN2A bases 2078 to 2944	From AMI bases 1746 to 2237
From SPEXOA bases 54 to 881	From AMI bases 2304 to 3800
From SPLYS bases 207 to 1622	From AMI bases 3800 to 4726
From SPLYTPN bases 201 to 1157	From AMI bases 4735 to 5802
From SPPBPX bases 253 to 2505	From AMI bases 5813 to 6739
From SPPENA bases 233 to 2272	From COMA bases 1324 to 3478
From SPPOLA bases 120 to 2753	From ENDA bases 133 to 327
From SPSULA bases 245 to 1195	From ENDA bases 366 to 1190
From SPHEXA bases 971 to 3505	From HEXB bases 499 to 2448
From SPHEXA bases 474 to 920	From RECP bases 1 to 1971
From SPHEXAORFL bases 1 to 324	From SULD bases 930 to 1742
	From UNG bases 40 to 693

REFERENCES

1. **Alff-Steinberger, C.** 1984. Evidence for a coding pattern on the non-coding strand of the *E. coli* genome. *Nucleic Acids Res.* **12:**2235–2241.

2. **Alloing, G., M. C. Trombe, and J. P. Claverys.** 1990. The *ami* locus of the gram-positive bacterium *Streptococcus pneumoniae* is similar to binding protein-dependent transport operons of gram-negative bacteria. *Mol. Microbiol.* **4:**633–644.

3. **Dowson, C. G., A. Hutchison, and B. G. Spratt.** 1989. Nucleotide sequence of the penicillin-binding protein 2B gene of *Streptococcus pneumoniae* strain R6. *Nucleic Acids Res.* **17:**7518.

4. **Garcia, P., J. L. Garcia, E. Garcia, and R. Lopez.** 1986. Nucleotide sequence and expression of the pneumococcal autolysin gene from its own promoter in *Escherichia coli*. *Gene* **43:**265–272.

5. **Hui, F. M., and D. A. Morrison.** 1991. Genetic transformation in *Streptococcus pneumoniae*: nucleotide sequence analysis shows *comA*, a gene required for competence induction, to be a member of the bacterial ATP-dependent transport protein family. *J. Bacteriol.* **173:**372–381.

6. **Lacks, S. A., J. J. Dunn, and B. Greenberg.** 1982. Identification of base mismatches recognized by the heteroduplex-DNA-repair system of *Streptococcus pneumoniae*. *Cell* **31:**327–336.

7. **Lacks, S. A., B. M. Mannarelli, S. S. Springhorn, and B. Greenberg.** 1986. Genetic basis of the complementary *DpnI* and *DpnII* restriction systems of *S. pneumoniae*: an intercellular cassette mechanism. *Cell* **46:**993–1000.

8. **Laible, G., R. Hakenbeck, A. M. Sicard, B. Joris, and J.-M. Ghuysen.** 1989. Nucleotide sequence of the *pbpX* genes encoding the penicillin-binding proteins 2x from *Streptococcus pneumoniae* R6 and a cefotaxime-resistant mutant, C506. *Mol. Microbiol.* **3:**1337–1348.

9. **Lopez, P., M. Espinosa, B. Greenberg, and S. A. Lacks.** 1987. Sulfonamide resistance in *Streptococcus pneumoniae*: DNA sequence of the gene encoding dihydropteroate synthase and characterization of the enzyme. *J. Bacteriol.* **169:**4320–4326.

10. **Lopez, P., B. Greenberg, and S. A. Lacks.** 1990. DNA sequence of folate biosynthesis gene *sulD*, encoding hydroxymethyldihydropterin pyrophosphokinase in *Streptococcus pneumoniae*, and characterization of the enzyme. *J. Bacteriol.* **172:**4766–4774.

11. **Lopez, P., S. Martinez, A. Diaz, M. Espinosa, and S. A. Lacks.** 1989. Characterization of the *polA* gene of *Streptococcus pneumoniae* and comparison of the DNA polymerase I it encodes to homologous enzymes from *Escherichia coli* and phage T7. *J. Biol. Chem.* **264:**4255–4263.

12. **Méjean, V., I. Rives, and J. P. Claverys.** 1990. Nucleotide sequence of the *Streptococcus pneumoniae ung* gene encoding uracil-DNA glycosylase. *Nucleic Acids Res.* **18:**6693.

13. **Priebe, S., S. Hadi, B. Greenberg, and S. A. Lacks.** 1988. Nucleotide sequence of the *hexA* gene for DNA mismatch repair in *Streptococcus pneumoniae* and homology of HexA to MutS of *Escherichia coli* and *Salmonella typhimurium*. *J. Bacteriol.* **170:**190–196.

14. **Prudhomme, M., B. Martin, V. Méjean, and J. P. Claverys.** 1989. Nucleotide sequence of the *Streptococcus pneumoniae hexB* mismatch repair gene: homology of HexB to MutL of *Salmonella typhimurium* and to PMS1 of *Saccharomyces cerevisiae*. *J. Bacteriol.* **171:**5332–5338.

15. **Puyet, A., B. Greenberg, and S. A. Lacks.** 1989. The *exoA* gene of *Streptococcus pneumoniae* and its product, a DNA exonuclease with apurinic endonuclease activity. *J. Bacteriol.* **171:**2278–2286.

16. **Puyet, A., B. Greenberg, and S. A. Lacks.** 1990. Genetic and structural characterization of EndA, a membrane-bound nuclease required for transformation of *Streptococcus pneumoniae*. *J. Mol. Biol.* **213:**727–738.

17. **Radnis, B. A., D.-K. Rhee, and D. A. Morrison.** 1990. Genetic transformation in *Streptococcus pneumoniae*: nucleotide sequence and predicted amino acid sequence of *recP*. *J. Bacteriol.* **172:**3669–3674.

18. **Walker, J. A., R. L. Allen, P. Falmagne, M. K. Johnson, and G. J. Boulnois.** 1987. Molecular cloning, characterization, and complete nucleotide sequence of the gene for pneumolysin, the sulfhydryl-activated toxin of *Streptococcus pneumoniae*. *Infect. Immun.* **55:**1184–1189.

Appendix 3

Transformation of Streptococci and Related Organisms by Electroporation

At the 1986 streptococcal genetics conference, S. Harlander presented one of the first reports (5) on a new method for transforming bacteria with plasmid DNA. This technique, termed electroporation, involves passing a high-voltage electric pulse through an aqueous suspension of bacteria mixed with DNA. This method offers great potential for introduction of DNA into noncompetent strains of streptococci and related organisms that have been refractory to such procedures in the past. A number of participants at the 1990 meeting have developed and published methods for using electroporation to transform streptococci, lactococci, and enterococci. One thing that will be apparent to anyone reading this appendix is that there are a number of different methods by which organisms can be prepared successfully for electroporation. However, some important parameters have been found to be critical by several different groups, and investigators developing a method for their own organism should focus on these. Some of the most important variables seem to be the growth medium (including the use of glycine to inhibit cell wall synthesis), growth phase, electroporation solution, DNA concentration, and electroporation conditions (high field strengths of 5,000 to 10,000 V/cm). In this appendix, several of these protocols are presented, along with a list of some key references relating to electroporation of these organisms. Hopefully, these procedures will be useful starting points for investigators developing electroporation techniques for their own organisms.

GARY M. DUNNY
Bioprocess Institute and
    Department of Microbiology
University of Minnesota
St. Paul, Minn.

SELECTED REFERENCES

1. **Chassy, B. M., and J. L. Flickinger.** 1987. Transformation of *Lactobacillus casei* by electroporation. *FEMS Microbiol. Lett.* **44:**173–177.
2. **Cruz-Rodz, A. L., and M. S. Gilmore.** 1990. High efficiency introduction of plasmid DNA into glycine-treated *Enterococcus faecalis* by electroporation. *Mol. Gen. Genet.* **224:** 152–154.
3. **Dunny, G. M., L. N. Lee, and D. J. LeBlanc.** 1991. Improved electroporation and cloning vector system for grampositive bacteria. *Appl. Environ. Microbiol.* **57:**1194–1201.
4. **Fielder, S., and R. Wirth.** 1988. Transformation of bacteria with plasmid DNA by electroporation. *Anal. Biochem.* **170:**38–44.
5. **Harlander, S. K.** 1987. Transformation of *Streptococcus lactis* by electroporation, p. 229–233. *In* J. J. Ferretti and R. Curtiss III (ed.), *Streptococcal Genetics.* American Society for Microbiology, Washington, D.C.
6. **Holo, H., and I. F. Nes.** 1989. High-frequency transformation, by electroporation, of *Lactococcus lactis* subsp. *cremoris* grown with glycine in osmotically stabilized media. *Appl. Environ. Microbiol.* **55:**3119–3123.
7. **Luchansky, J. B., P. M. Muriana, and T. R. Klaenhammer.** 1988. Application of electroporation for transfer of plasmid DNA to *Lactobacillus*, *Lactococcus*, *Leuconostoc*, *Listeria*, *Pediococcus*, *Bacillus*, *Staphylococcus*, *Enterococcus*, and *Propionibacterium*. *Mol. Microbiol.* **2:**637–646.
8. **McIntyre, D. A., and S. K. Harlander.** 1989. Genetic transformation of intact *Lactococcus lactis* subsp. *lactis* by highvoltage electroporation. *Appl. Environ. Microbiol.* **55:**604–610.
9. **Powell, I. G., M. G. Achen, A. J. Hillier, and B. E. Davidson.** 1988. A simple and rapid method for genetic transformation of lactic streptococci by electroporation. *Appl. Environ. Microbiol.* **54:**655–660.

High-Efficiency Transformation of Lactococci by Electroporation

HELGE HOLO[1] AND INGOLF F. NES[2]

Laboratory of Microbial Gene Technology, NLVF, N-1432 Ås-NLH,[2] and
Norwegian Dairies Association,[1] Oslo, Norway

1. Use exponentially growing *Lactococcus lactis* subsp. *cremoris* BC101 in GM17 medium (M17 medium [Oxoid Ltd., London, England] supplemented with 0.5% glucose) to inoculate (1%) in SGM17 medium (GM17 medium with 0.5 M sucrose) containing 3% glycine; grow the culture overnight at 30°C.
2. At an optical density at 600 nm of 0.2 to 0.7, harvest the cells by centrifugation (10 min at $5,000 \times g$).
3. Wash the cells twice in ice-cold 0.5 M sucrose containing 10% glycerol.
4. Suspended the cells in 1/100 culture volume of washing solution and freeze them at $-85°C$. Both fresh and frozen cells are highly competent for electroporation with plasmid DNA.
5. Electroporation is carried out with the Gene Pulser (Bio-Rad Laboratories, Richmond, Calif.). Frozen cells are thawed on ice. One microliter of DNA (in 10 mM Tris-HCl–1 mM EDTA, pH 8) is mixed in the cuvette with 40 µl of competent cells immediately prior to electroporation. Electroporation is carried out at 10 to 12.5 kV/cm.
6. The discharged cell suspension is immediately diluted in 0.96 ml of reconstitution medium SGM17 containing 20 mM $MgCl_2$ and 2 mM $CaCl_2$ (the temperature of the reconstitution medium does not seem to be critical for this strain) and then incubated for 1 to 2 h at 30°C.
7. The cells are then spread on SR plates (per liter: 10 g of tryptone, 5 g of yeast extract, 200 g of sucrose, 10 g of glucose, 25 g of gelatin, 15 g of agar, 2.5 mM $MgCl_2$, and 2.5 mM $CaCl_2$, pH 6.8) or GM17 plates containing appropriate antibiotics to select for transformants.

The transformation frequencies obtained with *L. lactis* subsp. *cremoris* BC101 were 5.7×10^7 and 2×10^7 per µg of pIL253 and pSA3 DNAs, respectively, when the plasmids were isolated from lactococci. The following *L. lactis* subsps. *cremoris* strains have been successfully transformed with this procedure: NCDP 495, NCDO 504, NCDO 607, NCDO 893, NCDO 924, NCDO 1986, NCDO 1997, NCDO 2004, GS, and BC101 (plasmid-free). *L. lactis* subsps. *lactis* LM2336 was transformed with a frequency of 1.4×10^7 transformants per µg of pIL 253 DNA. The optimal glycine concentration to obtain competent cells was usually the highest concentration that allowed growth of the strain of interest. For the strains tested, the glycine concentration varied between <0.5 and 4% and may be even higher for other strains. It is therefore crucial to determine the MIC of glycine for any new strain to be used for transformation. Variation electrocompetence with different amounts of inoculum of a particular strain has been observed. This variation was reduced by inoculating the glycine-containing medium with a culture that had been grown exponentially for many generations. In the case of strain BC101, exponentially growing cells were more sensitive than stationary-phase cells to growth inhibition by glycine.

High-Efficiency Electrotransformation
of *Streptococcus pyogenes*

DANIEL SIMON AND JOSEPH J. FERRETTI

Department of Microbiology and Immunology, University of Oklahoma Health Sciences Center, Oklahoma City, Oklahoma 73190

1. Inoculate 20 ml of Todd-Hewitt broth supplemented with 5% horse serum (TH-HS) with 0.4 ml of an overnight culture. Grow the cells at 37°C for 3 h to an optical density at 560 nm of 0.25.

2. Centrifuge the cells at 7,000 × g for 5 min at 4°C.

3. Discard the supernatant. Resuspend the cells in 1 ml of ice-cold, sterile solution of 0.5 M sucrose in distilled water and transfer them to an Eppendorf tube.

4. Sediment the cells for 15 s in a microfuge, discard the supernatant, and resuspend the cells in the same volume of 0.5 M sucrose.

5. Wash the cells two more times as described above.

6. Resuspend the cells in 100 μl of 0.5 M sucrose.

7. To 1 to 10 μl of DNA solution, add suspension to obtain a final volume of 100 μl. Mix the solution and transfer it to a chilled Gene Pulser cuvette (0.2-cm electrode gap).

8. Expose the cells to a single electric pulse (peak voltage, 2.5 kV; capacitance; 25 μF; pulse controller, 200 Ω) in a Bio-Rad Gene Pulser and pulse controller.

9. Add immediately 0.9 ml of TH-HS. Incubate the preparation at 37°C for 1 h.

10. Mix the cells with TH-HS agar containing appropriate antibiotic. Incubate the cells at 37°C for 24 to 48 h.

By using this procedure, it was possible to transform *Streptococcus pyogenes* NZ131 with plasmid pIL252, a plasmid vector derived from pAMβ1, or pSA3, a *Streptococcus-Escherichia coli* shuttle vector. Transformation efficiencies of up to 7 × 10^7 CFUs/μg of DNA were obtained with recipient strain NZ131. It is important to note that this procedure was optimized for strain NZ131; electrotransformation of other strains was also observed with varying efficiencies. The most important parameters for successful electroporation are the growth stage of the recipient organism and the field strength. Preliminary results obtained with use of a 0.1-cm electrode gap cuvette suggest that the optimal field strength is 20 kV/cm. Linear DNA has been used with the electrotransformation procedure described above; however, efficiencies of transformation decreased by about 4 logs.

Electroporation of Glycine-Treated *Enterococcus faecalis*

ARMANDO L. CRUZ-RODZ AND MICHAEL S. GILMORE

Department of Microbiology and Immunology, University of Oklahoma Health Sciences Center, Oklahoma City, Oklahoma 73190

1. Prepare an overnight culture of *Enterococcus faecalis* JH2-2 by inoculating 2 ml of M17 medium, using a single colony from a fresh M17 plate containing rifampin (25 µg/ml) and fusidic acid (50 µg/ml).
2. Dilute the overnight culture 1/100 into 10 ml of fresh SGM17 (see below) and incubate it for 21 h at 37°C.
3. Harvest cells by centrifugation at 1,000 × *g* at room temperature for 10 min.
4. Wash the cells twice with 1 volume of ice-cold electroporation solution (0.5 M sucrose and 10% glycerol).
5. Resuspend the cells in 1/100 the original volume of electroporation solution, divide them into 40-µl aliquots, and freeze them at −70°C.
6. Thaw a 40-µl aliquot of frozen cells on ice for approximately 2 min. Mix the cells with 1 µg or more of plasmid DNA dissolved in TE (10 mM Tris-HCl, 1 mM EDTA, pH 7.5) on ice. Transfer the suspension to a prechilled electroporation cuvette (0.2 cm).
7. Apply a single pulse (peak voltage, 2.5 kV; capacitance, 25 µF; resistance, 200 Ω). Immediately following the discharge, mix the cell suspension with 0.98 ml of ice-cold SGM17MC (see below) and keep it on ice for 5 min.
8. Make appropriate dilutions in SGM17MC and incubate them at 37°C for 2 h.
9. Spread 100-µl aliquots of diluted transformed cells on SR medium (see below) containing appropriate antibiotics. Incubate the cells at 37°C for 2 days. Microcolonies will be visible within 24 h.

Medium compositions (per liter) are as follows:

SGM17
37.25 g of M17 (Difco)
0.5 M sucrose
8% Glycine

SGM17MC
37.25 g of M17 (Difco)
0.5 M sucrose
8% Glycine
10 mM $MgCl_2$
10 mM $CaCl_2$

SR
10 g of tryptone
5 g of yeast extract
200 g of sucrose
10 g of glucose
25 g of gelatin
15 g of agar
2.5 mM $MgCl_2$
2.5 mM $CaCl_2$
pH 6.8

This procedure yielded 2.8×10^6 transformants per µg of plasmid DNA when *E. faecalis* JH2-2 was used. However, a lower efficiency of 8×10^3 transformants per µg of plasmid DNA was observed when strain UV202 was used.

Transformation of *Enterococcus faecalis* and *Enterococcus faecium* by Electroporation

STEFAN FIEDLER AND REINHARD WIRTH

*Lehrstuhl für Mikrobiologie, Universität München, D-8000 München 19,
Federal Republic of Germany*

1. Grow cells overnight in Todd-Hewitt Broth (cells should be in *stationary* growth phase!).
2. Harvest cells by centrifugation and wash them in 10% glycerol with:
   (a) 1/1 original volume of growth medium.
   (b) 1/2 or original volume of growth medium.
   (c) 1/10 original volume of growth medium.
3. Resuspend cells of 1 liter of original growth medium in 1 ml of 10% glycerol and store them in small aliquots at 70°C.
4. For the actual transformation, mix 40 µl of this cell suspension with up to 200 ng of plasmid DNA, transfer the preparation to a chilled electroporation cuvette (2-mm inner width), and incubate it for 5 min at 0°C. Apply one pulse at 2,500 V (12,500 V/cm), using the Bio-Rad Gene Pulser 25-µF capacitor and the 400-Ω resistor of the pulse controller. A time constant of 5 to 15 ms should be obtained by this method; the optimal killing rate is about 50%.
5. After the pulse, transfer the suspension into 1 ml of Todd-Hewitt broth, incubate it for 30 to 90 min at 37°C for phenotypic expression of antibiotic resistances, and place the cells on selective plates.

Transformation efficiencies for various enterococci with use of this protocol range from 10^3 to 10^5/µg of DNA. Recently electroporation cuvettes with 1-mm inner width were made available by Bio-Rad. These offer the advantage that a maximum field strength of 25,000 V/cm can be applied. Because of Joule heating, we could use only cell suspensions having 1/10 the cell concentration used for previous experiments. The results show that the use of higher field strength does not seem to offer an advantage over results obtained with 2-mm cuvettes.

Electroporation of Enterococci, Streptococci, and Bacilli

Bioprocess Institute and Department of Microbiology, University of Minnesota,
St. Paul, Minnesota 55108

1. Grow cells for 12 to 15 h in BYGT or M9-YE medium (see below) plus glycine at various concentrations. Use a range of glycine concentrations and select the one that gives 70 to 90% reduction in the A_{660} of the culture in comparison with a control grown for the same period of time in the absence of glycine. In our hands, useful glycine concentrations for most strains grown in BYGT are in the range of 0.5 to 6.0%. (Group A streptococci and *Bacillus anthracis* transformed without glycine, and actually lysed during electroporation, if grown in glycine.)

2. Dilute the culture from step 1 into fresh medium (containing the same or a slightly higher glycine concentration) to bring the A_{660} to 0.05 to 0.08. Incubate the cells for 60 min at 37°C (90 min for slowly growing strains).

3. Chill the culture on ice; harvest the cells by centrifugation; wash the cells in 1/3 volume of chilled EP solution (EP solution is 0.625 M sucrose and 1 mM $MgCl_2$, adjusted to pH 4.0 with 1 N HCl).

4. Harvest cells from the wash and resuspend them in 1/30 to 1/100 the original volume of EP solution. Incubate the cells on ice for 30 to 60 min. (Cells may be frozen in a dry ice-ethanol bath and stored for at least 1 year at −70 to −85°C at this point and thawed in an ice water bath just prior to use.)

5. Add cells from step 4 to the electroporation cuvette (800 μl if using 0.4-cm cuvettes; 50 to 100 μl if using 0.2-cm cuvettes). Add 0.3 to 1 μg of DNA (<10 μl in H_2O or low-salt buffer; gradient purified for maximum efficiency).

6. Electroporate cells immediately, using the 25-μF setting on the Bio-Rad Gene Pulser. When 0.2-cm cuvettes and the pulse controller unit are used, the resistance should be set to 200 Ω and the field strength should be 8,750 to 10,000 V/cm.

7. Place cells on ice for 1 to 2 min, dilute them into 2 volumes of medium (plus inducing concentrations of antibiotics if an inducible resistance gene is used as a selective marker), and incubate them for 90 to 120 min at 37°C. Spread the cells on selective agar plates containing 0.25 M sucrose. Frequencies of transformation range from 2 × 10^5 transformants per ug of plasmid DNA for *Enterococcus faecalis* to 5 × 10^3 for *Streptococcus agalactiae*.

Medium compositions are as follows:

M9-YE
Yeast extract, 3 g/liter
Casamino Acids (Difco), 10 g/liter
10× M9 salts,[a,b] 1/10 volume
20% Glucose,[a] 1/100 volume
1 M $MgSO_4$,[a] 1/500 volume
1 M $CaCl_2$,[a] 1/10,000 volume

BYGT
Brain heart infusion, 19 g/liter
Yeast extract 5 g/liter
Glucose 2 g/liter
1 M Tris, pH 8.0 1/10 volume

---

[a]Add after autoclaving.
[b]Formula for 10× M9 salts (grams per liter): Na_2HPO_4, 60; KH_2PO_4, 30; NaCl, 5; NH_4Cl, 10.

Appendix 4

Cloning Vectors for Lactococci, Enterococci, and Streptococci

J. KOK

Department of Molecular Genetics, University of Groningen, 9750 NN Haren, The Netherlands

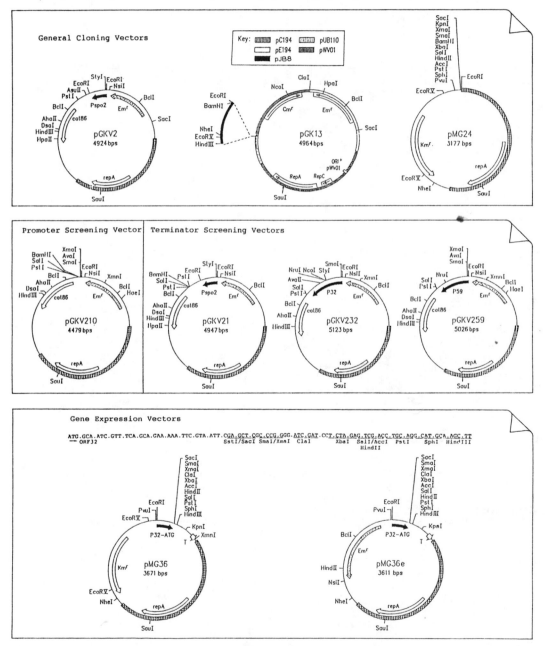

Continued on next page

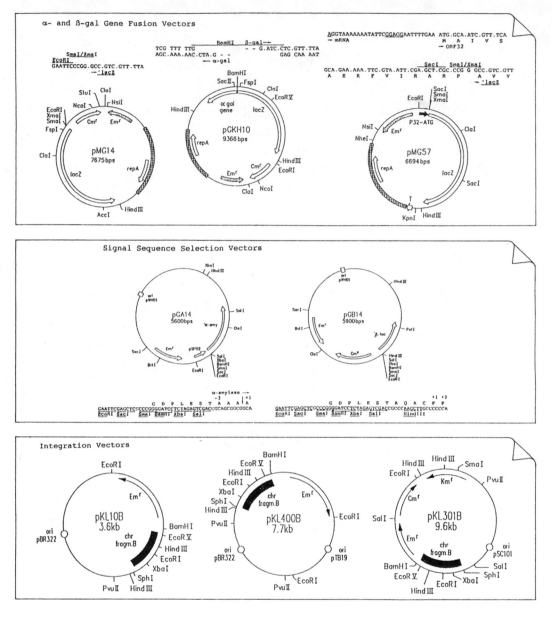

References: 1. J. Kok, this volume.
2. M. van de Guchte, Ph.D. thesis, University of Groningen, Haren, The Netherlands, 1991.

Cloning Vectors with Streptococcal Resistance Genes and an *Escherichia coli* Origin of Replication Which Can be Used for Cloning of Streptococcal Origins of Replication or for Gene Inactivation in Streptococci

L. TAO AND J. J. FERRETTI

Department of Microbiology, University of Oklahoma Health Science Center, Oklahoma City, Oklahoma 730731

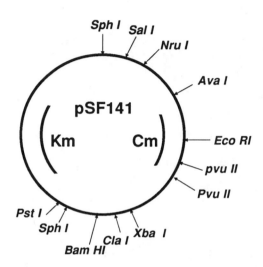

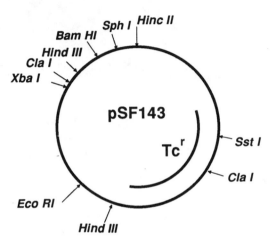

Vector:	pSF141
Size:	7.6 kb
Single sites for gene inactivation:	Cm^r: *Eco*RI, *Pvu*II
Other single sites:	*Sal*I, *Nru*I, *Ava*I, *Xba*I, *Cla*I, *Bam*HI, *Pst*I
Other hosts:	*E. coli*
Reference:	L. Tao and J. J. Ferretti, unpublished data

Vector:	pSF143
Size:	5.7 kb
Single sites:	*Hinc*II, *Sph*I, *Bam*HI, *Xba*I, *Eco*RI
Other host:	*E. coli*
Reference:	L. Tao and J. J. Ferretti, unpublished data

A Shuttle Vector Containing a Polylinker Region Flanked by Transcription Terminators Which Replicates in *Escherichia coli* and in Many Gram-Positive Cocci

D. J. LeBLANC,[1] L. N. LEE,[1] AND G. M. DUNNY[2]

Department of Microbiology, University of Texas Health Sciences Center, San Antonio, Texas 78284,[1] and Bioprocess Technology Institute, University of Minnesota, St. Paul, Minnesota 55108[2]

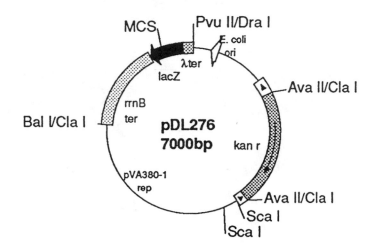

Reference: G. M. Dunny, L. N. Lee, and D. J. Le Blanc, Improved electroporation and cloning vector system for gram-positive bacteria, *Appl. Environ. Microbiol.* **57:**1194–1201, 1991

Author Index

Subject Index